KAPLAN

D1378395

Related Kaplan Books

DAT
MCAT 45
MCAT Comprehensive Review
MCAT Practice Tests
PCAT

ORGANIC CHEMISTRY EDGE

Second Edition

By Glenn Croston, Ph.D. and
the staff of Kaplan, Inc.

Simon & Schuster

NEW YORK · LONDON · SYDNEY · TORONTO

Kaplan Publishing
Published by Simon & Schuster
1230 Avenue of the Americas
New York, New York 10020

For bulk sales to schools, colleges, and universities, please contact: Order Department, Simon and Schuster, 100 Front Street, Riverside, NJ 08075. Phone: 800-223-2336. Fax: 800-943-9831.

Contributing Editor: Albert Chen
Project Editor: Larissa Shmailo
Production Editors: Maude Spekes, Rodolfo Robles
Interior Production: Laurel Douglas
Cover Design: Cheung Tai
Production Manager: Michael Shevlin
Managing Editor: Déa Alessandro
Executive Editor: Del Franz

Special thanks are extended to Karl Lee, Sara Pearl, and Michael Wolffe.

Manufactured in the United States of America
Published simultaneously in Canada

December 2003

10 9 8 7 6 5 4 3

ISBN 0-7432-4100-2

Table of Contents

ABOUT THE AUTHOR

Since receiving his Ph.D. in 1992 at the University of California, San Diego, Glenn E. Croston has worked as a research scientist in the biotechnology industry performing pharmaceutical research and publishing extensively. As the Director of In Vitro Pharmacology at the Ferring Research Institute in San Diego, he leads a group doing peptide and small molecule drug discovery and molecular pharmacology. In addition to research and teaching at the U.C. San Diego extension, he is also the author of Kaplan's *SAT II Biology E/M* and *AP Biology* and has written and edited a wide range of MCAT preparative materials. Currently he is writing the *Everyday DNA Handbook*, a practical guide to the biotechnology revolution in everyday life. The father of two young daughters, he loves spending free time with his favorite family and jogging on the beach.

Introduction

Organic chemistry is the chemistry of a single element out of the periodic table - carbon. What is so special about carbon that an entire course is devoted to it?

For one, the chemistry of carbon is the foundation of life on earth. The trees, insects and human beings would not be here without carbon. Life on another planet may be silicon-based, but on earth we are walking, living, breathing examples of sophisticated organic chemistry. Learning and understanding organic chemistry helps us to understand more about humans and our living planet, and is the foundation for learning biological sciences such as biochemistry and pharmacology.

In addition to providing the foundation to understanding life, organic chemistry has provided an enormously useful tool that has helped create much of the modern world, including plastics, polymers, medicines, dyes, fuel, lubricants, Teflon, and a lengthy list of other staples of contemporary life. For those planning a career in chemical engineering, materials science, medicinal chemistry or other similar fields, a solid grounding in organic chemistry is essential.

In fact, understanding chemistry is useful for everyone. Many environmental problems facing society are addressed most effectively with an understanding of chemistry in hand. The staples of our lives are results of chemical laboratories, from the plastic products we use to the medicines we take to the food additives we eat. Furthermore, a major trait learned in organic chemistry is analytical problem solving, an important skill in a variety of careers (and a major reason why medical schools require orgo as a pre-requisite). The logical thought processes involved in thinking through an organic chemistry problem can be similar to the process of thinking through a legal question or a medical situation.

As a college course, organic chemistry builds on inorganic chemistry and, depending on the text or the course, it may place great emphasis on the memorization of reactions. However, this book is an introductory text that emphasizes general concepts. With a solid foundation, it is much easier to understand the details for specific reactions, functional groups and strategies. Individual facts that may seem pointless on their own become part of a greater whole. Rather than memorizing a thousand independent pieces of information, it becomes possible to memorize a few general principles and then apply them. Specific reactions and molecules will be discussed, but primarily to illustrate a principle.

The first four chapters describe molecules, starting with the atom. Chemistry involves the interactions of electrons between atoms to form chemical bonds, making an understanding of electrons a key to understanding chemistry (Chapter 1). We can then move on to understanding how molecules are put together, looking at the variety of functional groups found in organic molecules and naming these molecules (Chapter 2). Molecules are not static, but have movement and interaction within molecules and with other molecules (Chapter 3). Organic molecules are often not flat but have three-dimensional structure essential to understanding the behavior of the chemical (Chapter 4). Chapters 5-8 present chemical reactions between molecules and an overview of the variety of organic molecules that are commonly encountered in the laboratory, the body or the lecture hall. Next we can deal with the practical aspects of organic chemistry, including lab techniques used to perform reactions and methods used to analyze the products of reactions and determine their molecular structure (Chapter 9). Finally, we will deal with organic chemistry as the foundation for the building blocks of life, with a brief survey of biological chemistry (Chapter 10).

This completes the explanation of the fundamental concepts of organic chemistry.

Then in Part II, we switch gears and go to review mode. In chapters 11-25 you can review all the essential content covered in an organic chemistry course. Here you'll also find practice problems you can use to build your organic chemistry skills or review for an exam. All problems are accompanied by complete explanations that tell you not only what the right answer is and why, but also why the other choices are wrong. You can use this section to prepare for the big test, whether it's the midterm, final, or a standardized test like the MCAT or DAT.

We hope that this text will enrich your learning of organic chemistry, and prove a useful reference. Chemistry is not a subject apart from life that you will only encounter in a college textbook or a professional school entrance exam. Chemistry is everywhere around us and inside us. What could be more interesting?

Periodic Table of the Elements

1 H 1.0																	2 He 4.0
3 Li 6.9	4 Be 9.0											5 B 10.8	6 C 12.0	7 N 14.0	8 O 16.0	9 F 19.0	10 Ne 20.2
11 Na 23.0	12 Mg 24.3											13 Al 27.0	14 Si 28.1	15 P 31.0	16 S 32.1	17 Cl 35.5	18 Ar 39.9
19 K 39.1	20 Ca 40.1	21 Sc 45.0	22 Ti 47.9	23 V 50.9	24 Cr 52.0	25 Mn 54.9	26 Fe 55.8	27 Co 58.9	28 Ni 58.7	29 Cu 63.5	30 Zn 65.4	31 Ga 69.7	32 Ge 72.6	33 As 74.9	34 Se 79.0	35 Br 79.9	36 Kr 83.8
37 Rb 85.5	38 Sr 87.6	39 Y 88.9	40 Zr 91.2	41 Nb 92.9	42 Mo 95.9	43 Tc (98)	44 Ru 101.1	45 Rh 102.9	46 Pd 106.4	47 Ag 107.9	48 Cd 112.4	49 In 114.8	50 Sn 118.7	51 Sb 121.8	52 Te 127.6	53 I 126.9	54 Xe 131.3
55 Cs 132.9	56 Ba 137.3	57 La* 138.9	72 Hf 178.5	73 Ta 180.9	74 W 183.9	75 Re 186.2	76 Os 190.2	77 Ir 192.2	78 Pt 195.1	79 Au 197.0	80 Hg 200.6	81 Tl 204.4	82 Pb 207.2	83 Bi 209.0	84 Po (209)	85 At (210)	86 Rn (222)
87 Fr (223)	88 Ra 226.0	89 Ac† 227.0	104 Unq (261)	105 Unp (262)	106 Unh (263)	107 Uns (262)	108 Uno (265)	109 Une (267)									

*	58 Ce 140.1	59 Pr 140.9	60 Nd 144.2	61 Pm (145)	62 Sm 150.4	63 Eu 152.0	64 Gd 157.3	65 Tb 158.9	66 Dy 162.5	67 Ho 164.9	68 Er 167.3	69 Tm 168.9	70 Yb 173.0	71 Lu 175.0
†	90 Th 232.0	91 Pa (231)	92 U 238.0	93 Np (237)	94 Pu (244)	95 Am (243)	96 Cm (247)	97 Bk (247)	98 Cf (251)	99 Es (252)	100 Fm (257)	101 Md (258)	102 No (259)	103 Lr (260)

PART I

FUNDAMENTAL CONCEPTS

Understand the fundamentals. This section provides an easy-to-understand explanation of the fundamental concepts of organic chemistry. Use it before you take the course to get a head start so that the course becomes easier and more understandable. Or use it as a review before the midterm or final. You can also return to this section when you begin to prepare for a standardized test like the MCAT or DAT.

Atom and Eve

CHAPTER ONE

THE ATOM

The Greeks believed that if you took an extremely sharp knife and cut something (an apple, rock or bread) into smaller and smaller pieces, you would eventually have a piece so small that you could not divide it any further. The Greeks had pretty much the right idea. It is an idea that may seem pretty obvious today, but to the Greeks was controversial. There are indeed indivisible pieces called atoms, and they are the fundamental currency of ordinary matter that we see, breathe and feel every day.

People also used to think that there were four fundamental substances that made up the world: fire, earth, air and water. This part is a little bit off. There are not four elements, but over 112 and counting as short-lived man-made radioactive elements are added onto the table over the years. Substances made of different elements have different properties. Hydrogen is the lightest element, usually forms a gas, and comprises most of the universe. As a metal, gold is shiny, shapeable (malleable) and able to conduct electricity. Carbon forms diamonds and graphite and is the central element in organic chemistry. The way the atoms of each element are put together determines this dazzling array of different properties.

Atoms are comprised of a nucleus that is very dense and heavy, with a cloud of electrons moving around it. The nucleus is composed of two types of particles: neutrons, which have no electric charge, and protons, which have a positive charge. Electrons have a negative electric charge and 1/1840 as much mass as a proton or neutron, which have equal masses. The nucleus contains almost all of the mass of the atom, but it is extremely small in size; most of the atom is actually empty space. You can shoot a beam of neutrons (if you have a particle accelerator) through a thin sheet of solid material and have most of the neutrons come out the other side untouched, sailing through the empty space that comprises most matter. In fact, neutrons, protons and electrons are themselves composed of smaller particles, with perhaps smaller particles inside of these. But these particles inside particles are the domain of the physicists and understanding electrons is sufficient for organic chemistry.

Each type of atom, each element, has a specific number of protons in the nucleus, and an equal number of electrons around the nucleus. It is the number of protons (called the atomic number) that defines each element. The nucleus usually has roughly the same number of neutrons as protons, and atoms with the same number of protons but different numbers of neutrons are called isotopes of an element.

As scientists learned more about elements, they started to see that there was an order to the way that different elements behaved that depended on the weight of the elements, or the number of protons and electrons they had. A major step toward bringing order to chemistry was the creation of the periodic table of the elements. Going across the table in a row from left to right, each element has one more proton in the nucleus plus one electron to match, as well as neutrons. Elements in each column in the table tend to behave chemically in similar ways. Understanding how atoms are put together helps to understand the nature of the periodic table and chemistry.

Instructor question: Now that we've covered the basics, does anyone know how the atom is actually structured?

Student response: Sure. Everyone's heard about the model of the atom as proposed by Niels Bohr in 1913. Basically the atom is like a little solar system. The big heavy nucleus is in the middle like the sun, with the electrons spinning around it in defined orbits like the planets. The positive charge of the nucleus keeps the negatively charged electrons from spinning off into space.

Instructor response: Well...that used to be the theory, but not anymore. It turns out that electrons are not little planets spinning around in orbit, but are in fact both particles and waves, depending on how you look at them. As particles, electrons have mass and can hit something with momentum, as a discrete particle. Surprisingly, electrons also behave like a wave. As waves, electrons are described by "wave functions", solutions to complex equations in quantum mechanics that describe regions where electron-waves have a defined probability of being located around the atomic nucleus. According to the Heisenberg Uncertainty principle, it is not possible to know precisely both the position and momentum of an electron at any moment. What is possible to determine is the *probability* of an electron occupying a region in space, by using the wave function. These different regions in space where electrons are likely to be found in an atom are called **orbitals**. We'll discuss these further later.

ELECTRON SHELLS AND ORBITALS

All chemistry, including organic chemistry, is about electrons, so understanding electrons is of fundamental importance. It is the electrons on the exterior of the atom (farthest from the nucleus) that are accessible and available to interact with other atoms in chemical reactions. Electrons fill shells around the nucleus at different energy levels extending outward from the nucleus. The shell nearest the nucleus, shell #1, has the least energy; increasingly energetic shells lie further out. Sometimes electrons can receive energy from an outside source that excites them to a higher energy level. They can also lose this energy in the form of light or other energy when falling back to lower energy levels. There are seven shells in all as you move through the periodic table, and seven corresponding rows to the periodic table. Within each shell are different types of orbitals that electrons occupy.

The different types of orbitals an atom can have are given by letters, including s, p, d, and f. Orgo deals mostly with the first two, s and p orbitals, which is fortunate since these are the simplest. The most common atoms used in organic chemistry have only s and p orbitals. Different orbitals have different shapes. S orbitals are spheres centered around the nucleus while p orbitals are a little more complicated with two halves (lobes) centered around the nucleus like a barbell. One lobe is + and the other is -, with a **node** in the middle where the electron can never be found (there is a 0% probability of finding electrons at the node). The plus and minus signs given to the two lobes of the p orbital do not indicate electric charge, but opposite spin of the electrons in the orbital, where spin is one of the properties of electrons. There are three p orbitals in each shell along three different axes in space (x, y and z). S orbitals are closer to the nucleus and have less energy than p orbitals.

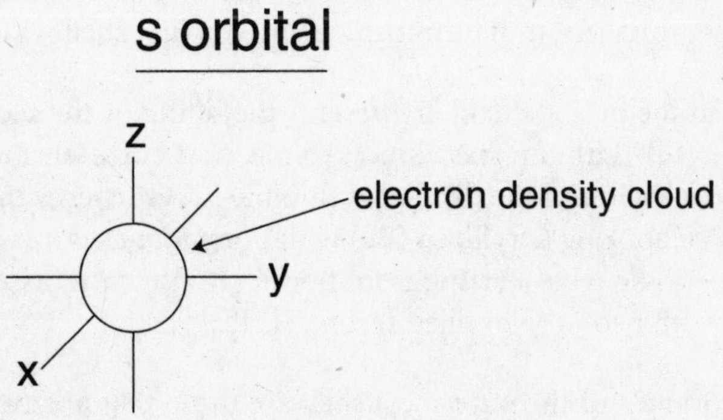

s orbital

z
electron density cloud
y
x

An orbital can be empty, with no electrons, or it can have one or two electrons in it. If an orbital has two electrons in it, the electrons must have different spin. When drawing electrons as arrows in orbitals, the spin of electrons is sometimes indicated by an arrow pointing up or pointing down. See for example the organization of the electrons of carbon into orbitals below. Each electron is indicated as an arrow in the different orbitals. The different spin of the two electrons in an orbital together is indicated by placing one arrow up and one arrow down.

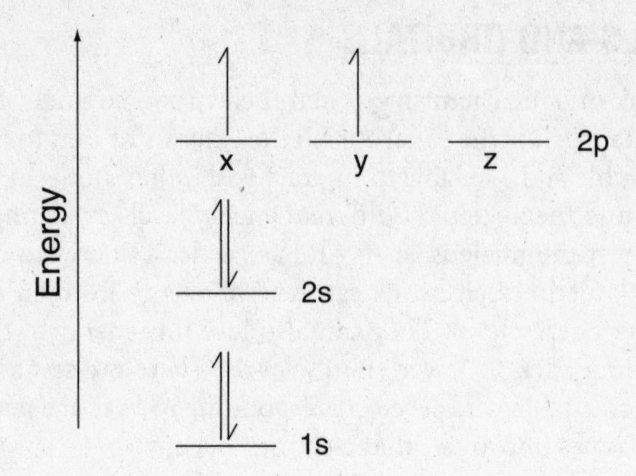

ELECTRONS AND THE PERIODIC TABLE

As atoms get larger, electrons fill the shells and orbitals of an atom in a certain order, filling in lower energy shells and orbitals first. This rule is called the Aufbau principle, and can be seen in the periodic table (see page xii). Going across the periodic table from left right, more and more electrons are added to the outer shell until it is filled (the lowest energy state). Hydrogen (H) has only one proton, and one electron. With only one electron, it will go into the lowest energy orbital, 1s. This orbital, like all orbitals can accommodate two electrons, as long as their spin is different. The next element is helium (He), with two protons. With two protons, helium also has two electrons. The second electron (with opposite spin) also goes into the lowest energy orbital, 1s, to fill this orbital and the first shell with 2 electrons.

The next level down in the periodic table represents the filling of the second electron shell, starting with lithium (Li). Lithium has three protons and three electrons. The first two electrons satisfy the first shell, and the third goes into the lowest energy orbital in the second shell, the 2s orbital. Neighboring beryllium (Be) with one more electron has a filled 2s shell. Traveling across the periodic table starting from boron (B), the next six electrons fill the 2p orbitals, culminating with neon's (Ne) filled second shell.

The electrons in the orbitals of the outermost shell are those that are available for bonding with other atoms. Electrons in inner shells are not available since they are tucked close to the nucleus; they are not reactive, since these shells are filled. Atoms "like" to have their outer shells filled, to complete an **octet.** This allows them to exist in a stable or low energy state. The word octet refers to having eight electrons in the outer shell, a full outer shell for the atoms we will discuss, except for helium (which has a complete outer shell of 2 electrons). Completing octets is essential to organic chemistry.

The varying degrees to which elements give up electrons or attract electrons to complete their octets describes and predicts their chemical behavior. The order in which orbitals are filled brings order to the periodic table. In the right hand column are the noble gases, including helium, neon, and argon.

Instructor question: Which elements would you predict to be most or least reactive?

Student response: I'd say the noble gases would be the least reactive chemical elements, because they already have complete octets, and have no energy driving the sharing of electrons with other atoms.

Instructor response: Exactly. How about over on the far left side of the table, in that first column—the alkali metals. Or the halogens in the second to last column on the right?

Student response: I guess I'm supposed to think about octets. Lithium and the elements in its column have only one electron in their outer shell and want to give up one electron to leave a filled inner shell. Fluorine and the other halogens (chlorine, bromine, etc), with seven valence electrons tend to draw one electron from other atoms to complete their octet. So both these columns must be pretty reactive.

Instructor response: You got it. An element's position in the periodic table indicates how many valence electrons it has in its outer shell available for bonding, and what type of bonds it is likely to form.

To summarize the rules for how electrons fill shells and orbitals:

1. Aufbau principle: Electrons are added first to the lowest energy orbitals, and then to new orbitals in order of increasing energy.
2. Pauli Exclusion principle: Electrons can only occupy the same orbital if they have opposite spin, and even then only two electrons can be in the same orbital.
3. Hund's rule: For orbitals that have the same energy, one electron is placed into each orbital before two electrons are placed in each orbital. For example, for the three p orbitals in the second electron shell, one electron will be added to each of these before any of them receives a second electron.

WHAT'S SO SPECIAL ABOUT CARBON?

Organic chemistry is essentially the chemistry of carbon, and to a large extent the basis of life. What are the unique properties of carbon that allow it to play such a key and versatile role?

Carbon is element number 6 in the periodic table, with six protons and electrons, and is relatively abundant in the universe compared to most elements. Generally, the more abundant elements in the universe are those with lower atomic weights (helium and hydrogen being the most abundant) with heavier elements increasingly rare. The abundance of elements in the universe is related to their rate of creation through nuclear fusion in stars, with elements heavier than helium forming when stars consume their hydrogen fuel and fuse heavier nuclei together as the star dies. When a star becomes a nova and explodes, the elements are dispersed to form new planets and stars, including our own humble little planet and all of its inhabitants. Oxygen, hydrogen, and nitrogen are also common elements in organic chemistry, and these are also relatively light, abundant elements.

Carbon is in the middle of the periodic table, with four electrons in its outer shell available for bonding. It can share these four electrons along with four electrons from other atoms to complete its octet, thus forming up to four stable covalent bonds with other atoms. The ability to form four covalent bonds increases the variety of structures carbon can form compared to a molecule like oxygen that forms two covalent bonds. Imagine a set of tinker toys, with beads and rods that connect them. You can make a lot more things using beads that have four connections than beads with two or three connections. Carbon can form long chains (octane), rings (benzene, glucose, cyclohexane), sheets (graphite) networks (diamond), or even soccer-ball shaped molecules called buckyballs. Being in the middle of the periodic table, carbon makes stable bonds with a lot of different atoms, further extending the variety of carbon-containing compounds.

CHEMICAL BONDS

So far we have talked about atoms, but atoms alone do not make chemistry. Chemistry is the interaction between individual atoms to form chemical bonds and molecules. Different types of bonds form depending on the elements involved and how they share electrons to fill octets.

A key factor in bond formation is the **electronegativity** of the atoms involved. Electronegativity is a characteristic of elements that describes how strongly an atom holds onto the electrons in its outer shell. An element that draws electrons toward itself and holds them very tightly is very electronegative. Another means to describe the electronegativity is the **ionization potential** of an atom, the amount of energy required to remove an electron from an atom to create an ion (see below). The greater the electronegativity, the greater the ionization potential. In other words, the stronger an atom holds on to its electrons, the more energy is required to yank it away.

$$Na + \text{energy} \longrightarrow Na^{+} + e^{-}$$

In the periodic table, there are two trends that affect electronegativity, one going across the table and the other going down. Going across the table from left to right, the elements in a row contain more and more protons, with more positive charges in the nucleus, while the electrons are still filling the same energy shell at roughly the same distance from the nucleus. This puts a greater positive charge in the nucleus pulling the electrons in more strongly and making elements more electronegative as one travels from left to right in the table. Going down the table, the elements become larger and less electronegative; elements further down in the table are filling outer shells that are further away from the nucleus, and so are held more loosely. Also, in heavier atoms, electrons in the inner shells act like a shield that reduces the hold of the nucleus on the electrons in the outer shell. Elements on the left side of the table that are less electronegative are more likely to donate electrons in a bond while elements on the right side of the table that are more electronegative hold on tightly to their own electrons and are more likely to accept electrons, filling their own octet. Carbon, in the middle of the periodic table, has moderate electronegativity, another feature of carbon that allows it to play a versatile role in chemistry.

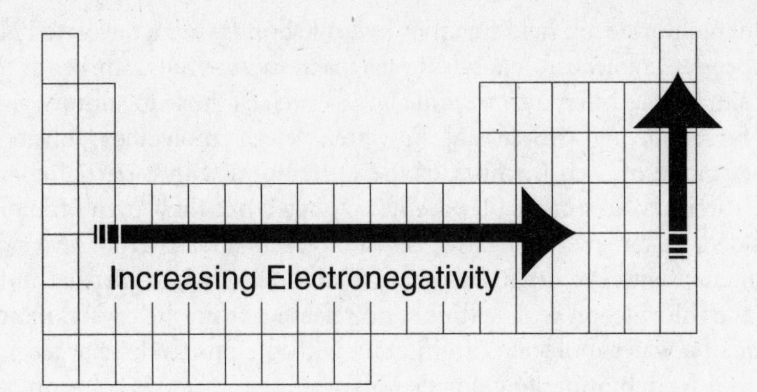

IONIC BONDS

Atoms can form different types of bonds depending on how they share electrons. One type of bond that forms between atoms is called the **ionic bond.** In the ionic bond an electron is transferred completely from one atom to another. The atom that donates the electron becomes positively charged since it now has one more proton than electrons. Atoms with a charge are called **ions** and positively charged atoms are called **cations.** The atom that accepts the electron now has one more electron than protons and is a negatively charged ion, or **anion.** Ionic bonds occur between one atom that is very electronegative and another that is not (also called electropositive).

An example of an ionic bond is the formation of sodium chloride, table salt. Sodium, an alkali metal on the left side of the periodic table with a single electron in its outer shell, is weakly electronegative, meaning that it holds its single valence electron loosely. Chlorine, a very electronegative atom, has seven electrons in its outer shell and "wants" to attract an electron to complete its octet. A bond between sodium and chloride is a match made in heaven. When sodium chloride is formed, sodium gives an electron to chlorine, leaving each sodium atom with a positive charge and each chlorine atom with a negative charge. The positive and negative charges attract each other and hold the atoms together. In the process of forming the ionic bond, sodium is left with the filled octet of the inner shell left behind and chlorine also has a filled octet. Everybody gets a complete octet and everybody wins.

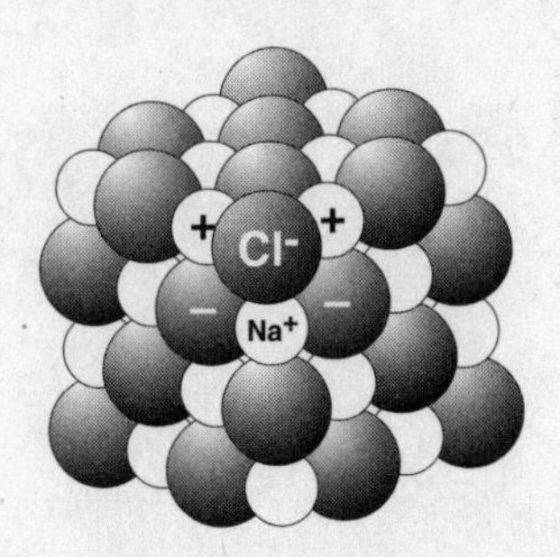

Crystals of sodium chloride are held together by ionic bonds. In fact, isolated NaCl does not exist as an independent molecule. In a salt crystal, each sodium atom and each chloride atom exist as part of a matrix of other atoms. Ionic bonds such as those in sodium chloride can be broken easily by dissolving the crystal in water. Water molecules solvate each atom, neutralizing the charge of each member of the ionic bond with a partial charge that each water molecule carries. Water molecules are not charged, but they are a little bit positive on the hydrogen side and somewhat negative on the oxygen side. The positive side of a water molecule can interact with the negatively charged chloride ions, to disrupt and displace the interactions of the chloride ion with sodium and release it from the crystal matrix. Similarly, the negative side of a water molecule can interact with the positively charged sodium in the crystal, to release it from binding to chloride ions, and thereby enter solution.

Bonds involving carbon, on the other hand, are never ionic. Because carbon has moderate electronegativity, it always shares electrons rather than completely donating or accepting them. Carbon can have a transient charge in some reactions, but this state is highly reactive and unstable. Which brings us to…

COVALENT BONDS

In ionic bonds, electrons are fully transferred from one nucleus to another. In a covalent chemical bond, electrons are shared by two atoms to complete the octet in the outer shell for both atoms involved. Atoms without filled octets are highly unstable chemically and rarely exist for long in nature. Covalent bonds occur between atoms that are not too far apart in their electronegativity, so that neither atom completely draws electrons away from the other atom in the bond. As a result of sharing, an electron can fill orbitals associated with both atoms, killing two birds with one stone.

In some covalent bonds, one of the atoms involved is more electronegative than the other to the extent that it draws the electrons in the bond partially but not completely toward itself, creating a partial charge in the bond. These bonds, in which electrons are shared unequally between the atoms in the bond, can be called covalent polar bonds.

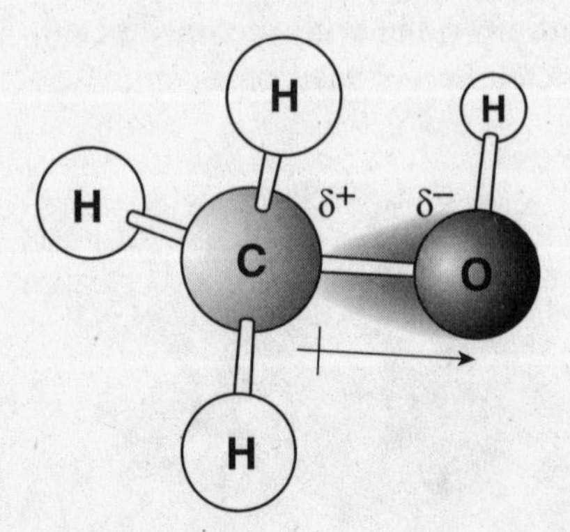

In a carbon-oxygen covalent bond, seen above, oxygen is more electronegative than carbon, drawing electrons toward the oxygen to create a partial negative charge on the oxygen and a partial positive charge on the carbon. Partial charges in bonds play an important role in the interaction between biological molecules, such as the hydrogen bonding in proteins and DNA which helps hold these molecules together.

Other covalent bonds are nonpolar. Carbon and hydrogen have almost the same electronegativity, so C-H bonds are non-polar, with electrons shared equally between both atoms in the bond. Compounds with only carbons and hydrogens have virtually no **dipole moment**. A dipole moment is due to unequal distribution of electrons in a molecule. Higher electron density in one area over another induces charge and polarity within the molecule, even if the molecule is not charged overall.

Sometimes the polar covalent bonds result in a partial charge in a molecule as a whole, creating a dipole moment.

Instructor question: Can a molecule with polar bonds have no net charge?

Student response: No, because polarity indicates dipole moments, right?

Instructor response: Actually, it all depends on the geometry of the molecule in question. Let's look at two examples. Chlorine is highly electronegative, and draws electrons away from carbon in a C-Cl bond. In trichloromethane, the presence of three chlorines creates an overall polarity in the molecule.

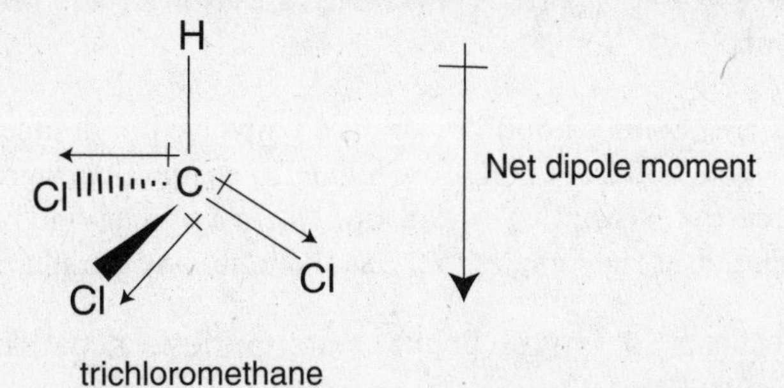

trichloromethane

If the molecule is carbon tetrachloride, on the other hand, the bonds are still polar, but equally so in all directions, resulting in no dipole moment on the molecule as a whole even though the bonds are polar.

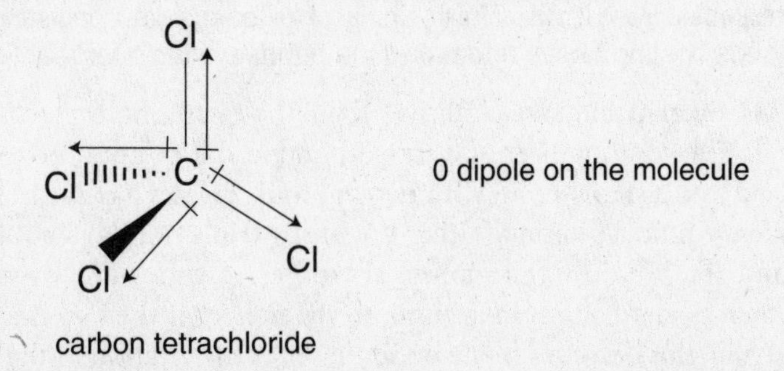

carbon tetrachloride

As noted earlier, electrons behave like waves when it comes to forming covalent bonds. When waves of any sort have the same sign and overlap, the waves are additive and reinforce each other. Waves of the opposite sign exhibit destructive interference. When two atomic orbitals overlap, their waves can overlap and add together to create a **molecular orbital** that is not an orbital for one atom or another, but an orbital for two atoms combined. A molecular orbital with two electrons in it shared between two atoms is a covalent bond. The molecular orbital with shared electrons draws the atoms together and counteracts the normal repulsion between positively charged atomic nuclei. Covalent bonds between atoms are basis of carbon chemistry and all covalent bonds are the result of molecular orbitals with shared electrons between atoms.

When two atomic orbitals join to form a molecular orbital, they actually form two orbitals, one of which is the molecular orbital discussed above that is involved in bonding. The other molecular orbital is called an antibonding orbital, but is not important for most chemistry. It is mainly of interest to note that orbitals are conserved in this way.

Both s and p orbitals are involved in forming molecular orbitals, and molecular orbitals occur in two different flavors, **sigma bonds** and **pi bonds**. All single bonds are sigma bonds. The greater the overlap between the atomic orbitals that create the molecular orbital, the stronger the bond will be. Sigma bonds with s orbitals are shorter, stronger and more stable than bonds with p orbitals, since the s orbitals are closer to the nucleus and contain lower energy electrons. One example of a sigma bond occurs when two hydrogen molecules form a hydrogen molecule when their 1s electrons form a molecular orbital together. When two p orbitals overlap end to end, or when an s and a p orbital overlap, these also form a sigma bond. Different sigma bonds can have different strengths and lengths depending on the energy of the electrons involved. Shorter sigma bonds are stronger and take more energy to break than longer bonds.

Whereas sigma bonds create single bonds, pi bonds are involved in double bonds and triple bonds, together with sigma bonds. Pi bonds are higher in energy than sigma bonds, weaker, less stable, and more reactive. When two carbon atom form a sigma bond between them, this leaves p orbitals unbonded, at right angles to the sigma bond and parallel to each other.

> **Instructor question:** So are pi bonds more or less accessible as participants in reactions?
>
> **Student response:** I'd say electrons in pi bonds are more easily involved in reactions than those in sigma bonds because they lie above the plane of the atomic nuclei.
>
> **Instructor response:** Yes. It's helpful to think of pi bonds as consisting of electron clouds floating above and below the bond, available for reactions.

If parallel p orbitals overlap and are additive, then they can share electrons to form a molecular pi orbital. The combination of a sigma bond and a pi bond between two atoms creates a double bond and a sigma bond with two pi bonds creates a triple bond. There is no rotation of molecules around a pi bond (either double or triple bonds) since the molecule is held in place around the bond by the molecular orbital. If you hold two pieces of wood together with one nail, a board can spin around on the nail, as if the nail is a sigma bond; if you put in a second nail, this becomes the pi bond that prevent rotation of the boards around the sigma bond.

HYBRID ORBITALS

Sometimes s and p orbitals around an atom can combine to form orbitals that have characteristics that are a mixture of s and p orbitals. These orbitals are called **hybrid orbitals.** The number of hydrid orbitals formed is equal to the number of atomic orbitals that hybridized to form them (orbitals are conserved). An s orbital can hybridize with one p orbital to make two sp orbitals:

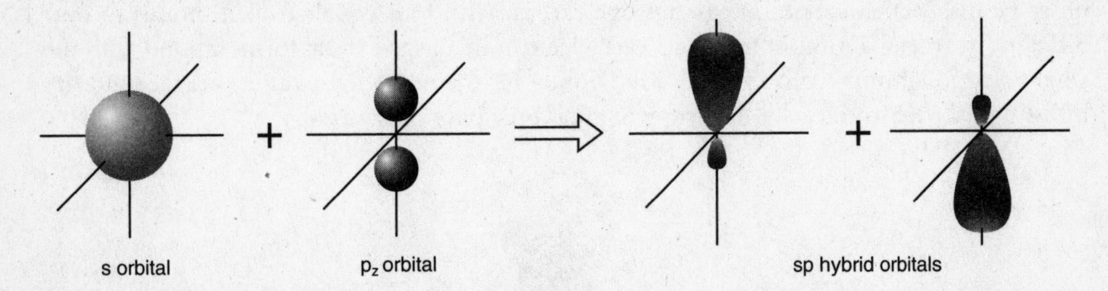

Hybridization of an s with 2 p orbitals creates 3 sp^2 orbitals and hybridization of one s with 3 p orbitals creates 4 sp^3 orbitals. The hybrid orbitals are all equal to each other and have an energy level and shape that are a mixture of those of the original orbitals:

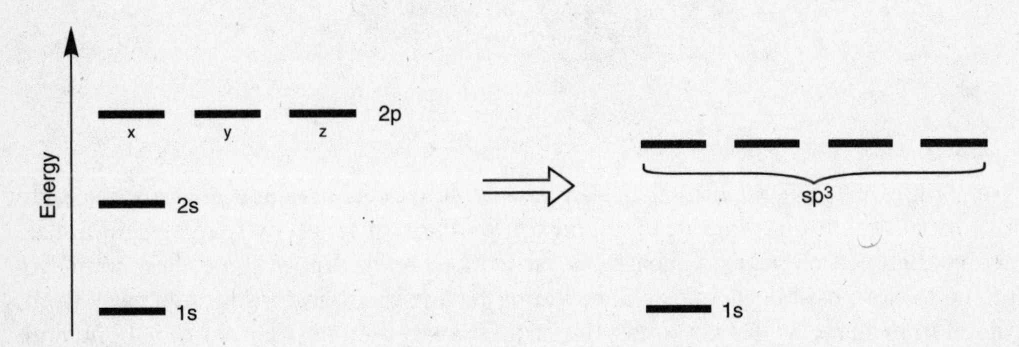

An sp orbital, for example, has characteristics that are part s and part p:

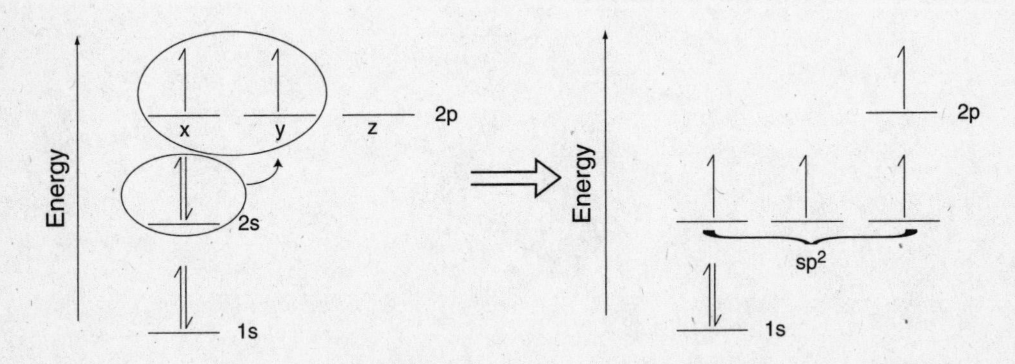

Carbon has 2 electrons in the 1s orbital, and 2 electrons in the 2p orbitals. If carbon were stuck with its electrons this way for bonding, then it would not form the four equivalent covalent single bonds that we know it does, so there must be something wrong with this plan. Turns out carbon uses hybrid orbitals to form covalent bonds using the 2s and 2p valence electrons in its outer shell. The s orbital and the three p orbitals are used to form four equivalent sp^3 hybrid orbitals, each containing one electron. By forming four equal sp^3 hybrid orbitals from the 1s and the 3 p orbitals, carbon forms four equal covalent bonds with other atoms. Methane, natural gas, has one carbon with four equal covalent bonds to four different hydrogen atoms. In methane, each electron in an sp^3 orbital forms a bond with the single electron from hydrogen. The four bonds have equal lengths and energies and are indistinguishable from each other, showing that they have hybridized.

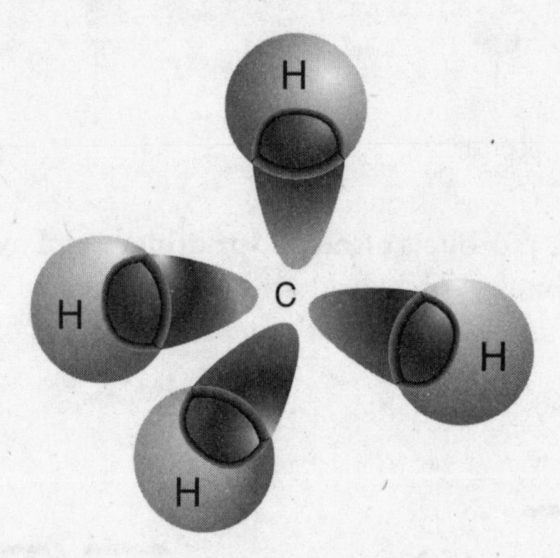

The four bonds in sp^3 bonded methane are also at equal distances and angles from each other. Normally bonds are arranged to maximize their distance in space to minimize interactions between electrons, which repel each other. As we know, molecules "want" to occupy the lowest possible energy state, including having their electons located away from each other. In methane, with four bonds, the furthest apart they can be is for each bond and hydrogen to be arrayed at the four points of a tetrahedral pyramid with 109.5 degree angles between bonds:

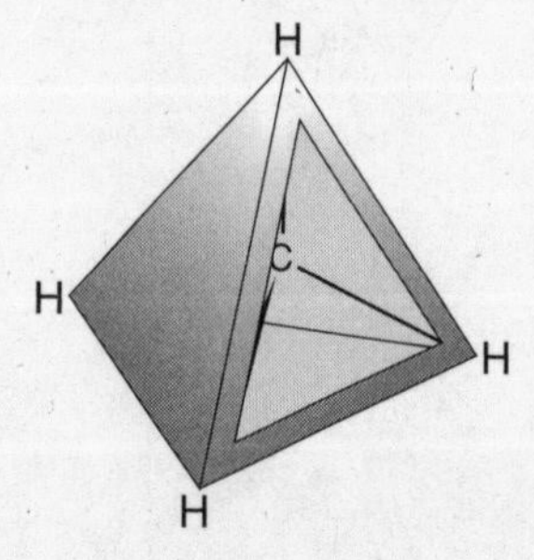

Molecules with sp^3 orbitals involved in bond formation tend to have tetrahedral shape, with 109.5 degree angles, though they can deviate from this value if the bonds are strained. In fact, this energetically favorable tetrahedral shape cannot occur without hybridization.

Ethylene, a plant hormone that makes fruits ripen, is a simple molecule with a double bond between two carbons:

H H
120° C=C
H 120° H

Ripe apples release ethylene and will make other fruit nearby ripen as well. Fruits that are picked before ripening are shipped to the supermarket and then sprayed with ethylene to make them ripen at the store. The double bond of ethylene includes a sigma bond and a pi bond between the carbons. In each carbon, two p orbitals are hybridized with the s to make three sp^2 hybrid orbitals that are intermediate between s and p.

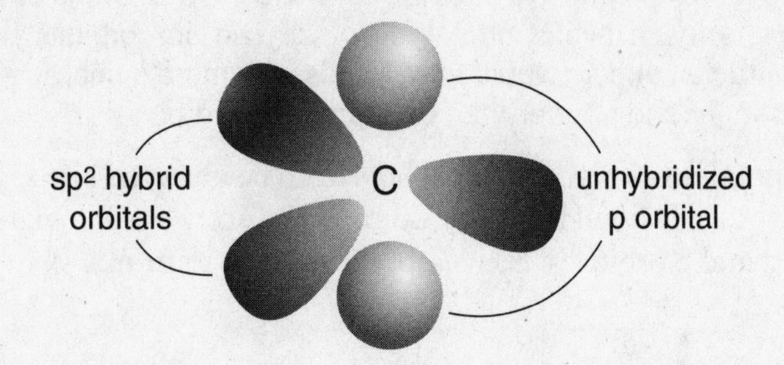

The sp^2 hybrid orbitals all lie in a plane together, spread at 120 degree angles from each other, with the remaining p orbital projecting out of the plane at right angles. With three orbitals in a plane, 120 degrees is the greatest possible angle of separation between the bonds (3 x 120 degrees = 360 degrees). Hybrid sp^2 orbitals from each carbon form a sigma bond between the carbons, while the remaining sp^2 orbitals form molecular orbitals with hydrogens. The figures below represent the double bond in ethylene in different ways.

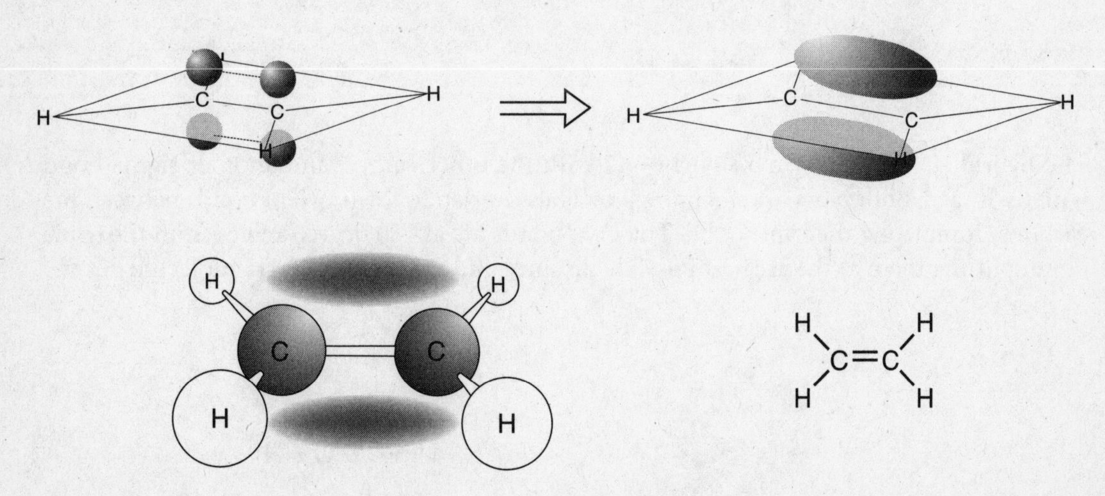

The unhybridized p orbitals above the plane overlap to form the pi bond of the double bond found in ethylene. Like all pi bonds, the pi bond in ethylene is weaker than a sigma bond since it has less overlap and involves p orbitals rather than s orbitals.

Hybrid orbitals have characteristics that are a combination of the atomic orbitals that go into the hybrid, in proportion to the contribution of each.

> **Instructor question:** Comparing sp^2 and sp^3 orbitals, which will create stronger bonds?
>
> **Student response:** Well, sp^3 orbitals are more stable, right? I'd say they would be stronger.
>
> **Instructor response:** Not quite. Think about the ratio of the s and p orbitals that go into making both types of hybrid orbitals. In sp^2 orbitals there is one s for two p orbitals (the hybrid is 1/3 s), while in sp^3 there is one s for three p orbitals (each hybrid orbital is 1/4 s). The sp^2 orbitals have a greater s character, and so would form bonds with more s character, and would be stronger and shorter than bonds with sp^3 orbitals.

The last example of bond hybridization, the sp bond, can be observed in acetylene (technically called "ethyne"), with a triple bond between two carbons. Acetylene is a highly flammable gas used in welding. An s and p orbital in each carbon hybridize to form 2 sp orbitals.

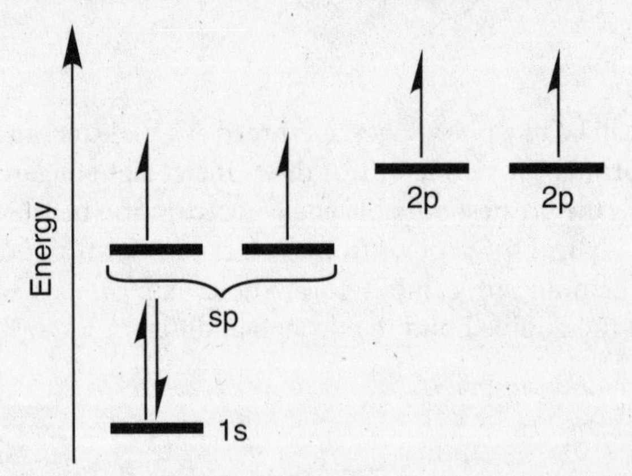

The hybrid sp orbitals form a sigma bond with the other carbon and another sigma bond with hydrogen. Both pairs of remaining p orbitals overlap to form two pi bonds between the carbons, completing the triple bond. The C-H bonds are at 180 degree angles from the triple bond, putting them at the greatest possible distance and making the entire molecule linear.

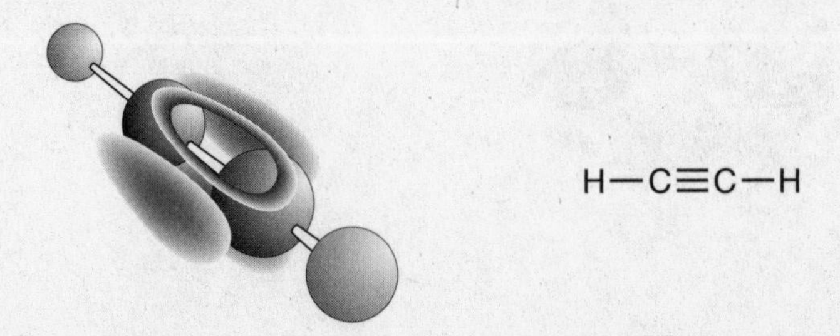

HOW TO DRAW BONDS AND MOLECULES

Molecular Formulas

Molecular formulas indicate which types of atoms are in a molecule and how many of each are involved. For example, $C_6H_{12}O_6$ is the molecular formula of glucose, indicating that each glucose molecule has 6 carbons, 12 hydrogens, and 6 oxygens. Unfortunately, many other molecules consist of the same atoms and each molecule can behave differently. The molecular formula alone does not indicate how the atoms in the molecule form bonds with one another.

Lewis Structures

Orbitals are the most accurate way to depict electrons, and it is important to understand orbitals, but they are a bit hard to draw on paper and not a convenient way to represent most of organic chemistry. There are easier ways to draw the bonds between atoms, starting with drawings called Lewis structures, named after the chemist who invented this method of drawing molecules. In a Lewis structure, each atom is drawn on the page with all of its valence electrons, those in its outer shell, drawn in pairs around the atom. Electrons are represented by a dot. To draw a Lewis structure:

- Draw all atoms, arranging the atoms around each other in a way that makes sense.
- Draw all valence electrons around each atom, correcting for any electrons lost or gained to create an ionic charge.
- Arrange electrons between atoms to complete the octets of all atoms. Two electrons drawn between atoms represent the shared electrons in a molecular orbital that form a bond between atoms. Four electrons shared between two atoms represent a double bond and six are a triple bond.
- Show any charges present on atoms by dividing shared electrons between atoms and comparing the number of valence electrons assigned to each atom in the structure with the atomic number of the atom. For example, if an atom has six valence electrons in its outer shell, but five electrons in a structure, the atom has a formal charge of +1.

In the drawing of methane drawn as a Lewis structure, carbon is drawn first with its four valence electrons, and then four hydrogens are arranged around it, each with one electron of its own. In total, there are eight electrons, and all of the atoms in the molecule have filled outer shells. Lewis structures are much more convenient to represent bonds and to track electrons than trying to draw molecular orbitals. To make things simpler still, the shared electrons in covalent bonds between atoms are commonly drawn as lines.

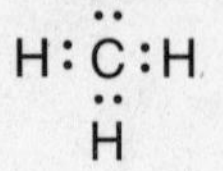

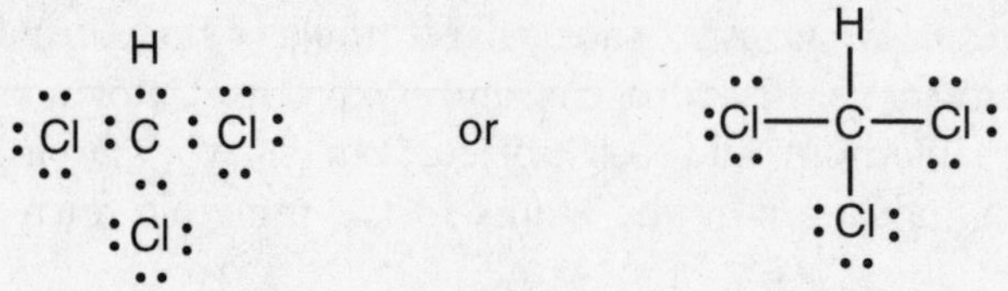

H
H:C:H
H

H
:Cl:C:Cl:
:Cl:

or

H
:Cl—C—Cl:
:Cl:

Each line is the equivalent of two electrons in a shared molecular orbital. A single line between atoms represents a single sigma bond, a double-line represents a double bond, and a triple line represents a triple bond. In line structures, unbonded electrons in the outer shell as sometimes still drawn as dots, especially when sketching the movement of electrons in a reaction. These electrons are often omitted from a drawing to leave only the bonds shown in the structure. Sometimes in organic molecules, these structures are simplified even further, leaving out the hydrogen atoms and bonds, making these assumed, and not specifying carbon atoms, leaving these assumed as well.

Instructor question: Look at the following structures for dimethylsulfoxide below. What is the formal charge on the sulfur?

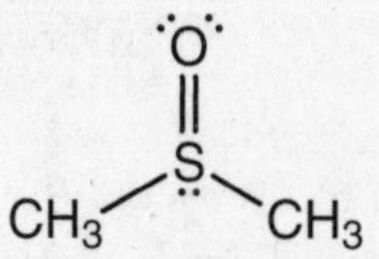

O
S
CH_3 CH_3

Student response: First of all, the equation for formal charge is:

Formal charge = (# valence e−) − [(1/2 # bonding e−) + (# nonbonding e−)]

Here, we know sulfur is a group VIA element so it has six valence electrons. In this molecule, it has four bonding pairs of electrons (or 8 bonding electrons) and one pair of nonbonding, or 2 lone, electrons. So the formal charge is 6 – (1/2 * 8) – 2 = 0. It has no formal charge.

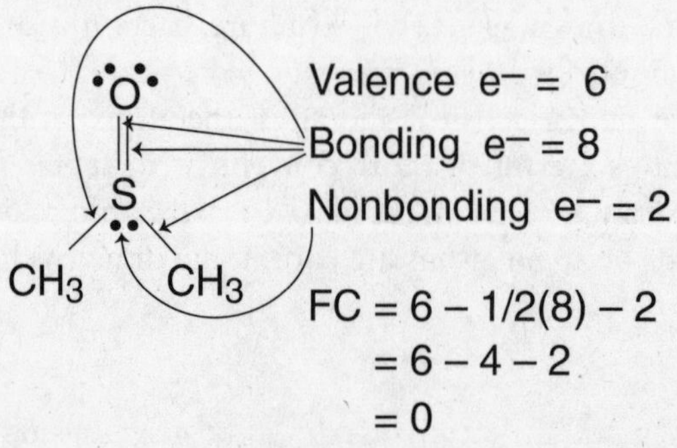

O
S
CH_3 CH_3

Valence e− = 6
Bonding e− = 8
Nonbonding e− = 2

FC = 6 – 1/2(8) – 2
= 6 – 4 – 2
= 0

Molecules themselves are too small to take pictures of, one reason why we draw them, but models can represent molecules in a way that helps us to understand them. University bookstores often sell molecular model kits that you can use to supplement a text by actually holding three-dimensional molecules in your hands. (This will come in handy in chapters 3 and 4.) Models will usually represent atoms as balls and bonds as sticks connecting them. They don't demonstrate the true nature of bonds and orbitals, but they do allow an easy way to look at the connectivity and geometry of organic chemicals.

RESONANCE

Some molecules can be drawn in more than one way. An example below is acetic acid, the chemical that makes vinegar smell and taste sour.

$H_3C—C$ $H_3C—C$

Instructor question: Compare the two drawings above – Which is correct? Which structure would be favored?

Student response: They look pretty much the same to me. How do you know which one is right?

Instructor response: You have the right idea. Both of these drawings are equally correct and indistinguishable, and neither of them exactly represents the true structure of acetic acid. The electrons in the pi bond between carbon and oxygen are not truly located between one or the other C-O, but are delocalized between the carbon and both oxygens. One way to represent this sharing of electrons between bonds is with both line drawings enclosed in large brackets, representing extremes in the movement of electrons between one bond and the other. In this case, both form are equivalent and would exist equally as resonance structures of acetic acid. The C-O bonds are both equivalent, and are intermediate between a single bond and a double bond in length. The movement of electrons within the bonds can be represented with an arrow, as in the following:

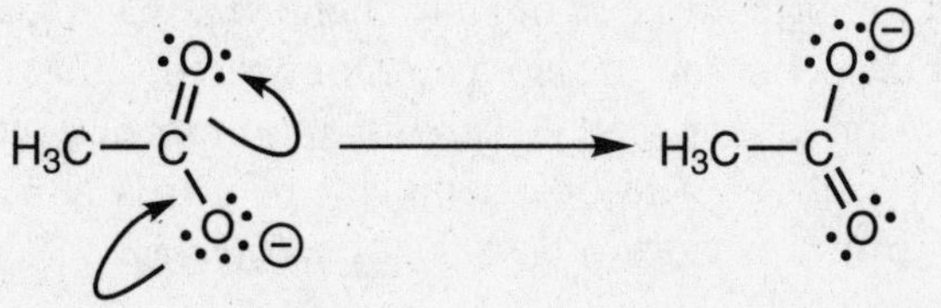

$H_3C—C$ → $H_3C—C$

This notation for the movement of electrons is often used when talking about reactions as well. The bonds in which the electrons are delocalized are hybrids between double and single bonds and are equal in length and strength in acetic acid. The delocalization is sometimes drawn as a dotted line across both bonds rather than specifying the location of a double bond in one location or the other. In molecular orbital terms, the delocalized electrons could be shown as:

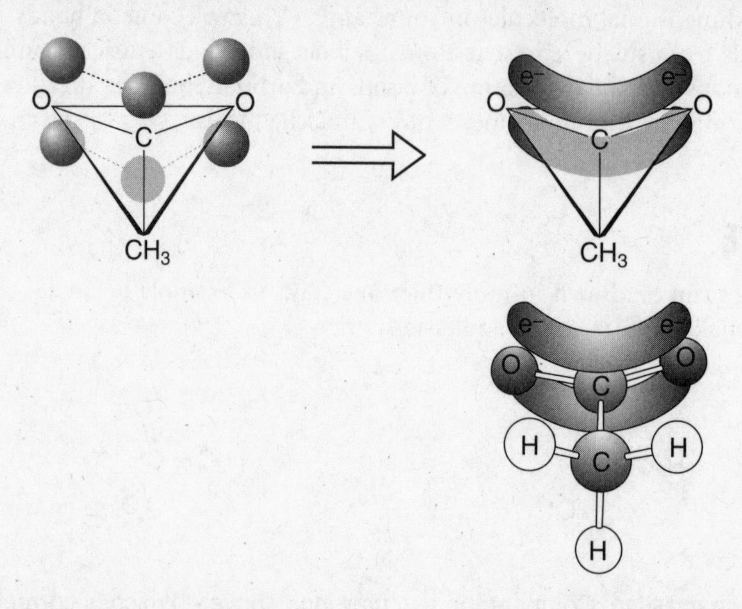

Student question: So resonance structures always contribute equally?

Instructor response: Not at all! Sometimes resonance structures are not identical to each other and contribute unequally. For example, if the movement of electrons creates a formal charge in one resonance form but not another, then the one with the charge is a less stable, higher energy structure that will contribute less to the overall structure. Also, if a resonance structure puts a charge an atom that normally would not support that kind of charge, it is probably not a great resonance structure. For example, if one structure has a charge of +1 on an oxygen and another structure with a charge of –1 on that oxygen, the latter is clearly a more significant contributor, because oxygen is an electronegative atom and can stabilize a negative charge much more easily than a positive one.

ACIDS AND BASES

Acids and bases play an important role in organic chemistry, making a brief review of acids and bases worthwhile. One definition of acids commonly used is that of Bronsted-Lowry, in which an acid is a proton donor, and a base is a proton acceptor. Every acid has a corresponding base that forms an acid-base conjugate pair. In a typical acid-base reaction, the transfer of a proton from an acid to a base can be written as:

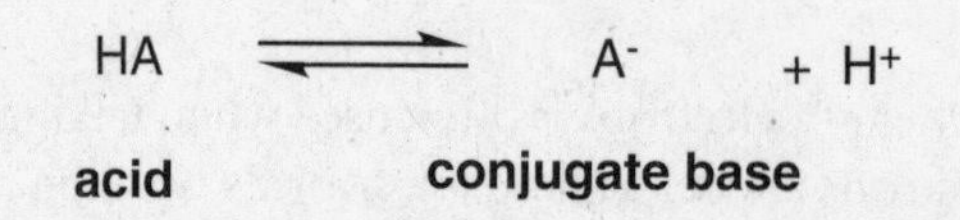

In this system, A- is the conjugate base and HA is the acid. An acid-base reaction does not need to look exactly like this however. Anything that accepts a proton is a base. Water can be either an acid or a base, depending on whether it is gaining or losing a proton in a given circumstance.

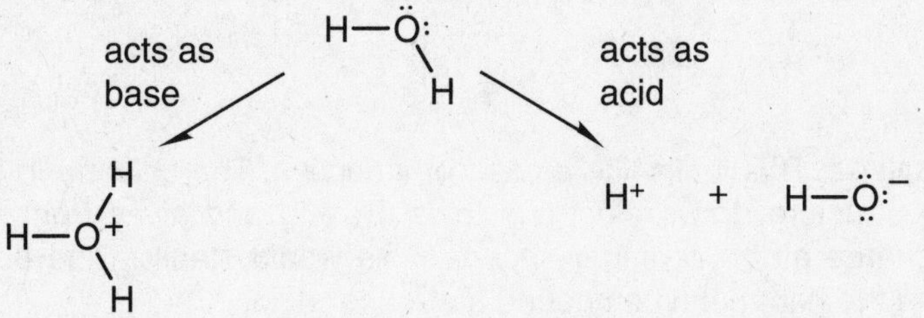

acts as base
acts as acid
H_3O^+ ← H_2O → H^+ + HO^-

Acid-base reactions exist in equilibrium in solution, described by the equilibrium constant K_a.

$$K_a = \frac{[A-]\,[H+]}{[HA]}$$

The K_a of compounds is a measure of the relative acidity of compounds. If an acid is strong and tends to completely dissociate in solution, then the equilibrium is driven to the right of the dissociation reaction, and the K_a is large. For a compound with a more basic character, it will tend to hold onto its protons, leading to a very low concentration of the conjugate base at equilibrium and a very small K_a equilibrium constant. The K_a of compounds can be compared as a measure of the relative acidity of compounds, just as K_b, less commonly used, measures basicity.

In a conjugate acid/base pair, if the acid HA is very strong, what does this indicate about the strength of the conjugate base A^-? If the K_a is large, and dissociation of HA is favored (it is a strong acid), then the conjugate base A^- is a very weak acceptor of protons. The structure of acids and bases determine the strength with which they donate or accept protons. If a conjugate base is very stable, then the acid will be a strong acid, so anything that stabilizes a conjugate base will make an acid stronger. Two of the factors that stabilize conjugate bases are resonance and induction.

Induction occurs when a electronegative substituent draws electrons toward itself, away from the negative charge created in the conjugate base. The electrons in nearby bonds are drawn in the same direction, and the negative charge of the conjugate base is partially spread out toward the electronegative substituent. Anything that diffuses or spreads out a negative charge will make it more stable. The more electronegative the substituent is, the greater the stabilization of the conjugate base.

Instructor question: In the following pair of compounds, which is the stronger acid?

$$\mathrm{CH_3COOH} \qquad \mathrm{CH_2ClCOOH}$$

Student response: This looks like a case of induction. The chlorine in the right-hand structure draws electrons toward itself, and away from the negative charge of the conjugate bases. This would stabilize the base and making the right-hand compound a stronger acid.

Instructor response: Yup. The greater electronegativity of a group, the greater the induction and the greater the acidity of the compound. For instance, a fluorine compound will be more acidic than a compound with a chlorine in the same position. Fluorine is higher in the periodic table than chlorine, so more electronegative. It will draw electrons more strongly toward itself, causing greater stabilization of the conjugate base through induction. Adding more than one electronegative group will also draw more electrons away to stabilize the base and increase acidity. Trichloroacetic acid is more acidic than acetic acid with one chlorine, for example. Also, the strength of induction and acidity depends on how far away electronegative groups are from the basic group they stabilize. The farther away, the weaker the induction, and the weaker the acid.

To summarize induction, stronger acids have:

- greater electronegativity of substituents
- a larger number of electronegative substituents
- closer electronegative substituents to basic group

Resonance also helps to stabilize conjugate bases and increase acidity. Resonance can delocalize the negative charge of the base, making it more stable if the negative charge is in a position where the electrons can move within the molecule. Similar molecules without resonance in which the acidic charge of the base is fixed are much less acidic. Resonance delocalization stabilizes the conjugate base of carboxylic acids such as acetic acid, to increase the acidity of these compounds. Phenol, a benzene ring with an alcohol on it, is an acidic compound because resonance delocalization of the negative charge stabilizes the base:

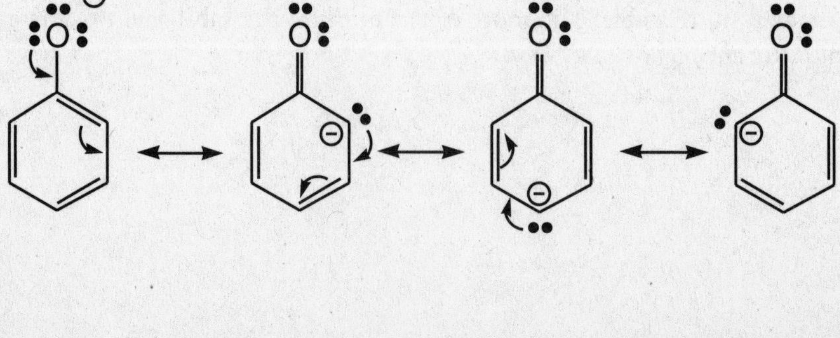

A different way to think about acid-base reactions is looking at reactions of **Lewis acids** and **Lewis bases.** A Lewis base donates outer shell electrons that are not involved in bonding. A Lewis acid accepts electrons. Note that this is different from the definition of Bronsted-Lowry acids as proton donors, though the two definitions say essentially the same thing (a base accepts a proton or donates an electron.) Traditional acid-base reactions can be understood in terms of Lewis acids and bases. For example, if sodium hydroxide reacts with hydrochloric acid, the hydroxide ion contains a pair of electrons in the outer shell that are not bonded that it can share. The hydrogen ion is a proton with no electrons, so it can act as an acceptor for the pair of electrons from the hydroxide ion. The reaction involves the movement of electrons from the hydroxide ion to the hydrogen ion, to share these electrons in a molecular orbital (a sigma bond) and form water.

$$^{-}OH + H^{+} \longrightarrow H_{2}0$$

Other reactions that do not appear as acid-base reactions can also be described as the movement of electrons between Lewis acids and Lewis bases. For example, unbonded electrons from methanol are donated in the reaction below, making the methanol the Lewis base and the alkane the Lewis acid.

$$CH_3-\overset{\oplus}{C}(CH_2CH_3)-CH_3 + CH_3\ddot{O}H \longrightarrow CH_3-C(CH_2CH_3)(\overset{\oplus}{O}H-CH_3)-CH_3$$

Lewis acid   Lewis base

Later on we will see that Lewis acids are called electrophiles and Lewis bases are called nucleophiles. Almost all chemical reactions can be thought of as the movement of electrons between Lewis acids and Lewis bases, making these concepts important to return to later.

Nomenclature and Functional Groups

CHAPTER TWO

When you travel, knowing a foreign language allows you to make your way around town with ease. When you sit at a café, you may use your language skills to order with confidence and impress the local inhabitants. Traveling now into the land of organic chemistry, it is time to learn its own special language. Every chemical has its own unique name that it shares with no other chemical. Rather than trying to memorize the names of the millions of possible chemicals, the thing to do is to learn a system that allows all molecules to be named using a few simple rules. A systematic way to name things like chemicals is called **nomenclature**. With a chemical nomenclature in hand, along with knowledge of the basic building blocks of organic chemistry, you can precisely name and intelligently discuss specific chemicals with others. The chemists in the cafes will think that you are native chemistry speaker and will be most impressed.

The international standard set of rules for naming organic molecules is called IUPAC, which stands for the International Union of Pure and Applied Chemistry. Chemists around the world can be assured that if they are using the IUPAC set of rules for organic nomenclature to name their molecules, all of their chemist friends (and you, after you read this chapter) will know what they are saying. IUPAC provides the rules of grammar for chemistry-ese. The different types of chemical groups found in organic chemicals, functional groups, are the parts of speech. Note that there are many common names for certain molecules that do not adhere to IUPAC rules, but you will learn the prevalent ones in class as you go. All compounds with common names have IUPAC names as well that are recognized as perfectly valid.

ALKANES

The simplest organic molecules are called **alkanes**, and are composed entirely of carbon and hydrogen atoms with single bonds (sigma bonds) connecting each atom in the molecule. A molecule with a double bond is not an alkane. Most alkanes have the molecular formula C_nH_{2n+2} and are **saturated**, meaning that the molecule has as many hydrogens in it that it could possibly have for that number of carbons. The smallest alkane and the simplest organic chemical is methane, with a single carbon forming covalent bonds with four hydrogens. Many alkanes are combustible liquids and gases like methane in natural gas, propane sold for heating and octane in gasoline.

Alkanes can have carbons connected to each other in long straight chains joined by carbon-carbon bonds, or branching structures.

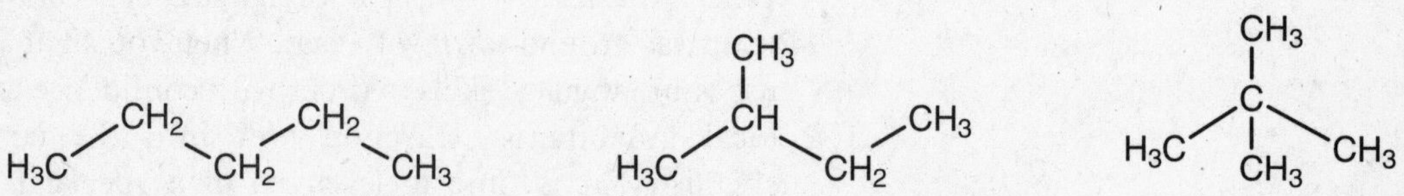

H_3C–CH_2–CH_2–CH_2–CH_3 H_3C–$CH(CH_3)$–CH_2–CH_3 $C(CH_3)_4$

For example, the molecule formula for these three structures is the same, although the molecules are quite different in structure and physical behavior. How can you name these chemicals in a way that you can tell someone else the name of a compound and they will know without a doubt what you are talking about?

Molecules are named according to the number of carbons they have, starting with methane, ethane, propane and butane. The prefixes meth- (1 carbon), eth- (2 carbons), prop- (3 carbons), and but- (4 carbons) appear frequently in the naming of a variety of organic molecules with 1-4 carbons. It is necessary to memorize these prefixes since they are used so frequently. Molecules with the suffix "–ane" are alkanes. The prefix "n-" appears before straight chain alkanes, which stands for "normal" and indicates that the carbons are in a straight line connected together, not branched. Sometimes alkyl chains are drawn with just sticks to show the bonds, assuming the reader knows that the intersection of the bonds shown represents the location of a carbon. Simplifying the structure yet further, the hydrogens can be eliminated from the drawing, making these assumed as well (see drawings in table). The naming of the simple straight chain alkanes is fairly straightforward—just count the carbons, and use the correct prefix, followed by the suffix -ane. For example, if the molecule has six carbons it is hexane (hex-ane).

NOMENCLATURE AND FUNCTIONAL GROUPS

n	Name	Formula	Representations
1	**Meth**ane	CH_4	
2	**Eth**ane	CH_3CH_3	
3	**Prop**ane	$CH_3CH_2CH_3$	
4	**But**ane	$CH_3CH_2CH_2CH_3$	
5	**Pent**ane	$CH_3CH_2CH_2CH_2CH_3$ or $CH_3(CH_2)_3CH_3$	
6	**Hex**ane	$CH_3CH_2CH_2CH_2CH_2CH_3$ or $CH_3(CH_2)_4CH_3$	
7	**Hept**ane	$CH_3CH_2CH_2CH_2CH_2CH_2CH_3$ or $CH_3(CH_2)_5CH_3$	
8	**Oct**ane	$CH_3CH_2CH_2CH_2CH_2CH_2CH_2CH_3$ or $CH_3(CH_2)_6CH_3$	
9	**Non**ane	$CH_3CH_2CH_2CH_2CH_2CH_2CH_2CH_2CH_3$ or $CH_3(CH_2)_7CH_3$	
10	**Dec**ane	$CH_3CH_2CH_2CH_2CH_2CH_2CH_2CH_2CH_2CH_3$ or $CH_3(CH_2)_8CH_3$	

Naming branched-chain alkanes is a little more complicated and requires some rules that everyone must follow if they are going to speak the same language. The rules are the following:

1. Find the longest continuous chain of carbons in the molecule. This will be the backbone of the molecule, with everything else hanging off of it. The main part of the name will be determined by this part of the molecule. If the longest continuous carbon chain has six carbons and is an alkane, the molecule is a hexane. The longest chain may not lie in a straight line in a drawing! This can be hard to see.
   a. Substituents are the groups that branch out from the longest carbon chain.

2. Number the carbon atoms in the longest carbon chain.
   a. Start numbering at the end of the carbon chain that results in the lowest possible numbers for the carbons with substituents.

3. Name the substituents.
   a. When a substituent is an alkane, it is called an alkyl group.
   b. The alkyl substituents are named in a manner similar to independent alkanes (methyl, ethyl, etc.) with a few special prefixes.
   c. The special prefixes for substituents with more than two carbons are:
      i. n- (n is for normal, unbranched)
      ii. iso- (a specific arrangement of carbons in a "T" formation)
      iii. sec- (sec stands for secondary)
      iv. tert- (or t-, stands for tertiary)
   d. When more than one of the same substituent is attached to a carbon, prefixes can be added on to indicate this: di-, tri-, tetra-, etc. For example, "di-methyl" means that two methyls are attached to that carbon.

4. Identify the number of the carbon each substituent is attached to. Numbers must be repeated when more than one substituent is attached to one carbon. If two methyls are on the 4th carbon, it would be noted as 4,4 dimethyl-.
   a. Each substituent will have a name, and the number of the carbons in the main chain it is attached to.

5. Put the entire name together.
   a. First list the substituents, in alphabetical order, with the number of each followed by its name.
      i. Prefixes like tert- or tri- are ignored when alphabetizing.
   b. Put dashes between numbers and words.
   c. Put commas between numbers.
   d. At the end of the name list the longest chain. These rules apply to the moment to alkanes, but with slight modification will apply to other types of organic molecules as well. The best way to learn the rules is by using them, so keep practicing.

Instructor: Let's try naming the molecule below.

$$\begin{array}{c} CH_3 \\ | \\ CH_2CHCH_3 \\ | \quad\quad CH_3 \\ CH_3CH_2CHCHCHCH_2CH_3 \\ | \\ CH_3 \end{array}$$

Student response: First, we look for the longest carbon chain. The longest chain of carbons is seven across, so—

Instructor response: Wait! Remember, don't get fooled by that straight chain across. The longest chain can sometimes wrap around like a pretzel.

Student response: Oops. I see it now. The longest chain is eight carbons long, making this molecule an octane, with alkyl substituents. Now that I've identified the basic structure of the branched alkane, the substituents need to be named. In this case, the substituents are simple, either with one carbon (methyl-), or with two carbons (ethyl-). In addition to naming the substituents, their location along the alkyl chain needs to be identified by the number of the carbon they are attached to. The carbons in the longest alkyl chain are numbered sequentially starting at #1 to allow the positions of substituents to be indicated. The numbering could start from either end of the chain. I want the carbons numbered so that the substituents have the lowest possible numbers. If I compare the two possible ways to number our molecule, and the resulting substituent numbers either way…

$$\begin{array}{c} CH_3 \\ |\ ⑧ \\ ⑥CH_2CHCH_3 \\ |\quad ⑦\ CH_3 \\ |\ ④\quad | \\ CH_3CH_2CHCHCHCH_2CH_3 \\ ⑤\ |\ ③\ ②\quad ① \\ CH_3 \end{array}$$

WRONG

$$\begin{array}{c} CH_3 \\ |\ ① \\ ③CH_2CHCH_3 \\ |\quad ②\ CH_3 \\ |\ ⑤\quad | \\ CH_3CH_2CHCHCHCH_2CH_3 \\ ④\ |\ ⑥\ ⑦\quad ⑧ \\ CH_3 \end{array}$$

RIGHT

Now the substituents can be named and numbered: 2-methyl, 4-ethyl, 5-methyl, 6-methyl.

Instructor response: Good. Keep it in pieces until you're ready to put it all together. The substituents are placed in the name alphabetically, so an ethyl substituent appears in the name before methyl. Also, if a substituent appears more than once, it will be preceded by a prefix such as di-, tri-, etc. These prefixes are not used as part of the alphabetization process (so triethyl would come before dimethyl in the name of a compound since 'e' is before 'm'). In the case of the molecule above, 4-ethyl comes first since ethyl is alphabetically before methyl. Next, the methyl groups are listed together as 2,5,6-trimethyl, followed by the name of the longest carbon chain, octane. The structure as a whole is named: 4-ethyl, 2,5,6-trimethyloctane.

Note that there is no space or hyphen in the last segment of the name (it's not "trimethyl-octane.") The way that carbons (or other atoms) are connected to the longest carbon chain is sometimes indicated by the use of the terms primary, secondary, and tertiary. If there is one carbon attached to the carbon being discussed, it is called primary. A primary carbon is on the end of a carbon chain. A secondary carbon is attached to two carbons, such as a carbon in the middle of a straight chain alkane. A tertiary carbon is attached to three carbons, such as the carbon in a chain with an additional alkyl group attached. Whether a carbon is primary, secondary or tertiary can affect how a chemical behaves in a reaction.

Some alkane substituents have common names, relating to the arrangement of carbons in branched structures. These common names contain the prefixes n- for normal for straight chain structures, iso-, sec- for secondary, and tert- for tertiary. Use of these common names for these substituents is usually acceptable, and it might be helpful to memorize them.

		CH_3	CH_3	CH_3
CH_3-CH-	CH_3CH_2CH-	CH_3-C-	CH_3-C-CH_2-	$CH_3-CH-CH_2-$
CH_3	CH_3	CH_3	CH_3	
isopropyl	*sec*-butyl	*tert*-butyl	neopentyl	isobutyl

Using these rules, every compound structure should have a unique name and every name corresponds to one compound. If the name is ambiguous, and can be used to draw more than one structure, then the name is incomplete or wrong. Common names for some molecules do not follow IUPAC rules, but are still acceptable. The rules as stated here are simplified, and do not account for all possible situations. They are enough to cover most situations however, and are essentially the same for molecules with other functional groups, as we will see.

Instructor comment: Note that if there are two alkyl substituents on an alkane equidistant from the ends, the substituent that comes first in the alphabet gets the lower number.

CYCLOALKANES

In addition to straight chain or branched structures, alkanes can form ring structures, in which carbons are joined together in a ring. Ring alkanes are named according to the number of carbons in the ring. A ring with three carbons is cyclopropane, followed by cyclobutane (four carbons), cyclopentane (five carbons) and cyclohexane (you can take a stab at how many carbons here). Cyclic alkanes do not have the same number of hydrogens as straight chain or branched alkanes. Since the carbons are in a ring, cycloalkanes have two fewer hydrogens in their molecular formula than a regular alkane with the same number of carbons.

Rings made of 5 or 6 carbons are the most stable. Cyclopropane and cyclobutane are strained rings, since they have smaller angles in the structure than normal sp^3 hybrid orbitals normally form in covalent bonds in carbon. Sp^3 orbitals are normally at a 109 degree angle, while the interior angle of cyclopropane is 60 degrees (think of a triangle). This strain makes these molecules relatively high in energy. Cyclopentane or cyclohexane are more stable since these rings allow the normal 109 degree bond angles in the ring that sp^3 orbitals are normally involved in. Rings with more than 6 carbons are strained as well, if forced to be planar, and would contort to seek a more suitable angle.

In naming molecules with substituents branching off of the cycloalkane, the ring carbons are numbered so that the number one carbon is the one with most substituents. If there is only one substituent on a ring alkane, you do not have to give it a number. It could only be #1, making it unnecessary to write or state the number. Remember these are 3-D molecules that can be looked at from any point in space. An example would the naming of chlorocyclobutane, drawn below:

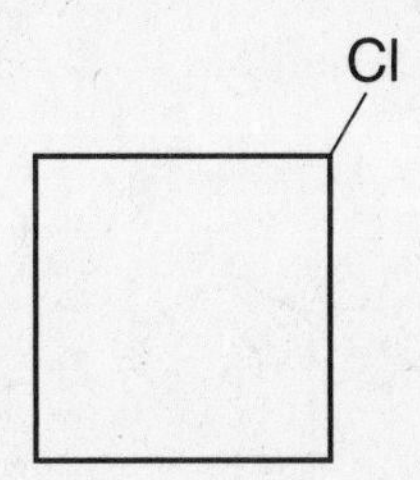

ALKENES AND ALKYNES

Functional groups are chemical groups that differ from alkanes. One type of functional group is the carbon-carbon double bond. If a carbon-carbon double bond is added to an alkane, the molecule becomes an **alk*ene***, an unsaturated hydrocarbon. An alkene is named with the same prefixes as alkanes, depending on the number of carbons in the longest carbon chain: butene, pentene, hexene, etc. Ethene and propene would the proper IUPAC way to name the two carbon and three carbon alkenes respectively, but these are more commonly called ethylene and propylene. Alkenes can change the rules for naming the carbon backbone of a branched chain molecule. With alkanes, the longest continuous carbon chain is the backbone. In alkenes, the longest chain should include the double bond, even if this is not the longest carbon chain. The double bond position is indicated with the number of the carbon where the double bond starts. In the carbon chain, the numbering should place the double bond (or double bonds) with the lowest possible numbering.

Instructor question: What is the name of the following molecule?

$$CH_3CH{=}CHCH_3$$

Student response: The molecule has four carbons, giving it "but-" as the root of the name. The presence of the double bond makes it an alkene, with the –ene suffix: butene. The carbons are numbered to indicate the position of the double bond, and we want to give the double bond the lowest number. In this case, the number will be the same if you start numbering at either end: carbon 2 is the first carbon in the double bond. The molecule is 2-butene.

Instructor question: Nice. How about this one?

$$CH_3CH_2CH{=}CHCH_3$$

Student response: This molecule has five carbons and has a double bond, making it a pentene. Now we deal with the double bond. Numbering from either end of the molecule, we see that if we number from the left, the double bond would be at carbon 3, but starting from the right the double bond starts at carbon 2. Since numbering from the right produces a lower number, this is the correct answer: 2-pentene.

When alkenes are substituents, they are named by the same rules with alkane substituents, with a few exceptions for some common names used with simpler double-bond substituents: vinyl, allyl, and methylene (see below).

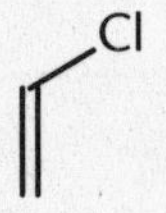

Cl

Br

chloroethene
(vinyl chloride)

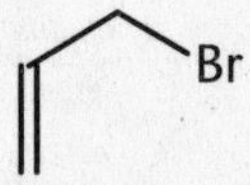

3-bromo-1-propene
(allyl bromide)

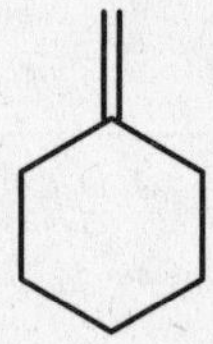

methylene cyclohexane

Alkenes are planar, without rotation of the molecule around the double bond due to the pi bonds between p orbitals. The pi bonds along with the sigma bonds form multiple connections between atoms, holding two atoms in place. With the groups on both sides of the double bond held in place, they must be in one of two orientations, called cis and trans. If all the groups are the same, then one cannot distinguish cis and trans, but if they are different, then cis and trans become important distinctions in structures. The cis and trans forms have the same molecular formula and even the same connectivity, but they still differ in their three dimensional structure, making these molecules isomers of each other. Cis and trans refer to the positions of substituents on the two ends of the double bond. If the substituents are on the same side, then they are cis, but if they are on opposite sides then they are trans. Isomers and stereochemistry will be discussed in Chapter 4 as an important element of chemical structure.

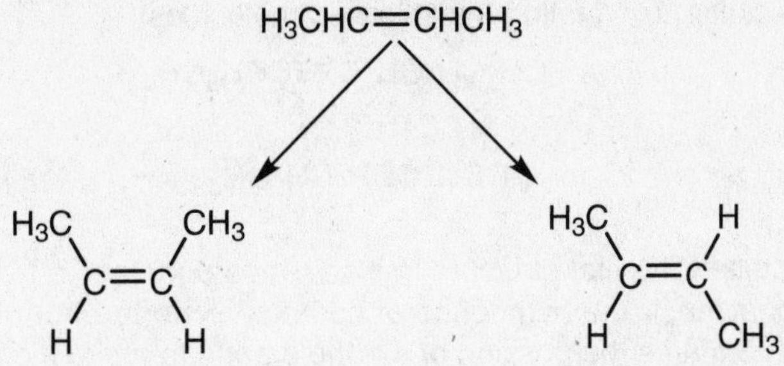

$H_3CHC{=}CHCH_3$

cis-2-butene

trans-2-butene

Student comment: You can think of it like this: your *cis*-ter is always *on your side*. And *trans* is – like *trans*portation, to move away. Get it?

In molecules that have two or more double bonds spaced one bond apart from each other, the double bonds can be **conjugated**, with electrons partially delocalized. Conjugated double-bonds are often found in odor-causing molecules and are also common in dyes. The conjugated double bonds in dyes absorb light at a specific wavelength(s), leaving the remaining light reflected. The reflected light produces the color of the molecule. The wavelengths of light absorbed depend on the extent of the delocalized electrons in the conjugated double bond system. Indigo, used to color jeans, is a common example of a dye with a conjugated system that reflects blue most strongly and so appears blue.

A diene

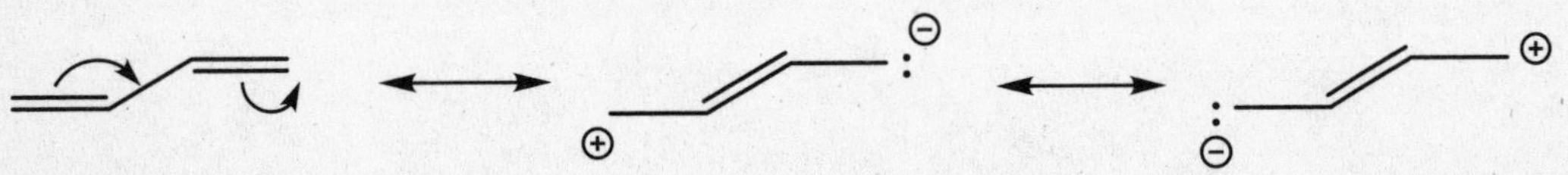

Cycloalkenes could be made from cycloalkanes by the introduction of a double bond in the ring. The resulting structure is named in the same way as cycloalkanes. Putting a double bond in cyclohexane, and removing two hydrogen atoms makes cyclohexene. If the double bond is the only modification, then there is no need to number it.

Alkynes are alkanes with a triple bond, making them more unsaturated than either alkanes or alkenes. The addition of a triple bond results in the removal of four hydrogens compared to the molecular formula of the corresponding alkane. Alkynes are named in the same way as alkenes, starting out by identifying the number of carbons in the chain containing the triple bond to name the molecule and adding the suffix "–yne" on the end. Finally, the carbons are numbered to indicate the position of the triple bond in the carbon chain, giving the triple bond the lowest possible number.

Instructor question: Try naming the following molecules:

1. $CH_3CH_2CH_2C{\equiv}CCH_2CH_3$

2. $H_3CC{\equiv}CCH_2CH_2CH_3$

Student response: The first structure has seven carbons and a triple bond, making it heptyne. If the numbering of carbons is started from the right, then carbon 3 marks the position of the triple bond to make the structure 3-heptyne. The second molecule has six carbons and a triple bond, making it hexyne. Numbering the carbons from the left gives the triple bond the lowest number, making the molecule 2-hexyne.

BENZENE AND THE AROMATIC RING

One of the most commonly encountered structures in organic chemistry is the benzene ring, also called an aromatic ring. The term aromatic derives from the fact that members of this family are responsible for some of the common smells found in foods like cinnamon or vanilla. Benzene rings are also found in a huge variety of medicines, dyes, and plastics, and are present in many biological molecules such as DNA and proteins. For the current definition of aromaticity, read on.

Benzene is a ring structure with six carbons, with the molecular formula C_6H_6 usually drawn with three double bonds alternating with three single bonds. The molecule can easily be drawn as its resonance structure by moving the electrons in each double bond around the ring to the next bond.

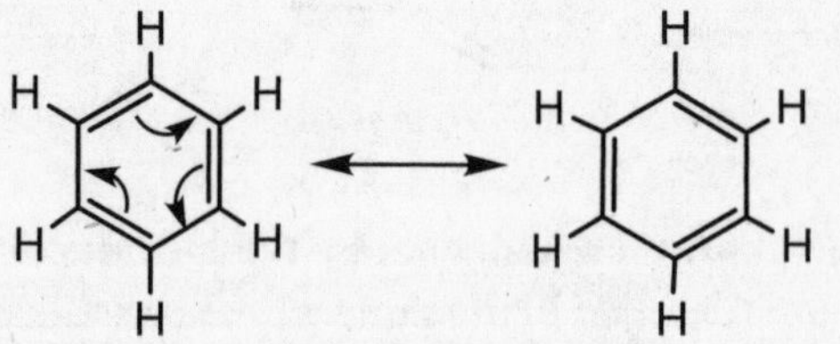

The true story is that neither of these resonance forms is exactly right. Benzene is a flat molecule, making the p orbitals for each carbon stick out from the plane of the carbons and lie parallel to all of the other p orbitals. This is an ideal situation for electrons to be shared among molecular orbitals. Rather than existing as electrons localized in a specific pi bond however, electrons in a benzene ring are delocalized among the orbitals above or below the ring of carbons. Three pairs of electrons are shared equally among all six carbons in the ring, in ring-shaped orbitals above and below the plane of the ring like donuts. Therefore, a more accurate way to draw benzene is a circle inside a hexagon, indicating the delocalized pi-orbital electrons. Sometimes the regular line-bonds are helpful to draw to keep track of electrons during reactions involving the benzene ring or in drawing resonance structures.

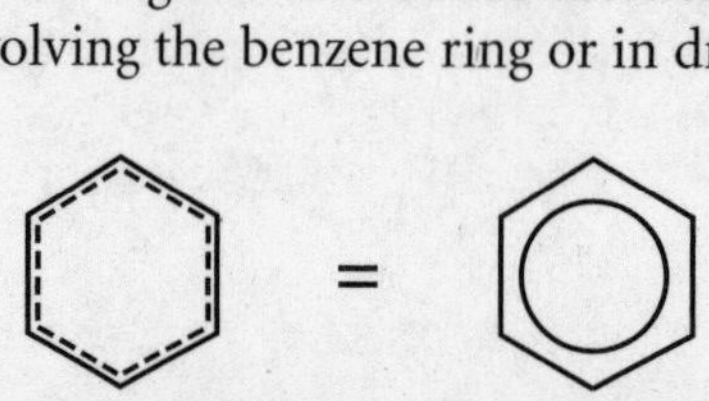

The delocalization of the pi orbital electrons in benzene explains some of the properties of this molecule. Benzene is more stable, and less reactive, than would be predicted without delocalization. If the bonds in benzene were actually alternating single and double bonds, this would affect the lengths of the bonds and the shape of the molecule. What is observed however is that all six C-C bonds are the same in length, and about halfway between single and double bonds. The ring is flat as a result of the electron clouds above and below the plane of the carbon ring, and all of the internal angles between the carbons are equal as well.

Benzene is one example of a broader class of compounds that are called aromatic. Aromatic compounds do not all have six carbons like benzene, but they do all follow rules that can help to determine if a molecule is aromatic or not:

- They are all cyclic.
- They must be planar (flat).
- The p orbitals are perpendicular to the plane of the ring (and parallel to each other).
- They obey Huckel's rule, which states that aromatic compounds must have 4n+2 pi electrons, where n is a number (0,1,2,3, etc). In other words, an aromatic molecule could have 2 pi electrons (plugging n = 0 into the equation), 6 (n = 1), 10 (n = 2), 14 (n = 3), and so on.

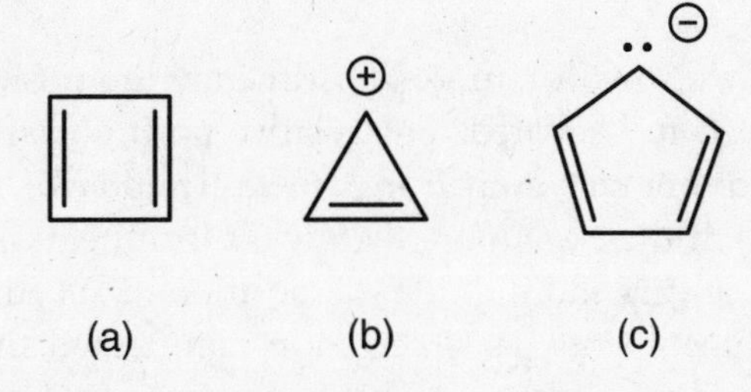

Instructor question: Of the three structures above, which one(s) are aromatic?

Student response: Let's see. They are all cyclic. In the first structure (a), all four carbons will be sp^2 hybridized, and the molecule will be planar. The molecule has two pi bonds, with two electrons in each, this makes 4 electrons in pi orbitals. I need to see if this fits Huckel's rule. Um, how does that work again?

Instructor response: You're doing fine. Think of it this way: can 4n + 2 = 4?

Student response: If n = 0, it equals 2. If n = 1, it equals 6. So no, it doesn't work. This molecule is not aromatic. Choice (b) is cyclic, and with three carbons must also be planar. How many pi electrons does it have? It has one double bond, meaning it has 2 pi electrons. If n = 0, then 4n+2 = 2, making 2 pi electrons acceptable for Huckel's rule. This molecule meets all of the criteria to be aromatic! The last example (c) is cyclic, and will be planar. Huckel's rule will also be satisfied, with four pi electrons from the two double bonds and two more electrons unbonded on the fifth carbon, for a total of six. If n = 1, then 4n+2 = 6. Choice (c) is aromatic too.

AROMATIC NOMENCLATURE

The great variety of aromatic molecules and their broad utility derive from the variety of functional group substituents added to the benzene ring to change the behavior of molecules. A benzene ring with a single substituent is named with the substituent name, then the name benzene. For example, a benzene with a chlorine is named chlorobenzene. Some of the simpler benzene derivatives, with a single functional group as a substituent, have common names that are often used. For example, a benzene with a hydroxy group is usually called phenol rather than hydroxybenzene. Phenol is a component in many cough syrups, acting as an anesthetic in small doses and as a caustic agent in more concentrated forms. Benzene with a single methyl substituent is commonly called toluene, and benzene with a single amine group is commonly called aniline.

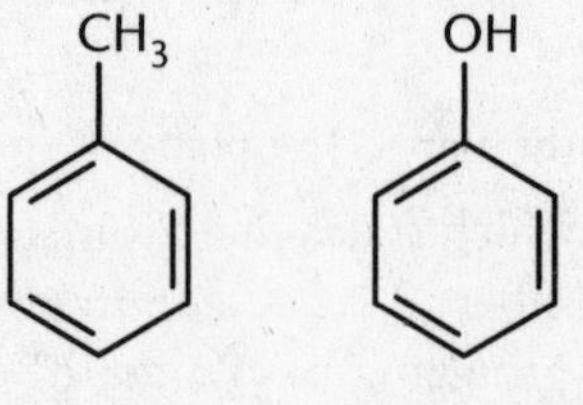

CH_3 OH

Toluene Phenol

In a benzene-derivative with two or more substituents, the positions of the substituents relative to each other need to be stated. The relative positions are often indicated by the prefixes ortho- (next to, or one step away from), meta- (middle of, or two steps away from), and para- (farthest from). If there are more than two substituents, it is necessary to number the carbons in the benzene ring and indicate the position of substituents according to their position. If one of the substituents has a common name, then this will serve as the base structure to start numbering the carbons. The common name for a benzene with a single methyl substituent is toluene. The common name could be used to name derivatives of toluene with another substituent added. If a bromine is added, the molecule becomes ortho-, meta- or para- bromotoluene, depending on the position of the bromine relative to the methyl group around the ring. Alternatively, the IUPAC name of each molecule could be figured out, indicated below the common name equivalent.

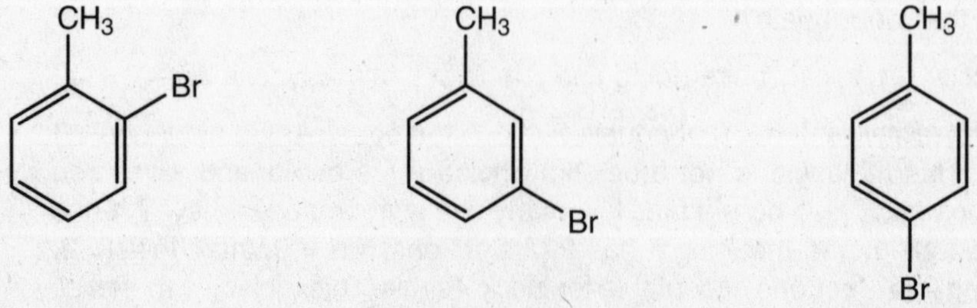

CH_3 Br CH_3 Br CH_3 Br

o-bromotoluene (1-bromo-2-methylbenzene)

m-bromotoluene (1-bromo-3-methylbenzene)

p-bromotoluene (1-bromo-4-methylbenzene)

More complicated molecules with multiple benzene rings are found in many molecules, such as naphthalene and anthracene. Naphthalene is what gives moth balls their pungent odor and hopefully keeps moths out of your clothes. Like benzene, these larger ring systems are planar, but in these larger aromatic ring systems the electrons are delocalized across the entire ring system. The flat, planar nature of some aromatic compounds makes them a health hazard, allowing them to lie inside DNA and disturb its structure to cause mutations and cancer.

FUNCTIONAL GROUPS – THE REST OF THE STORY

The alkanes, alkenes, alkynes and aromatic rings are all hydrocarbons, composed of carbon and hydrogen. The world of organic chemistry is much broader and more interesting than hydrocarbons, however. It is now time to introduce the other components of the anatomy of organic compounds, the wide and wonderful world of functional groups. The basic carbon backbone of organic compounds is modified to include many different types of groups, including halogens, oxygen-containing groups (ens, aldehydes, ketones, esters, ethers and carboxylic acids), and nitrogen-based groups (amines, amides). This assortment of functional groups added onto hydrocarbons provides the full range of functions and behavior of organic compounds, altering physical traits like melting and boiling point, changing the way they interact with other molecules, making them acids or bases, and changing their reactivity. The functional groups will be encountered throughout organic chemistry.

The naming of organic compounds, even with this variety of functional groups, follows essentially the same rules as alkanes. However, the identification of the carbon backbone and its numbering are influenced by which functional groups are present, and how a functional group affects the naming of a molecule depends on its priority. The most oxidized functional groups have the highest priority, with priority decreasing with the state of oxidation. The order of priority for functional groups when naming compounds goes as follows, from highest priority to lowest:

Carboxylic acid>esters>amides>nitriles>aldehydes>ketones>alcohols>amines>ethers.

Carboxylic acid groups are the most oxidized, and the highest priority in naming compounds, and ethers are the least oxidized and the lowest priority. Oxidation indicates the relative energy state of the bonds in molecules, where the most oxidized compounds have the least energy and the least oxidized molecules have the most energy. The priority determines how the carbon backbone is selected, and how the molecule is named. The highest priority group is indicated by the suffix at the end of the name, and the other lower priority groups are named as substituents.

Family	Structure	Naming System	Example	IUPAC name (common name)
Alcohol	$R_3C{-}OH$	hydroxy- / -ol	$HO{-}CH_2{-}CH_3$	ethan**ol** (ethyl alcohol)
Aldehyde	$R{-}CHO$	oxo- / -al	$CH_3{-}CHO$	ethan**al** (acetaldehyde)
Alkene	$R_2C{=}CR_2$	alken(yl)- / -ene	$H_2C{=}CH_2$	eth**ene** (ethylene)
Alkyne	$R{-}C{\equiv}C{-}R$	alkyn(yl)- / -yne	$H{-}C{\equiv}C{-}H$	eth**yne** (acetylene)
Amide	$R{-}C(=O){-}NR_2$	-amide	$CH_3{-}C(=O){-}NH_2$	ethan**amide** (acetamide)
Amine	R_3N	amino- / -amine	$H_3CH_2C{-}N(CH_3){-}CH_3$	N,N dimethylethan**amine** (ethyldimethylamine)
Arene	C_6R_6	phenyl- / -benzene	$C_6H_5{-}CH_2CH_3$	ethyl**benzene**
Carboxylic Acid	$R{-}C(=O){-}OH$	-oic acid	$CH_3{-}C(=O){-}OH$	ethan**oic acid** (acetic acid)
Ester	$R{-}C(=O){-}O{-}R$	-oate	$H_3C{-}C(=O){-}O{-}CH_3$	methyl ethan**oate** (methyl acetate)
Ether	$R{-}O{-}R$	alkoxy- / -ether	$CH_3CH_2{-}O{-}CH_2CH_3$	**ethoxy**ethane (diethyl**ether**)
Haloalkane	$Br{-}R$	halo- / alkyl halide	$Br{-}CH_2{-}CH_3$	**bromo**ethane (ethyl brom**ide**)
Ketone	$R{-}C(=O){-}R$	oxo- / -one	$CH_3{-}C(=O){-}CH_3$	propan**one** (acetone)

HALOGENATED ALKANES

One of the common substituents to modify alkanes is the halogen group, the periodic table family of elements that includes chlorine, fluorine, bromine and iodine. The uses of halogenated organic molecules have included dry-cleaning agents, refrigerants, aerosol propellants, and insecticides. The loss of ozone from the polar atmosphere has been attributed to the wide spread use of certain chloro-fluorocarbons such as aerosols and refrigerants, now being removed from use by world-wide agreement.

The numbering of an alkane with halogen substituents follows the same rules as with alkyl-group substituents, numbering the backbone to give the substituents the lowest possible numbering. The alkyl halide is primary, secondary, or tertiary, depending on how many carbons are attached to the carbon with the halogen. Alkanes with halogens are named with the prefixes chloro-, fluoro-, bromo- or iodo-, depending on the halogen.

ALCOHOLS

Alcohols have the general structure R-OH, where R represents any alkyl group. Alcohols are common in nature, including the ethanol generated by yeast as a metabolic by-product that produces wine and beer. Alcohols are like water, with an alkyl group substituted for one of the hydrogens in the water molecule. Like water, the R-O-H bonds are bent, and the oxygen has unbonded valence electrons. Oxygen is more electronegative than carbon or hydrogen, pulling electrons toward itself in alcohols and creating a partial charge that plays an important role in interactions between molecules through hydrogen bonds. The hydrogen bonding of water and alcohols is responsible for many of the physical properties of these molecules.

Alcohols are slightly acidic, able to donate protons very weakly. The conjugate base is very strong; the equilibrium constant for dissociation, the K_a, is usually on the order of 10^{-15} or smaller. One important exception is phenol (benzene with an alcohol group) in which the negative charge of the conjugate base is stabilized within the aromatic ring (see Chapter 1). In other words, phenol is acidic because it can easily donate a proton and delocalize the resulting charge.

As with other functional groups, the alcohol group is primary, secondary or tertiary depending on the carbon it is attached to. For simple alcohols, change the name of the longest carbon chain from its alkane name to have an –ol suffix on the end. For example, ethane with an alcohol group becomes ethanol, followed by propanol, butanol, etc. In more complicated alcohols, it becomes necessary to number the carbons and to specify the number of the carbon with the alcohol group. For example, 1-propanol is different than 2-propanol, where the number indicates the carbon that the –OH alcohol group is attached to. The carbons are numbered to give the alcohol group the lowest possible substituent number. Molecules with more than one alcohol group can be indicated with the suffix –diol (two alcohols), -triol (three alcohols), etc.

If double or triple C-C bonds are present, these are indicated by changing the root name of the longest chain from –an (for alkanes) to –en (alkenes, double bond) or –yn (for alkynes, triple bond). For example, in 2-penten-1-ol, the two refers to the position of the double bond, and the 1 to the position of the alcohol. The carbon with the alcohol is number 1 because alcohols take precedence over double or triple bonds in the numbering of carbons.

Instructor question: Now, try to name the following molecule:

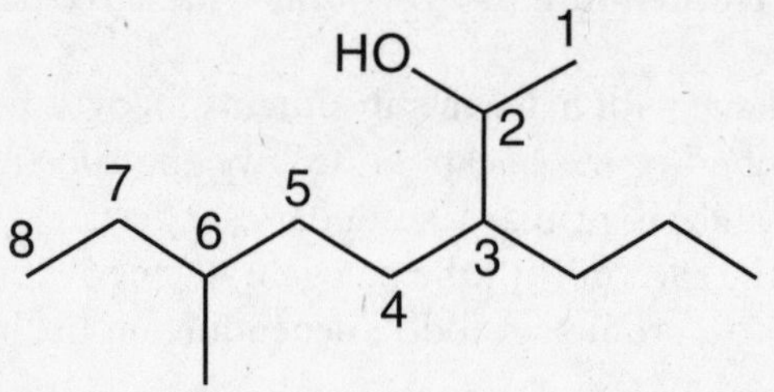

HO 1 2 7 5 8 6 3 4

Student response: We know that the –OH gives priority to the carbon chain it is attached to. Starting with the end of the carbon chain with the –OH group, the longest carbon chain has eight carbons, and since it is an alcohol with the –OH group on carbon 2, the base structure is 2-octanol. This leaves two alkyl substituents, a methyl group on carbon #6 and a propyl group on carbon #3. The methyl group comes first in the alphabet, and first in the name, making the molecule: 6-methyl-3-propyl-2-octanol.

Instructor response: Jolly good show! Let's get a little trickier this time:

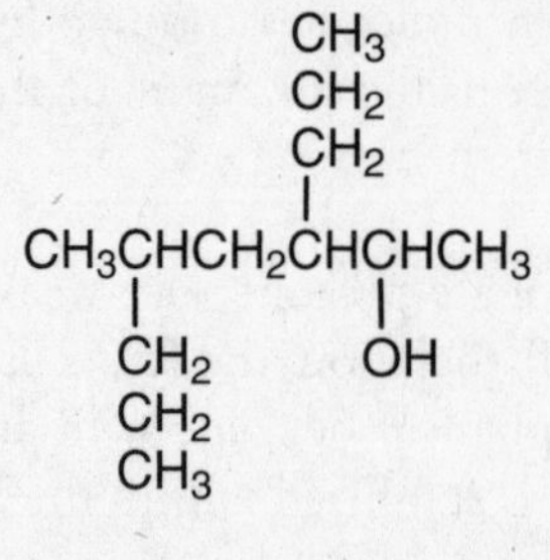

CH_3
CH_2
CH_2
$CH_3CHCH_2CHCHCH_3$
CH_2 OH
CH_2
CH_3

Student response: Well, the longest carbon chain has nine carbons, so the molecule is an nonane. Now we—

Instructor response: Hold up there, chief. You have the right idea, but the priority group in the molecule is the alcohol group, and the base structure for naming should always include the priority functional group. The longest carbon chain that contains the alcohol as a direct substituent has eight carbons, so it is an octanol. The carbons should be numbered from the end that gives the alcohol group the lowest number, #2 in this case to make the base structure 2-octanol. The base structure has two substituents, a propyl on carbon #3 and a methyl on carbon #5. M(ethyl) comes before p(ropyl) in the alphabet and in the name, so putting it all together, the molecule is named: 5-methyl-3-propyl-2-octanol.

ETHERS

Ethers are compounds with the general structure where R1-O-R2 where R1 and R2 indicate two alkyl chains that are joined by an oxygen group. The letter R is often used when drawing an organic structure where R can be any type of alkyl group. The naming of ethers begins by finding the alkyl group on one side of the oxygen that is larger. In IUPAC, ethers are named as alkoxyalkanes, with the larger alkyl group named as the second part, and the other smaller alkyl group on the other side of the oxygen named with the suffix –oxy. An ether with a methane on one side of the oxygen and ethane on the other side would be named methoxyethane.

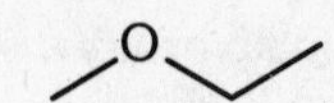

methoxyethane

Diethyl-ether (a common name) often called just "ether", was at one time commonly used as an anesthetic in surgery and is still used in the organic chemistry lab as a solvent. Diethyl-ether is made of oxygen, with an ethyl group on both sides. Ethers, like alcohols, are a bit like water, but with two R groups instead of one. One important difference between ethers and alcohols is that ethers do not have an –OH for hydrogen bonding, which alters behavior like boiling points.

Ethers can form ring shaped structures, including three membered rings commonly called epoxides. The epoxide ring is strained, as with other three membered rings like cyclopropane. Cyclic ethers with more than five members in the ring are more stable, as might be predicted by their larger bond angles that allow the normal sp^3 tetrahedral structure most stable for each carbon.

ALDEHYDES AND KETONES

Aldehydes and ketones are functional groups that contain an oxygen with a double bond to a carbon: C=O. This group is called a **carbonyl** group and is also found in carboxylic acids. Aldehydes have the carbonyl group on the end of an alkane chain, while ketones have the carbonyl group in the middle of an alkane chain. As with most molecules, the simpler aldehydes and ketones have common names that are acceptable in naming. Formaldehyde, used to preserve anatomical specimens, is the simplest aldehyde, with only one carbon, and acetone, fingernail-polish remover, is the simplest ketone. Octanal, the eight-carbon aldehyde, has a potent fruity smell.

In naming molecules, aldehydes are indicated by the suffix –al, and ketones are indicated by the suffix –one on the end of a molecule's name. As you might expect, the carbon backbone is selected to have the carbonyl group in it, and numbered to give the carbonyl group the lowest possible number. Since the carbonyl group is by definition on the end of the carbon chain in an aldehyde, it is not necessary to number the group in the name (it will be on carbon #1 most of the time). Aldehydes and ketones are more oxidized and have a higher priority in naming than alcohols and ethers.

CARBOXYLIC ACIDS

Carboxylic acids are the most oxidized functional groups with oxygen, and the highest priority functional group in naming molecules. The carboxyl group has a carbonyl and an attached –OH group, such as in the common carboxylic acid acetic acid (found in vinegar). Carboxylic acids have a sour taste, such as the sour taste of lactic acid in spoiled milk.

The carboxylic acids are fairly strong acids (much more so than alcohols) with a pK_a of approximately 5-6. They are weaker acids than the inorganic acids like hydrochloric or sulfuric acid, but are orders of magnitude more acidic than most other organic compounds. The dissociation of carboxylic acids to donate protons occurs readily because the conjugate base is stabilized by delocalization of electrons between the two carbon-oxygen bonds (see Chapter 1).

Carboxylic acids are named in IUPAC by placing the carbon chain with the carboxylate as the carbon backbone, finding the longest carbon chain that contains this group, numbering the carbons, and then naming and ordering the substituents. The carboxylate group is indicated by the suffix –oic acid. The IUPAC name for acetic acid would be ethanoic acid.

Instructor question: Let's practice a name once again, for the following structure:

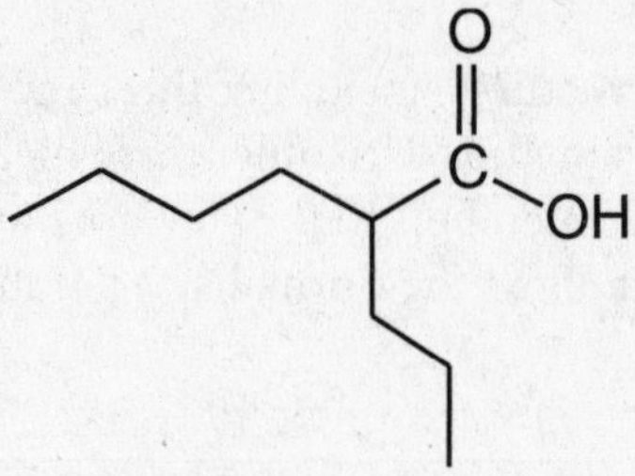

O
C
OH

Student response: OK, here goes. In this case, the position of the carboxylate gives priority to the carbon chain. This is not the longest carbon chain in the molecule, but since it has the carboxylate on it, it will be the foundation of the name and the numbering of the molecule. The longest carbon chain that has the carboxylate has six carbons, making it hexanoic acid. The carbon of the carboxylate is the #1 carbon, with a propyl substituent on carbon #2. Putting it all together now, the name is (drum roll, please): 2-propylhexanoic acid.

Instructor response: How about one more? That's the spirit! Try naming this beast:

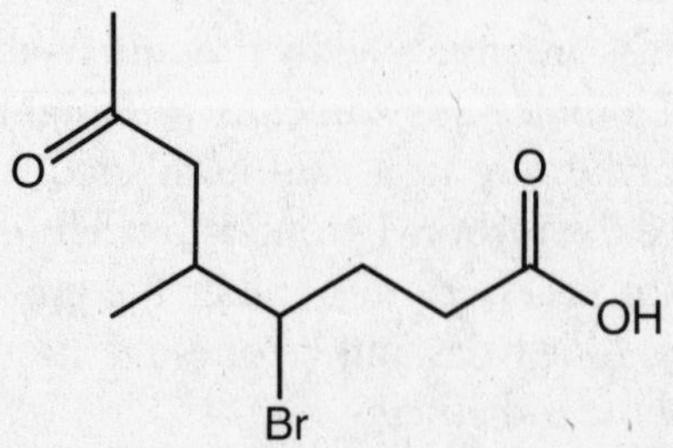

O
O
OH
Br

Student response: I surrender.

Instructor response: You can do it. First, prioritize the functional groups.

Student response: Well, this molecule has a bromo-, a ketone, a methyl, and a carboxylic acid. The most oxidized, and the highest priority functional group, is the carboxylic acid. The carbon in the carboxylate will be the number one carbon and the root name of the structure will have the suffix of the carboxylic acids (-oic acid). With eight carbons, the root structure is going to be octanoic acid.

Instructor response: Good! The other substituents are 7-oxo (oxo is the prefix used for aldehydes and ketones when they are present as lower priority groups), 5-methyl, and 4-bromo. Ordering the substituents according to the alphabet (b then m then o), the name of the molecule is: 4-bromo-5-methyl-7-oxooctanoic acid. Not so scary after all, is it? Um....hello?

DERIVATIVES OF CARBOXYLIC ACIDS

There are many structures that are not themselves carboxylic acids, but are closely related derivatives, including esters, amides, and anhydrides. Esters are common in nature, with the general structure:

$$R-C(=O)-O-R_2$$

In which R2 replaces the hydrogen in the corresponding carboxylic acid. Making the ethyl ester of butanoic acid produces the molecule that smells like pineapples and other simple esters also have fruity smells. Esters are named with the suffix –oate and the alkyl group names as a prefix. The compounds that produces the pineapple smell is named ethyl butanoate.

Amide groups form the backbone of all proteins, and their general structure consists of an amine group that replaces the –OH in carboxylic acids:

$$R-C(=O)-N(R)R$$

Amides are named with the suffix –amide.

AMINES

Amines are, broadly, nitrogen-containing compounds. Many organic chemicals, both naturally occurring and produced by man, are amines, including many bad smells. Compounds like dimethylamine have a strong fishy smell that is (usually) perceived as unpleasant. The nomenclature of amines is founded in the naming system used for alkanes. Amines are often referred to as primary, secondary and tertiary, although in this case these names refer to the substitution of the nitrogen involved, and not the carbons it is bound to. A nitrogen atom attached to one carbon is primary, attached to two carbons is secondary and attached to three carbons is tertiary.

Some amines have common names, including the simplest aromatic amine, aniline, and simple alkyl-amines, such as ethylamine. In IUPAC naming of amines, the suffix –amine is added to the root name of the corresponding alkane name, making the IUPAC name for ethylamine, ethanamine. As with other functional groups, the position of the amine substituent on the longest continuous carbon chain is indicated by the number of the carbon it is attached to, numbering the carbons to give the nitrogen the lowest possible substituent number.

Like carbon, nitrogen commonly forms sp^3 hybrid orbitals that are tetrahedral in shape. However with five valence electrons rather than four, nitrogen can fill its octet by forming just three covalent bonds, leaving one electron pair unbonded. Quaternary amines are sometimes formed, with a positive charge, and are usually good proton donors.

The N-H bond is polar since N is fairly electronegative, although less so than oxygen. Thus, amines that can hydrogen bond have increased boiling points. This would not include tertiary amines, that have no N-H bond, only N-alkyl bonds.

One of the key physical properties of amines is that they are weak bases. In water, the water molecules will donate a hydrogen to an amine. The strength of this reaction, and of the basicity of an amine, is indicated by the equilibrium constant K_b, which is described by the reaction:

$$A^- + H_20 = HA + OH^-$$

Larger values of K_b indicate more basic amines, and smaller numbers are less basic. The K_b is related to K_a, since the stronger the base, the weaker the acid.

All in all, you should now be a fluent speaker of organic chemistry, fully versed in all the parts of speech and the rules of grammar. Venturing onward, these skills will allow us to refer easily to molecules involved in a dizzying array of reactions. Isn't this exciting?

Molecular Dynamics and Intermolecular Forces

CHAPTER THREE

Organic molecules contain atoms connected to each other with covalent bonds. We have learned how atoms and bonds form molecules and how to draw and name these molecules. Do atoms and bonds tell the whole story though?

Once we have figured out how the atoms in a molecule are held together by bonds, and we can draw and name the molecule, we must know everything about that molecule, right? It turns out that is not the case. The drawings represent organic molecules as static entities, unmoving and unchanging. Even a molecular model held in your hand does not do much on its own other than rest quietly. Real molecules, however, are never still. Within a molecule, bonds vibrate, bond angles fluctuate, electrons move in molecular orbitals, groups rotate around single bonds, and the molecule as a whole moves through space. Every molecule is moving in many different ways all the time. In addition, molecules can interact with other molecules, either of the same type or of different kinds. The bigger the molecule the more complicated the story becomes. The drawing on paper of atoms and lines is a simplified story, but the true story is far richer.

INTERACTIONS BETWEEN MOLECULES

The structure of organic chemicals affects their physical properties. The melting and boiling points, as well as solubility in water or other solvents, are important properties of a molecule that depend on interactions of the molecule with other molecules. The melting point represents the transition of a substance from a solid to a liquid and the boiling point is the change from a liquid to a gas. In both liquids and solids, materials are held together by interactions between adjacent molecules. In a solid, the molecules are held rigidly in place through interactions with other molecules to form a crystal structure. To become a liquid, the molecules in a solid need enough energy to overcome the rigid interactions in the crystal to move about more loosely in a liquid. This energy comes in the form of kinetic energy, or heat. As heat is added to a solid, the molecules absorb the energy until at the melting point they have enough energy to break free of the interactions that hold them in the crystal and move about in liquid form. A liquid still has interactions and structure, but the molecules move about dynamically forming and breaking interactions among themselves. Adding heat energy to a liquid increases the movement of the molecules in the liquid. At the boiling point molecules have enough kinetic energy to break free of their interactions and escape into the gas phase, where the individual molecules are separated in space.

Clearly, energy is required for molecules to break their interactions to go from a solid to a liquid or a liquid to a gas. The stronger the interactions between molecules, the more energy is required to break the interactions. Temperature represents kinetic energy, so the stronger the interactions between molecules, the higher the melting temperature and the boiling temperature of a chemical. What types of interactions hold molecules together in solids and liquids? The three main types of interactions between molecules are hydrogen bonds, dipole-dipole interactions and Van der Waals forces. The strength of these interactions is: hydrogen bond>dipole-dipole>Van der Waals forces. All three of these forces are caused by electrostatic interactions between molecules arising in different ways and of different strengths.

DIPOLE-DIPOLE INTERMOLECULAR INTERACTIONS

An important force in interactions between molecules is dipole-dipole interaction. In covalent bonds between atoms with different electronegativities, electrons are drawn to the more electronegative atom to create polarity in the bond and partial charges on the atoms. In a bond between carbon and bromine, for example, the bromine is more electronegative than carbon and draws electrons toward itself in the bond, creating a partial negative charge on the bromine and a partial positive charge on the carbon. The molecule as a whole is not an ion, but simply has unequal charge distribution with a concentration of positive charge in one part of the molecule and negative charge in another part of the molecule. This unequal charge distribution in molecules is the dipole moment that draws molecules to interact with each other in dipole-dipole interactions (see figure).

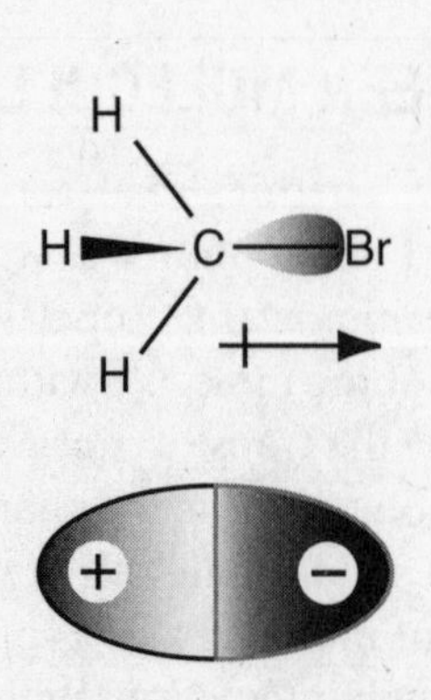

A dipole in a bond does not always result in a dipole moment in the molecule that contains the bond. Take the example of the two forms of 1,2-dichloroethene, called the cis and trans forms. These two molecules have the same molecular formula and the atoms are connected in similar ways, but in the cis form, the two chlorines are on the same side of the molecule, while in the trans form the two chlorines are on opposite sides of the molecule. The cis and trans forms are a type of isomer called geometric isomers, which we will talk about later. The carbon-chlorine bonds will be the same in these two molecules, with chlorine drawing electrons toward itself and creating a dipole in the bond with carbon. In the cis form, the dipole negative on both chlorines is on the same side of the molecule and will result in a negative charge on this side of the molecule as a whole. In the trans form, however, the dipoles in the two C-Cl bonds are pointed in exactly opposite directions, 180 degrees apart, and will cancel each other out so that while both carbon-chlorine bonds have a dipole, the molecule as a whole has no dipole moment at all.

cis- 1,2- dichloroethene:

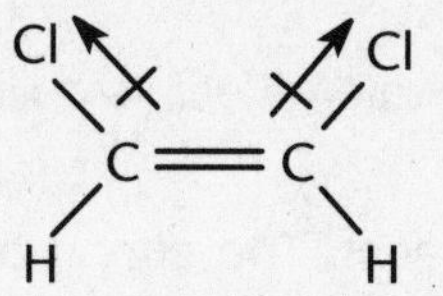

trans- 1,2- dichloroethene:

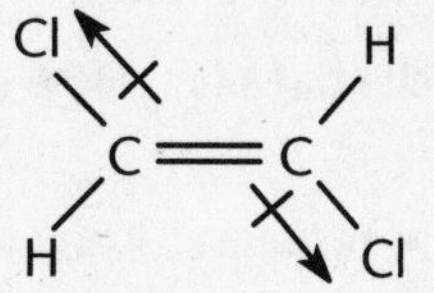

Instructor question: Which of the above two forms of 1,2-dichloroethene will have a higher boiling point, or will it be the same for both?

Student response: They would have the same properties, since they're same molecule. Right?

Instructor response: Not necessarily. Recall that the stronger two molecules are held together, the more energy it takes to pull them apart. Molecules with strong interactions need more energy (more heat) to escape these interactions in the liquid and fly off into the vapor phase. If a molecule has a dipole moment, then the positive and negative charges on different molecules will interact and hold the molecules together, increasing the amount of heat that must be added for boiling to occur. The trans form of the dichloroethene has no dipole moment and no partial charges holding molecules together, so it will boil easily. The cis form has a dipole moment, so the partial charges will interact between molecules and increase the boiling temperature, making the cis form the isomer with the higher boiling point.

HYDROGEN BONDING BETWEEN MOLECULES

Another important interaction between molecules is the **hydrogen bond**. Again, relative electronegativities play a vital role. Hydrogen is less electronegative than halogens, oxygen, or nitrogen, so in covalent bonds with these atoms, electrons are drawn to the other atom and away from hydrogen. The partial positive charge on the hydrogen interacts with non-bonding electrons on the electronegative atom of another molecule, creating a strong interaction between molecules. This is the famous hydrogen bond. The hydrogen bond is a very strong form of dipole-dipole interaction, and you will see it throughout organic chemistry. Water is a great example of hydrogen bonding (see figure). In the –O-H bond in water, electrons are drawn strongly toward the oxygen. The partial positive charge on hydrogen in one molecule is attracted to the partial negative charge on oxygen in another molecule to form a hydrogen bond between water molecules. In ice, the hydrogen bonds create a rigid matrix of interactions between water molecules, creating crystals. In liquid water there are still many hydrogen bonds between neighboring water molecules, but the interactions are dynamic and transient, rapidly forming and breaking again as individual molecules move within the solution.

Instructor question: How do the hydrogen bonds in water affect its boiling point?

Student response: They have a pretty important effect. The strong and numerous interactions caused by hydrogen bonding between molecules in water give water an unusually high boiling point.

Instructor response: Exactly. And good thing for us, since this unusual property of water keeps liquid water on the planet and makes life possible.

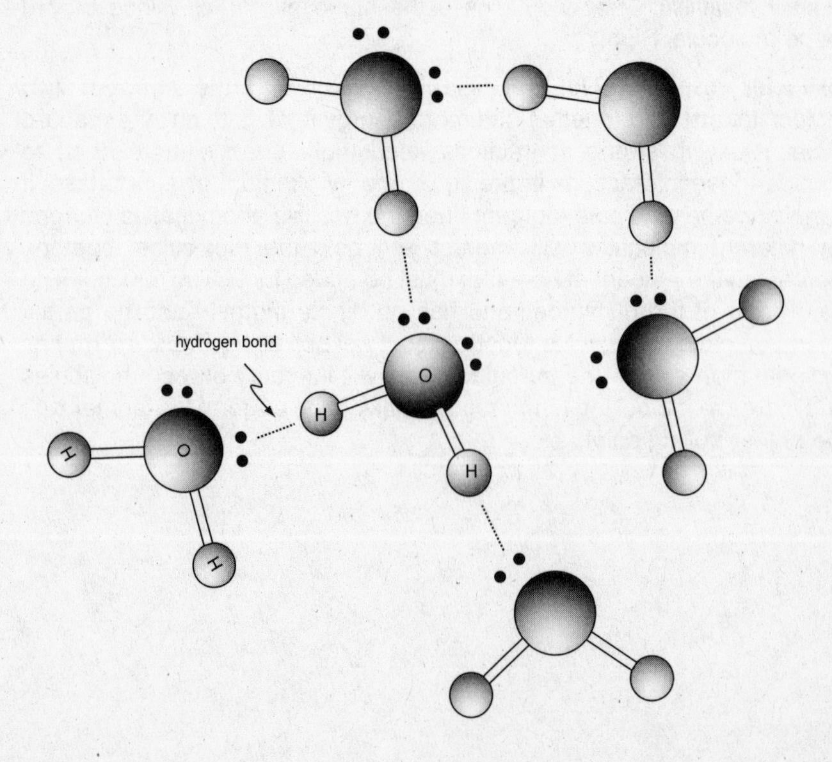

Organic molecules that can be involved in hydrogen bonding include alcohols, amines and carboxylic acids. As with water, the ability to form hydrogen bonds in many of these molecules can greatly increase the boiling point of a compound. Hydrogen bonding also plays a role in solubility of molecules in solvents like water that are polar and capable of hydrogen bonding. A molecule that can form hydrogen bonds will be much more soluble in water than a molecule that cannot.

VAN DER WAALS FORCES

Dipole-dipole interactions and hydrogen bonds are much weaker than covalent bonds. But the weakest of them all are **Van der Waals forces** or **London dispersion forces**. These weak interactions arise from unequal electron distribution in molecular orbitals, even in non-polar covalent bonds like a carbon-hydrogen bond. The electrons in a non-polar bond like a C-H bond will on average over time be evenly distributed between the two atoms, but at any given moment the electrons can be located anywhere between the two atoms and might randomly be located more on one side of the bond or the other. These random fluctuations in electron location create transient dipoles in covalent bonds and the molecules therein.

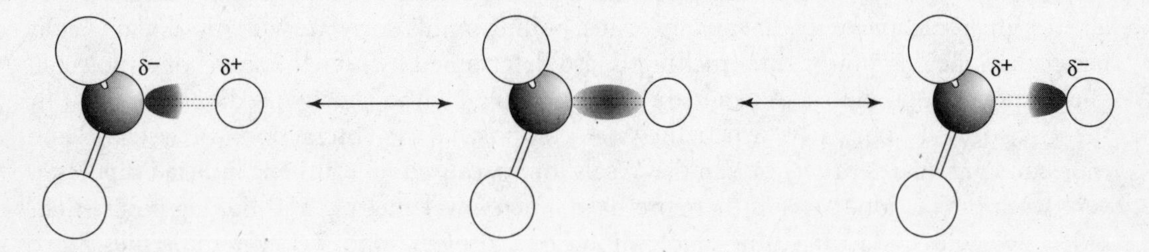

This transient dipole creates a small charge that can influence the position of electrons in other neighboring molecules. If a transient positive charge forms in one molecule, it can draw electrons in a neighboring molecule toward it, inducing a transient dipole in this neighbor molecule as well. The result can be a transient dipole-dipole interaction between molecules, all originating from momentary transient random inequalities in electron distributions in bonds. As quickly as the interaction arises it can disappear again, to reappear elsewhere perhaps in another part of the same molecule.

δ− δ+ δ− δ+ δ− δ+

electrostatic attraction

London dispersion forces or Van der Waals forces can occur in any molecule, but they are so weak compared to regular dipole interactions or hydrogen bonds that they are only observed when they occur in a non-polar setting like the C-H bond. As such, they are the main interaction between non-polar molecules like alkanes.

PHYSICAL PROPERTIES OF ALKANES

Different molecules have personalities, patterns of behavior that can be measured by certain factors such as boiling point, melting point and solubility. A chemical's personality can be predicted based on how the chemical is put together and how this geometry will affect interactions between molecules.

> **Instructor question:** As a general rule, bigger molecules have higher melting and boiling points. True or false?
>
> **Student response:** Well, bigger molecules have a larger surface area, so they will have more interactions between molecules. It would take more energy to break them up. I'd say true.
>
> **Instructor response:** You're right. Also, since bigger molecules weigh more, it takes more kinetic energy for them to speed up enough to overcome the interactions between molecules. For example, it takes less energy to speed up a methane molecule than an octane molecule.

This trend is very clear in straight chain alkanes. At room temperature (about 30 degrees Celsius), methane and ethane are gases, hexane and octane are liquids and straight-chain alkanes greater than twenty carbons are solids. Boiling points correlate with molecular weight better than melting points since melting is also determined by how the shape of a molecule packs into a crystal matrix. Alkanes are very non-polar, with virtually no dipole moment in the C-C or C-H bonds of which they are composed. The interactions between alkane molecules are made entirely of Van der Waals forces, caused by transient induced dipoles, a very weak interaction. Alkanes therefore have much lower melting and boiling points than other molecules that form strong dipole-dipole or hydrogen bonds between molecules. Since interactions between alkanes depend on Van der Waals forces, the greater the Van der Waals interactions between alkane molecules, the greater the boiling point. Increasing the length of an alkane increases the boiling point, in part, because the longer the molecule is, the more Van der Waals interactions it can form.

Another structural feature that affects the melting point and boiling point is the degree of branching within the molecule. More branched molecules are more compact, with less surface area available to interact with other molecules, so the more branched a molecule is, the lower its boiling point is. However, the trend for melting temperature is the opposite of this. More branched molecules are less flexible, and make better crystals, so it takes more energy to melt them from solid to liquid. Increased branching thus increases the melting temperature yet lowers the boiling point.

Instructor question: Let's try applying some of this information. Try to arrange the following three molecules in order of increasing boiling point:

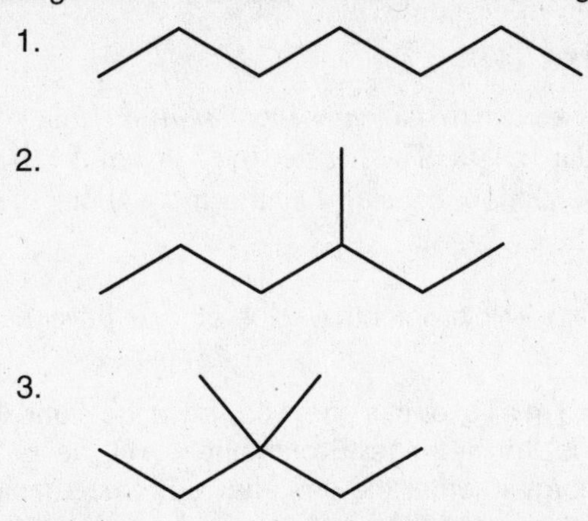

Student response: Hmm. All three of these molecules have the same number of carbons and hydrogens: 7 carbons and 16 hydrogens, with the molecular formula C_7H_{16}. Increasing size generally increases the boiling point, but here, all three molecules are the same size.

Instructor response: Correct, but they do differ in the degree of branching. We know that branching decreases the boiling point by decreasing the surface area available for Van der Waals interactions...

Student response: So the most branched molecule should have the lowest boiling point, which would be molecule #3, followed by #2, and then #1, the least branched, which should have the highest boiling point. I'm brilliant!

Solubility is another important part of a molecule's personality and is also related to structure. What does the structure of alkanes predict about their solubility in water? Solubility is determined by how well a molecule interacts with molecules of the solvent, rather than how well a molecule interacts with more copies of itself, which is the case with boiling point. If a molecule can form strong interactions with the solvent molecules, then it will be very soluble in that solvent. Salt dissolves in water because polar water molecules interact well with the sodium and chloride ions.

We know that water is a very polar solvent, with a strong dipole moment. But alkanes are very non-polar molecules, without the ability to form hydrogen bonds, and are not soluble in water. The interaction of water molecules with other water molecules is much more favorable than interactions with alkane molecules, so rather than mixing and dissolving each other, alkanes form a separate layer when they are mixed with water. We've all observed the concept of oil and water not mixing, and now you know why. Alkanes will, however, dissolve in non-polar solvents like other alkanes or benzene, since these molecules can interact favorably with one another.

Student comment: A cute one-liner you may hear in class is "like dissolves like," which sums up this concept. Polar substances dissolve other polar substances; non-polar substances dissolve non-polar counterparts.

PHYSICAL PROPERTIES OF OTHER ORGANIC CHEMICALS

Halogenated Alkanes

As with simple alkanes, halogenated alkanes increase in melting and boiling temperature as they increase in molecular weight. The heavier the halogen involved, the higher the boiling point, since molecular weight is one factor that affects the ability of molecules to gain enough kinetic energy to enter the gas phase.

Instructor question: Will bromoethane or ethane have a higher boiling point?

Student response: I'm all over this one. They have the same length, and the only difference is the bromine. Bromoethane will be heavier, and will require more energy to enter the gas phase. Bromoethane will therefore have a higher boiling point than ethane.

Instructor response: Bingo. Also, the electronegativity of bromine will create a dipole moment in bromoethane, increasing interactions between molecules and further elevating the boiling point of bromoethane compared to ethane.

Alcohols and Ethers

As with alkanes, alcohols of increasing molecular weight have higher boiling points. However, the boiling point of an alcohol is much higher than the boiling point of an alkane of similar weight. Why? As you might expect, strong hydrogen-bonding interactions between alcohol molecules increase the amount of energy required for each molecule to leave the liquid phase and enter the gas phase. As in water, the –O-H bond in alcohols is bent and is quite polar, causing a strong partial negative charge on the oxygen and partial positive charge on the hydrogen. In a hydrogen bond, the H in one alcohol molecule will bond with the O in another. This network of interactions between molecules helps to hold the molecules together in the liquid and raise the boiling point. Alcohols can also readily form hydrogen bonds with water molecules when they are mixed with water, making low molecular weight alcohols like methanol, ethanol and propanol very water soluble and high in boiling point.

Ethylene glycol, or 1,2-ethanediol according to IUPAC, has two hydroxyl groups. The hydrogen-bonding capacity of this molecule is large and gives it such a high boiling point that it is used for anti-freeze to prevent boiling in car radiators. Ethylene glycol is also poisonous, requiring care in its handling and storage.

Ethers are like alcohols, but with two alkyl groups bound to oxygen instead of one. With two alkyl groups bound to the oxygen, the oxygen does not have a hydrogen bound to it and cannot form hydrogen bonds. How might the boiling point of ethers compare to the boiling point of alcohols? Without hydrogen bonds to hold ether molecules together, they have boiling points closer to those of alkanes than alcohols.

Carboxylic Acids and Amines

One of the most important traits of carboxylic acids is their strong ability to form hydrogen bonds, even more so than alcohols. This means that carboxylic acids have higher boiling points than molecules like alkanes that do not form hydrogen bonds, and they have even higher boiling points than alcohols. The strength of the hydrogen bond derives from very strong polarity of the –OH bond in carboxylates, and the ability to form two hydrogen bonds in one molecule. Carboxylic acids have such a strong tendency to hydrogen bond that they will even form dimers amongst themselves, with two molecules forming two hydrogen bonds between them. The strong hydrogen bonding potential of carboxylates makes them very water soluble as well.

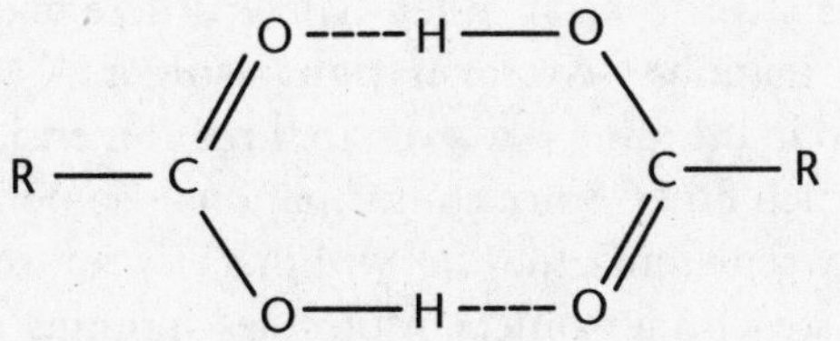

two hydrogen bonds

Amines tend to be bases rather than acids, but they share in common with carboxylates the ability to form hydrogen bonds. Nitrogen, like oxygen, is fairly electronegative compared to hydrogen. The hydrogen on one amine molecule will therefore form hydrogen bonds with the non-bonding electron pair on the nitrogen of another amine molecule. Nitrogen is less electronegative than oxygen (you can tell this because it is to the left of oxygen in the periodic table) so it forms weaker hydrogen bonds than oxygen. Amines have increased boiling point compared to alkanes, but lower boiling points than alcohols for this reason.

CONFORMATIONAL ISOMERS

Molecules are constantly moving in many different ways. As we talked about earlier, kinetic energy (heat) causes the molecule as a whole to move faster in solution or to leave the solution phase to boil and become a gas. We refer to these interactions as *inter*molecular. But movements also occur within molecules themselves, called *intra*molecular interactions. For instance, there is rotation within a molecule around a sigma bond. Sigma bonds are single bonds, molecular orbitals that form head on from overlap between atomic orbitals (see figure). Rotation of this bond does not disrupt the molecular orbital in any way, so molecules easily and rapidly rotate around single bonds. Bond rotation with a pi bond present would require disrupting the pi bond, which would take a lot of energy. Pi bonds therefore prevent bond rotation where they are present, in double and triple bonds. The different orientations of a molecule that occur as a result of rotation around a single bond are called **conformational isomers.**

Ethane is a good place to start to look at bond rotation and conformational isomers since it is a fairly simple molecule. A molecule of ethane can easily rotate around the single sigma bond between carbon atoms:

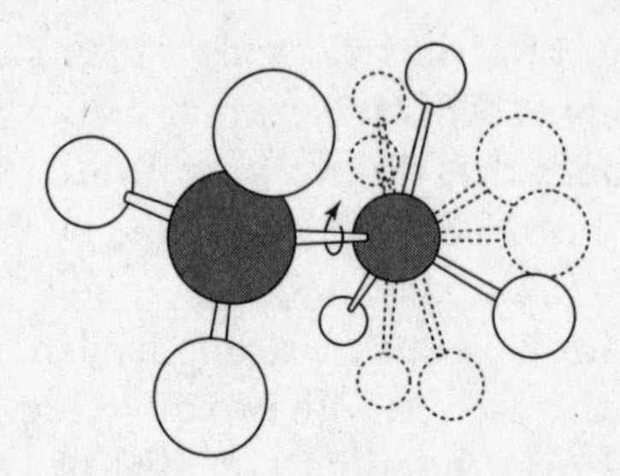

When ethane rotates around the carbon-carbon bond, the hydrogens on the two carbons change their positions relative to each other. These different orientations have different energy and distinct conformations—**conformational isomers**. Conformational isomers have the same atoms connected in the same way, even in three-dimensions, but are rotated around a single bond relative to each other. Since conformational isomers have the same atoms and the same connections between atoms, they fall within a class of isomers called stereoisomers, which are different than structural isomers. Molecules spinning around a bond are rapidly and constantly interconverting from one conformational isomer to another, so that it is not normally possible to isolate a specific conformational isomer in the laboratory, unlike structural isomers.

> **Instructor question:** How can we remember the difference between stereoisomers and structural isomers?
>
> **Student response:** The way I remember is that the stem "stereo" always refers to spatial or distance relationships, whereas the word "structural" calls to mind actual architecture. So stereoisomers have the same connections but different ways of being arranged in space, whereas structural isomers totally differ in the way they are connected, though they are comprised of the same building blocks.

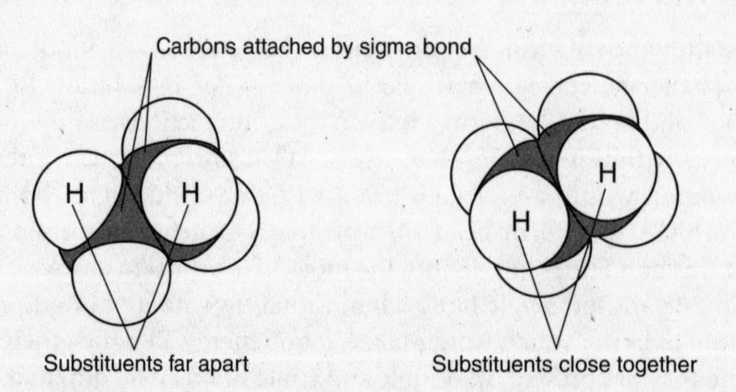

As always, it can be hard to visualize three-dimensional structural questions in two dimensions, so a special way to look at rotation on paper has been devised. The conformational isomers of ethane or other molecules can be visualized by drawings called Newman projections. To draw a Newman projection, focus on the bond that is rotating in the molecule. There may be more than one bond that rotates, but focus on rotation around one bond at a time. Now, visualize the molecule as if you are looking at the rotating bond head-on rather than sideways. If you visualize ethane this way, viewing the carbon-carbon bond end on rather than sideways, so that one C is behind the other, the hydrogens will project out away from the C-C bond in the middle, like two hydrogen wheels attached to a carbon-carbon bond axle. As the molecule rotates around the carbon-carbon bond, the Newman projection is a good way to see how the substituents on each carbon are positioned relative to each other.

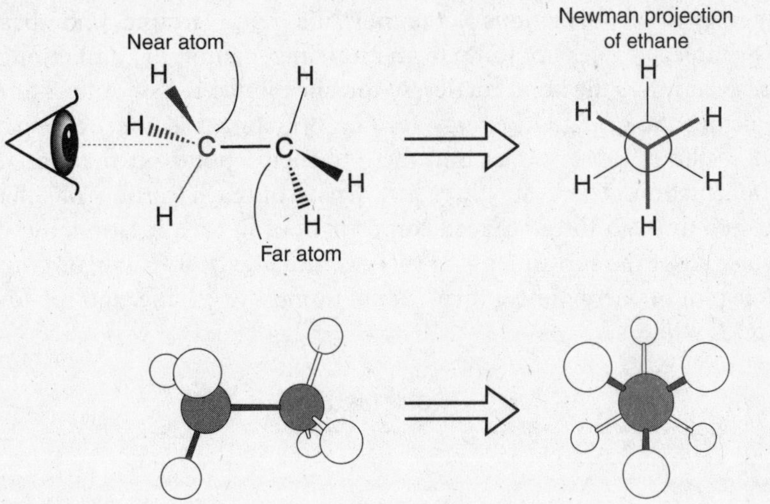

Near atom

Far atom

Newman projection of ethane

As ethane rotates around the C-C bond, the different conformations can be visualized with Newman projections. The different conformations of ethane are more or less stable, with more or less energy, depending on how close the hydrogens are to each other. The energies and stabilities of different conformational states are determined by the degree of interaction between groups bound to each carbon. The closer the groups, the less stable and the higher energy the conformation is. The farther apart the groups become as the molecule rotates around the bond, the more stable and lower energy of the conformation. When the hydrogens are lined up in the Newman projection, they are at their closest point. Each front hydrogen is directly in front of each back hydrogen, so you see only 3 spokes of the wheel, with each 'spoke' representing two hydrogens. Often we draw the front and back slightly separated so we don't forget that both exist. This conformation is called **eclipsed**, since one group is hidden behind the other; think of a solar eclipse. The eclipsed conformation is less stable than other conformations, since the electrons in the bonds with carbon (C-H bonds in this case) are at their closest distance and therefore repel each other, causing **torsional strain**. Also, just in terms of space, the closer the hydrogens are, the more they bang into each other. This is called **steric strain** and such a molecule would be high in energy if forced to remain this position. If the bond rotates further, then the hydrogens are spaced maximally apart between each other in the drawing, a lower energy state. This conformation is called **staggered**, and is the most stable conformation of alkanes.

maximum distance between substituents

staggered

eclipsed

The relative energy of conformations as the molecule rotates around a bond can be plotted on a chart. The molecule does not jump from one conformation to another but must makes a continuous transition as the bond rotates, so the line plotted is a smooth continuous curve. In a plot depicting the change in energy for conformational isomers of ethane, the high points would be the eclipsed isomer and the low energy point on the plot would be the staggered conformation. There are three hydrogens on each carbon in ethane, and the molecule must go through three eclipsed conformations in each rotation, but each of these is the same since all of the substituents in this case are identical. The high points will all be the same in the plot of energy in conformational isomers of ethane, and the low points will all be the same as well.

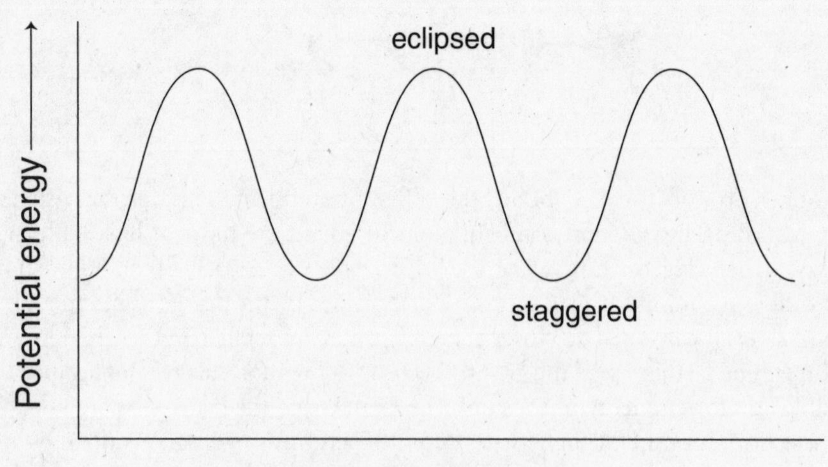

CONFORMATIONAL ISOMERS OF BUTANE

Ethane is a very simple example of conformational isomerism since all of the groups rotating on both ends of the bond are hydrogens. In other cases, larger and bulkier groups are bound as substituents and can have larger and more complicated affects on the energy of conformational isomers. Bigger groups on each side of the rotating bond, or more electronegative groups, will come closer as they rotate past each other and exhibit more repulsion between electrons, increasing the torsional strain. Increased torsional and steric strain caused by bulky groups increase the energy of the eclipsed state compared to rotation of two hydrogens past each other. Lets look at the rotation of butane around the carbon-carbon bond in the middle of the molecule as an example of the affect of bulky groups on conformational isomers. If butane is drawn as a Newman projection, it might look like the following:

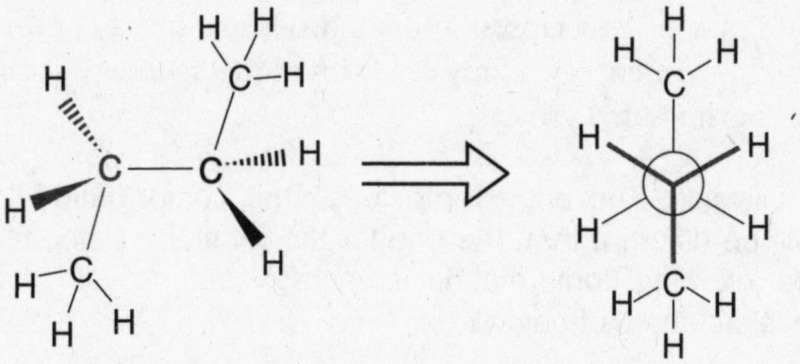

In this representation, the C-C bond in the middle of butane is viewed head on. This leaves one methyl substituent in the front and another in the back, along with two hydrogens on the carbons at both ends of the rotating bond. The methyl groups are bulky substituents that will repel each other if their orbitals are brought close together. The most stable state then is for the bulky groups to be as far apart as possible in the molecule. The conformation in which the bulky groups are the farthest apart and butane has its lowest energy is called **anti**, as depicted in the figure above. From the anti position, butane will rotate 60 degrees until the methyl substituents overlap with the adjacent hydrogens (see figure).

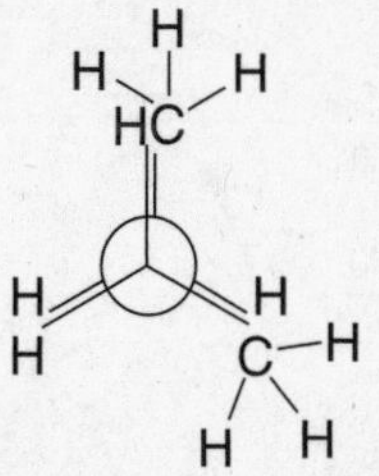

The above eclipsed state has more torsional strain than the anti state, and higher energy than an eclipsed state in ethane as well, since the methyl group is bulkier. Continuing the bond rotation another 60 degrees, the molecule arrives at a staggered conformation in which the two bulky groups are nearer than in the anti position, although still staggered. This position is called gauche (see figure). The methyl groups are close enough to show torsional strain in the gauche position, increasing the energy of this conformation compared to the anti state. Nonetheless, the gauche conformation is still staggered, and has lower energy than any eclipsed state.

repulsion between methyl groups

Finally, rotating the groups another 60 degrees, a total of 180 degrees rotation compared to the anti conformation, brings butane around to an eclipsed conformation in which the two bulky methyl groups are at their closest approach. This eclipsed conformation will have the most strain and the highest energy of any conformational isomer of butane due to repulsion of the electrons on both methyl groups.

Instructor question: The energy profile for the conformational states of butane will be different than the one for the rotation of ethane. Can you match the following conformation isomers of butane with their correct energy levels in the plot below?

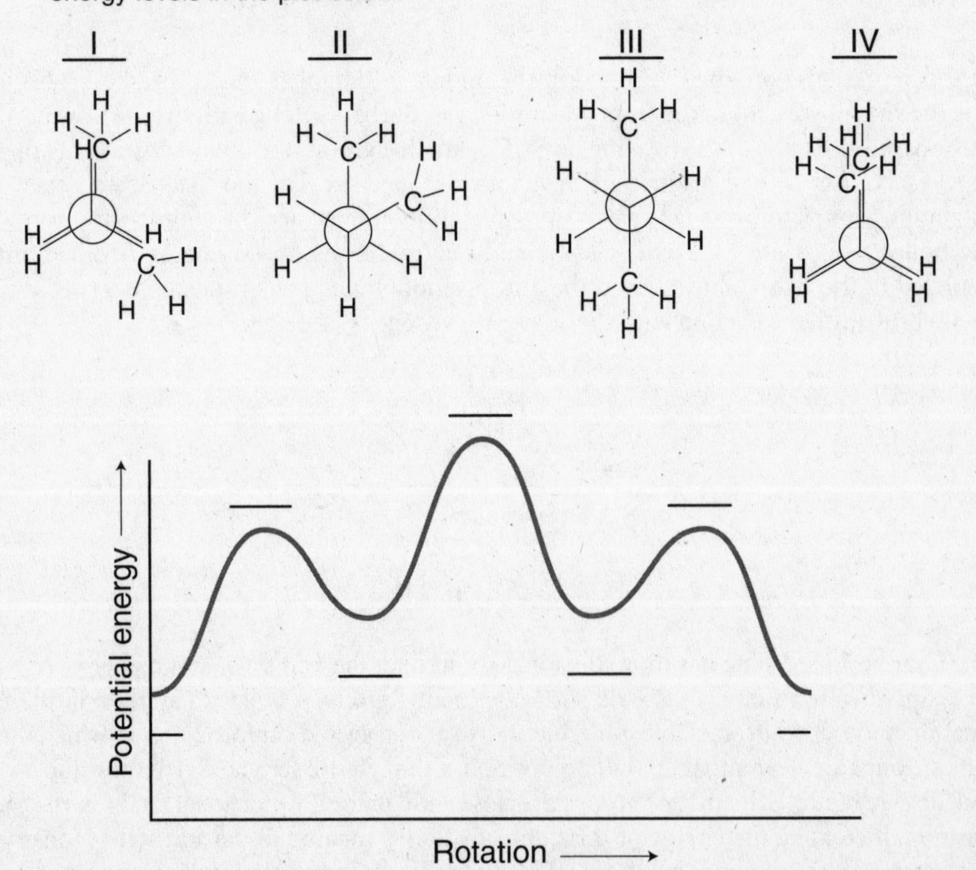

Student response: Let's see. The lowest conformation for butane is for the anti orientation, III. The anti state would be found at either the extreme left or right of the plot, where the energy is the lowest.

Student #2: I agree. The next lowest energy conformational isomers is the other staggered orientation, gauche, which is isomer II. This isomer will appear in both of the valleys between the peaks in the plot.

Instructor response: Yep. And the highest energy state will be the conformation with the greatest strain, caused by the closest approach of the bulky methyl groups and their electrons. This conformation is IV, with the two methyl groups eclipsed. This is the center peak. The remaining isomer, I, is another eclipsed state, in which the methyls are eclipsed by hydrogens. This conformation will still be high energy, since it is an eclipsed conformation, but will be lower energy than the eclipse of the methyl groups with each other. This conformation will be found at the two lower peaks.

DYNAMICS OF CYCLOHEXANE

Cyclohexane is a six-membered carbon ring. The bond angles in cyclohexane are generally the 109.5 degree angles predicted from the tetrahedral geometry of sp^3 hybridized orbitals of carbon, and the covalent bonds they help to form. How does cyclohexane assume this state?

Other smaller cyclic hydrocarbons are unable to assume the ideal tetrahedral geometry and as a result are less stable (higher energy). Cyclopropane has angles of 60 degrees and cyclobutane is essentially a square in shape, with 90 degree angles. These molecules are both strained and have much higher energy than would otherwise be the case if they were straight chain alkanes. The strained state of cyclopropane is also caused by the lack of flexibility in this molecule, forcing the hydrogens to be in an eclipsed conformation, near each other. The ability to assume a staggered conformation between adjacent carbons, even in a cyclic molecule, would allow more favorable, lower energy conformations.

How can a molecule assume different conformations if it is cyclic? Cyclohexane, with six carbons in its ring, has a fair degree of flexibility, enough to allow some bond rotation between carbons, although not full 360 degree rotation around bonds since the groups are attached on both sides to other ring carbons. This flexibility allows cyclohexane to be somewhat floppy, moving in equilibrium between different conformational states. These different conformational states and the flexibility of the molecule allow the carbons in cyclohexane to have the favored tetrahedral bond formation around each carbon.

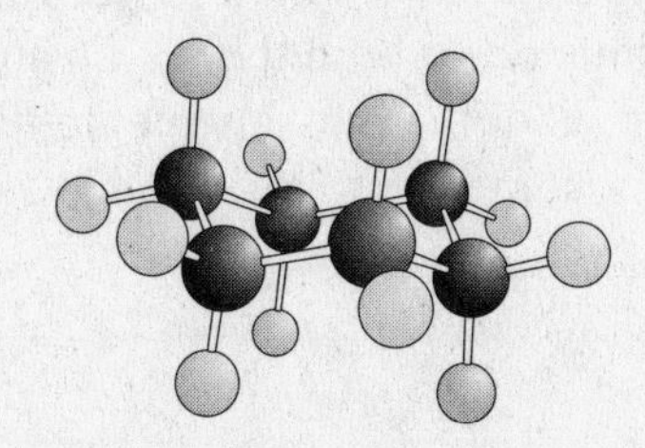

Bond rotation and flexibility produce some distinct conformational states, or conformational isomers, represented in the figure below. These isomers have distinct names, starting with the **chair** and **boat** conformations, named for their fairly clear resemblance to their namesakes. In this drawing, the molecule is simplified to represent different sections of the molecule as if they are hinged. In the chair conformation, the carbons on opposite sides of the ring are on opposite sides of the molecule, one up and the other down. In the boat, the two opposite carbons are both facing up. The molecule is said to "**flip**" between these conformational isomers, existing in solution in an equilibrium between these states, and a full range of other intermediate states, with individual molecules constantly changing their conformation through rotation around carbon-carbon bonds. As with other conformational isomers, these isomers are rapidly and continually interconverting between each other, and it is not possible at ordinary temperatures to isolate the different conformational isomers.

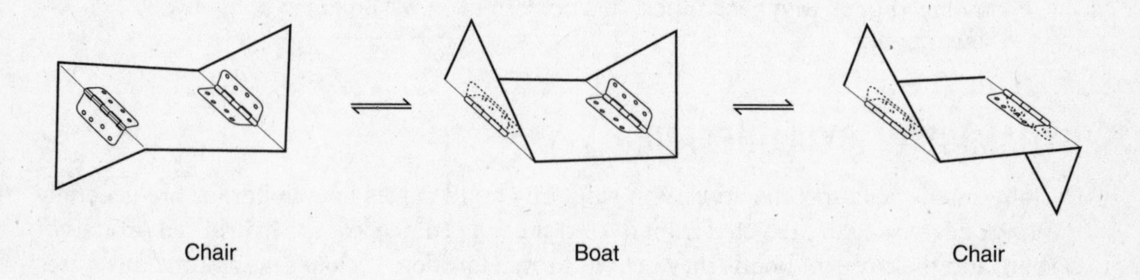

Fleshing out the diagram for cyclohexane with the wedges and dashes to indicate the shape of the molecule, the different states can be represented as:

The chair and the boat conformations are not equivalent in energy or stability. In the chair conformation, the shape of the molecule allows 109.5 degree bond angles, tetrahedral geometry, and the substituents are in a staggered conformation. In the drawing below, the chair form of cyclohexane is drawn in a Newman projection, looking at the molecule with the #2 carbon in front, and looking down both the C1-C6 and C3-C4 bonds in the Newman projection. In this conformation, all of the substituents are staggered, with minimal torsional strain, helping to make the chair isomer very stable and low energy.

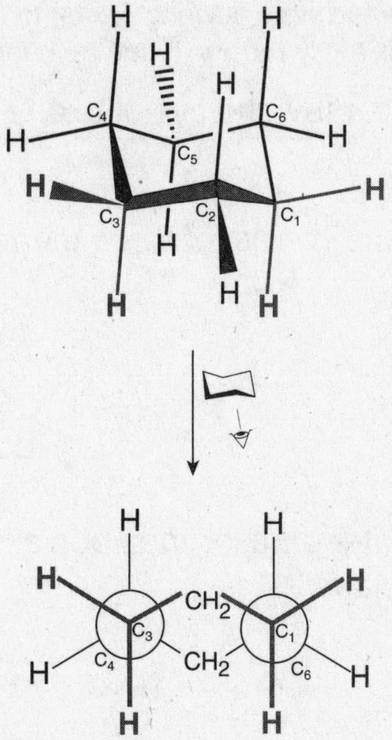

Now let's look at the boat conformation to compare the situation. In the boat, with the carbons up on both ends, the substituents all line up in an eclipsed conformation if we do the same Newman projection as above. This eclipsed conformation increases the energy and reduces the stability of the boat isomer compared to the chair. In the boat isomer, the C1 and C4 carbons and the electrons of their hydrogen friends are brought into close enough proximity that they repel each other because there is not enough room for both groups on top of the molecule. As noted earlier, this affect is called steric strain or hindrance, when bulky groups in a molecule interfere with each other. The torsional strain and the steric affects make the boat conformation less stable than the chair. Think about it this way – what would be a more comfortable and stable place for you to sit, in a rowboat or a nice comfy chair?

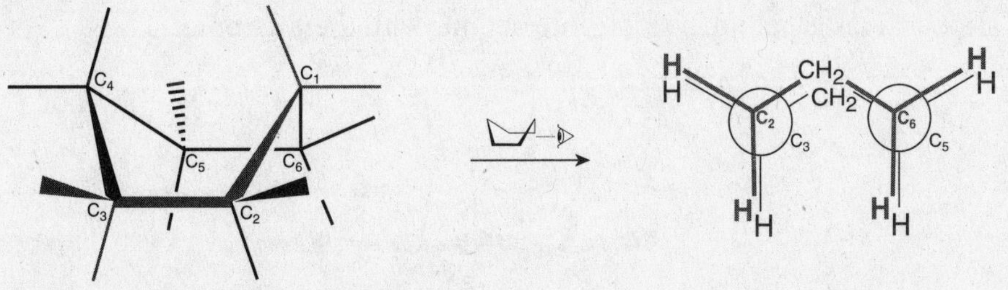

Student question: This may sound silly, but I'm having a heck of a time drawing the chair conformation.

Instructor response: Everyone does at first. One way of doing it is to draw two slightly slanted parallel lines, say pointing downward from Maine to Texas. Then on the Maine end, draw the two lines coming down to a point to the right, and on the Texas end to the left make the two lines come up, converging at a point. Simply redraw the lines slanting in the opposite direction (from Washington state down to Texas) to "flip" to the other chair conformation. It sounds tricky but you'll get the hang of it.

How to draw the cyclohexane chair

1. The conformations consist of three planes that alternate pitch

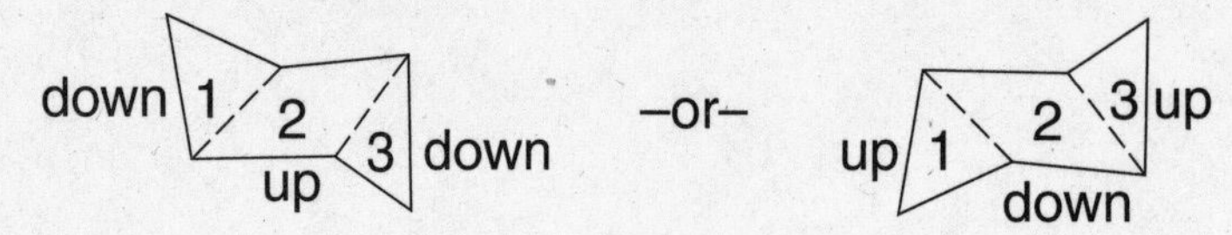

2. Draw two off-center parallel lines at a slight angle to the horizontal (plane 2)

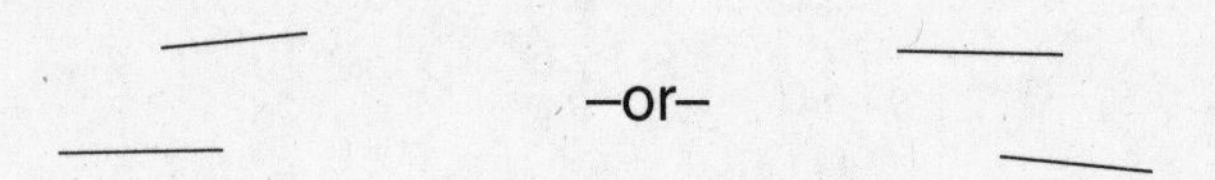

3. Connect both ends with a wedge. Keep in mind #1

In the three dimensional structure of cyclohexane, the hydrogens line up either point up and down, or stick out more or less flat, in the plane of the ring. The hydrogens sticking out away from the molecule around the edge are called **equatorial**, since they project out around the middle like the equator. Sticking out and away from the carbons in the ring, groups in the equatorial positions avoid unfavorable interactions with these carbons.

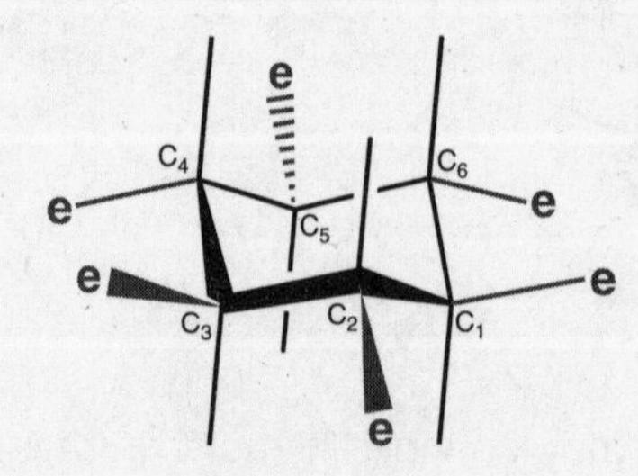

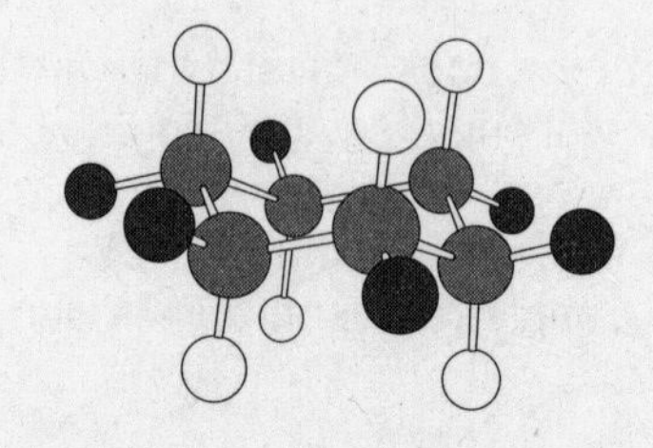

The bonds pointing straight up or down perpendicular to the ring in cyclohexane are in the **axial** position. There are three axial hydrogens pointing up, and three pointing down, alternating in the up and down axial positions going from carbon to carbon around the ring. The groups in the axial positions interact with one another above and below the ring; they therefore have more steric hindrance than groups in the equatorial positions which just jut out into space.

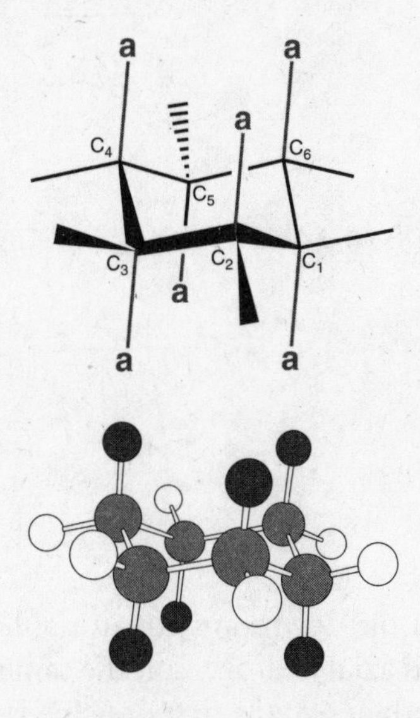

In an earlier figure, the cyclohexane ring was drawn interconverting from one chair form to the boat, and then to another chair form. When one chair converts to the other chair isomer, all of the equatorial groups become axial and all of the axial groups become equatorial as a result of the bond rotation that causes flipping between the conformational isomers. Any given substituent can therefore exist in equilibrium between being in an axial or equatorial position. You can verify this by using a model and flipping it back and forth.

> Instructor comment: An important fact is that though a substituent will go from equatorial to axial and vice versa, if it starts out "up" it will remain "up," and the same thing if it is "down." What I mean is, let's say in one chair conformation a hydrogen is in the axial position pointing straight up. If you flip the chair, it will be equatorial, but still pointing slightly up rather than down. All equatorial positions are drawn either slightly up or slightly down, representing that they generally point out into space, but not straight out. Similarly, if a hydrogen in one conformation is equatorially slightly downward facing, in the flipped chair it will be drawn as axially straight down. If a point on the chair is up, draw the axial line up and the equatorial line slightly down. If the point on the chair is down, draw the line straight down and the equatorial line out and slightly up.

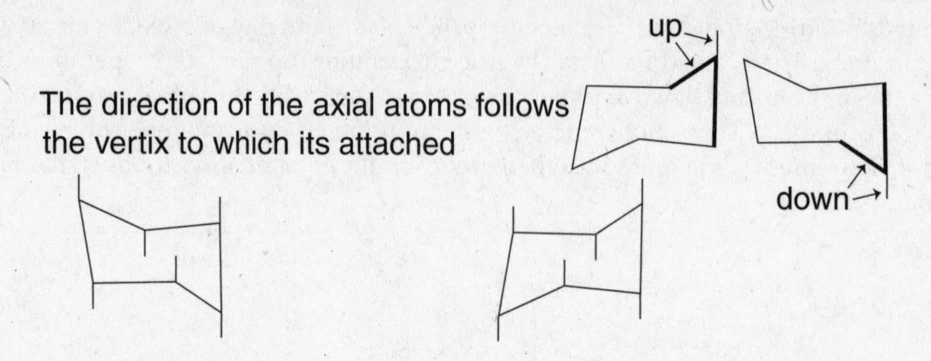

The direction of the axial atoms follows
the vertix to which its attached
up
down

The equatorial atoms are drawn in approx. parallel to horizontal

Note in the diagram below that the downward equatorial hydrogen on C6 on the left hand drawing becomes the downward axial hydrogen on the same C6 on the right. It may be hard to see at first that the upward carbon 6 in the first drawing is the same carbon pointing down in the third drawing, but with time you will grow more familiar with this transformation.

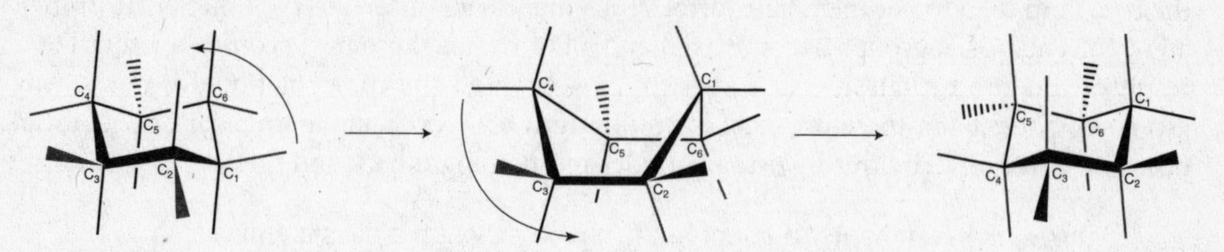

In cyclohexane, all of the substituents on the carbon ring are hydrogens. Hydrogen is the smallest substituent, and as observed in the case of conformational isomers of butane, the presence of bulkier substituents affects the energy of different conformational isomers. Putting a bulky group like a methyl group on the cyclohexane ring changes the energy states of the conformational chair and boat isomers. With all hydrogens in cyclohexane, the two different chair isomers were identical in energy. If a methyl group is added to the molecule (see figure) then one isomer will be energetically favored more than the other, the one with less interaction and steric hindrance of the methyl with other groups. Axial groups are always closer and have more hindrance, while bulky groups in the equatorial spots are freer in open space and other molecules are not there to bump into them. The chair isomer with the methyl in the equatorial state is the more favored, lower energy isomer. In general, always draw bulky substituents in the equatorial position.

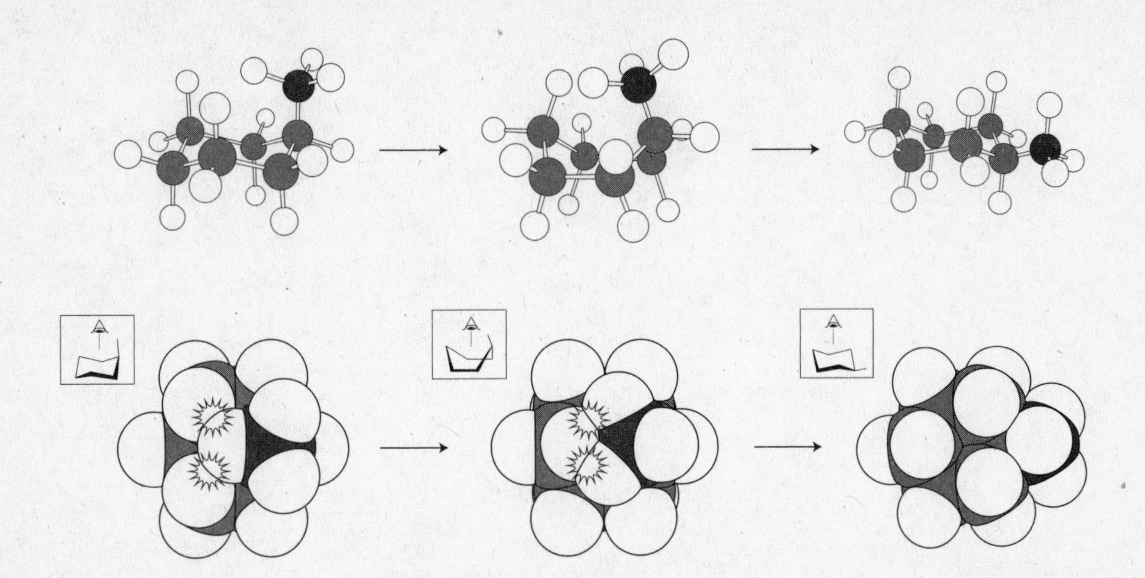

In the flipping of the methylcyclohexane between isomers, a plot of energy against conformation can be drawn, as was done for the energy of other conformational isomers as the bond rotated. In this case, the diagram starts out at the chair form in which the methyl is in the axial position. Moving into the boat form, with steric hindrance, the boat will be the highest energy and least stable isomer. Finally, the molecule can assume the most stable chair state, with the methyl in a nice open equatorial spot. The boat is the least favored state, but the molecule must pass through the boat to achieve the most stable, lowest energy chair state.

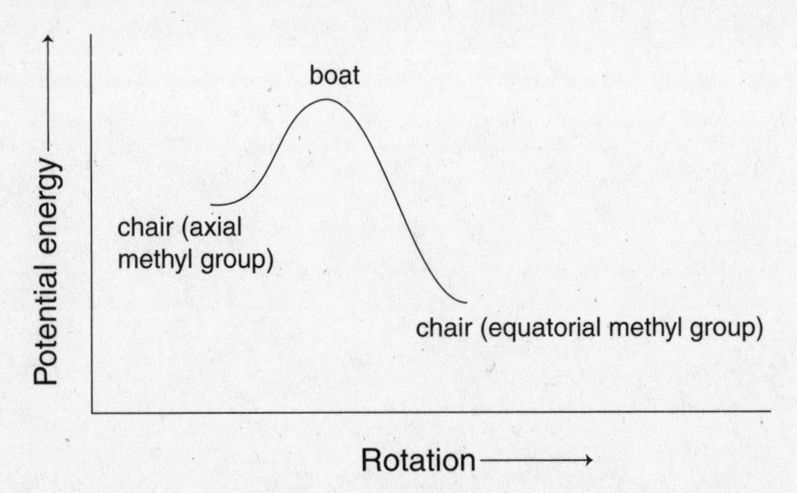

Chapter 3 has focused on the interactions and movements between molecules and within molecules and how these affect molecular behavior. Interactions between molecules affect the melting point, boiling point and stability. Another result of molecular personality is the conformational isomers resulting from bond rotation. Conformational isomers are isomers of organic molecules, in the class of isomers called stereoisomers, leading us onward into another facet of molecular structure and behavior, stereochemistry, presented in Chapter 4.

Three-Dimensional Molecular Structure: Isomers and Stereochemistry

CHAPTER FOUR

Organic molecules are complex creatures. The more you learn about them, the more questions arise. There are lots of different ways to describe molecules on paper, starting out with simple methods like the molecular formula. One of the problems with the molecular formula is that it tells you what atoms are present but it does not tell you how the atoms are connected. Indeed, many different molecules can have the same molecular formula. **Isomers** are compounds that all share the same molecular formula, but are different in some other way. Molecules with the same molecular formula but with different connections between atoms are called **structural isomers**, one type of isomer. Other isomers called **stereoisomers** are molecules that put together with the same connections, but have different 3-dimensional orientations. The conformational isomers encountered in Chapter 3 that differ from each other by way of rotation around a bond are one type of stereoisomer. Other types of stereoisomers are geometric isomers, enantiomers and diastereomers, all to be revealed in this chapter.

STRUCTURAL ISOMERS

The molecular formula tells you which atoms are in a molecule and how many there are of each. The atoms indicated by the molecular formula can often be connected to each other in many different ways, and the way they are joined affects important qualities like boiling point and reactivity. Connecting things differently can create entirely different functional groups even.

> **Instructor question:** Starting with the molecular formula C_3H_6O, what are some different molecules that can be drawn?

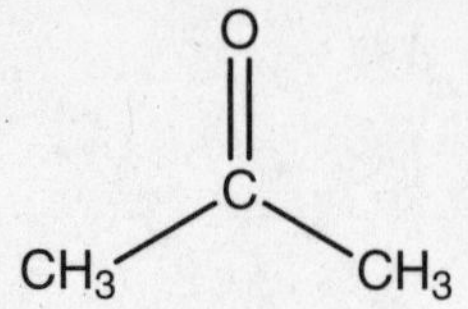

O
C
CH_3 CH_3

H OH
C
H_2C—CH_2

CH_3—O—CH=CH_2

propanone (acetone) 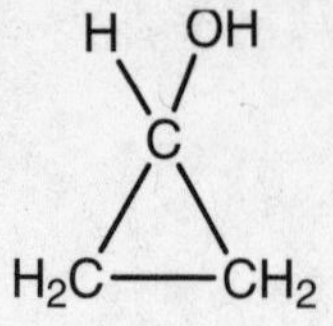 | cyclopropanol 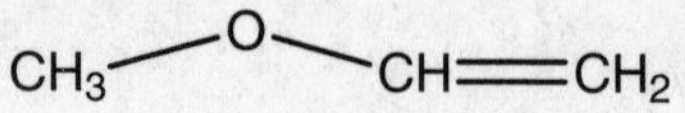 | methoxyethene

Instructor response: All of these molecules have the same molecular formula but they look quite different from each other. Even with this relatively simple molecular formula, the atoms can be connected differently to produce very different molecules, including different types of functional groups. In this case a ketone, an alcohol, and an ether were all drawn based on the same molecular formula. How will the behavior of these molecules vary?

Student response: Boiling point, for one. Only the alcohol can hydrogen bond, giving it a much higher boiling point than the others. The alcohol will also be more water soluble than the ether or the ketone since only the alcohol can hydrogen bond with water.

Structural isomers have the biggest difference between isomers, both in physical behavior and in chemical reactivity. The differences between stereoisomers, on the other hand, are generally much smaller.

STEREOISOMERS

The names and drawings we have discussed so far should be able to fully describe the atoms and connections in any organic molecule. Do the atoms and connections tell you everything about that molecule? We already saw one case in which this was not true; the different conformations produced by rotating a molecule around a bond are an additional aspect of structure not contained in a simple description of atoms and bonds. Conformational isomers are an example of stereoisomers. Isomers that have the same molecular formula and the same connections between atoms but have distinct three-dimensional shapes in one way or another are called stereoisomers.

Structural isomers are pretty easy to distinguish, since the atoms are connected differently in each isomer. Drawing out a simplified Lewis structure can reveal the differences between two structural isomers. In stereoisomers, the atoms may be connected to each other in the same way when drawn in two dimensions using a Lewis structure, but they may still be different in three dimensions, only differing from each other by the orientation of groups to each other in space.

The difference between one stereoisomer and another may look trivial, but stereochemistry can have very important consequences. Biological molecules are often extremely stereospecific (restricted to one stereoisomer). There are two stereoisomers of amino acids, the molecules that serve as the basic building block of proteins. Chemically, the two stereoisomers of amino acids differ only in their optical activity, but all life on earth uses only one of the stereoisomers. Potato starch and the cellulose in wood are very different in appearance, but chemically are very similar, differing mainly in their stereochemistry. The stereospecificity of the chemistry of life is the envy of chemists who struggle in the laboratory for months or years to reproduce the complex stereospecific molecules that nature can produce with apparent ease.

CONFORMATIONAL ISOMERS

The type of stereoisomers that are the most similar to each other are the conformational isomers, presented in Chapter 3, with the same molecular formula, the same connectivity, and the same orientation of atoms around each carbon. The only differences between two conformational isomers are the changes in orientation between groups in the molecule produced by rotation around single bonds. There are differences in energy between the different conformational isomers, such as staggered compared to eclipsed. The differences in energy are usually small compared to the thermal energy of molecules at room temperature however, so that the conformational isomers are constantly and rapidly interconverting.

It is usually not possible to isolate one conformational isomer, but there are two ways that this can happen. Heat energy contributes to the energy of bond rotation, meaning that if the heat is reduced, the molecules may have insufficient energy to rotate. If the temperature is reduced to extremely low temperatures, there is not enough energy for the bonds to overcome unfavorable high-energy eclipsed conformations and the molecule gets stuck in the conformations found in low-energy valleys. Another way to increase the abundance of a conformational isomer is by putting bulky groups onto a molecule like cyclohexane. If the group is bulky enough to create steric hindrance, certain conformations may be so high in energy that the molecule will be forced to assume lower energy conformations. For example, a bulky alkyl chain like a tert-butyl group may not fit into an axial position due to repulsion with other groups, forcing the molecule into a specific chair isomer. A bulky group may also block the molecule from the boat formation for similar reasons.

ENANTIOMERS

Another type of stereoisomer is the enantiomer. Some of the characteristics of enantiomers are the following:

- Two enantiomers have the same atoms and the same connectivity as each other, but they differ in three dimensional structure.
- An enantiomer cannot be rotated around a bond to produce one enantiomer from the other.
- Enantiomers are mirror images of each other that cannot be superimposed on each other by any rotation of the molecule as a whole or parts of the molecule.
- Since enantiomers are mirror images of each other, they always come in pairs.

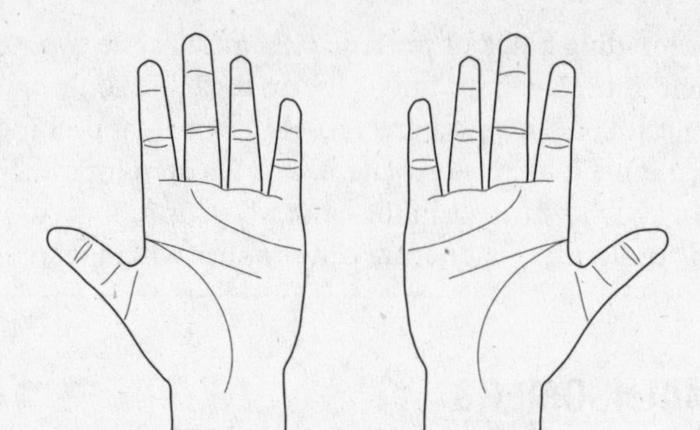

Two enantiomers are related to each other a lot like your hands are related to each other. Looking at your two hands, they are essentially the same, with the same parts, and put together in basically the same way. They are not the same however – each hand is the mirror image of the other, so that the left hand cannot simply be rotated in space to be the same as the right hand. In the figure below, for example, molecule A and B have the same atoms connected to each other, and present mirror images of each other. Are A and B the same molecule? If you get the two medium gray balls facing the same way, the two dark ones are facing opposing directions. No matter how you rotate the two molecules, they cannot be superimposed on top of each other without breaking a bond. Since the molecules 1) have the same molecular formula, 2) have the same connections, and 3) are mirror images of each other, but 4) are not superimposable, A and B are enantiomers of each other.

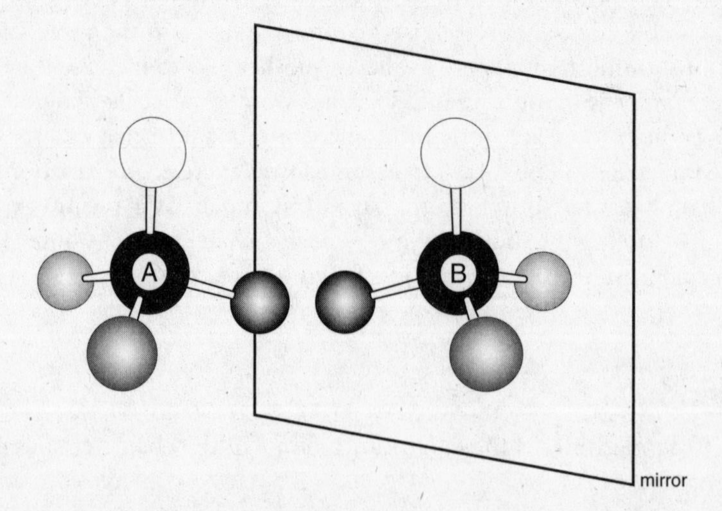

Molecular models can be very useful to understand stereochemistry. Drawings work well in two dimensions, but are sometimes hard to understand when you try to show three-dimensional aspects of structure without 3-D glasses.

An easy way to see if a molecule might be an enantiomer is to look for something called chiral carbons. A carbon that has four different groups bound to it is a **chiral** carbon and serves as a chiral center in the molecule. The carbon below has four different groups bound to it: H,

Cl, F and I. This carbon is a chiral carbon, sometimes called a stereocenter, and the molecule will be chiral. Chiral molecules have no plane of symmetry.

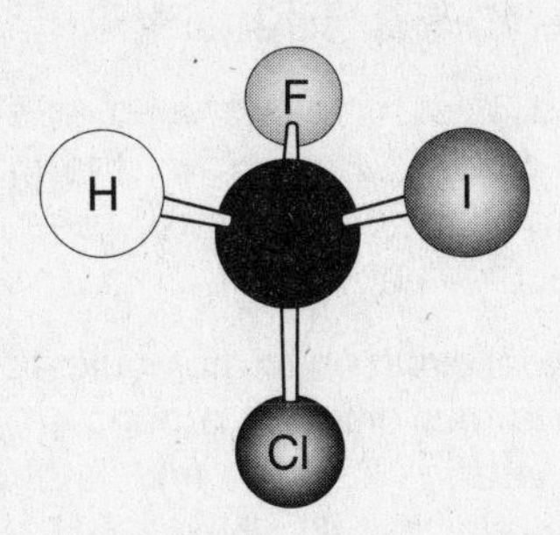

It follows that one way to rule out chirality in a molecule is to look for an internal plane of symmetry. An internal plane of symmetry would allow something to match its mirror image. For example, if you slice a plane vertically through a coffee mug, splitting the handle down its center, you can see that the two sides of the mug on either side of the plane are mirror images and are symmetrical, giving the coffee mug an internal plane of symmetry. This coffee mug is therefore achiral (achiral means not chiral). If a molecule has an internal plane of symmetry, then it is achiral, and will not have an enantiomer. As with learning to name molecules, practice is the best way to learn how to spot chiral centers in molecules.

> **Instructor question:** Let's have a go at looking for chiral centers in the following molecules, using the rules presented. Hint: Not all of these molecules will necessarily have a chiral center.

1. $CH_3CH_2C(OH)(H)CH_2CH_3$ (OH above and H below the central C of $CH_3CH_2CCH_2CH_3$)

2. $CH_3C(H)(H)CO_2H$ (H above and H below the central C of CH_3CCO_2H)

3. $H_2N—CH(CH_3)—C(=O)—OH$

4. Six-membered ring containing O, with OH on the carbon adjacent to O

5. HO— attached to a six-membered ring containing O, at the carbon opposite the O

Student response: OK, to be a chiral center, a carbon must have four different groups bound to it. None of the carbons in molecule #1 meet this criteria, because the hydrogens on the carbons are all equivalent and that central carbon has a CH_2CH_3 on either side.

Instructor response: And this molecule also has an internal plan of symmetry: if you draw a vertical line down the middle of the molecule, splitting that –OH down its middle, the two sides are a mirror image of each other, so the molecule is not chiral.

Student response: Gotcha. Moving on to molecule #2, there are once again no carbons with bonds to four different groups. In molecule #3, at last a chiral carbon is found. Carbon #2, in the middle of this molecule, has the following four groups bound to it:-H, $-NH_2$, $-CH_3$, and –COOH, making this carbon a chiral center. Molecule #4 does not seem to have a chiral center, because that carbon only has an O, an –OH, and a carbon attached to it. A chiral carbon has 4 different things on it.

Instructor response: Don't forget the implicit H that is attached to that carbon! Most of the carbons in the ring have two hydrogens attached and cannot be chiral. The ring carbon with the –OH substituent however has four unique groups bound to it, -H, -OH, -O-Ring, and –C-ring. This carbon is bound to the ring on both sides, but the ring is not the same in both directions, making this a chiral center.

Student response: And last, I would have thought molecule #5 had a chiral center, since it has a carbon attached to –H, -OH, and two carbons in a ring; but on second thought, both sides of the ring are identical. Molecule #5 has a clear plane of internal symmetry, drawn horizontally through the middle of molecule, to produce a top and bottom half that are mirror images of each other. Therefore this is not a chiral molecule and has no chiral centers.

We can also say that two molecules are enantiomers of each other if two groups on a chiral carbon can be snapped off of their bonds and switched with each other to produce one molecule from the other.

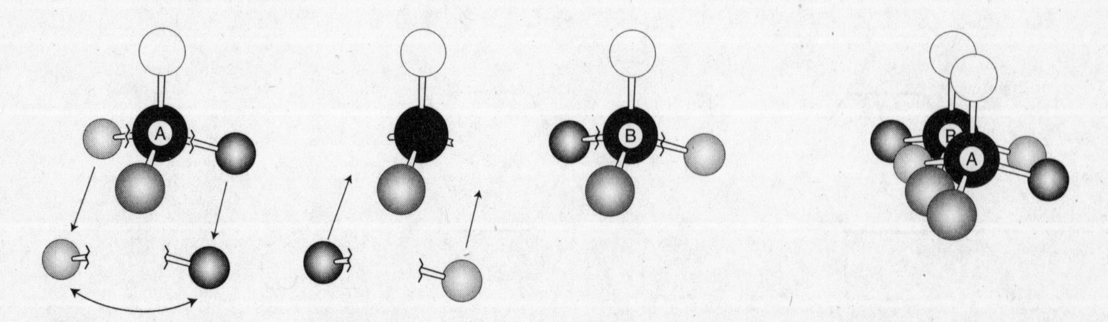

One way to draw chiral molecules on paper uses a dash and wedge style, designed to indicate which groups would be sticking out of the page and which would be behind the plane of the page. The solid wedges signify coming out of the page, and the dashed wedges represent projecting behind the page, away from the reader. This may sound trivial, but many a student has had points taken off an exam for failing to use the correct dashes and wedges. This style of drawing is sometimes replaced by Fischer projections, in which the groups are drawn with lines, without wedges, with the position of the groups indicating if the groups are coming out of the page or are behind the page. In Fischer projections, the convention is that the groups on the left and right of the structure are coming out of the page, and the group on top is behind the page.

Student comment: Think of a bow tie, in which the line up and down is a man wearing a bowtie, the man being the vertical line behind the page, and the horizontal line as the tie, sticking out from the page.

The Fischer "bow tie"

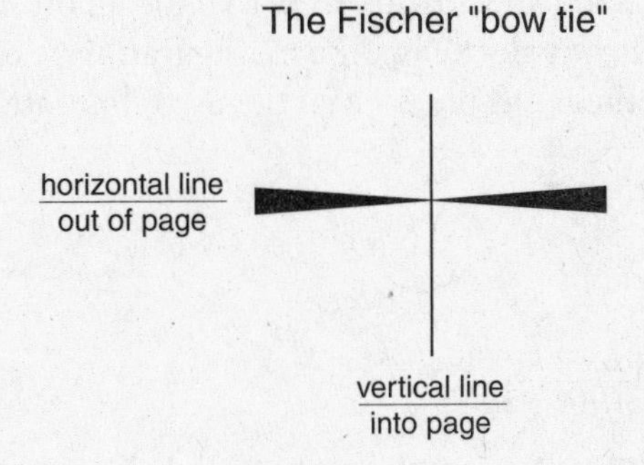

Unlike structural isomers which have very different properties, enantiomers have the same properties like boiling point and melting point. How do biological systems distinguish between enantiomers that are so similar chemically? Biological systems respond to chemicals with enzymes or other molecules that are made of proteins, large complex molecules. Proteins can bind to a molecule in multiple places, making contact with enantiomers in three or more places in space. Proteins are themselves composed of stereospecific components, the amino acids, so they are able to distinguish very effectively between two enantiomers by forming a three-dimensional recognition site that matches one enantiomer and not the other. Three-dimensional recognition in sensory receptors allow the nose to distinguish one enantiomer of the molecular carvone as the smell of peppermint, and the other enantiomer as the smell of dill. Stereospecific biological reactions are essential to life.

NAMING OF STEREOISOMERS: ABSOLUTE CONFIGURATION

As with other molecules, it is important for chemists, and yourself as a budding chemist, to be able to name organic molecules that are stereoisomers. Otherwise, you might walk into the café and try to order one molecule, and end up with the enantiomer – such a faux pas! The naming must include a way to communicate the arrangement of the groups around a chiral carbon.

Instructor question: If you name a molecule chorofluoroiodomethane (see discussion ahead), does this tell you the full story about this molecule?

Instructor response: No, it does not! This molecule has a chiral center and will have an enantiomer, meaning that the name given would apply to either enantiomer and is therefore not complete.

For a chiral molecule, the arrangement of the atoms around the chiral center needs to be a part of the name for the name to be complete. The naming of the organization of a molecule around the chiral center is called the **absolute configuration**. R and S are the names given to describe the absolute configuration of the groups around a chiral center in two enantiomers. The system to name the absolute configuration starts out by assigning a ection to the four different groups attached to the chiral carbon using rules we will go over in a minute. When the priority of each group is known, draw the structure with the lowest priority group (priority #4) behind the plane of the paper (dashed wedge). The three other groups then are in the plane of the paper, and numbered 1, 2, and 3. Draw a circle starting at group priority #1, connecting it then to #2 and #3 with an arrow to indicate the direction of the line around the circle. If the circle is drawn clockwise, the configuration is called R, and if the circle is drawn counter-clockwise, the molecule is given the S-configuration.

Absolute Configuration Rule #1

The main rule about the priority of groups around the chiral center is based on the atomic number of the groups attached to the carbon. The heavier something is, the higher the priority of the group when you are determining the absolute configuration. Check the periodic table.

Absolute Configuration Rule #2

The second rule is that if the first atoms in groups attached to the chiral carbon are the same, go down the line to the next atoms connected, and so on. For example, if the atoms immediately adjacent to the chiral carbon are themselves carbons, then look at the next atoms down the line until different groups can be distinguished. At the first fork, make a designation as to which group "wins."

Absolute Configuration Rule #3

The third rule concerns double and triple bonds in groups when determining priority. If you are going down a line of atoms and get to a double bond, double the atoms attached to the other end to figure out priority. C=O would be considered as C with two oxygens bound to it. It is helpful to actually draw in your own lines attaching more oxygens. For a triple bond, triple the attached atoms.

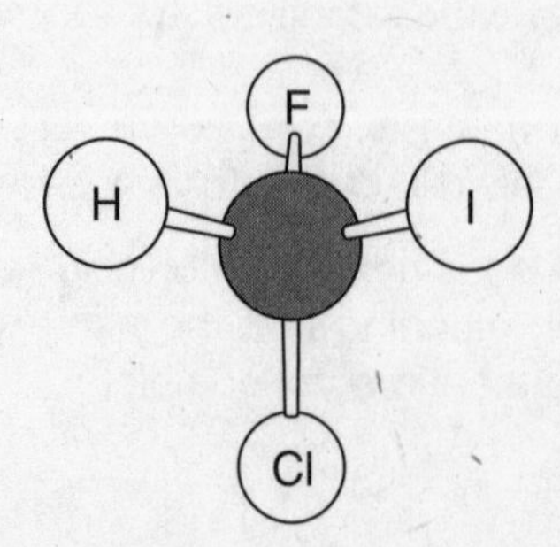

Instructor question: Let's take a crack at assigning priority first in a simple case, then the fun stuff. In the above molecule, what is the order of priority?

Student response: Looking at a periodic table, iodine is the heaviest (priority #1), followed by chlorine (#2), fluorine (#3), and hydrogen (#4). Next, we rotate the molecule so that the lowest priority group is pointing away, behind the page. The lowest priority group in this molecule would be hydrogen, the lightest element.

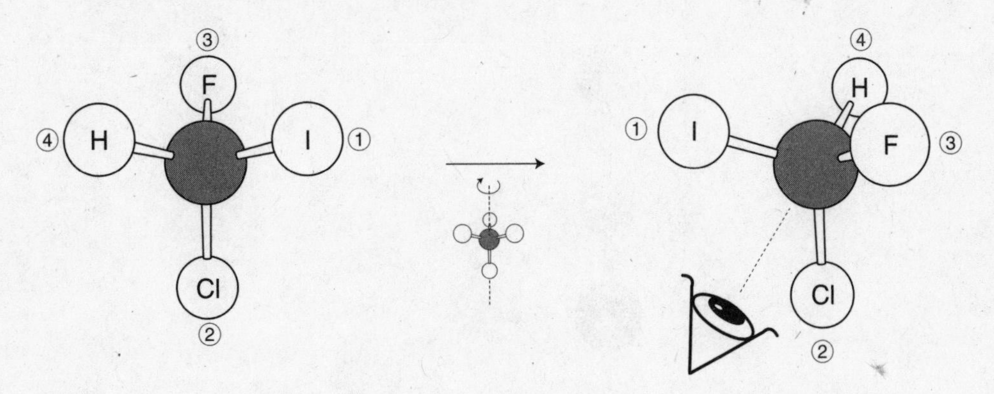

Instructor response: Good so far. Now that we have the lowest priority group oriented behind the page, we draw a line circling around to connect the remaining groups, from high to low priority. In this case, the priority goes from iodine, to chlorine, to fluorine in a counter-clockwise fashion. Counterclockwise order of priority groups means that the molecule has the S absolute configuration. Clockwise would have indicated the R isomer. Now you can place your order with confidence knowing the complete name: (S)-chlorofluoroiodomethane.

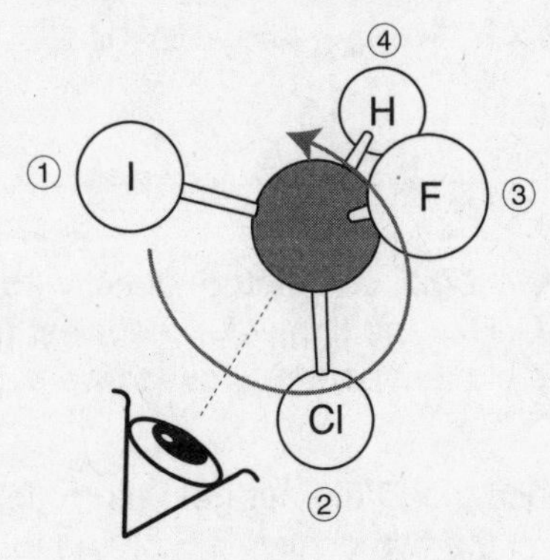

(S)–chlorofluroiodomethane

Student response: This may sound dorky, but here's how I remember R versus S: if your circle goes around clockwise, I think of an alaRm clock. Clockwise = R. And I just know that the other way has to be S. Or here's another way: when you turn the steering wheel clockwise, you're turning it to the Right.

Instructor response: Cute. Now, time for something a little more challenging. Determine the absolute configuration of this molecule, 1,2-butanediol.

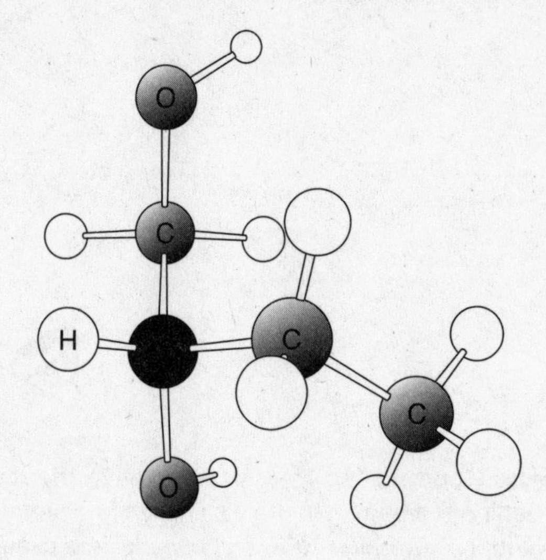

This molecule might be hard to figure out when drawn this way, so reduce it to a wedge and dash drawing.

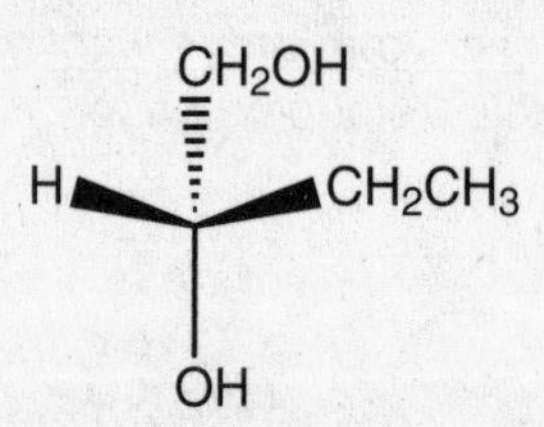

Student response: The atoms connected directly to the chiral carbon include: H, O, C and C. H is the lightest and lowest priority and O is the heaviest, making it the highest priority. How do we distinguish the two Cs again?

Instructor response: Since the C did not help us to determine the priority, look at the next atom in the two chains (according to rule #2 for priority). In one group the next atom is oxygen, and in the other chain it is carbon. Oxygen has a higher priority, since it's heavier, making the CH_2OH group a higher priority than the CH_2CH_3 group. The ordering of the priority of these groups becomes OH> CH_2OH> CH_2CH_3>H. Since H is the lowest priority, we want to rotate the molecule to put the H atom behind the rest of the molecule.

(2) CH_2OH
(4)H CH_2CH_3 (3)
OH (1)

Instructor response: Yes, but BE CAREFUL! To do this, you CANNOT just rotate everything to the right or left. Here's how we do it. Keep one substituent steady, and rotate the other three. Don't keep the H steady since that is the one we want to move. Here, for instance, keep the –OH in the downward position and rotate the other three clockwise so that the H goes on top, the CH_2OH goes right, and the CH_2CH_3 swings over to the left.

(2) CH_2OH
(4)H CH_2CH_3 (3)
OH (1)

rotate →

(4) H
(3) CH_3CH_2 CH_2OH (2)
OH (1)

Student response: Following the priority of the groups from 1–3 makes a counter-clockwise circle, making the molecule the S enantiomer, (S)-1,2-butanediol.

Another way to figure out the absolute configuration is by using Fischer projections. Let's revisit good old chlorofluoroiodomethane. First, convert the wedges and dashes, or stick and ball drawing, to a Fischer projection. How to do this? Remember the bowtie – the line across is the tie sticking out, and the vertical line is the man wearing the tie. Since the H and I are sticking out, these are on the left and right side of the Fischer projection, with F and Cl behind, and on the vertical line. The next step is the same as before, assigning priority to the groups, and writing these on the drawing.

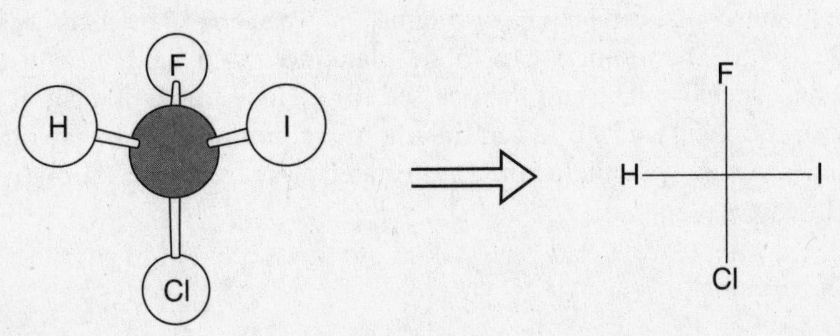

And finally, draw a circle connecting the prioritized groups. Recall that we usually want the lowest priority group sticking up (meaning into the page.) If the lowest priority group (#4) is indeed on the vertical line, then the R and S configuration are determined as before, with R for clockwise and S for counterclockwise. If, however, the lowest priority group is on the horizontal line, as drawn here, then after drawing the arrow from high to low configuration, you simply reverse the rule for how to assign R or S. For example, this molecule would be R if the H were on the vertical line. Since it's on the horizontal line, we just take the opposite: S.

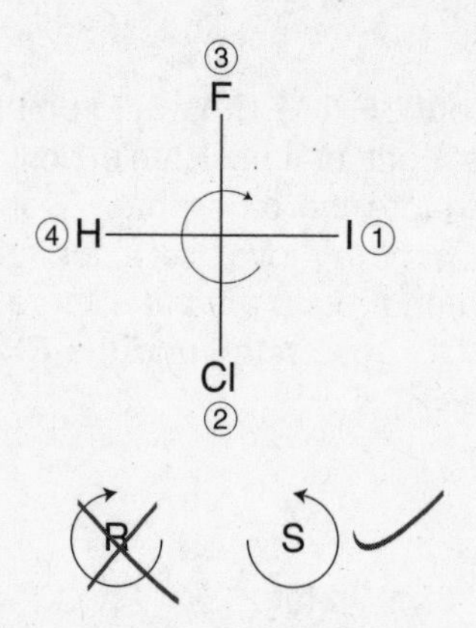

Student comment: Note that if you prefer, you can employ the switcheroo method we talked about earlier: you could keep one group steady, the Cl for instance, and rotate the other three to get the lowest priority group on top, then just assign R or S as usual. There are lots of equally valid ways of assigning absolute configuration.

PHYSICAL PROPERTIES OF ENANTIOMERS

As we mentioned earlier, the two enantiomers of a molecule are very similar in structure and physical properties. An alcohol with the hydroxyl group on a chiral carbon will have the same boiling point for both enantiomers – they will both hydrogen bond in the same way. The fact that enantiomers behave the same way can make isolating them from each other very difficult. If a mixture is crystallized, individual crystals might contain only one enantiomer or the other, and it might be possible to tell the crystals apart under the microscope based on the "handedness" of the crystal. Louis Pasteur did it, but it would not be easy.

One physical property that can help us distinguish between enantiomers is how they interact with polarized light. Enantiomers can rotate plane-polarized light. If a beam of light is polarized along a specific plane and then passed through a sample of a chiral molecule, the plane of polarization will rotate by a certain amount, either one way or the other. If the line goes in polarized vertically, it might come out rotated away from the vertical by an angle of a certain size.

Student question: So the way that a molecule interacts with polarized light can predicted from the absolute configuration.

Instructor response: Actually, no. The R or S name is based on rules of nomenclature as we just went over, not the physical property of how a molecule rotates polarized light. A given molecule will produce a specific amount of rotation, called the specific rotation. If the light is rotated clockwise, the rotation is positive, and if it is rotated in the other direction, the rotation is negative. A compound that alters the plane of polarization is **optically active**. And the S enantiomer of a molecule will always rotate light by the same magnitude but in the opposite direction from the R enantiomer.

SPECIFIC ROTATION

Light is a wave. Regular light, coming out of a light bulb or the sun, is a mixture of waves oriented randomly in different directions. It is possible however to select from this mixture light that is oriented in one particular direction. A device that does this is called a polarizer. Polaroid sunglasses, which you might have spied your grandmother wearing, are polarizers, meaning that they let through light of only one orientation. The light hitting the glasses is a mixture of waves, yet the light reaching your eyes is all oriented in the same way (it is polarized). You can pass polarized light through your sample of a chemical to see if it is optically active. Simply measure the plane of polarization of light after it goes through the sample and compare this to the polarization before passing through the sample; if the angle changed when light went through the sample, you have an optically active compound in your beaker.

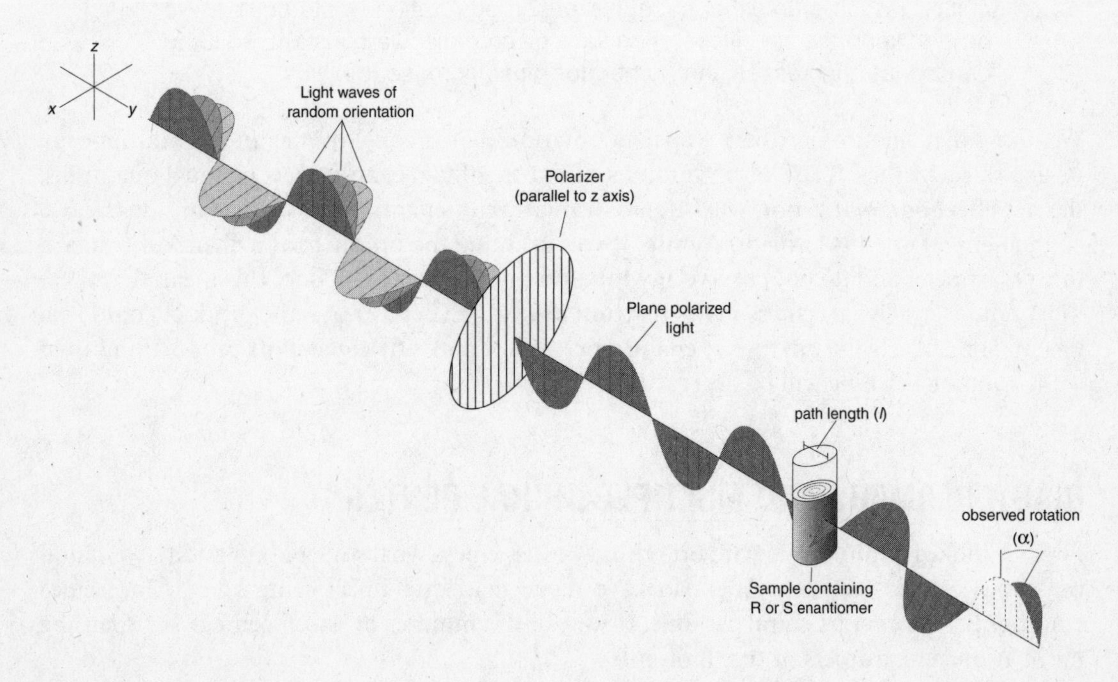

The polarimeter determines the angle of rotation. The angle of rotation is influenced by how much of the chemical is in the solution that the light passes through (the concentration of the sample), how far the light must pass through the sample (the farther it travels, the more molecules it will encounter along the way), and how each different chemical affects polarization. Each different chemical affects polarization of light in a way that is constant for that chemical, and distinct from other chemicals. This constant value for a chemical is called its **specific rotation**, represented by that little alpha symbol shown below. The observed rotation will depend on the concentration of sample and the size of the beaker, but the specific rotation is always the same for a compound once you correct for the other factors involved. If a compound rotates polarized light clockwise, the specific rotation for that compound is positive, and if it rotates in the opposite direction, the specific rotation is negative.

$$\alpha_{specific} = \frac{\alpha_{observed}}{c \bullet l}$$

where
$\alpha_{specific}$ = specific rotation
$\alpha_{observed}$ = observed rotation
c = concentration of solution
l = path length

Instructor question: What will happen to the angle of rotation if the concentration of a chiral molecule is doubled? How about if the path length is doubled?

Student response: In either case, the observed rotation would double. We have to keep the specific rotation for a given molecule constant, so if the denominator increases, the numerator must increase too.

We know that enantiomers have a specific rotation that is equal in magnitude but opposite in sign to each other. If a solution contains a mixture of two enantiomers in equal quantities, the net effect on polarization will be zero. A mixture of enantiomers such as this, with 50% of each enantiomer, is termed **racemic**. If you examine the products of a chemical reaction in a polarimeter and do not observe any rotation of polarized light, does this mean that there must not be any chiral centers in the reaction products? Clearly not – the product could be a racemic mixture. If the mixture of enantiomers is not perfectly equal in its proportions, then some rotation of angle will be observed.

DIASTEREOMERS AND MULTIPLE CHIRAL CENTERS

First we looked at molecules with one chiral center. Guess what's next ? You got it, two chiral centers. Molecules can and often do have more than one chiral center. Sugar molecules commonly have several chiral carbons. If we call the number of chiral centers "n," then the number of stereoisomers of that molecule is 2^n.

Instructor question: Consider the molecule below, 1,2-dichloro-1-propanol: how many chiral centers does it have? How many stereoisomers of this molecule might there be?

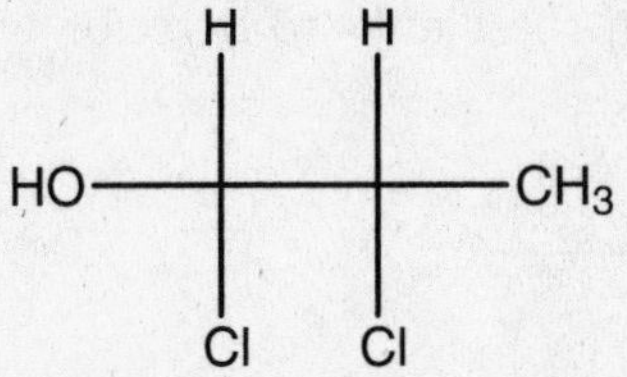

H H

HO—+—+—CH_3

Cl Cl

Student response: I see two chiral carbons, those in the middle of the structure, which is drawn here as a Fischer projection. To calculate the maximal number of stereoisomers, we know n = 2, and 2 squared is 4. There are 4 stereoisomers.

Instructor response: Well done. But just for kicks, let's try drawing out the stereoisomers below:

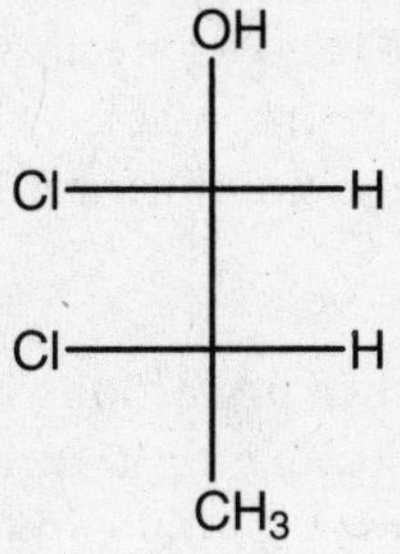

OH
Cl—+—H
Cl—+—H
CH_3
1

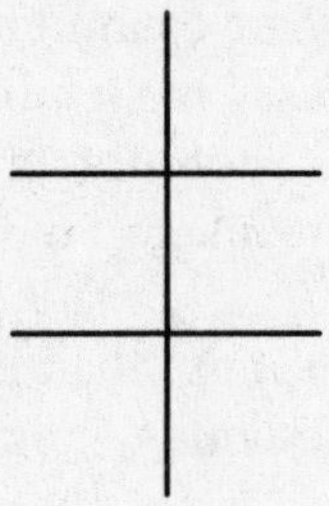

mirror

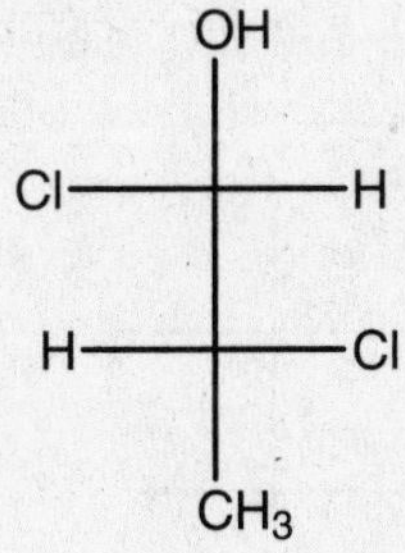

2

OH
Cl—+—H
H—+—Cl
CH_3
3

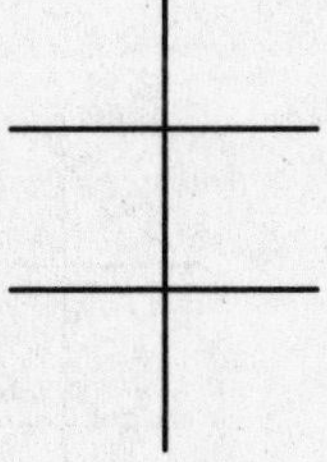

mirror

4

Student response: No problemo. These are easy to draw: just leave the top and the bottom the same and switch the groups on the left and right side of the vertical line, making them mirror images.

What do the structures mean? If two molecules are not superimposable, but are mirror images, then they are enantiomers of each other. If two molecules are stereoisomers with more than one chiral carbon, but they are not enantiomers (not mirror images), then they are called **diastereomers**. Fill in the blanks to describe what type of isomers each of the following are to each other:

i. 1 and 2 are ________________

ii. 1 and 3 are ________________

iii. 1 and 4 are ________________

iv. 2 and 3 are ________________

v. 3 and 4 are ________________

vi. 2 and 4 are ________________

Molecules #1 and #2 are enantiomers of each other and #3 and #4 are also enantiomers since they are in both cases mirror images that are not superimposable. The other pairs are stereoisomers and are not superimposable, but they are also not mirror images, meaning that they must be diastereomers and not enantiomers.

In compounds with more than one chiral carbon, Fischer projections become even more valuable to track absolute configuration. It might be possible to keep track of one chiral carbon using wedges and dashes, but tracking two or more chiral carbons this way is too much. The rules for drawing Fischer projections with two or more chiral centers are the same as the rules for molecules with one chiral carbon, with the groups on every horizontal line sticking out from the page. Instead of a man with one bowtie, imagine now that he is a creative dresser and has two (or more) bowties. The enantiomer of a molecule with more than one chiral center will have the opposite configuration at every carbon. If one enantiomer is R at both carbons, the other enantiomer is S at every carbon.

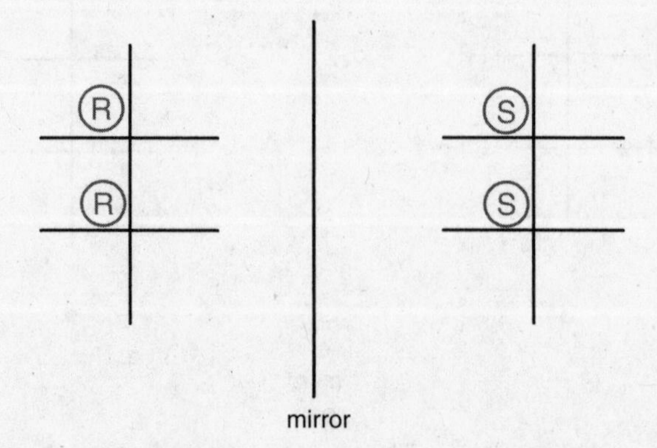

Diastereomers, however, have different combinations of chirality, matching perhaps the chirality at one carbon and not the other.

(R)
(R)

(R)
(S)

(S)
(R)

Instructor question: Let's take another crack at distinguishing enantiomers from diastereomers. In the molecules below, what is the relationship of the two molecules in #1 and the two molecules in #2 to each other?

1.

CHO
Br — — CH_2OH
H — — OH
CH_2OH

CHO
HOH_2C — — Br
H — — OH
CH_2OH

2.

CHO
Br — — CH_2OH
H — — OH
CH_2OH

CH_2OH
H — — OH
Br — — CH_2OH
CHO

Student response: With number one, we see that these molecules have the same groups and the same connections – they must be a type of stereoisomer, either enantiomer or diastereomer. These are Fischer projections, making it possible to determine the absolute configurations if necessary. But it should not be necessary, because for them to be enantiomers, they must be mirror images of each other. The positions of the groups on the upper chiral carbon are reversed, switching Br and CH_2OH to change the configuration of this carbon. The configuration on the lower chiral carbon in the Fischer projection is the same in both molecules. Since the molecules are not a mirror image of each other, they are not enantiomers and must be diastereomers.

Instructor response: Right. Now compare the two molecules in #2 now – how do they relate to each other? Are they superimposable? Try to rotate this molecule in your mind or on paper to see if one superimposes on other as a result of rotation. The answer is no, it does not. By rotating the molecule 180 degrees, the top and bottom groups match, but the other groups do not match. They are, however, a mirror image of each other. What type of isomers are non-superimposable but are mirror images of each other? Enantiomers, of course. By examination of the Fischer projection and the relation between positions of groups, it is possible to see that the molecules are enantiomers without going through the more difficult process of determining the absolute configuration for everything.

Molecules that have more than one chiral center and an internal plane of symmetry are called **meso** compounds. Meso molecules lack optical activity in the molecule as a whole, even though they contain chiral carbons. Examine this molecule – would it be optically active?

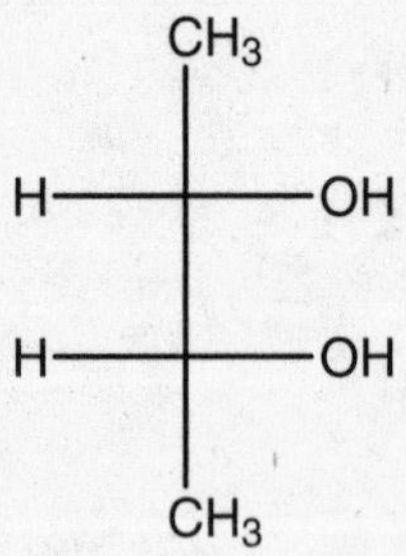

This molecule has two chiral carbons, and it also has an internal plane of symmetry. The upper half of this compound is a mirror image of the lower half. The rotation of light by one chiral carbon in a meso molecule cancels out the rotation from the other chiral carbon, resulting in a lack of optical activity for the molecule even though it contains chiral carbons.

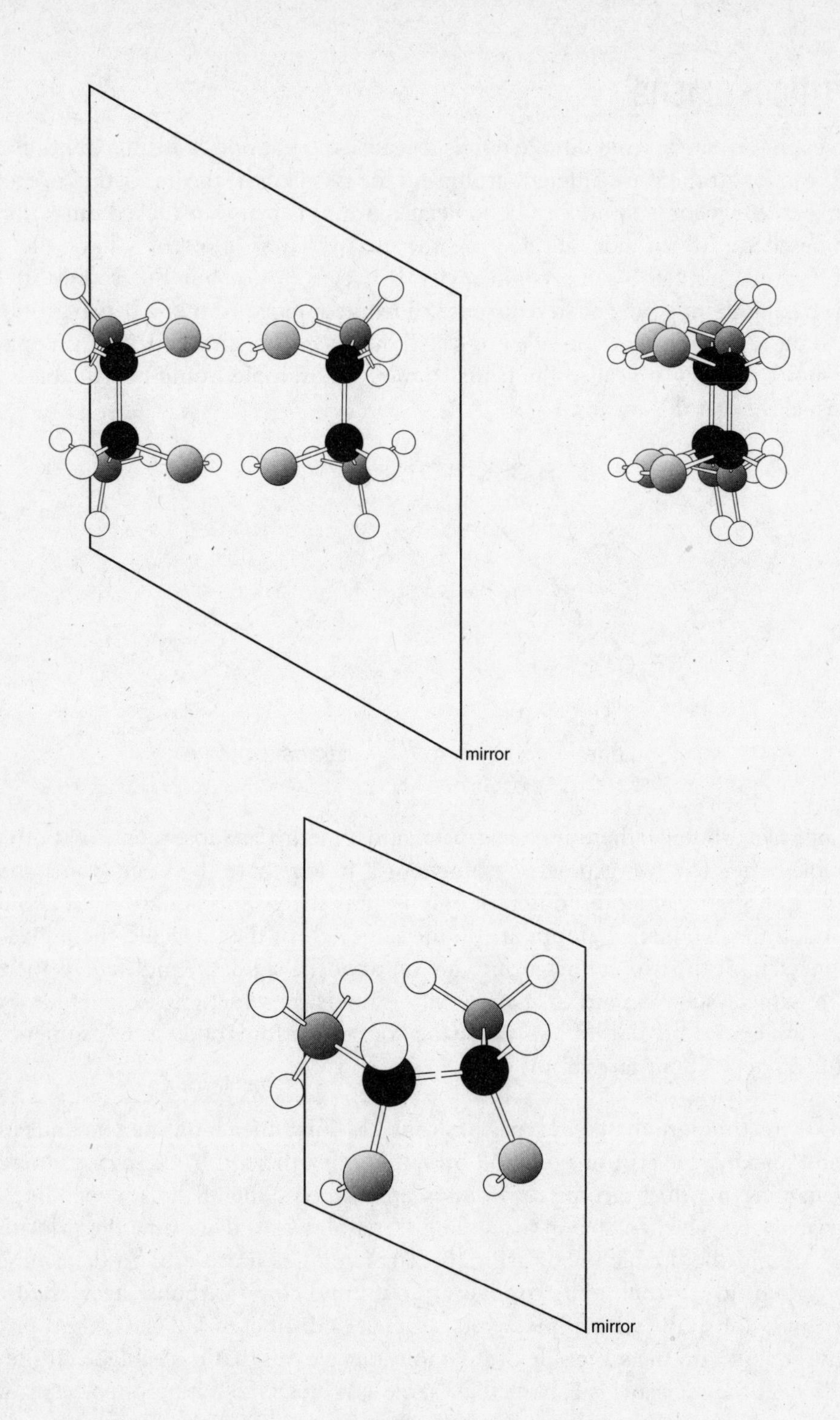
mirror
mirror

GEOMETRIC ISOMERS

Molecules cannot rotate around double bonds, because the pi bond holds the bond in place to prevent rotation. If there are different groups on the two ends of the bond, they are locked in place in a specific geometry. In a cyclic molecule groups can also be locked into a specific geometry based on which side of the ring they exist. These molecules have the same molecular formula, and the same connectivity between atoms, but differ only in their geometry, making them a type of stereoisomer called **geometric isomers**. If two groups are locked onto the same side of a molecule it is called the **cis** isomer and if they are on opposite sides of a molecule, they are called the **trans** isomer. An example would be valuable at this point to visualize what this means.

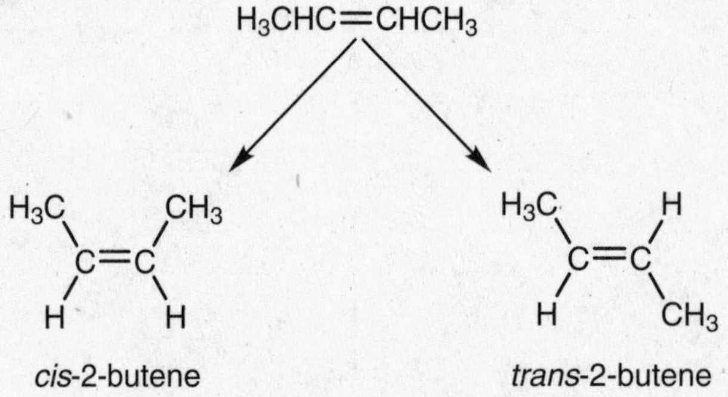

$H_3CHC{=}CHCH_3$

H_3C CH_3 C=C H H

cis-2-butene

H_3C H C=C H CH_3

trans-2-butene

In a molecule like 2-butene, there are a hydrogen and a methyl as substituents on both ends of the double bond. The two geometric isomers of 2-butene have the same atoms and the same connections between atoms, differing only in how the groups are arranged about the double bond. If the two methyl groups are on the same side of the molecule, the molecule is the cis isomer, and if the two methyl groups are on opposite sides, the molecule is the trans isomer. The same system is used to name cyclic isomers, in which cis isomers have both substituents on the same side of the ring (the top or bottom) and trans isomers have substituents on opposite surfaces of the ring.

Sometimes there are more than two types of groups as substituents on the two ends of the double bond, making the cis/trans method of naming insufficient. In these cases there is a system of naming in which geometric isomers are named either E or Z depending how groups are arranged. The first step in using the E/Z naming is to determine the priorities of the groups around the double bond, using the same rules as those used to determine the priority of groups to name absolute configuration around chiral carbons (heaviest atom = highest priority; stop at the soonest place you can make a distinction). If the highest priority substituents are both on the same side of the molecule, we call that molecule Z. If they are on opposite sides the molecule is E. Note that "same side" refers to both groups' being above or below the double bond, not both to the right or the left of it. For instance, in the molecule below, R1 and R3 are in the Z position, *not* R1 and R2.

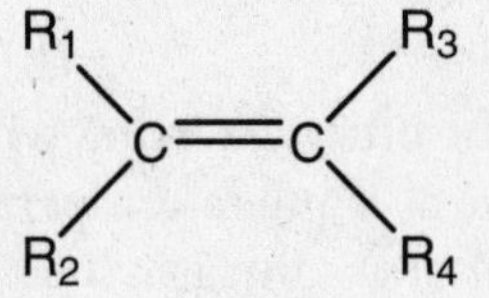

R_1 R_3
C=C
R_2 R_4

Student comment: You can remember the difference between Z and E as, "these groups are on zee zame zide." Get it? Z = same side.

The following present examples of the assignment of configuration for geometric isomers using the E/Z naming system.

1. 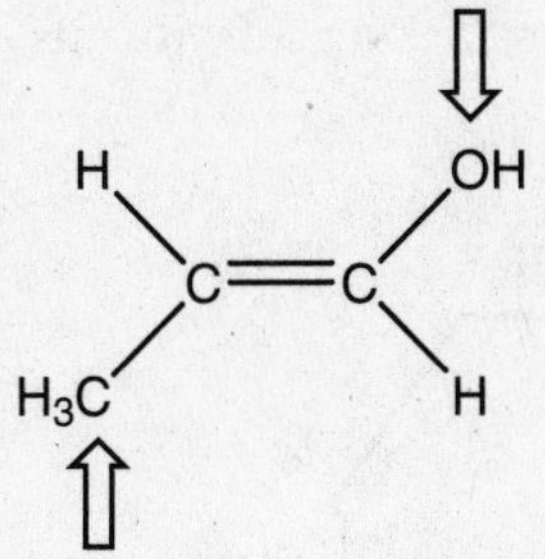 H, OH; C=C; H_3C, H — (E) isomer

2. 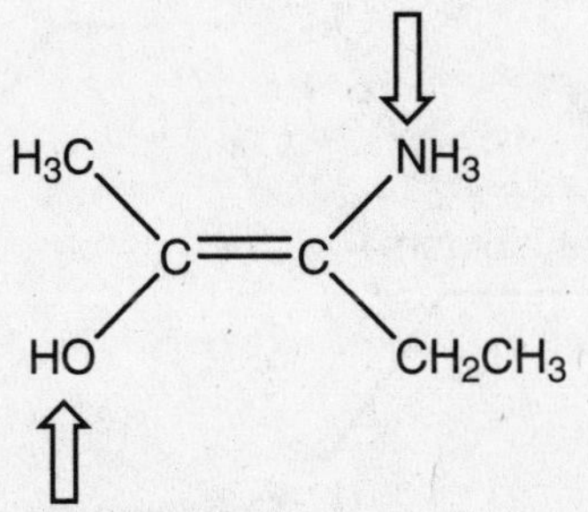 H_3C, NH_3; C=C; HO, CH_2CH_3 — (E) isomer

3. 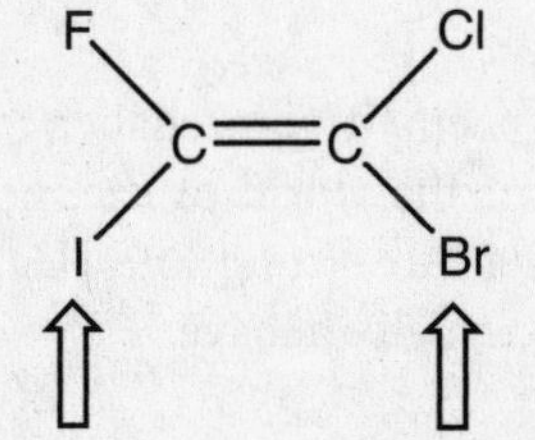 F, Cl; C=C; I, Br — (Z) isomer

4. 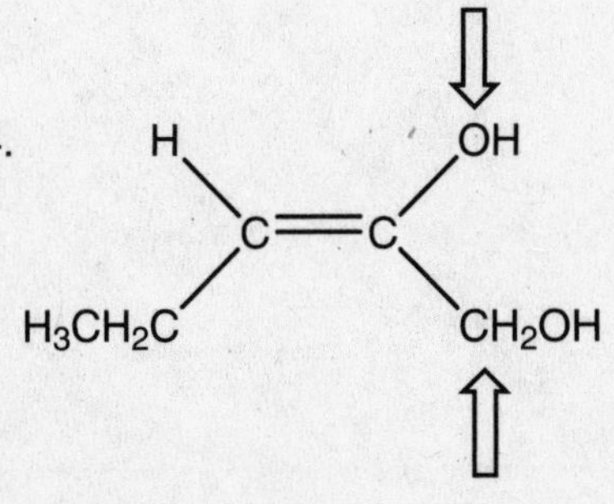 H, OH; C=C; H_3CH_2C, CH_2OH — (E) isomer

SUMMARY OF ISOMERS

There are several different types of isomers, starting with the major divisions of structural isomers and stereoisomers. Structural isomers are perhaps the easiest to name and draw because they are the most different and can easily be represented in two dimensions with Lewis structures. They also differ the most in their physical properties. Stereoisomers can also be subdivided, into geometric isomers, conformational isomers, enantiomers and diastereomers. Conformation isomers differ only through bond rotation and rapidly convert between forms. Enantiomers and diastereomers both have the same connections between atoms but differ in their orientation around chiral carbons. Enantiomers and diastereomers are both not superimposable on each other but enantiomers are mirror images of each other while diastereomers are not. Geometric isomers occur where a molecule can be locked into a configuration by double bonds or rings, locking substituents in place on one or two sides of a molecule.

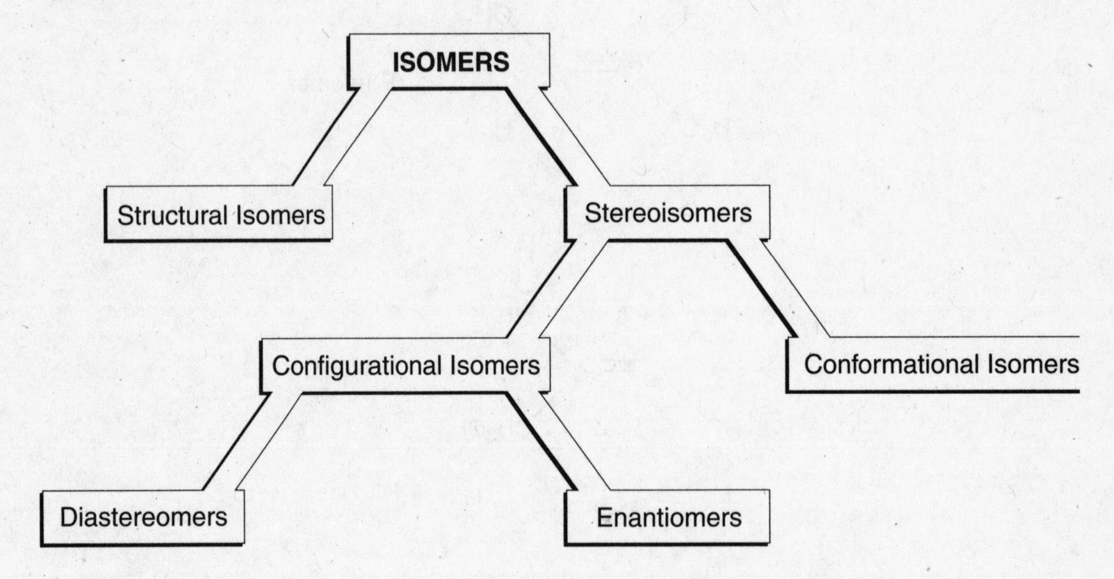

The naming and description of stereospecificity are challenges, but practicing extensively helps. This chapter completes our tour of the world of the molecule. Next we will journey into the mystical land of reactions, using our strong knowledge of molecular structure to learn the principles of how these things actually behave.

Generalities about Organic Chemistry Reactions

CHAPTER FIVE

So far we have explored molecular structure, including refinements such as where the electrons are, how molecules rotate around bonds, and the three-dimensional structure of isomers. This is only the beginning. Now that we have learned the language of molecular structure it is time to expand into a broader realm, chemical reactions. We have already seen that molecules are not static unchanging bodies drawn on a page or carved in stone. Life would be pretty boring if that were the case. Molecules can and do change. Chemistry happens. Chemistry occurs when electrons move their molecular orbitals to break old covalent bonds and form new ones. That is the whole ballgame right there. Now all we have to do is figure out the details about when, where, how, and why bonds break or form. The details of how molecular orbitals (covalent bonds) change are determined by the energy and positions of electrons, including the type of functional groups, delocalization, electronegativity, the type of bonds (pi or sigma), stereochemistry, strained bonds…all the topics we have been discussing. There was, in fact, a reason for all that stuff. The more we understand about reactions, the better we can use reactions in the lab to make what we want.

Organic chemistry is amazingly useful for humans, producing neat things like medicines and plastic grocery bags, and it is also amazingly versatile. The versatility of organic chemistry is one reason why it is the chemistry of life and why you are here to read this. The millions of possible reactions in organic chemistry may make it seem desperately complex, but luckily there are some common themes that tie them all together. It is not necessary to memorize millions of reactions. There are a few basic types of reactions that can be applied in many different cases and there are rules that predict what how reactions will proceed. With some knowledge of the building blocks, an understanding of bonds that hold the functional groups together, and some basic knowledge of how reactions work, it becomes possible to piece together the information about reactions in a powerful way.

THE REACTION PROFILE

What is a reaction? Every reaction starts out with molecules called **reactants** and ends up with **products**. The reactants and products can be drawn in a reaction equation, as presented for the reaction below. The reaction equation will not tell you how the reaction happened, but it will tell you where the reaction started and where it ended up.

> **Instructor question:** In the reaction below, what are the reactants and products?

$$HO^- + CH_3Cl \xrightarrow{CH_3OH} HO{-}CH_3 + Cl^-$$

> **Student response:** The reactants, on the left side of the equation, are the hydroxide ion and the chloromethane, and the products, on the right side of the equation, are methanol and chloride ion.

Two of the key questions to address with every new reaction that you encounter walking down the street (as you size up the expression on its orbitals) are:

- Will this reaction occur spontaneously?
- How quickly will it occur?

As we will see, the first question is asking if the reaction will proceed on its own, without any external energy. This is determined by the energy of the reactants and products, and it involves thermodynamics. The second question, on the other hand, involves kinetics. Even if a reaction occurs spontaneously, it may still take a long time to occur. Spontaneity says nothing about rate of a reaction since thermodynamics and kinetics are very different. We will return to these questions soon.

In the passage from reactants to products, the products can either have more energy or less energy than the reactants. If the products have less energy than the reactants, this means that energy is given off if the reaction proceeds.

> **Instructor question:** In the case of the example above, do the reactants or the products have more energy?
>
> **Student response:** Well, OH^- obviously wants to react, but then again so does Cl^- . . . .

Instructor response: You're on the right track. Think about which is *more* reactive, hydroxide ions or chloride ions? Which is a stronger electron donor (Lewis base)? Between these two choices, the hydroxide ion is a much stronger electron donor, and is much more reactive than chloride ion. Chloride ions are pretty content in water floating around, but as a strong base hydroxide ions will react rapidly with many different things. Hydroxide ions are less stable, very reactive, and occupy a higher energy state.

The energy of molecules as they traverse the course of a reaction is represented by the **reaction profile** or reaction energy diagram. The y-axis in the reaction profile plots the relative potential energy of different components of the reaction. High-energy molecules are higher on the y-axis of the plot. The x-axis is called the reaction coordinate, and depicts the route through which the pairing of reactants and products must travel through the course of the reaction. Reactants are on the left side of the reaction, the starting line, and products are on the right. For the reaction equation presented above, we can begin to fill in a reaction profile with the information we have deduced.

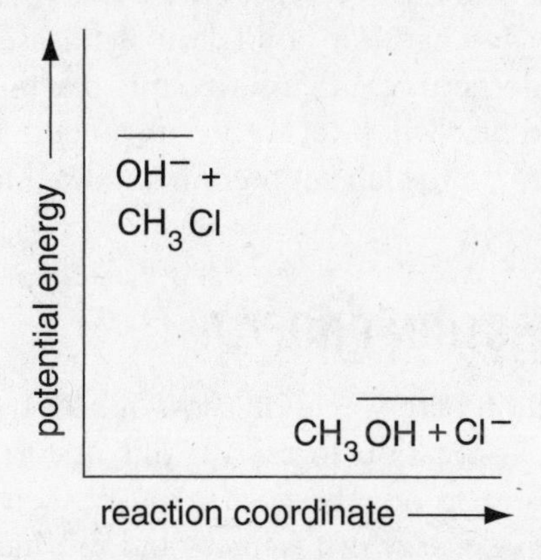

The reactants are on the left side and the products are on the right side of the profile. Since the reactants in this case have more energy than the products, they are higher up on the y-axis than the products. We can already tell something important about this equation, the answer to one of our key questions. The reactants have more energy than the products. In going from reactants to products, energy will be released. In the land of thermodynamics (which is to say, the universe that we inhabit), any process that goes from higher energy to lower energy, releasing energy into the surrounding system, is a favorable and spontaneous process. If the reactants are at a lower energy than the products then a reaction will not be favorable. The reaction presented in the above example is spontaneous and will occur on its own, without the addition of any outside energy.

If a reaction is spontaneous, what does this indicate about the equilibrium of a reaction? The reaction equation can be placed into an equilibrium equation, which in this case would look like:

$$K_{eq} = \frac{[CH_3OH][Cl^-]}{[CH_3Cl][OH^-]}$$

In the equilibrium equation, the top of the ratio contains the concentrations of products, and the bottom contains the concentrations of reagents. K_{eq} is the equilibrium constant.

> **Instructor question:** What does it mean if the K_{eq} is large?
>
> **Student response:** Well, according to the equation, a large K_{eq} means the numerator must be larger than the denominator, and since the numerator is the products, more product than reactant must exist at equilibrium. In other words, the reaction favors products at equilibrium, so the reaction is favored thermodynamically.
>
> **Instructor response:** Well done. Conversely, if the K_{eq} is very small then the reaction favors the reactants at equilibrium.

The difference in energy between the reactants and the products is called the heat of reaction, or enthalpy. Reactions in which the products have less energy than the reactants are **exothermic** – some of the energy will be released as heat during the reaction. A beaker in which an exothermic reaction takes place can get quite hot. Where does the heat come from? It is heat that was contained in chemical bonds but is released during the reaction. The opposite of exothermic is endothermic. In **endothermic** reactions, the products have more energy than the reactants, and heat will be absorbed during the reaction. If a beaker with an endothermic reaction is placed on the lab counter and allowed to proceed it will get cold.

KINETICS AND THE REACTION PROFILE

The other question we started off with was about how quickly a reaction moves. Do we have enough information about our reaction to answer this question yet? If the reactants will break and form bonds to become products spontaneously, then a reaction must go quickly, right? Actually, no—the relative energy of reactants and products tells you how favorable a reaction will be, but it says nothing about how a reaction will get there or how fast it will go. This is the realm of kinetics. For example, it is very thermodynamically favorable for wood to burn, since heat and energy are released. If a stick of wood is left out in the counter however, it will not spontaneously degenerate or burn (at least not usually). In fact, thermodynamics favors the oxidation of almost any biological material to carbon dioxide, and yet we all manage to walk around quite stably, except for the unfortunate victims of spontaneous combustion.

What determines the rate of reactions? The path that is taken from reactants to products determines the rate of reactions. Molecules do not leap directly from one state to the next – there is something in between. What happens in the middle stage between reactants and products? What is the energy state of the molecule when they are in between reactants and products? One way to draw the reaction profile would be to connect the dots drawing a straight line from one stage to the other (see below).

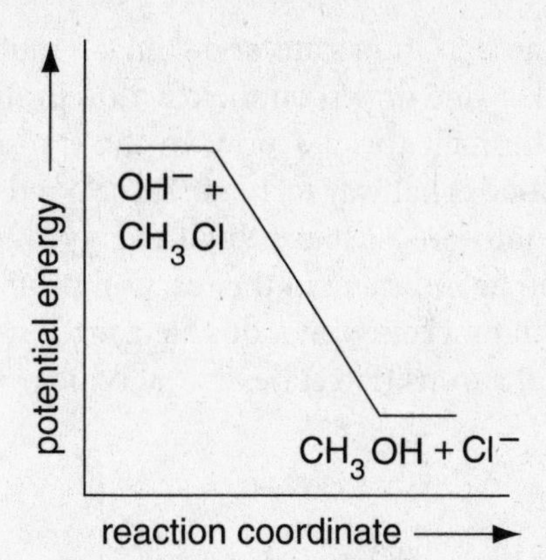

However, this is not the true picture. This reaction profile is WRONG. To go from reactants to products, electrons must move from one molecule to the next, bonds must break and new bonds must form. The movement of electrons in reactions is often represented by the addition of arrows in the reaction equation, where the arrows show how electrons will move to form and break bonds.

$$HO^- + CH_3Cl \xrightarrow{CH_3OH} HO{-}CH_3 + Cl^-$$

In the middle is something called the **transition state** which is between reactants and products. Energy that is put into the reactants, generally through heat or kinetic motion, gives the molecules enough energy for electrons to leave their comfortable homes in molecular orbitals to move to homes in new orbitals. As we all know, moving is never easy and takes a lot of energy. The transition state is unstable, and does not exist as a species floating in solution that a chemist can isolate, no matter how skilled she is. The structure of the transition state is hypothetical. What can be said is that since it is unstable and takes energy to create, the transition state will be at a *higher* energy state than either the reactants or the products. The correct reaction profile in this case must therefore look like the following:

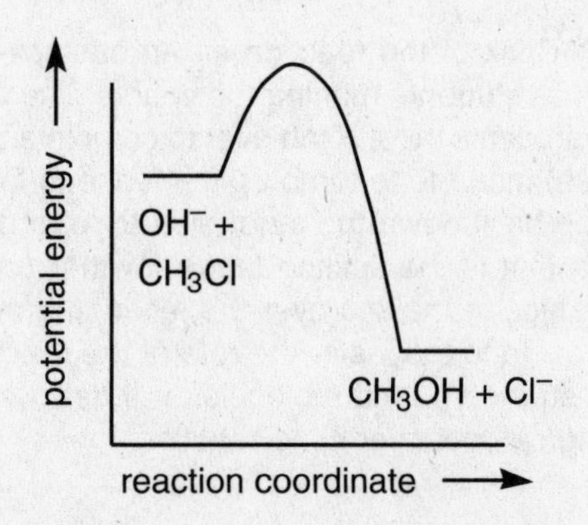

The energy increases to form the transition state, and then the molecules lose energy to become the products in this reaction. The line drawn on the reaction profile is the path that molecules must travel as they progress through this reaction. In the transition state the C-Cl bond is halfway broken and the C-O bond is halfway formed. These bonds in the state of transition are called partial bonds. These bonds are unstable, high-energy transient creatures, gone in an instant. The transition state can be indicated in the reaction profile at the top of the hill where it would be present, enclosed in brackets with a double dagger superscript outside to indicate that it is not a stable species but a transition state. Partial bonds are drawn as dashed lines.

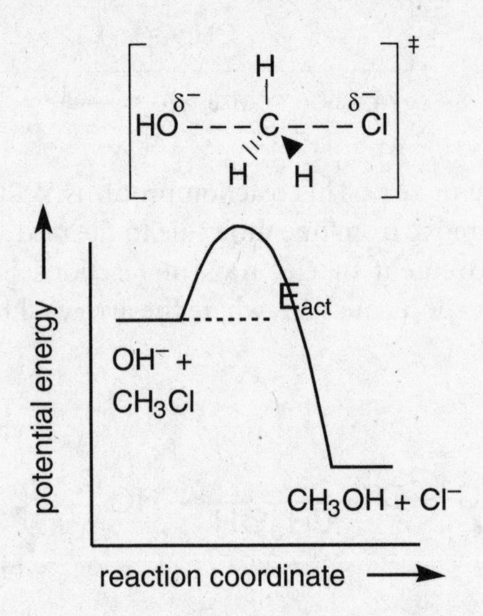

In this diagram, a partial negative charge is indicated in the transition state on both the hydroxyl and the chloride ions. These partial charges come from the charge in the reactants moving as the electrons move from one state to another. As the reaction proceeds, the charge on the oxygen decreases and the charge on the chloride ion increases. Like electrons, the charge will be conserved over the reaction course, even if the way it is distributed in the molecule changes.

To go from reactants to products, the reactants must first form the high-energy transition state. The energy to form the transition state comes from thermal energy, generated by molecules knocking about in solution. The energy required for reactants to form the transition state is called the **activation energy**, labeled E_{act} in the reaction profile.

> **Instructor comment:** Think of the reaction as an obstacle course and the molecules involved as students running the course. The activation energy is a high fence the students must climb over to complete the course. Once students (molecules) manage to climb up the fence to form the transition state, they can drop back down the other side to form products. It is the climb up the fence that is the limiting factor for the reaction. The more molecules that manage to make it over the fence, the more that can roll down to form product, and the faster the rate of the reaction. If that fence is low, it's easy to surmount, and molecules will get over it quickly. If it's so high that nothing can get over it, no reaction.

The rate of a reaction, then, is not determined by the difference in energy between reactants and products; this is thermodynamics and determines spontaneity of a reaction, not rate. The rate of a reaction is determined by the size of the energy barrier between reactants and products in the reaction profile, otherwise known as the activation energy. The lower the activation energy, the more rapid the reaction will be.

Temperature is one way to affect the reaction rate. Most reactions go more quickly at higher temperatures. In solution, or a gas, not every molecule has the same kinetic energy at a given temperature. The temperature indicates the average kinetic energy, but individual molecules might have energy over a range that is somewhat bell-shaped. With the students on the obstacle course, not every student will run and tackle the fence with the same energy – some are faster than others. The faster students will make it easily over the fence, while some of the slower ones may never make it over. If you jazz up the students with a reward at the end of the course (beer? No more laps?), then you will increase the energy of the students so they attack the wall with more energy and more of them can make it over. With molecules, at a low temperature only a few molecules will have enough energy to make it over the activation energy and the reaction will proceed slowly. Adding more heat increases the speed of molecules and the energy of collisions, allowing more molecules to reach the high energy transition state. If more molecules reach the high-energy transition state, then more will go forward to form product.

Note that changing the temperature does not change the activation energy, but just gives more molecules enough energy to reach the activation energy. However, another way to increase the number of students who make it over the fence is to lower the fence so that even the less energetic students can make it over. Catalysts help to lower the fence, or reduce the activation energy. Catalysts speed up reaction rates without themselves being consumed in the reaction, as reactants are. One important class of catalysts are the enzymes in your body that play myriad roles to keep the body up and running. Enzymes increase the reaction rate by bringing reactants together and stabilizing the transition state to lower the activation energy.

The kinetics of reactions are influenced by several things, as described by the rate equation. Along with temperature, adding more of one reactant or another can increase the rate. The equation also contains a rate constant, k, not to be confused with the equilibrium constant. The bigger the k, the faster the reaction. Example of a rate equation:

- rate = $k[OH^-][CH_3Cl]$

> **Student comment:** Note that although it is called a rate *constant*, implying it does not change, it is only constant for a given temperature. Each temperature has its own k.

Reactions that involve a single reactant are called unimolecular and reactions with two reactants are called bimolecular. The kinetics of these reactions will vary depending on which type they are. The rate of unimolecular reactions varies with changes in only the single reactant, while the rate of a bimolecular reaction has second-order kinetics, depending on the concentrations of both reactants involved.

Some reactions pass from reactants to products through a single transient transition state. The transition state is not a stable species, existing only in the instant that it takes for bonds to break and reform in a single step process. It exists only as a part of the change that occurs, and cannot be isolated independently in a test tube. Other reactions occur in two or more steps. In reactions with two steps, there may be two separate transition states between reactants and products, with a reaction intermediate in the middle (see below). Passage through the two transition states can be thought of as two separate reactions, one from the reactants to the intermediate, and the other from the intermediate to the final products. Each step has its own activation energy. The heat of reaction is still the difference in energy between the reactants and products, as in a one-step reaction. In a reaction like this, the intermediate is theoretically something you could isolate in the lab, unlike the transition state. Practically speaking, the intermediate is generally not very stable and is highly reactive. The deeper the energy well the intermediate sits in, the more likely that it could be isolated. The ability to overcome the activation energy limits the rate of the reaction, and is the rate-limiting step in the reaction. If a reaction has more than one step in it, then the step with the highest activation energy will be the rate-limiting step for the reaction as a whole.

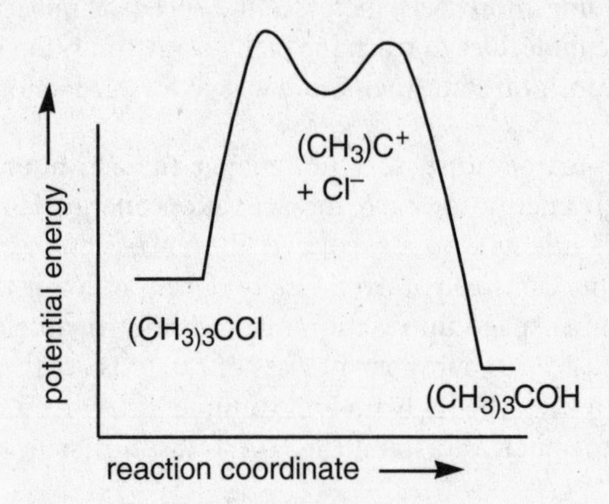

A FEW COMMON REACTION INTERMEDIATES

In organic chemistry, there are some very common reaction intermediates that are encountered over and over again in reaction mechanisms. These do not represent stable compounds normally found in molecules, but highly reactive species generally present for only a brief period of time. These species represent carbons with varying numbers of valence electrons. Carbon normally exists with eight valence electrons shared in four covalent bonds, meaning that four of the electrons are contributed by carbon and the other four by atoms it forms bonds with. If a carbon atom has only three covalent bonds, meaning that it has three electrons that are contributed by the carbon atom, then if it has no additional electrons in its outer shell it will have a positive charge. Carbon atoms with a positive charge are called **carbocations** (cations in general are positive ions). Lacking a filled outer shell, carbocations are highly reactive and do not generally stick around too long, reacting easily with negatively charged species. An example of a carbocation intermediate follows, and will be explored further as part of the S_N1 substitution reaction mechanism discussed later.

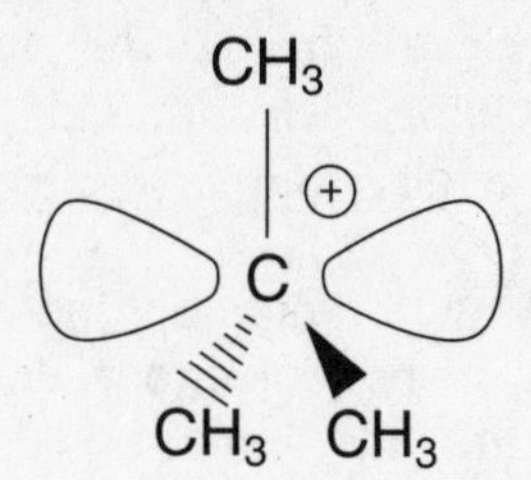

Another common intermediate in organic reactions contains carbons with three bonds again, but with two additional electrons in the outer shell; this is called a **carbanion**. Carbanion carbons have eight electrons in outer shell, three contributed by other atoms in bonds and five associated with the carbon atom. With five electrons on a carbon, the atom has a negative charge and will be quite reactive, despite its filled outer shell.

Finally, a reaction intermediate with three bonds and a single unbonded electron is called a **free radical**. The carbon has no charge, but it does have an unfilled shell and will be highly reactive.

Reaction Mechanisms

Once you know what the players in the reaction are, you can begin to think about how it will happen. Reactions can often occur by many different routes. Was it Colonel Alkene in the parlor with the hydrochloric acid? Going through the details about what happens during a reaction to break and reform bonds helps to understand the general features of different types of reactions.

In general when going through a reaction, you should look for the following:

- What is the reaction equation, with reactants and products?
- What type of reaction is it?
- What can we predict about the energy of reactants and products and the equilibrium status of the reaction?
- What is the main product that can be expected?
- What is the reaction mechanism?
- Does the probable reaction mechanism indicate anything about the reaction rate?

Reactions are all about electrons. One thing that affects the reactivity of electrons is the type of bond they are involved in. Electrons that are more accessible will be more available for reactions. In sigma bonds, electrons are less exposed than the pi bonds where the electrons lie up out of the plane of the molecule. Pi electrons are therefore generally more reactive.

The movement of electrons to form or break bonds occurs between parts of molecules that are electron-rich and those that are electron-poor. Molecules or parts of molecules that attract electrons because they are electron-poor are called **electrophiles**. The flip side of the coin, the partner in crime of the electrophile, is the electron-dense molecule called the **nucleophile.** The nucleophile has lots of electrons seeking a nucleus to call home and will donate its electrons in a reaction to an electrophile to form a covalent bond.

> **Student comment:** Here's a way to remember these terms: "phile" means "lover of." Electrophiles are lovers of electrons, since they are hungry for electrons. Nucleophiles, on the other hand, would love to get their hands on a positively-charged nucleus.

Examples of potential electron donors (nucleophiles) include double or triple bonds, in which the electrons in the pi bonds are rich and accessible. Polar covalent bonds also represent an enrichment of electrons on the more electronegative member of the bond. The negatively charged member of one polar covalent bond, the nucleophile, might donate electrons to the positively charged electrophile of a polar bond in another molecule.

The definitions of Lewis acids and Lewis bases also correspond to electrophiles and nucleophiles, respectively. Lewis acids are electron acceptors, making them electrophiles. A hydrogen ion, with no electrons, is a good Lewis acid and electrophile. If reacted with hydroxide ions, hydrogen ions will accept electrons to form a covalent bond with the hydroxide nucleophile, forming water. The non-bonding electron pairs in oxygen and nitrogen help to make Lewis bases in organic molecules.

Most reactions can be classed into one of three different types: elimination reactions, addition reactions, and substitution reactions. Identifying which type of reaction is involved helps to predict the reaction mechanism and rate of a reaction. It's all pretty straightforward at this level – the devil is in the details. We will meet the devil in Chapters 6, 7 and 8.

Nucleophilic Substitution Reactions

CHAPTER SIX

One of the main types of organic reactions is substitutions, exchanging a group in a molecule for a different group. Many substitution mechanisms involve nucleophiles though substitution in aromatic molecules usually involves electrophiles. Nucleophiles have abundant electrons and are looking for a good electron acceptor, while electrophiles are the opposite, deficient in electrons and looking for an electron donor. In this chapter we will deal with nucleophilic non-aromatic substitution. Electrophilic aromatic substitutions are discussed in Chapter 8.

Within nucleophilic substitutions, there are two basic types of reaction mechanisms, termed S_N1 and S_N2. Let's look again at the reaction presented in the last chapter:

$$\mathrm{HO^-} + \mathrm{H_3C{-}Cl} \xrightarrow{\mathrm{CH_3OH}} \mathrm{HO{-}CH_3} + \mathrm{Cl^-}$$

This is a substitution reaction. The hydroxide ion is substituted for the chloride group in this reaction. There are a few ways we can envision this reaction proceeding. In one mechanism, the hydroxide ion attacks the chloromethane and the chloride ion leaves simultaneously. Another mechanisms would involve the chloride leaving the molecule first, followed in a separate step by reaction of the hydroxide ion to form the final product. The first mechanism is called S_N2 and the second mechanism is called S_N1. After we go through the details of these two types of substitution reactions, we will come back at the end of the chapter to compare and contrast these two types of reaction mechanisms.

S_N2 REACTIONS

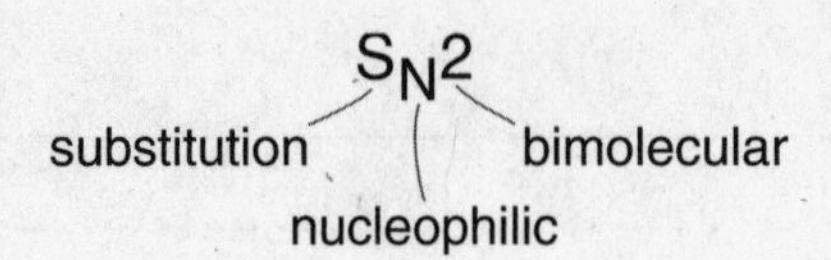

S_N2 reactions are bimolecular reactions resulting in the displacement of one group in a molecule by another one. The S_N2 name says it all – substitution, nucleophilic, and the 2 indicates that it is a bimolecular reaction. The reaction starts with the collision of a nucleophile with the molecule it reacts with, and ends with the release of a leaving group from this molecule.

Instructor question: In this figure, can you identify the nucleophile, the alkyl halide attacked by the nucleophile, and the leaving group?

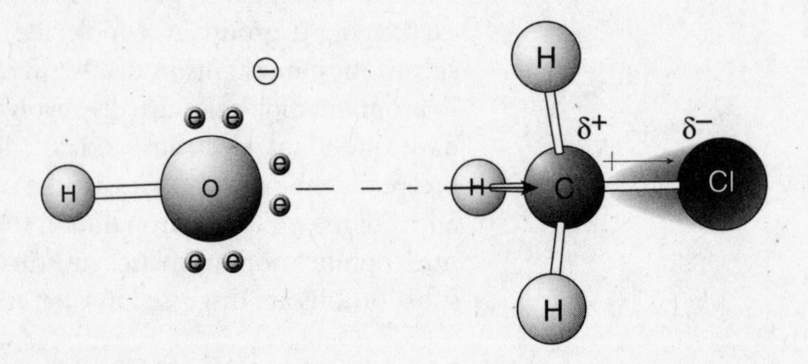

Student response: The hydroxide ion must be the nucleophile that provides the electrons to attack the alkyl halide (chloromethane), causing a new bond between C-O to form and breaking the old bond between C-Cl, with Cl^- as the leaving group. But I'm a little unclear as to how you recognize a nucleophile.

Instructor response: Good question. If you notice an abundant supply of electrons buzzing about its head, then it is probably a nucleophile. You do not necessarily need to be negatively charged to be a nucleophile, but you do need to have some electrons available and ready for action to go on the attack. Examples of nucleophiles include the following list of the usual suspects:

$:NH_3$ $\quad H\ddot{O}H \quad R\ddot{O}H$

Student response: I get it. All of these have one or more pairs of unbonded electrons.

The factors that determine how fast an S_N2 reaction will proceed include:

- strength of the nucleophile
- the nature of the solvent used for the reaction
- the identity of the substrate that is attacked by the nucleophile
- the nature of the leaving group

S_N2 - STRENGTH OF NUCLEOPHILE

Not every nucleophile is equal - some nucleophiles are stronger than others. What gives nucleophiles their strength – their hair? Charged nucleophiles are usually stronger than uncharged nucleophiles. Having a charge is not a prerequisite, but it helps. Strong nucleophiles are usually also strong bases. A strong Lewis base will have a strong propensity to donate electrons, just the sort of thing that you would look for in a good nucleophile. If the atom associated with the electrons is too electronegative, however, this can decrease the strength of the nucleophile; the atom will be unwilling to relinquish its electrons to its electrophile friends.

Instructor question: Let's examine the following molecules and rank them in order of decreasing nucleophilic strength, shall we?

(a) CH_3O^-

(b) CH_3COO^-

(c) $C_6H_5O^-$

Student response: No idea.

Instructor response: Okay, this is a bit hard to see at first. Think about the strength of each base and how well it can stabilize its negative charge. In each of these examples, it is the oxygen atom that has the nonbonding pair of electrons to attack the carbon and form a covalent bond. Each of the oxygens has a negative charge, so we can't distinguish among any of these three molecules based on charge. The difference is their degree of basicity, which could be determined from the material covered elsewhere in this book up to this point. The acetate ion CH_3COO^- is the conjugate base of a moderately strong acid, acetic acid, making it a poor base. Also the negative charge is delocalized over two oxygens, so it's pretty stable. Phenol (C_6H_5OH) is a weaker acid than acetic acid, but still with a fair degree of acidic nature that makes the conjugate base ($C_6H_5O^-$) a stronger base than the acetate ion. Finally, the methoxy ion is a very strong base, one you will see again and again. Its conjugate acid, methanol, isn't very acidic; alcohols generally aren't. Remember: weak acid = strong conjugate base. Methoxy ions can't delocalize their charge so they're itching to attack. The rank order of strongest to weakest base, and also strongest to weakest nucleophile, goes: $CH_3O^- > C_6H_5O^- > CH_3COO^-$. However, a warning: occasionally you will see that the strongest bases are not always the best nucleophiles.

S_N2 - SOLVENT EFFECTS

The solvation of nucleophiles influences the strength with which they act as nucleophiles. To act as a nucleophile, a molecule needs to leave behind its solvent molecules and bare itself to attack the molecule it will react with. Take the case of our old friend Mr. Hydroxide ion. Floating in solution he is surrounded by methanols that hydrogen bond with the hydroxide ion, including the nonbonding electron pair. If these electrons are to bond with anything other than solvent then they need to ditch the solvent molecules that are getting in the way. Small nucleophiles have a concentrated charge and are easily solvated; large nucleophiles have more diffuse electrons and are less strongly solvated, so a large nucleophile may have an advantage in an S_N2 reaction, to escape its solvent and attack.

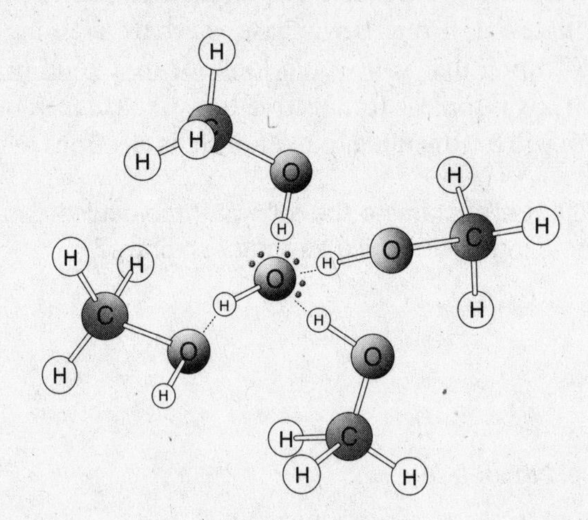

Let's examine the role of nucleophile solvation in one particular example. A substitution reaction starting with iodide ions and chloromethane has a faster rate of reaction than the reaction of hydroxide ions with chloromethane presented earlier, despite the fact that hydroxide ions are a stronger base than iodide ions. Which is a stronger nucleophile, iodide or hydroxide? Since everything else in these reactions is the same, the only difference is the nucleophile. The faster reaction must be the one with the stronger nucleophile, meaning that iodide is the stronger nucleophile.

> **Student question:** Wait a sec. Why is iodide a stronger nucleophile when we know it is a weaker base than the almighty hydroxide ion?

$$:\ddot{\underset{\cdot\cdot}{I}}^{-}: + H_3C{-}Cl \xrightarrow{CH_3OH} \left[\overset{\delta-}{I} \text{---} CH_3 \text{---} \overset{\delta-}{Cl} \right]^{\ddagger} \longrightarrow I{-}CH_3 + :\ddot{\underset{\cdot\cdot}{Cl}}^{-}:$$

transition state

> **Instructor response:** Yup, it's tricky. An important difference between iodide and hydroxide ions is the way they are solvated. The size of the nucleophile is one factor that determines how a nucleophile interacts with solvent and strong the nucleophile will be. A small nucleophile has electrons spread over a smaller area than a large nucleophile, and has solvent molecules bound more tightly to it. The tighter the solvent interactions, the harder it is for the molecule to escape bound solvent to perform its nucleophilic substitution magic. Unfortunately, memorizing a list of the strongest nucleophiles is probably a good idea.

The way that a solvent interacts with a nucleophile depends in part on the solvent. Solvents can be broken down into classes that indicate how the solvent will interact with solvated material. One type is **polar protic solvents**, which have a dipole moment, making them polar, and a hydrogen bound to F, O or N. This hydrogen is what puts the protic into polar protic solvents. Water and alcohols are examples of polar protic solvents. The hydrogen is electron deficient in polar protic solvents, since is bound to an electronegative partner, so it can interact with negative charges to solvate negatively charged nucleophiles. The electronegative atom in a protic solvent can solvate positively charged molecules as well. The ability of polar protic solvents to form hydrogen bonds is another aspect of how they interact with other molecules.

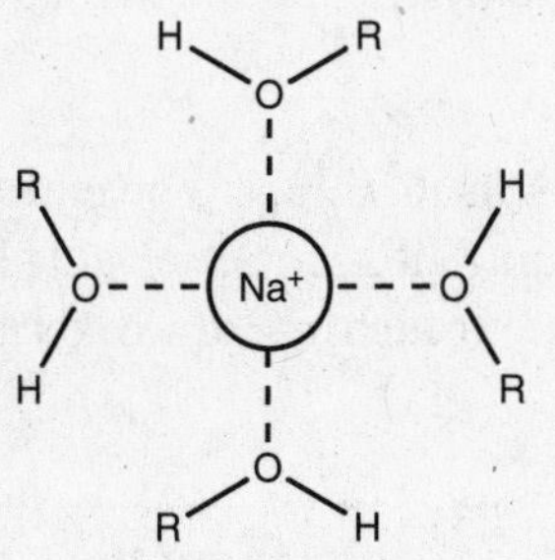

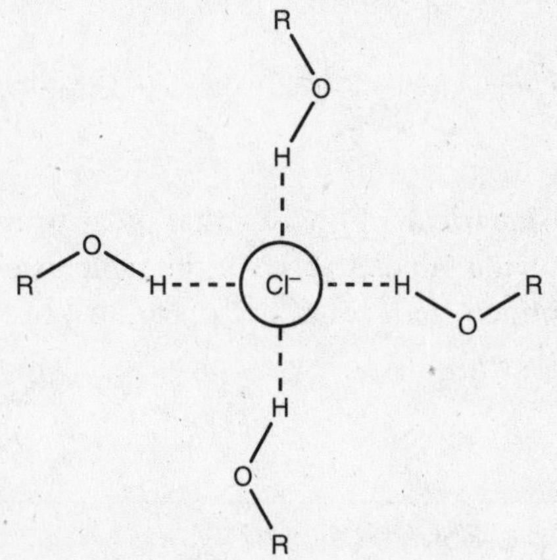

Student comment: If you see an –OH in the solvent, it's protic. Solvents abbreviated with capitalized letters (DMSO, etc.) are usually aprotic.

Another class of solvents is **polar aprotic solvents** that have a dipole moment but lack the hydrogen bound to an electronegative atom. An example of a polar aprotic solvent is acetone (see figure). These solvents lack the electron deficient hydrogen found in polar protic solvents, so they cannot solvate anionic nucleophiles as strongly. In acetone, the carbon end of the polar carbonyl group is also not very accessible, helping to reduce the strength of interaction with solvent. Polar aprotic solvents can still solvate cations pretty well by interaction with the electronegative part of their dipole, but they cannot form hydrogen bonds at all.

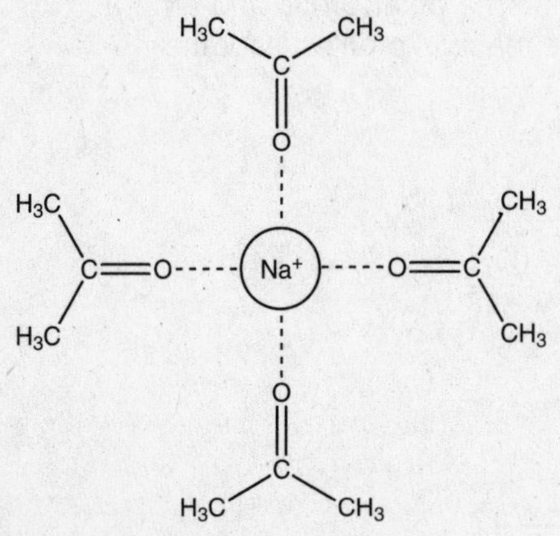

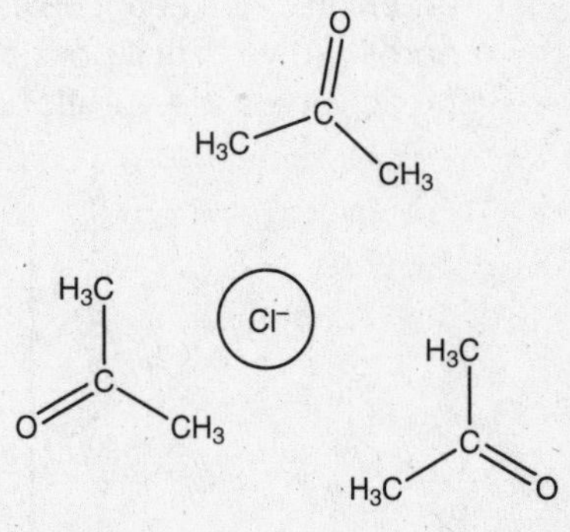

Acetone is just one example of a polar aprotic solvent. Some other examples include DMF, DMSO, and HMPT. What do these structures all have in common? They all have an electron withdrawing oxygen but lack a hydrogen bound to this oxygen that would make the solvent protic rather than aprotic.

N,N-Dimethylformamide (DMF)

Dimethylsulfoxide (DMSO)

Hexamethylphosphoric triamide (HMPT)

A comparison of the way that polar protic and aprotic solvents solvate things is given in the following table. A value in the table listed as "No" does not mean that aprotic solvents have absolutely zero solubility of anions, but that they solvate anions much more weakly than protic solvents.

	Polar protic	**Polar aprotic**
Solvate cations?	Yes	Yes
Solvate anions?	Yes	No
Form hydrogen bonds?	Yes	No

Back to S_N2 reactions for a moment. Would an S_N2 reaction occur more rapidly in a polar protic or aprotic solvent? Think about the information we have covered so far to answer this question. The stronger the interaction of solvent with a nucleophile, the more it prevents the nucleophile from attacking its target. The S_N2 reaction will go more quickly with the weaker solvent, the polar aprotic solvent, which is the solvent of choice.

Instructor question: Let's say we do an experiment testing the reactivity of an anionic nucleophile with an alkyl halide in both a polar protic and an aprotic solvent. Is it possible to tell which of the reaction profiles is for the protic versus the aprotic solvent?

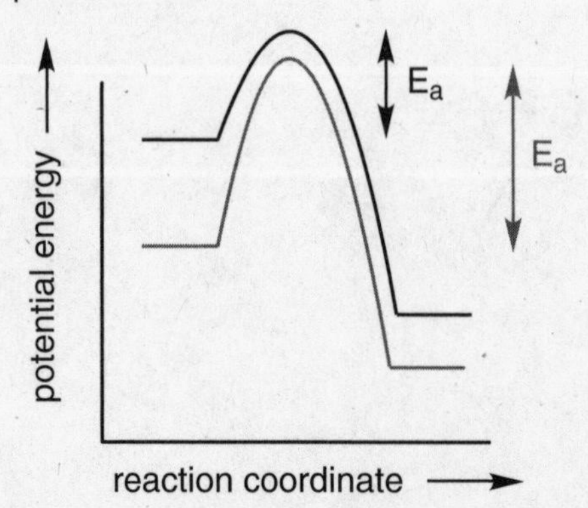

Student response: Well, according to the diagram, the two trial runs differ in the energy level they start with. We know the one starting at a lower energy level is more stable. And the better the solvation, the more stable something is. The nucleophile in this reaction is anionic (negatively charged) and must be better solvated in its solvent. The solvent that would interact strongly with it would be a polar protic solvent. Aprotic solvents do not solvate anions very well (see table). The better solvation and reduced energy in protic solvents means that the lower plot is probably the polar protic solvent and the upper plot is the aprotic solvent.

Instructor response: Exactly! You can see that the plot for the reaction with aprotic solvent has a smaller activation energy, since the compounds start at a higher point. With smaller activation energy, it takes less energy for molecules to make it over the top and more of them can do it, increasing the rate compared to the protic solvents with the larger activation energy.

S_N2 - SUBSTRATE EFFECTS

The molecule that is attacked in an S_N2 reaction is called the substrate, while the attacker is the nucleophile. The nature of the substrate can have an important influence on how well a reaction proceeds, even if everything else is the same. For example, compare the following two reactions:

$$CH_3Cl + OH^- \rightarrow CH_3OH + Cl^-$$

$$CH_3CH_2Cl + OH^- \rightarrow CH_3CH_2OH + Cl^-$$

These two reactions have the same nucleophile (OH-) and the same leaving group (Cl-), but because the substrate is different the reactions will proceed differently. An important difference among substrates is the connectivity of the carbon attached to the leaving group, the very same carbon that the nucleophile will attack. If the carbon is attached to one other carbon, then it is called primary, if it is attached to two other carbons it is secondary, and if it is attached to three carbons it is tertiary. In the figure below, the position of a halogen in each substrate is indicated by "X".

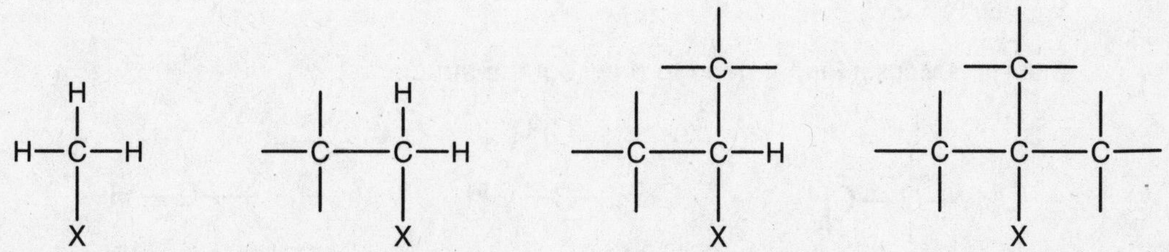

methyl halide | primary alkyl halide | secondary alkyl halide | tertiary alkyl halide

How does the substrate structure affect S_N2 reactions? The alkyl chains attached to the carbon are much larger and bulkier than hydrogen. As larger chains surrounding the atom of interest, they get in the way of the nucleophile, creating steric hindrance that makes S_N2 reactions less favorable. The bulkier the groups and the more there are of them, the slower the S_N2 reaction will be. This shows up in the reaction profile as an increase in the activation energy for reactions with bulky substituents. Compare in the figure the access of the hydroxyl nucleophile to the carbon in methyl chloride to the access of the nucleophile in t-butyl chloride. In the t-butyl chloride, the carbon is sterically blocked and the reaction does not proceed, while in methyl chloride the reaction proceeds readily. In fact, tertiary carbons are so sterically hindered that they are rarely involved in successful S_N2 reactions.

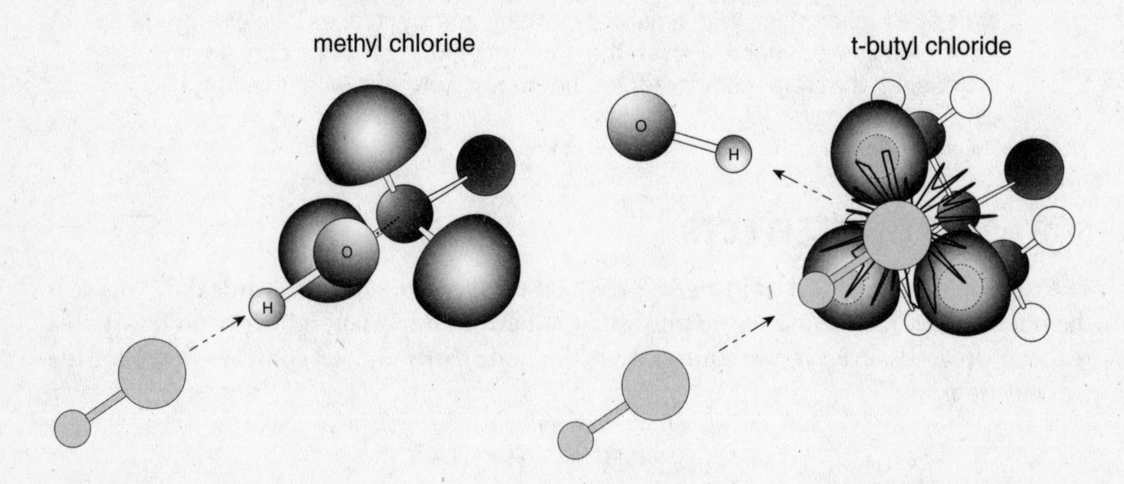

The affect of substrates on the rate of S_N2 reactions is (in order of decreasing reactivity):

methyl halide> primary halide > secondary halide > tertiary halide.

Instructor question: Compare the reactivity of the following three substrates in an S_N2 reaction and rank them from most reactive to least reactive.

(a) $CH_3CHBrCH_3$

(b) $(CH_3)_3CBr$

(c) $C_6H_5CH_2Br$

Student response: First, I need to draw out the structures:

a. $CH_3-CH(Br)-CH_3$ (H above C, Br below C)

2°

b. $CH_3-C(CH_3)(Br)-CH_3$ (CH3 above C, Br below C)

3°

c. $C_6H_5-CH(H)-Br$ (benzene ring attached to C bearing H, H and Br)

1°

Now, the leaving group is bromine in each case, so the only difference is the connectivity of the carbons the bromines are attached to. With the structures drawn out we can easily see the connectivity of the carbon of interest, the one attached to the bromine. Molecule (b) is tertiary so we know that's the worst for S_N2. But the other two are close…(a) is secondary for the carbon being attacked, whereas molecule (c) is primary. On the other hand, that primary one has that big old ring nearby. That's got to be bulky.

Instructor response: You're right, it can be a tough call. However, in this case the primary one wins out. Normally, straight chain primary carbons with no branching or other stuff around them are what we think of as ideal S_N2 candidates. Nonetheless, the rank order of reactivity for these three molecules in S_N2 will be:

primary (molecule c) > secondary (molecule a) > tertiary (molecule b)

S_N2 LEAVING GROUPS

The leaving group is the part of the substrate molecule that gets replaced by the attacking nucleophile. The leaving group carries with it a pair of non-bonding electrons that were previously in the bond between the substrate and leaving group. If a reaction is to proceed spontaneously (note, we must be talking about thermodynamics now, not rates), then the leaving group must be a weaker nucleophile than the attacking nucleophile. What would happen if this were not the case? The reaction would go backward instead of forward. What if we added a catalyst into this scenario? The reaction would still go backward, but would do so faster since the catalyst would reduce the activation energy and allow both forward and backward reactions to increase in rate.

A strong base, or a strong nucleophile, is a poor leaving group since it will tend to drive the reaction in the backward direction. A good leaving group is a weak base that will allow the reaction to proceed forward. A weak base is stable with the non-bonding pair of electrons it carries away from the substrate, and does not have a strong compulsion to act as a nucleophile or move its electrons somewhere else. A good leaving group is happy just the way it is. The better the ability of the leaving group to stabilize the electrons it carries, the better a leaving group it will be.

What constitutes a good leaving group? Think of things that stabilize electrons. Anything come to mind? How about resonance stabilization of electrons, delocalizing the negative charge around more bonds? Or induction by nearby electronegative groups?

The halogen leaving groups that we have already encountered are a good example of a plain, old-fashioned, no frills leaving group. The halogens, like chloride ions, are the weak base of strong conjugate acids (like HCl). Chlorine is fairly electronegative so it wears electrons and a negative charge comfortably, which helps to drive S_N2 reactions forward.

Another good leaving group contains sulfate or sulfonate ions. Why is this? Examine the resonance forms of a methanesulfonate ion, also called mesylate. The resonance forms stabilize the negative charge, spreading it around through delocalization to drive S_N2 reactions forward when chloride is the leaving group.

$$CH_3-S(=O)_2-O^{\ominus} \longleftrightarrow CH_3-S(=O)(-O^{\ominus})=O \longleftrightarrow CH_3-S(=O^{\ominus})(=O)-OH$$

Instructor question: Let's kick it up a notch. How would the triflouromethanesulfonate group below compare to methane sulfonate as a leaving group?

$$CF_3-S(=O)_2-O-$$

Student response: The trifluoro- part of the molecule will be very electronegative, withdrawing electrons from the rest of the molecule. The withdrawal of electrons will cause induction, allowing still more stabilization of the negative charge on top of the stabilization created by resonance delocalization.

How does a better leaving group affect the reaction rate and the reaction profile? We know two things about leaving groups that will affect the reaction profile. One is that the better the leaving group, the faster the reaction rate will be. How is a reaction rate increased? By reducing the activation energy. A reaction with a better leaving group will therefore have a lower activation energy. We can also deduce that the products will have less energy, since the better the leaving group, the more stable a reaction product it will be. The reaction profile comparing reactions with two different leaving groups, assuming that the reactants have equal stability, might look like this:

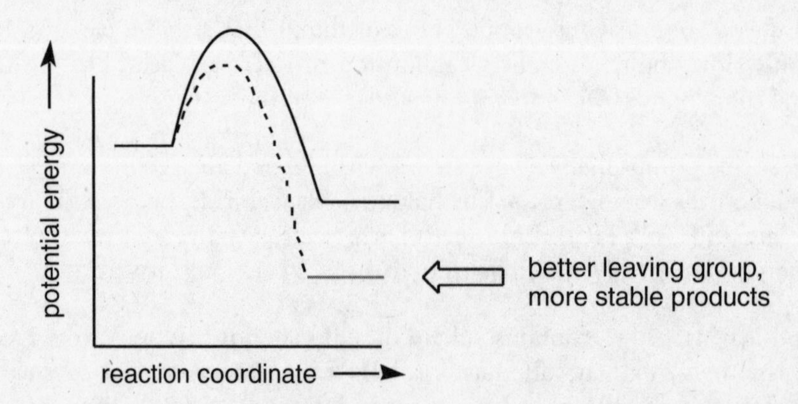

So the things that make for a strong S_N2 reaction are:

- strong nucleophile
- weak solvation of nucleophile (polar aprotic solvents)
- good leaving group
- lack of steric hindrance in the substrate

S_N1 REACTIONS

In an S_N2 reaction, the first step in the reaction is the nucleophile attacking the substrate, kicking out the leaving group. This works pretty well for some molecules, but not for others. Tertiary carbons are quite resistant to S_N2 reactions due to steric hindrance. S_N2 is not for everyone (different strokes for different molecules) but luckily there is an alternative – the S_N1 reaction. You knew that there had to be an S_N1 if there was an S_N2, right? S_N1 stands for substitution, nucleophilic, and unimolecular (1). How the unimolecular part works will become clear in a moment.

In an S_N1 mechanism, involving for example a tert-butyl chloride, the leaving group actually leaves of its own accord, without waiting around for the nucleophile to arrive. Nobody tells it to and nobody kicks it out—it just gets up and leaves all on its own. (So it's got to have a pretty good reason to want to take off, but we'll get to that later.)

What is left behind after it leaves? The nucleophile has not arrived on the scene yet, but the leaving group has carried a pair of electrons away from the carbon it was bound to. What's a carbon to do? Be positive. Positively charged, that is. The rest of the molecule forms a positively charged reaction intermediate called a carbocation. The first step then is for the leaving group to take off and leave behind a positively charged reaction intermediate called a carbocation.

What's the second step? The carbocation is not very stable and does not hang around for long. If there is a nucleophile in the house, the carbocation will react with it. For example, if hydroxide ions are around, they will react with the carbocation intermediate to form an alcohol.

$(CH_3)_3C{-}Cl \xrightarrow{\text{slow}} (CH_3)_3C^+ + Cl^-$

intermediate

$(CH_3)_3C^+ + {}^-OH \xrightarrow{\text{fast}} (CH_3)_3C{-}OH$

Carbocations are present as an intermediate in all S_N1 reaction mechanisms. The carbocation intermediate is very reactive, but it does exist as a real species if only for a very brief time. In contrast to transition states, the carbocation intermediate is theoretically something a chemist could isolate. Reaction intermediates do not appear in the overall reaction equation since there will not be any intermediate remaining when the reaction is over.

What does a carbocation look like? If you were a very speedy chemist, you might be able to catch one in the wild, and if so you would observe it to look somewhat like the following artist's rendition:

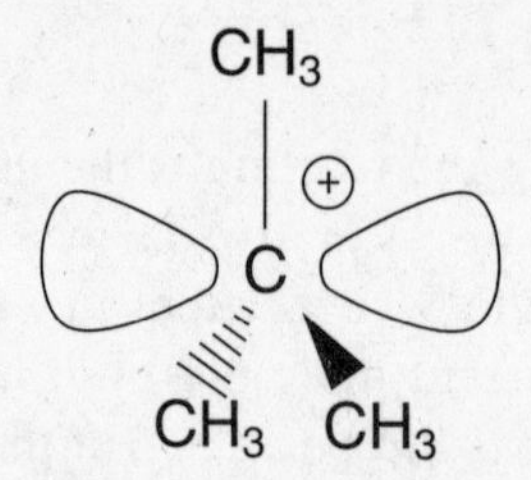

The carbocation intermediate has three covalent bonds, in this case with three other carbons, making it a tertiary carbocation. It will have the three bonds sp^2 hybridized and planar around the carbocation, at 120 degrees from each other. There are no unbonded electrons since the leaving group left town with these. An empty p orbital stands perpendicular to the sp^2 orbitals.

How will the reaction profile look for this S_N1 reaction? First consider the relative energies of the reactants, the intermediate, the products, and the transition states. Going from the reactants to the intermediate, the intermediate will be less stable (higher energy) than the reactants. Forming the intermediate requires breaking a covalent bond (C-Cl) and separating charges. Between the reactants and intermediate is a transition state with higher energy than either, in which the C-Cl bond is in the process of breaking.

Where do we go from here? In the second step, the intermediate must pass through another high-energy transition state in which it reacts with a hydroxide ion. The final product will be tert-butanol, and will have the lowest energy of all species in the reaction. If this were not true, the reaction would not proceed forward to products, but in the direction toward the lowest energy species, which would be backwards. Filling in the whole profile, then, will give us:

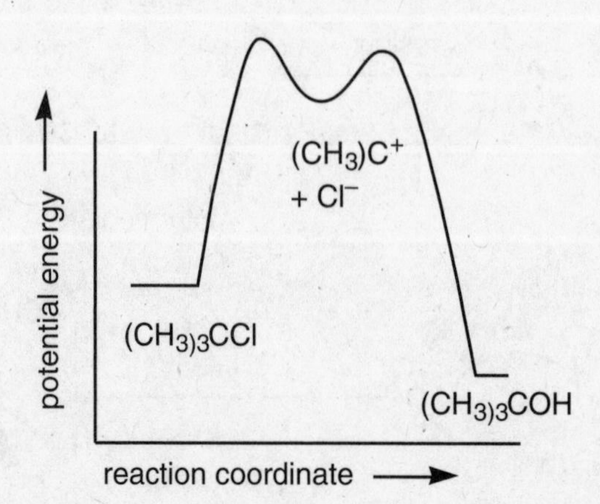

There are two steps in this reaction, each with its own transition state and its own activation energy. The first step is the reaction from reactants to intermediate, and the second step is the reaction from intermediate to product.

Instructor question: Which step in this reaction will have the highest activation energy? Hint: drawing in the activation energies in the reaction profile should make it pretty clear.

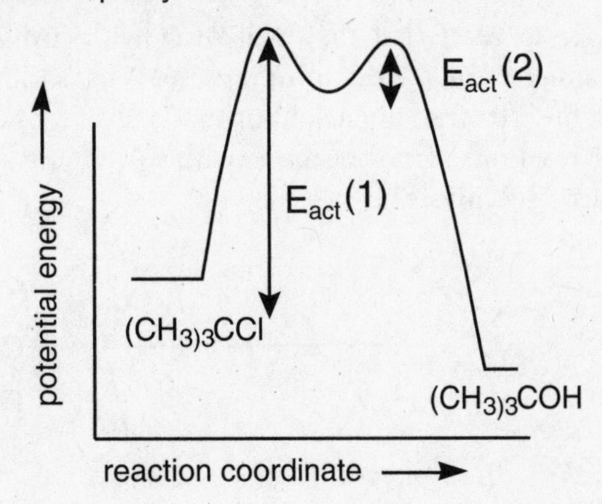

Student response: Going from reactants to the carbocation intermediate has a much higher activation energy than the activation energy for the second step for the carbocation to react with a nucleophile to form the product. With a higher activation energy, the first step is much slower than the second step. This means that the intermediate will react with nucleophile as rapidly as it is formed.

Instructor response: Correct. In fact, the first step is so much slower that it is a bottleneck in the overall reaction, also called the **rate-determining** step. The production of tert-butanol depends almost entirely on the rate of carbocation production, and very little on the rate of the second step.

Imagine an assembly line in which a toy is assembled in several steps along the way. If one step is very slow, and the other steps are very fast, then the toys will pile up at the slow step. Changing the fast steps will not change the overall rate of toy production since everything depends on the slow step, but making the slow step faster will speed up the whole reaction. The slow step in the toy assembly process is the rate-determining step.

If we add more hydroxide ions or if we increase the strength of the nucleophile we add to our tertiary S_N1 reaction, how will the rate of tert-butanol production be affected? Neither of these will affect the slow, rate-determining step of carbocation formation, so neither increasing the concentration or strength of the nucleophile will speed up the overall reaction. Since the hydroxide ions have no affect on the reaction rate, the rate equation looks like this:

$$\text{rate} = k[(CH_3)_3CCl]$$

This is a first order rate equation, in which the rate of product formation depends only on the concentration of the substrate reactant. Increasing the concentration of the substrate will increase the rate of product formation, but increasing the concentration of nucleophile will not.

THE ROLE OF SOLVENT IN S_N1

Carbocations are so eager to react that they will react with just about any excuse for a nucleophile that comes along, strong, weak or otherwise. What's the most likely nucleophile that the carbocation will find? Its friendly neighborhood solvent, of course. The carbocation is surrounded by it. S_N1 reactions of carbocations with solvent are so common that there is a word for these reactions – **solvolysis**.

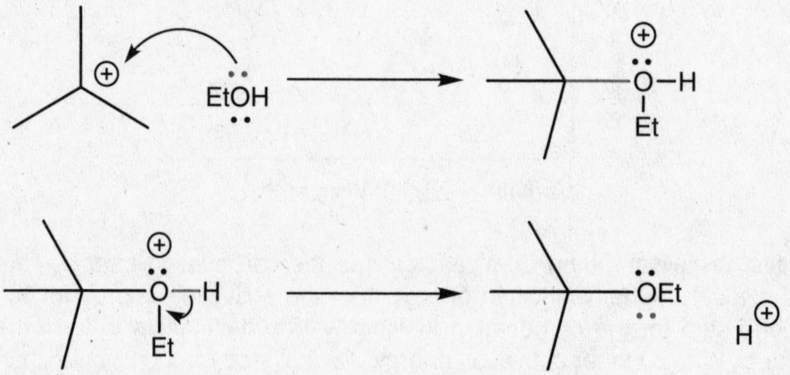

Let's say that you want to produce tert-butanol from tert-chlorobutane. In one experiment, you try to dissolve some potassium chloride in ethanol to provide the nucleophile. What are you likely to observe? The ethanol is not as strong as hydroxide ions, but as the solvent the ethanol molecules are so relatively abundant that you are likely to see a lot of ethanol reacting with the carbocation intermediate. It's hard to do chemistry without a solvent though, so what can you do? Use water as the solvent. The carbocation will react with a water molecule, and a hydrogen will leave to produce the desired product, tert-butanol.

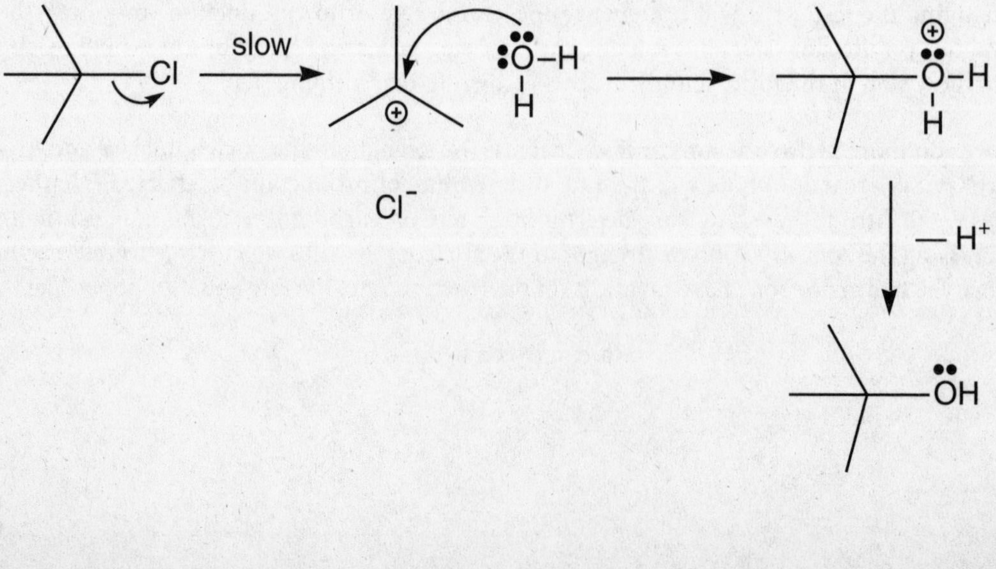

Perhaps the most important role solvent plays in S_N1 reactions is that of stabilizing the carbocation intermediate. Formation of the carbocation is the rate-limiting step, which makes sense: it takes energy to form that unstable carbocation. So helping the process along by stabilizing the carbocation is one way to make the reaction rate increase. Stabilizing the intermediate will decrease its energy and the activation energy required to form it, increasing the rate of carbocation formation and the reaction as a whole.

> **Instructor question:** What kind of solvent is up to this challenge?
>
> **Student response:** Well...the solvent should stabilize the intermediate carbocation and the chloride ions, which are charged. A polar solvent will be necessary, but I'm not sure if it should be protic or aprotic...
>
> **Instructor response:** Let's think about it. S_N2 reactions use aprotic solvents like acetone since weak solvation of the nucleophile helps the reaction to proceed. In this case, however, the reaction rate does not depend on nucleophile strength. It depends on stabilization of the carbocation intermediate, something a protic solvent can help out with, as well as stabilization of the Cl^- leaving group. A protic solvent also provides some nice hydrogen bonding backup.

S_N1 SUBSTRATE EFFECTS

The substrate in S_N2 reactions was very important, because of the affect on steric hindrance. In S_N1 however, there is less steric hindrance, because the surrounding groups are planar and more out of the way, and what hindrance there is does not affect the overall reaction rate, since the nucleophilic attack is not the limiting step.

The substrate still matters, but in a different way. The rate of formation of the carbocation is the rate limiting step and anything that speeds this up by stabilizing the carbocation will make the reaction go faster.

How would you build a more stable carbocation? It just so happens that alkyl groups can donate electrons to stabilize the carbocation by induction. An arrow pointing in the direction of induction is sometimes drawn to indicate the shift of electrons to mask the positive charge.

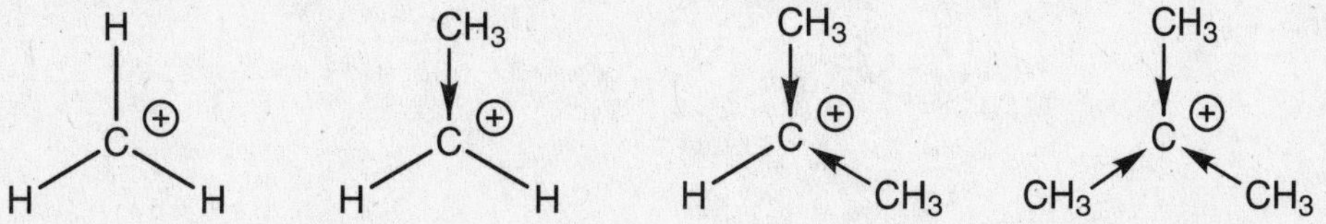

If one alkyl chain is good, then more must be better, and this is exactly right. The more alkyl chains bound to the carbocation, the more induction that occurs and the greater the stability of the carbocation. The order of stability of the carbocation due to inductive stabilization is:

tertiary > secondary > primary > methyl

Tertiary carbons form carbocations more readily, with lower activation energy and increased rate, due to increased stabilization of the carbocation. Conveniently, the order of preference for carbocation formation and therefore for S_N1 reactions is the exact opposite of the preference for S_N2 reactions, except for different reasons.

Resonance stabilization can also help to delocalize the positive charge of the carbocation and increase its stability. The rule we stated above is that primary alkyl halides are not good targets for S_N1 reactions, right? Usually. Yet the molecules below have primary alkyl halides that readily use S_N1 reaction mechanisms. How can this be? Resonance stabilization comes to the rescue again. In both of these molecules the positive charge of the carbocation can be delocalized and therefore stabilized.

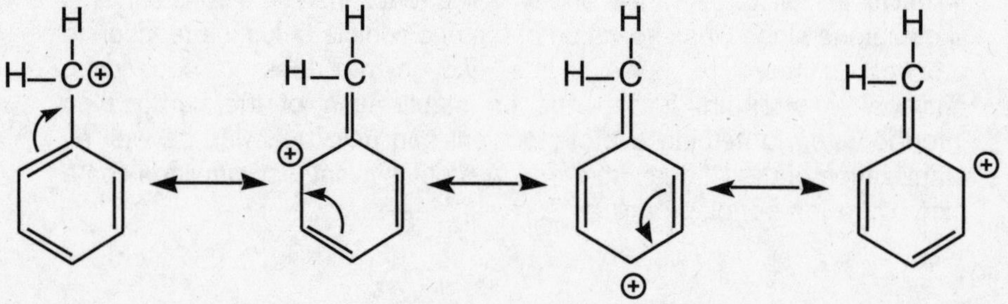

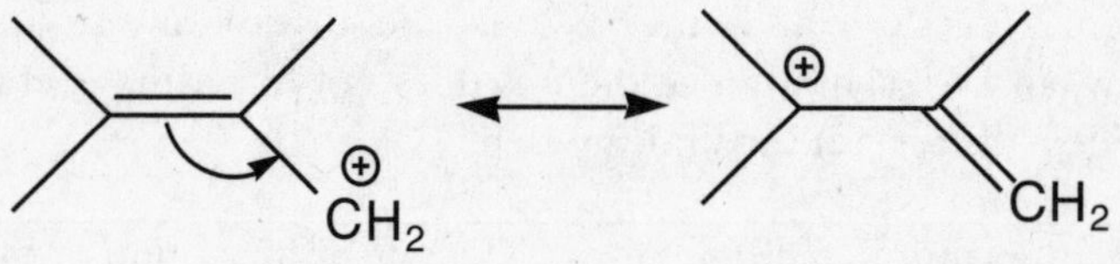

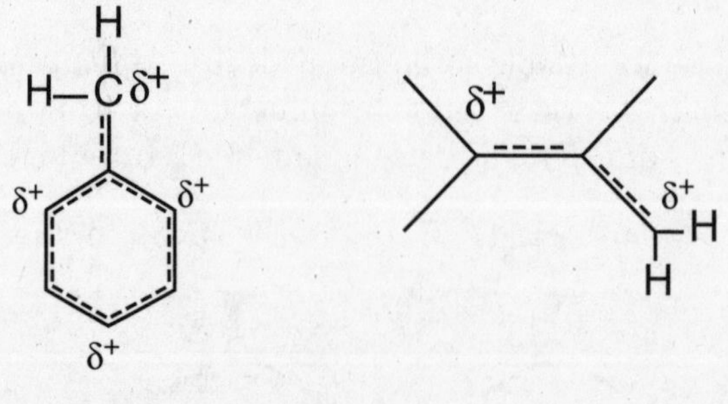

S_N1 AND THE LEAVING GROUP

The nucleophile and its strength may not be considerations for S_N1 reactions, but the leaving group still is. As with S_N2 reactions, a good leaving group is still a weak base. The better the leaving group, the more likely that it will up and leave on its own. A bad leaving group may never leave, slowing carbocation formation and.... Need we say more?

Perhaps just a bit more. What exactly makes a good leaving group for an S_N1 reaction? The same things that make a good leaving group for S_N2 reactions. A good leaving group is comfortable with itself as a solvated group, and does not feel a strong need to donate electrons as a nucleophile. A good leaving group is able to stand proudly on its own electrons, alone in the solvent. A chloride ion is a good leaving group for either S_N1 or S_N2, because it likes being an anion in solution. It has no pressing need to donate electrons or accept protons. The only difference is that in S_N1 the leaving group gets impatient and takes off without waiting around for the nucleophile to arrive on the scene.

NUCLEOPHILIC SUBSTITUTION STEREOCHEMISTRY

The reaction mechanisms for S_N1 and S_N2 reactions have stereochemical consequences for the products. If the substrate and the product of an S_N2 reaction, for example, are both optically active, how will the reaction affect the optical activity?

In an S_N2 reaction, the nucleophile attacks the carbon from the opposite side as the leaving group is attached. It comes in from the back side of the molecule, knocking the leaving group out through the front door. At the end of the reaction, the three other groups have flipped from one side to the other, and in the process they reverse the stereochemistry of the chiral center from R to S or vice versa. A commonly used analogy is that the molecule flips inside out like an umbrella caught in a strong wind. Note how the molecule in the diagram below is bent upward at first and then downward after the reaction.

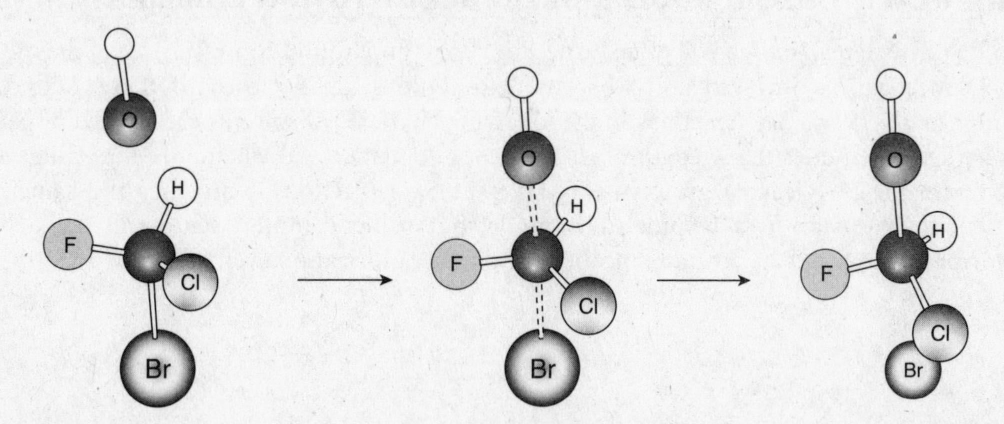

Experimentally, this is just what is observed, so these chemists must know what they are talking about. If you start out a reaction with (S)-2-bromobutane and react it with hydroxide ions by an S_N2 mechanism, the result is inversion of the molecule to produce (R)-2-butanol.

Student question: What about the stereochemistry of S_N1 reactions? Is the same inversion observed?

Instructor response: Actually no, not at all. The first step in the S_N1 is the loss of the leaving group to form the planar carbocation. At this point the nucleophile could (and does) attack the carbocation from *either side*. Remember, both sides are equivalent. The products will therefore not be one stereoisomer or the other, but an equal mixture of both stereoisomers. The products may still be chiral, but with an equal mixture of the two different enantiomers, optical activity will be lost. This is called what, by the way?

Student response: Yeah, yeah. A racemic mixture.

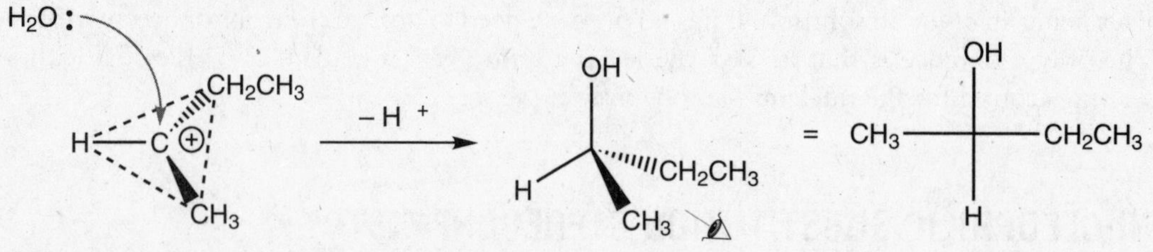

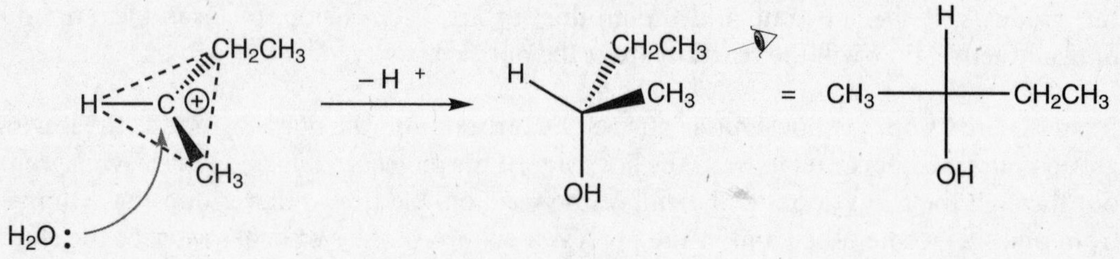

AND NOW...THE BIG NUCLEOPHILIC SUBSTITUTION SUMMARY

Going through reactions, it will often be necessary to compare and contrast some of the more likely mechanisms involved and what the expectations are for each of these. Let's try comparing S_N1 and S_N2 reactions in the following chart. What solvents are favored by each and why? When does the strength of the nucleophile matter and when does it not matter? Compare the best leaving groups in S_N1 versus S_N2 reactions – are they the same or different? The chart is a helpful study guide, but using common sense will make the information in the chart far more meaningful than memorization ever will.

NUCLEOPHILIC SUBSTITUTION REACTIONS

	SN1	SN2
kinetics	first order: rate = k[RX]	second order: rate = k[RX][Nu–]
mechanism	unimolecular, stepwise	bimolecular, concerted
solvent effect	favored by protic solvents	favored by polar aprotic solvents
substrate effect	determined by carbocation stability: methyl < 1° < 2° < 3° from inductive effects (but keep in mind the role that resonance can play!)	determined by steric effects: methyl > 1° > 2° > 3°
leaving group	weak base	weak base
nucleophile	strong nucleophile not necessary (usually solvolysis)	rate enhanced by a strong nucleophile
stereochemistry	racemization	inversion of configuration

Elimination and Addition Reactions

CHAPTER SEVEN

Pi bonds have a special place in the chemist's heart. The electrons in pi bonds are available for a range of reactions that do not occur elsewhere, including the reactions at the core of this chapter: elimination and addition reactions. Alkenes and alkynes are the beneficiaries of the unique properties of electrons in pi bonds. What is so special about pi bonds?

One unique feature of pi bonds is their accessibility, up out of the plane of the nuclei of a molecule, sticking out in the open. Pi bonds lock a molecule in place, preventing rotation around double and triple bonds and helping to create the geometry of a molecule. Also, the electrons in pi bonds can delocalize amongst themselves, spreading around within a molecule. Delocalization of electrons and charges through resonance in conjugated systems stabilizes molecules: the more spread out a charge is, the more stable it is (nature abhors concentrated charges). Electrons in sigma bonds do not have this distinction. The position of the pi bond, a molecular orbital formed from two parallel p orbitals, makes the electrons higher energy, less stable, and more reactive than electrons in sigma bonds. That reactivity can come in handy.

ELIMINATION REACTIONS

"What is an elimination reaction?" Funny you should ask. Consider the following reactions.

$$CH_3\underset{\displaystyle Br}{\underset{|}{C}}HCH_3 \longrightarrow CH_2{=}CHCH_3 + HBr$$

$$\text{cyclohexanol (OH)} \longrightarrow \text{cyclohexene} + H_2O$$

$$CH_3CH_2CH_2CH_2{-}\overset{\oplus}{N}{\equiv}N \longrightarrow CH_3CH{=}CHCH_3 + N_2 + H^+$$

$$CH_3{-}\underset{\displaystyle OH}{\underset{|}{CH}}{-}CH_2{-}\overset{\displaystyle O}{\overset{\|}{C}}{-}S{-}ACP \longrightarrow CH_3{-}CH{=}CH{-}\overset{\displaystyle O}{\overset{\|}{C}}{-}S{-}ACP + H_2O$$

What do these reactions all share in common? Have a look at the carbon-carbon bonds in the reactants and the products. Perhaps something about double bonds? The products of each reaction include a double bond that was not there in the reactants. Where did these double bonds come from? What else is present in the products? When the double bond is formed between two carbons, something else is lost to form a second product. Where did this second product come from?

In the first reaction, the formation of the double bond is accompanied by the appearance of HBr as a product. Where did the HBr come from? When the new pi bond was formed, the carbons went from being bound to four things with sp^3 hybridization, to being bound to three things with sp^2 hybridization. Something had to leave, and that was a bromine atom from one carbon and a hydrogen atom from the other carbon involved in the double bond. The formation of the double bond goes hand in hand with the departure of a hydrogen and a leaving group. Hence, certain groups leave or get "eliminated," giving elimination reactions their name.

In the second reaction, water appeared as a product of the reaction. This type of elimination reaction is called a dehydration reaction, and it acts as a means to create an alkene from an alcohol.

Now for the good stuff. As we go through elimination reactions, we will want to understand the mechanisms involved, deduce the outcomes of reactions, and predict the relative rates and thermodynamics therein. First, we have some background territory to cover to get us up to this point.

ALKENE ESSENTIALS

What makes a good (stable) alkene and a bad (less stable) alkene? Upbringing? Stability of alkenes is affected by the types and positions of substituents around the double bond.

Student question: What exactly is stability?

Instructor response: Stability has to do with thermodynamics – the relative potential energy of compounds. Let's try this. If two compounds, the infamous compound A and compound B, are both reacted to produce the same product, but different amounts of heat are produced, how do you know which compound was more stable?

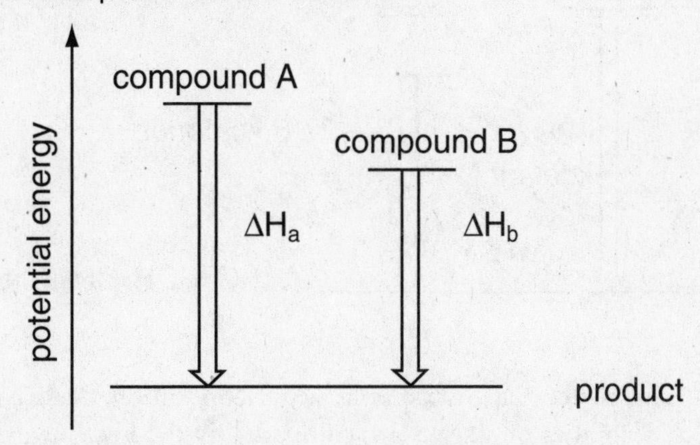

Student response: Well, compound A releases more heat to make product, so it must be more stable, right?

Instructor response: Want to think about that one a little more? Compound A must have started out with more energy than B, since it released more, so compound A must have more energy, and must be less stable than compound B. The reaction with compound A would release more heat than the reaction with compound B and A would be more exothermic than B. Remember, more stable compounds have less energy. Let's try another. Which of the following molecules is the more stable conformation of butene, the cis or the trans isomer?

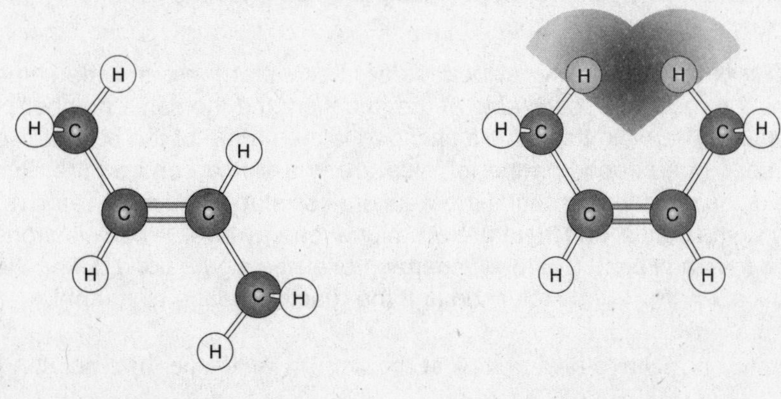

trans-2-butene

cis-2-butene

Student response: The shading around the hydrogens kind of gives it away. In the cis isomer, the hydrogens are close to each other and their electrons fight like cats and dogs, repelling each other way with steric hindrance. The trans isomer is much happier. The extra energy spent on steric hindrance in the cis isomer gives this molecule more energy and therefore less stability than the trans isomer.

A more quantitative way to look at the problem looks like this:

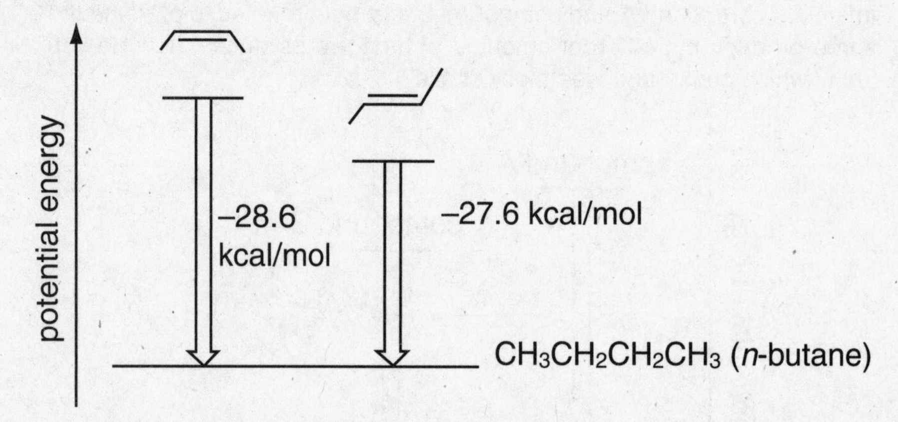

As in a lot of things in science, there is a specific way, a convention, that we use to express the stability of alkenes. Stabilities of alkenes are compared by the heat involved in addition of hydrogen to the double bond to create a single bond, also known as the heat of hydrogenation, the enthalpy involved in making the double bond saturated with hydrogen.

$$\mathrm{C{=}C} + H_2 \longrightarrow \mathrm{-\overset{|}{\underset{H}{C}}-\overset{|}{\underset{H}{C}}-}$$

Instructor question: Would this reaction be exothermic or endothermic? In other words, would energy be released or do you need to put energy into this reaction to make it proceed?

Student response: Looking at both sides, I see that there are the same number of bonds on each side of the equation and the same number of electrons. The main thing that's changed is the nature of the bonds. Two electrons in a pi bond on the left side are in a sigma bond on the right side of the equation. Pi bonds are weaker (less stable and more reactive) than sigma bonds, and therefore at a higher energy level. When electrons move from a pi bond to a lower energy orbital in a sigma bond during the reaction, energy will be released and the reaction will be exothermic.

Let's broaden our horizons a bit and look at the stability of alkenes in general. A general rule about alkene stability is that the more alkyl substituents around the double bond (not including hydrogens), the more stable the molecule will be. Also, as we already encountered, trans isomers are more stable than cis. Putting this together leads to the following trends in alkene stability.

$$CH_2{=}CH_2 < CH_2{=}C(H)(R) < (H)(R)C{=}C(H)(R)\ (cis) < (R)(H)C{=}C(H)(R)\ (trans) < (R)(H)C{=}C(R)(R) < (R)(R)C{=}C(R)(R)$$

Increasing stability

What does the relative stability of alkenes tell you if two alkenes are possible reaction products? Does it indicate the rate at which the alkenes will form? No, this is a function of the activation energy. It will, however, indicate the thermodynamic favorability of forming different reaction products and possibly the equilibrium status of the reaction. The trends in alkene stability will help to predict reaction products as we proceed.

E2 REACTIONS – BIMOLECULAR ELIMINATION

Elimination reactions can be either bimolecular or unimolecular, as with substitution reactions. In a bimolecular elimination reaction (E2) one molecule bumps into another one to cause the reaction to occur in a single step. In the case of the E2 reaction, the incoming molecule donates electrons to a hydrogen, and a leaving group packs its electrons and exits as well. A double bond is formed in the process.

$$B:^{\ominus} + H{-}C{-}C{-}X \longrightarrow C{=}C + HB + X^-$$

The occurrence of these two events at the same time is not a coincidence. Follow the bouncing electrons in the reaction above. The movement of the electrons, as always, is the key to the reaction mechanism. Where do the pi orbital electrons in the product come from? When the B atom in this reaction enters the scene with its unbonded electron pair, it donates the electrons to hydrogen. Hydrogen, with too many electrons on its hands, shuffles its electrons off to form the pi bond between the carbons. Finally, the carbon at the other end of the nascent double bond completes the electron toss by handing off to the leaving group, which heads out with an extra nonbonding electron pair. So, the pi bond electrons came from the C-H bond in the reactant. However, the movement of electrons does not really happen in a stepwise process, as outlined here. In the transition state the electrons are moving simultaneously all the way between the nucleophile and the leaving group, creating the following transition state:

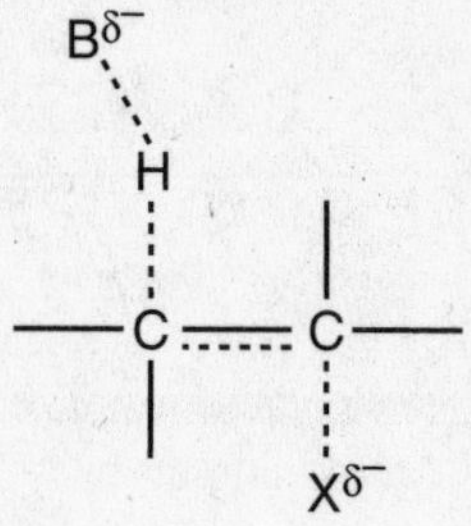

In the process of forming the pi bond, the hybridization of the carbons involved and the geometry of the molecule will change. A carbon with four single bonds is sp^3 hybridized. The appearance of a double bond, however, will cause the carbons to change to sp^2 hybridization, with three planar sp^2 orbitals arranged around the carbon. What is the bond angle between each sp^2 orbital? If the sp^2 orbitals are planar and are equally spaced, and there are three of them, the answer must be 360 degrees/3 = 120 degree angles between sp^2 bonds. The p orbitals that form the pi bond are located perpendicular to the plane of the sp^2 orbitals.

Can we deduce a rate equation for an E2 reaction? Sure thing. The reaction is bimolecular, so the concentration of both reagents will affect the rate. This, and throw in a rate constant for good measure, and you are all set: rate = $k[RX][B^-]$.

> **Instructor question:** What type of incoming molecule might donate electrons to hydrogen in an elimination reaction?
>
> **Student response:** Oh goody, it's our old friend Mr. Nucleophile making another guest appearance.
>
> **Instructor response:** It sure is. A strong nucleophile, such as a strong base like hydroxide ions, is ideal for donating some electrons in an elimination reaction, just as it was in a substitution. In fact, this can complicate things a bit. When a nucleophile happens across a substrate molecule, it could result in an elimination reaction, but it could also result in a substitution reaction. S_N2 and E2 reactions can and do compete with each other, like siblings, although they do have different characteristics which we will discuss in a bit.

In the following example, the hydroxide nucleophile donates electrons either to hydrogen to result in an E2 reaction or to the carbon to result in a substitution reaction.

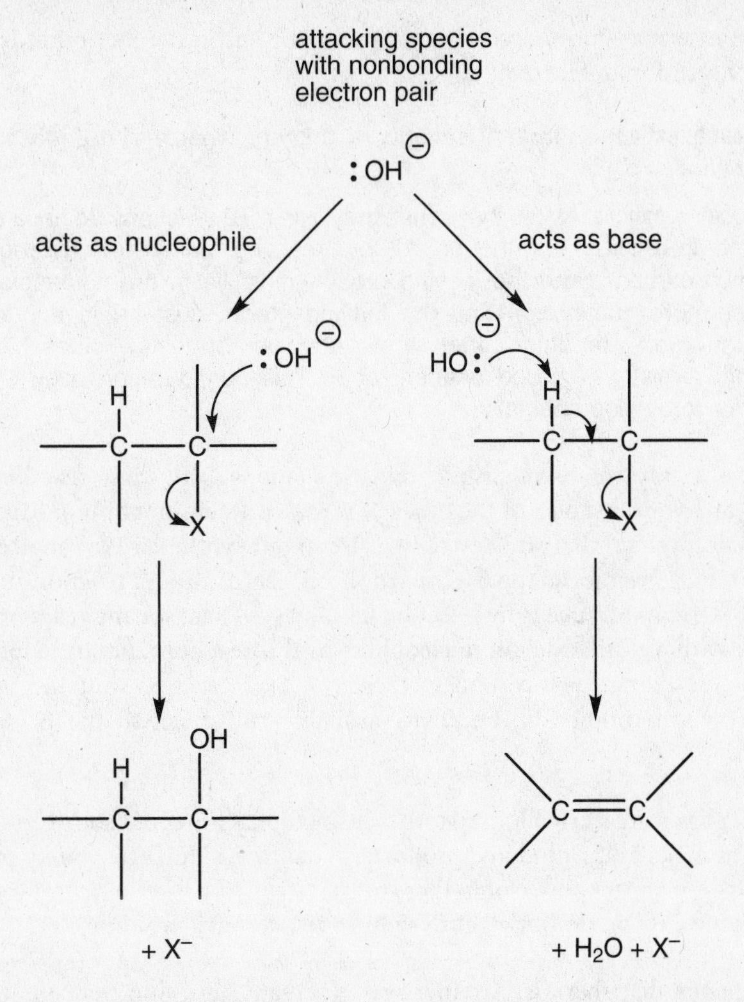

Wait a minute – the whole point of organic synthesis, usually, is to make a specific product that we have in mind. If reactions can go any which way, then is all chaos? Is there nothing that we can do to control the direction a reaction takes? Do not fear, there is a way. We can control whether a reaction goes to elimination or substitution. The answer is steric effects.

Recall if you will the importance of steric hindrance in S_N2 reactions. The more hindered the substrate, the less favored an S_N2 reaction is. If the nucleophile cannot get access to a carbon, then it cannot initiate an S_N2 reaction. The culmination of this is tertiary compounds, which are so hindered that they pretty much don't do S_N2 reactions at all.

Elimination may involve the same nucleophiles as S_N2, but it does not happen the same way. In an elimination reaction, the nucleophile does not need to get in close to the carbon involved in the double bond. The nucleophile in an E2 reaction only needs to contact the hydrogen it will donate electrons to and pluck off the lucky hydrogen, and it can do this quite nicely on the exterior—more accessible—part of the substrate. Steric hindrance is not as much of a concern for an E2 reaction.

> **Student question:** So tertiary, secondary, primary, whatever? E2 reactions don't care?
>
> **Instructor response:** Not in that sense they don't. E2 reactions do have one geometric requirement, though. All four reacting atoms—the hydrogen, the two carbons, and the leaving group—must lie in the same plane. Furthermore, the proton and the leaving group must be in an "anti" configuration, meaning that they are on opposite sides. This stereochemistry is called antiperiplanar. This configuration allows the orbitals to overlap optimally.

We now have a way to distinguish elimination and substitution reactions, by finding conditions that favor one and not the other. If we want an elimination product without any S_N2 reaction occurring, what strategy can we use in our synthesis? We can use conditions in which steric hindrance blocks the S_N2 reaction but leaves the E2 reaction unaffected. One way to create steric hindrance is by selecting a bulky substrate for the reaction, such as tert-chlorobutane with a hydroxide ion nucleophile. In this reaction, the nucleophile would not be able to access the tertiary carbon to create an S_N2 reaction, but the protons on the neighboring carbons would still be plenty available and open, so the E2 reaction would proceed.

If the only way to create steric hindrance is by using bulky hindered substrates, then you will only be able to make bulky hindered products. What if you don't like bulky products? They are not for everyone, you know. Could there be a strategy to create steric hindrance and select for the E2 reaction if you do not want to synthesize a tertiary product?

It just so happens that there is. Another way to create steric hindrance is to use a bulky hindered nucleophile rather than a bulky substrate. If the nucleophile is t-butoxide rather than hydroxide, then E2 will be favored once again since the size of the nucleophile will block S_N2 in the same way that the bulky substrate does. In this way the reaction mechanism can be selected without limiting the substrate used in the reaction.

When an elimination reaction occurs, introducing a double bond into an alkane, are there any preferred configurations of the alkane when the reaction occurs? As we noted earlier, yes. The nucleophile tends to remove the proton in an anti configuration from the leaving group, with the leaving group and the proton on opposite sides of the molecule and the remaining groups lying flat in a plane. The resulting alkene will be planar and held in place without rotation by the pi bond in the double bond. This preference for the anti configuration determines whether the cis or trans isomer will form as the product in the E2 reaction.

$B^{\ominus}$: H X → [$B^{\delta-}$ H $X^{\delta-}$]‡ → + HB + X^-

transition state

:$B^{\ominus}$ H X → [$B^{\delta-}$ H $X^{\delta-}$]‡ → + HB + X^-

SOME E2 PRACTICE

Instructor question: Let's try to predict what will happen in some E2 reactions. With each of the following substrates, predict the E2 product that will be produced.

a) tert-butyl bromide

Student response: As we now know, tertiary substrates are no problem for elimination reactions. The nucleophile just swoops in and sucks off a proton from one of the outer carbons, allowing the Br leaving group to proceed on its merry way and give 2-methyl propene as the product.

B^- H—CH_2 CH_3—C—Br CH_3 → CH_2=C(CH_3)CH_3 + HB + Br^-

Instructor response: Nicely done. Try another.

b) chlorocyclohexane

Student response: I'm frightened.

Instructor response: Fear not. There is nothing mysterious or special about cyclohexane in elimination reactions. The reaction will proceed as with any other alkyl halide substrate. The nucleophile will remove a proton from a neighboring carbon, and the chlorine will leave to produce cyclohexene.

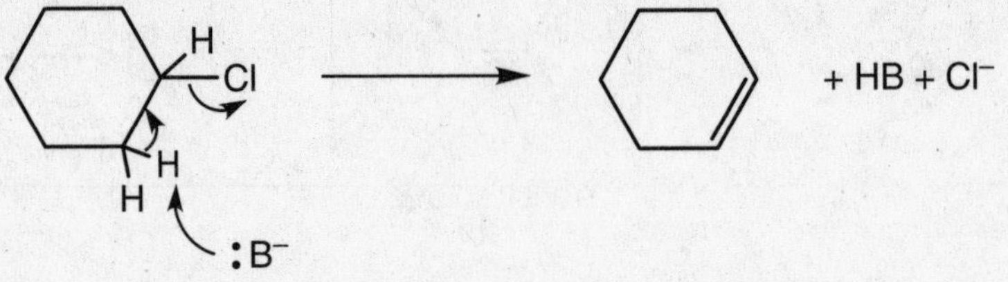

+ HB + Cl^-

Try another:

c) 1-bromoethylbenzene

Student response: The starting material has a secondary bromine on the benzene substituents. There is no hydrogen on the neighboring carbon that is in the benzene ring so the elimination must proceed by removing a proton from the other carbon in the ethyl substituents. The aromatic ring is untouched, with the creation of a double bond in the alkyl substituents to create ethenylbenzene.

Instructor response: Excellent. This molecule is also known commonly as styrene and is the precursor of polystyrene created when styrene units are polymerized together.

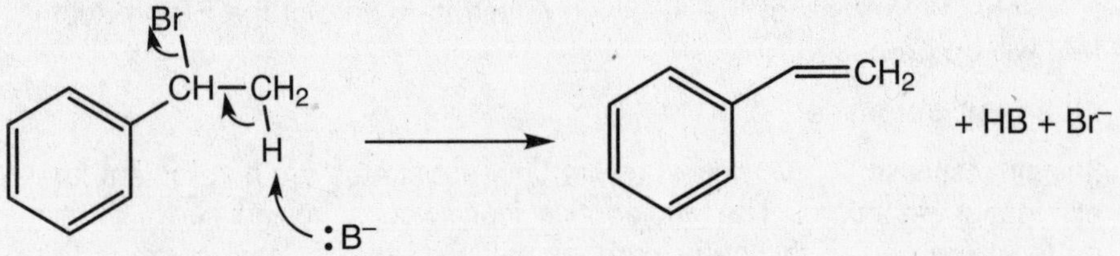

+ HB + Br^-

d) trans-1-chloro-2-methylcyclohexane

Instructor question: This last one may take some thought to draw the structure, but the pieces are all there. The molecule is a cyclohexane ring with two substituents, a chloro and a methyl group, in trans configuration of each other on neighboring carbons. As a cyclohexane, the molecule can exist in one of two different chair configurations. In one chair configuration the two substituents will both be in the axial position, while in the other chair state both will be equatorial. Is one of these chair configurations more favored than the other?

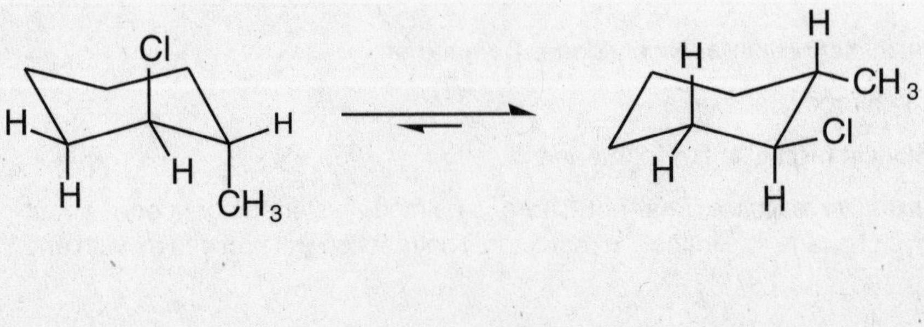

Student response: Thermodynamically, we know equilibrium will favor the configuration on the right with both groups in the equatorial position. The equatorial position for cyclohexane substituents is favored because the substituents and their electrons have more room to breathe isolated from the electrons of their repulsive neighbors.

Instructor response: True...but unfortunately, the more stable conformation is not in this case the one that is favored for an E2 elimination reaction. The key here is the chlorine that will be the leaving group. The leaving group and the hydrogen removed by the nucleophile *must* be in the antiperiplanar configuration. When the chlorine is in the equatorial position, it is not anti to anything. When the chlorine is axial, sticking straight up, it is anti to the methyl group on one side and anti to a proton on the other. The methyl substituent prevents elimination on that side since there is not a hydrogen in the correct anti position. This leaves only one possibility, for the double bond to form between the carbon bound to the chlorine and the next carbon further away from the methyl group.

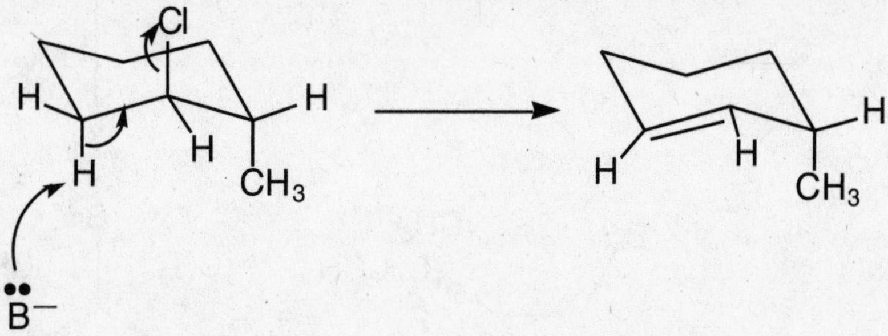

Another way to draw this is with Haworth projections, not worrying about the chair configurations of cyclohexane. This might make it easier to visualize why the double bond forms on one side and not the other in this reaction.

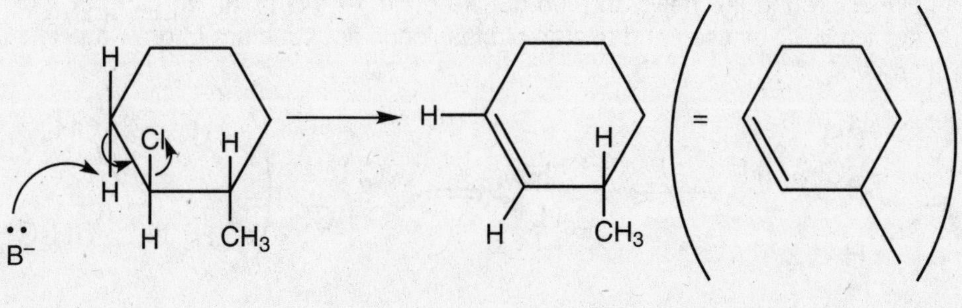

e)

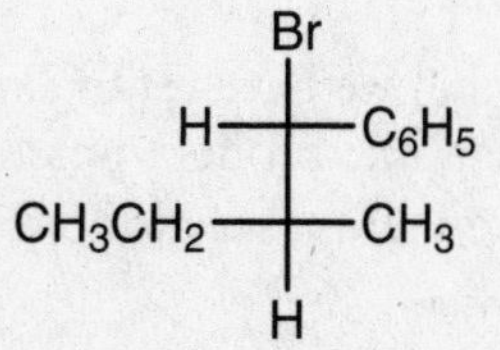

Student response: Yikes. Where to begin?

Instructor response: Let's try to break it down. First, draw the structure in a way that gives us more of a direct sense for the 3D structure of this puppy. Remember that in a Fischer projection, which is what we are starting out with here, the horizontal lines are the bowties, coming up out of the page, and the groups at the top and bottom are pointing out behind the page. Either dash and wedge drawings or Newman projections are handy to get a grip on the 3D structure, and to rotate the leaving group (Br) and the proton to be eliminated into the anti position.

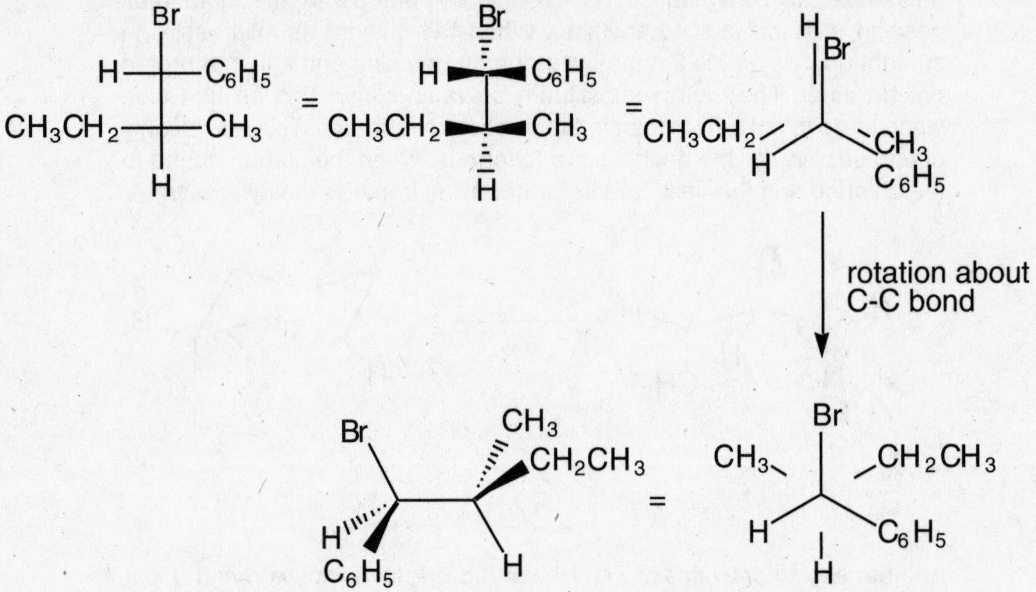

Br
H—C6H5
CH3CH2—CH3
H
=
Br
H C6H5
CH3CH2 CH3
H
=
H
Br
CH3CH2 CH3
H C6H5

rotation about C-C bond

Br CH3 CH2CH3
H C6H5 H
=
Br
CH3 CH2CH3
H C6H5
H

Once you get things into the anti configuration, the rest is downhill. Just remove the proton and the leaving group, leaving the remaining groups planar around the newly formed double bond. At this point we can put away our 3D glasses and use an old fashioned flat structure to draw the product.

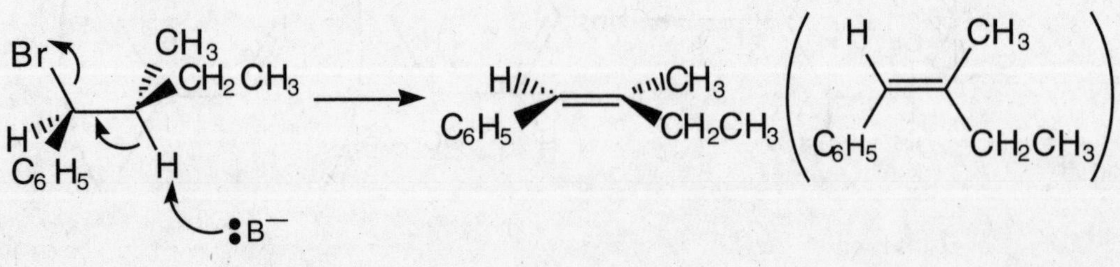

Br CH3 CH2CH3
H C6 H5 H
:B⁻
→
H CH3
C6H5 CH2CH3
(H CH3
C6H5 CH2CH3)

Are there any other complexities of E2 reactions that we have yet to explore? Here is a brain teaser for you: What happens if there is more than one product possible in a reaction? For example, if the following is the substrate in an elimination reaction, what products might be formed?

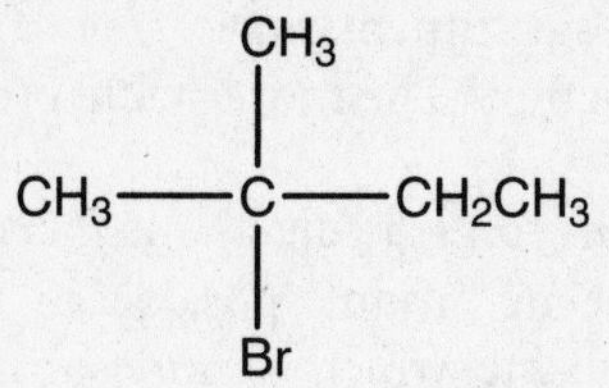

$$CH_3\text{—}C(CH_3)(Br)\text{—}CH_2CH_3$$

The key is to recognize what hydrogens might be involved in the elimination. The hydrogens need to be attached one carbon away from the leaving group (Br). This would make the hydrogens on the two methyl groups at the top and left of the molecule candidates to be removed in an elimination. Another possibility is the removal of a hydrogen from the secondary carbon in the ethyl substituents on the right side of the drawing. How many different products are possible in this E2 reaction and what are they?

The six methyl hydrogens are all equivalent and distinct from the two ethyl hydrogens (methylene hydrogens if you like). There are two possible products, which are:

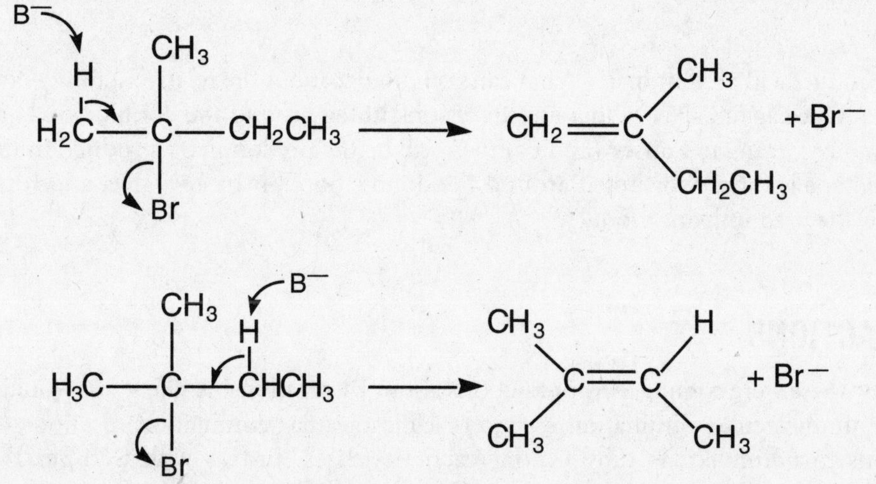

$$B^- + H\text{—}H_2C\text{—}C(CH_3)(Br)\text{—}CH_2CH_3 \longrightarrow CH_2{=}C(CH_3)(CH_2CH_3) + Br^-$$

$$B^- + H_3C\text{—}C(CH_3)(Br)\text{—}CH(H)CH_3 \longrightarrow (CH_3)_2C{=}C(H)(CH_3) + Br^-$$

Now for the good stuff: How much of the two products are formed?

First, let's try to answer the question assuming that it does not matter which methyl is removed, that they are all exactly the same as far as the nucleophile goes and the choice is entirely random. In this case you would predict a ratio of 6:2 (or 3:1) between the first product with removal of one of the six methyl hydrogens or two of the methylene hydrogens.

Is this what is observed in the lab? As it turns out, no. What is observed is that we get about 30% of 2-methyl-1-butene (product #1, made by removing a methyl hydrogen) and the remainder as the other product, 2-methyl-2-butene. How can this be?

Let's go back to the beginning of this chapter where we talked about the relative stability of different alkenes. Compare the stability of these two alkene products. Which one is more stable? Alkenes with more alkyl substituents have greater thermodynamic stability. Which product predominated in the reaction? The more stable product.

As it turns out, there just happens to be a rule about this called **Zaitsev's rule**. Zaitsev said that if there is more than one alkene product possible in a reaction, then the predominant product will be the more stable alkene, which is more substituted.

Let's look at another example of this. If you start with 2-bromopentane and work some E2 magic on it, what products will be observed and which one will predominate? First, determine the possible products, with the double bond formed on one side or the other of the leaving group.

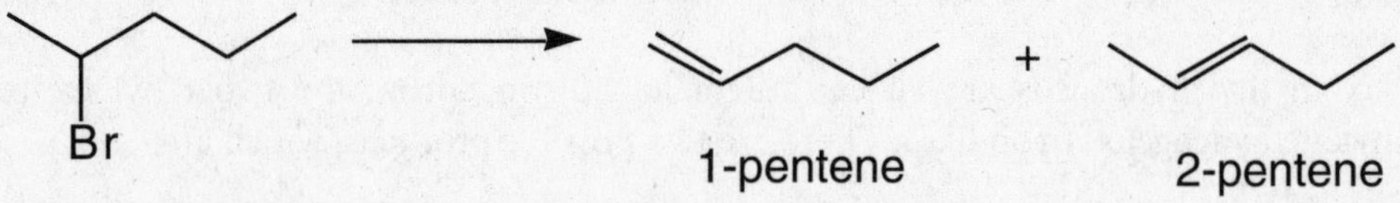

Br
1-pentene
2-pentene

Which product will predominate? What can you predict about the relative stability of the two products? Once again, alkenes that are more substituted around the double bond are more stable, and according to Zaitsev (and us) they will be the predominant product. In this case, the 2-pentene is more substituted around the double bond, is more stable, and (drum roll please) is the predominant product.

E1 REACTIONS

You knew these were coming. Any predictions about E1 reactions? A safe guess would be that they are unimolecular elimination reactions. Unimolecular elimination reactions (E1) are analogous to unimolecular substitution reactions (S_N1). Just as with S_N1 reactions, E1 reactions do not wait around for nucleophile to get things going. The substrate, all on its own, takes the initiative to lose a leaving group and create a carbocation reaction intermediate.

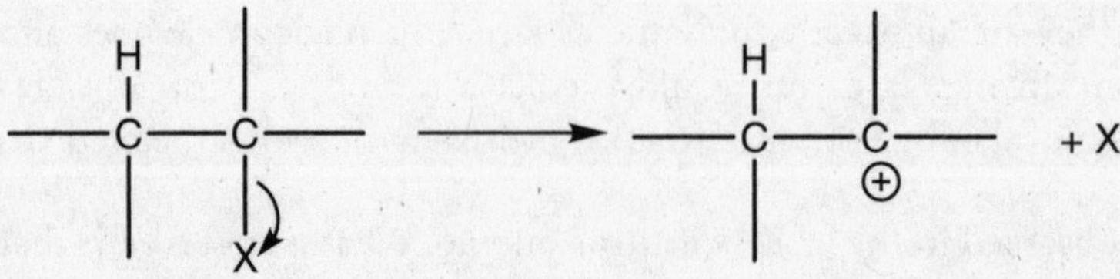

H
C
C
X
H
C
C
⊕
+ X^-

What would you predict about the rate law for E1 reactions? Being unimolecular, the rate will be proportional to the concentration of substrate, and a rate constant. The rate limiting step will be the creation of the carbocation, as with S_N1 reactions. The E1 reaction is completed in a second step by the removal of a proton by a nucleophile to form the double bond, like in E2.

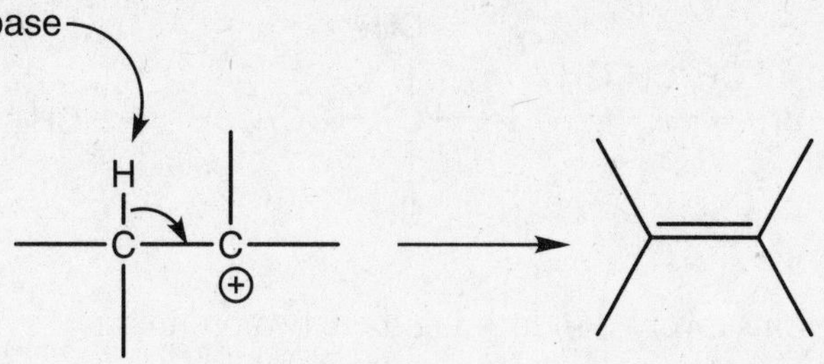

base

How will the strength of the nucleophile affect the rate of an E1 reaction?

As with S_N1 reactions, the rate-limiting step is the first one, the creation of the carbocation. The nucleophile does not appear in the rate equation. Once the carbocation forms, it will be so reactive that it will react with just about any nucleophile that it encounters, most commonly the solvent.

How can you speed up a E1 reaction? Anything that can be done to make our friend the carbocation more comfortable will help move things along. As with S_N1 reactions, polar protic solvents speed up E1 reactions by stabilizing both the charged carbocation and the leaving group. The substrate structure can also affect the stability of carbocations. More substituted alkanes form carbocations more readily in the order of carbocation stability: tertiary > secondary > primary.

Instructor question: Can you put the following molecules in order of decreasing rate of E1 reactions?

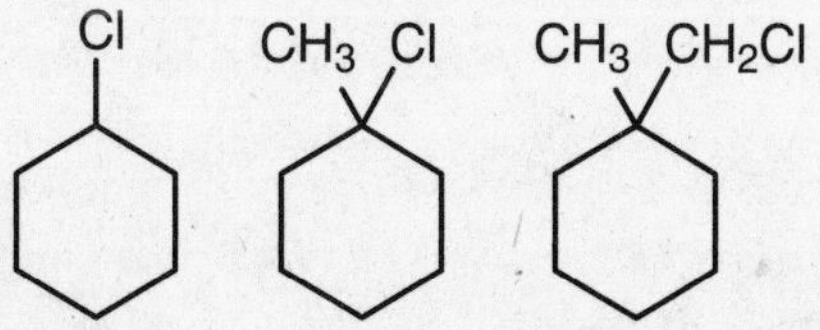

Cl CH_3 Cl CH_3 CH_2Cl

Student response: We want to rank the molecules in terms of their ability to form carbocations. The better molecules are at forming carbocations, the faster the reaction will be. We just said the order of reactivity will be tertiary > secondary> primary, so that gives us:

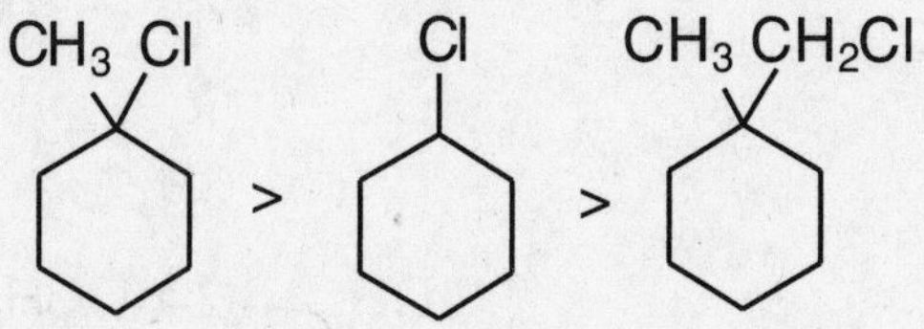

CH_3 Cl > Cl > CH_3 CH_2Cl

If E1 and S_N1 reactions are so similar, how can we know which reaction will occur if we add nucleophile to a substrate? The truth is that sometimes you get more than one reaction happening, whether you want it or not. If you mix t-butyl bromide with ethanol, for example, you get S_N1 and E1 reactions both occurring, producing two different products:

$$CH_3{-}C(CH_3)_2{-}Br \xrightarrow{CH_3CH_2OH} CH_3{-}C(CH_3)_2{-}OCH_2CH_3 + CH_2{=}C(CH_3)_2$$

What are the mechanisms involved in forming these two products?

The first step for the E1 and S_N1 reactions is the same: formation of the tertiary carbocation. If the ethanol acts as a nucleophile to donate protons to the carbocation carbon, then substitution occurs and an ether is produced. If, however, the ethanol donates electrons to a proton and removes it, then an alkene is formed by an E1 mechanism.

$$CH_3{-}C(CH_3)_2{-}Br \longrightarrow CH_3{-}C^{\oplus}(CH_3)_2 + Br^-$$

E1:

$$H{-}CH_2{-}C^{\oplus}(CH_3){-}CH_3 \quad CH_3CH_2\ddot{O}H \longrightarrow CH_2{=}C(CH_3)_2 + CH_3CH_2\overset{\oplus}{O}H_2$$

S_N1:

$$CH_3{-}C^{\oplus}(CH_3)_2 \quad CH_3CH_2\ddot{O}H \longrightarrow CH_3{-}C(CH_3)_2{-}\overset{\oplus}{O}(H)CH_2CH_3$$

$$\xrightarrow{-H^+} CH_3{-}C(CH_3)_2{-}OCH_2CH_3$$

Let's look at another example. If we react 2-iodo-3-methylbutane with butanol, what reaction products would you predict would be observed?

If you guessed that both E1 and S_N1 reactions would occur, starting with the loss of the iodide leaving group to form the secondary carbocation, then you guessed right. What is the second step? If you are talking S_N1, the butanol will act as a nucleophile to add to the carbocation and create an ether. If you are talking E1, then the butanol could remove a proton from either of the adjacent carbons to create two different E1 products.

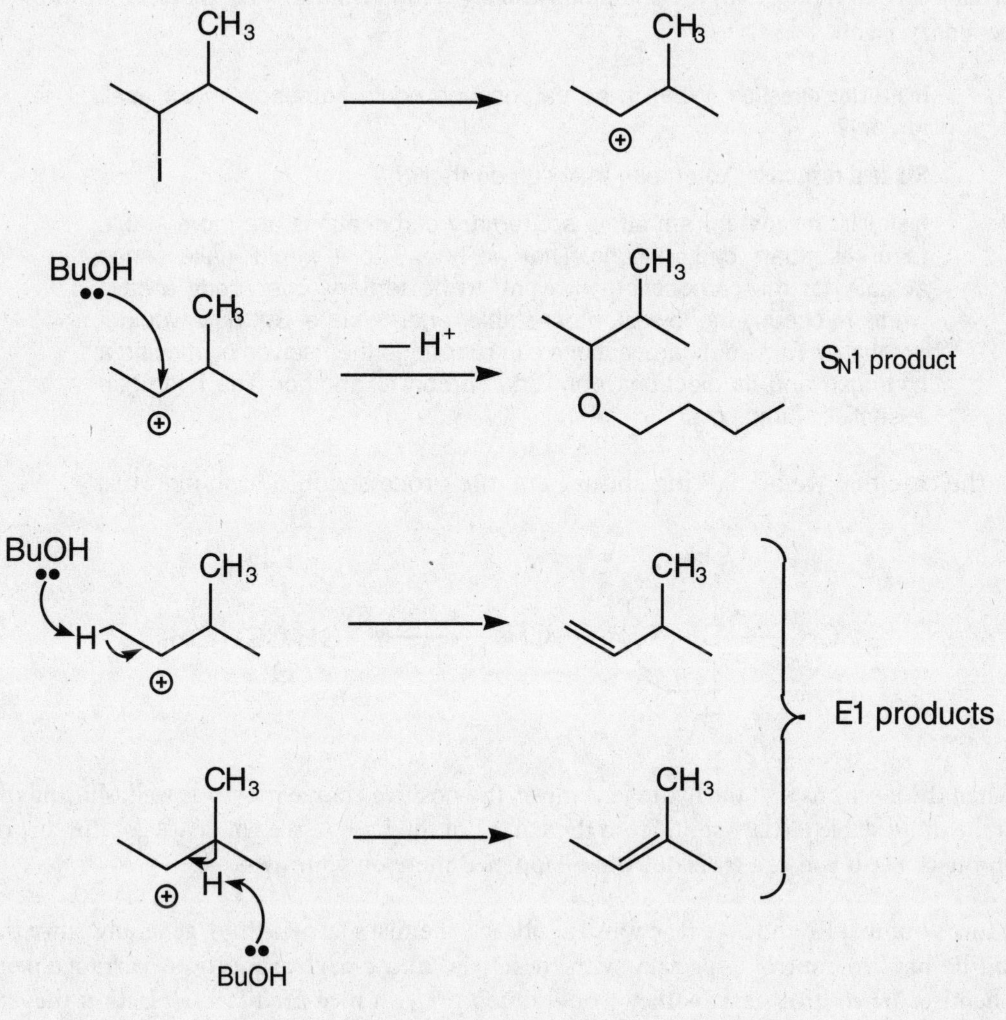

CH_3 CH_3

BuOH CH_3 CH_3 $-H^+$ S_N1 product

BuOH CH_3 CH_3

E1 products

CH_3 CH_3

BuOH

This is not the whole story, however. There are two more reaction products that are also observed. They are:

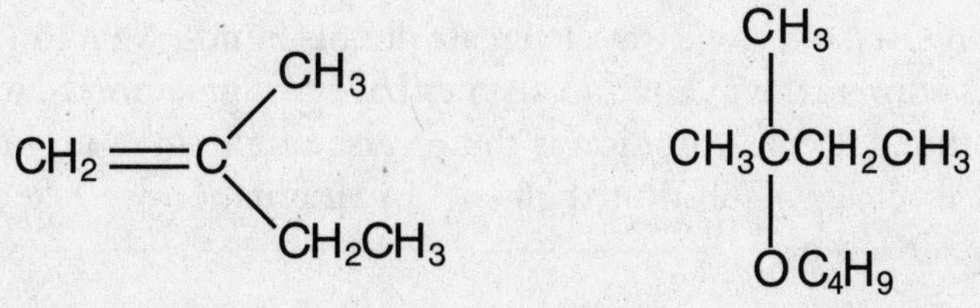

$CH_2{=}C(CH_3)CH_2CH_3$ $\qquad$ $CH_3C(CH_3)(OC_4H_9)CH_2CH_3$

Where in the world did these come from? Talk about left field. One looks like an elimination product and the other like a substitution product, but they've got things in the wrong places it looks like. No, we have not lost our minds quite yet. There is just something unusual going on here.

If you look at the second structure above, the ether, it seems as if the ether has added to the tertiary carbon rather than the secondary carbon. How could this be? The carbocation was secondary, right?

> **Instructor question:** What if the carbocation could somehow make itself tertiary?
>
> **Student response:** You mean they can do that??
>
> **Instructor response:** I am afraid so. Tertiary carbocations are more stable than secondary carbocations. That we know, so it would make sense actually for the carbocation to "want" to be tertiary. Everybody always wants to occupy the lowest, most stable, energy state. But how would it get there? Turns out carbocations can rearrange themselves by moving a hydrogen and its electrons from one carbon to another. The hydrogen essentially jumps over.

In the case that we are talking about here, the process would look like this:

$$CH_3-\underset{H}{\overset{CH_3}{C}}-\overset{\oplus}{C}H-CH_3 \longrightarrow CH_3\underset{\oplus}{\overset{CH_3}{C}}CH_2CH_3$$

When the electrons and the hydrogens move, the positive charge moves as well, shifting over to the more stable tertiary spot. From there it is not too hard to see where we got the surprise products. See if you can trace out what happened there on your own.

It turns out that E1 and S_N1 reactions are not the chemist's favorite tool, generally, since they can be hard to control, especially with these wild and crazy carbocation rearrangements. Chemists are control freaks – they would much prefer a nice predictable reaction they can control like an S_N2.

ELECTROPHILIC ADDITION REACTIONS

Now that we have figured out some ways to create double bonds, what can we do with them? Addition reactions convert the pi bond in alkenes back to sigma bonds, with the "addition" of two groups to the molecule. Addition is the reverse of elimination. In addition, we add stuff to an alkene to create a substituted alkane. In elimination we take away substituents from alkanes to create alkenes.

A key to addition reactions is the reactivity of pi bond electrons. What type of molecule is likely to react with the electrons in a pi bond to create an addition reaction?

So far we have discussed mostly the nucleophile, which wanders about with electrons that it has a penchant to donate. Is a nucleophile likely to react with the pi bond electrons? No, thank you, it is full of electrons already. A more likely scenario is that the electrons will react with something that is hungry for electrons, which would be an electrophile. Electrophiles are electron-deficient molecules that serve as electron acceptors in reactions. Examples would include protons (hydrogen ions), carbocations, or any other Lewis acids that accept electron pairs. An important detail to remember is that in electrophilic attack, we draw the electron-pushing arrow as originating from the pi bond, because that is where the electrons are coming from.

The first step in an addition reaction is for an electrophile to come along and take off with the electrons from the pi bond to create a new sigma bond between itself and one of the carbons. In the process a carbocation is formed. (These carbocations seem to be everywhere.) Just as in E1 and S_N1 reactions, the first step and the rate limiting step in electrophilic addition reactions is the formation of a carbocation, only this time it is formed by adding something to the substrate rather than losing something.

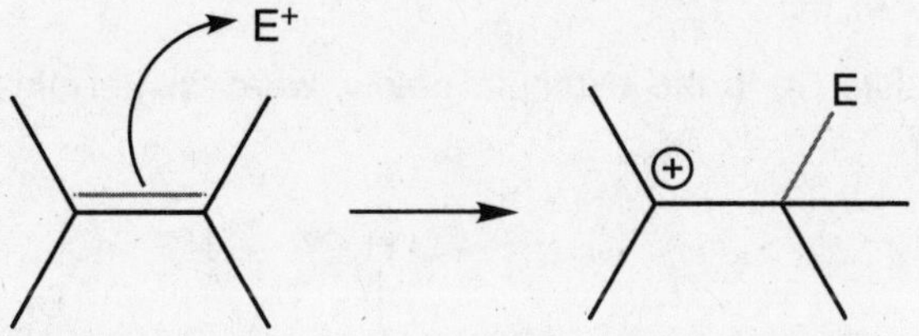

E^+ → E, ⊕

The second step is for something to add to the reactive carbocation intermediate. The carbocation will react with a negatively charged species to create a substituted alkane as the final reaction product. If the negatively charged species is a halogen ion, then the result is an alkyl halide. If the addition species is OH-, then the product of the addition will be an alcohol, using the reverse reaction of a dehydration reaction.

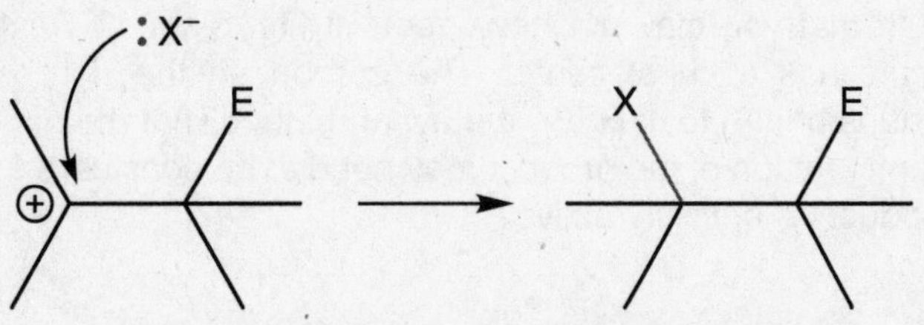

$:X^-$, E, ⊕ → X, E

The reaction profile for an electrophilic addition reaction will resemble other two step processes, including an energy minima in the middle of the reaction pathway for the carbocation reaction intermediate, surrounded on either side by two high-energy transition states. The order of energy if the reaction is to proceed forward is: reaction intermediate > reactants > products.

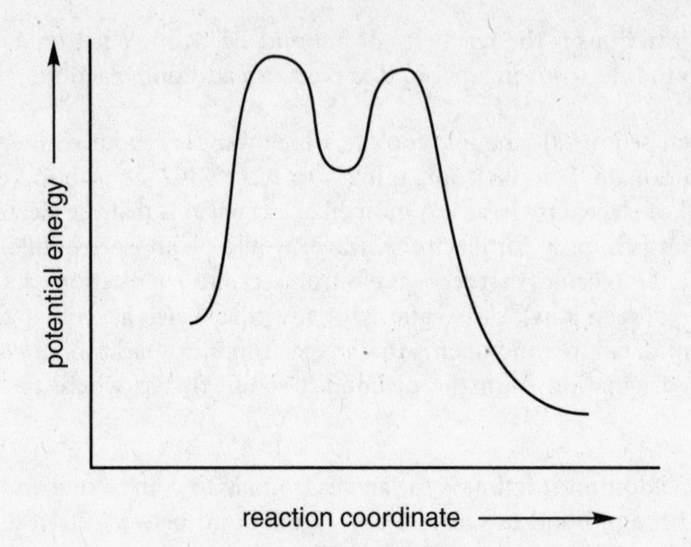

An example of an addition reaction would be the addition of HBr to trans 2-butene. In the first step of the reaction, hydrogen ions act as electrophiles to add to one carbon and break the pi bond, creating the carbocation intermediate. The second step is for a negative species to add to the intermediate, in this case Br^-.

Instructor question: So in the example below, what will the final reaction product(s) be?

$CH_3CH{=}CHCH_3 \xrightarrow{H^+} H{-}C(H)(CH_3){-}C^{\oplus}(H)(CH_3) \xrightarrow{Br^-}$

Student response: Is this a trick question? Don't you just add Br to the right side of the molecule?

Instructor response: Well, is there something more about the structure of this molecule that you may not have seen at first glance? Think chiral. This molecule has a chiral center. The carbon on the right has four different groups bound to it: H, Br, methyl and ethyl. That means that the absolute configuration of the product may need to be considered. Will the reaction product be optically active?

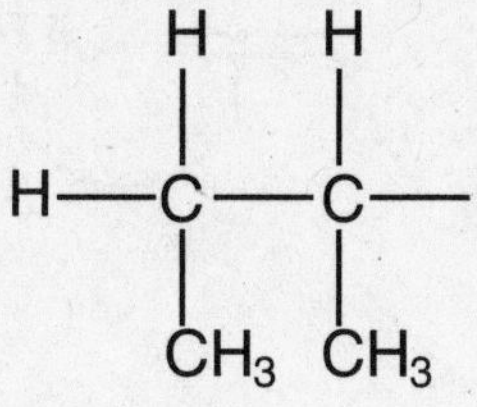

H H
H—C—C—Br
CH_3 CH_3

Student response: The way I see it, the reactants are not optically active, so the product should not be either.

Instructor response: You got it. Reactions do not create optically active products from substrates that lack optical activity. Another way of looking at this is that the carbocation intermediate has free rotation around the bond in the middle, and the three groups around the carbocation will be planar. The Br^- can and will add to the carbocation from either side of the carbocation. Both enantiomers will be synthesized in equal quantities and the product will be a racemic mixture, lacking optical activity.

If you started with the cis isomer of 2-butene, would you expect a different result in the above reaction? As it turns out, the end result is the same. Cis butene, being a cis alkene, has more energy than the trans isomer, so more energy will be released in the addition reaction with the cis isomer, but the same products will be observed. The same carbocation will be formed by either the cis or trans isomer, leading to the same racemic mixture at the end of the day.

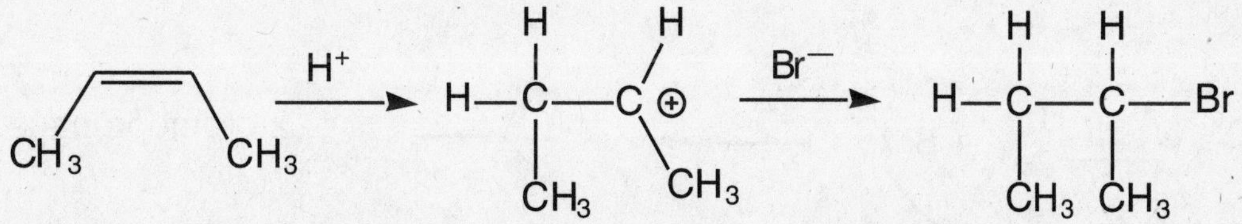

$CH_3CH{=}CHCH_3$ (cis) $\xrightarrow{H^+}$ H—C(H)(CH_3)—C⊕(H)(CH_3) $\xrightarrow{Br^-}$ H—C(H)(CH_3)—C(H)(CH_3)—Br

In some cases, two different products of an addition reaction might result, depending on how the addition reactant like HBr adds to the double bond. If H adds to one side of the double bond you get one product, and if it adds to the other side you get another product. Consider an addition reaction starting with propene and HBr. What two products would be observed? The proton can add to either side of the double bond to create two different carbocation intermediates. In one case, a primary carbocation is created; in the other case, you get a secondary carbocation.

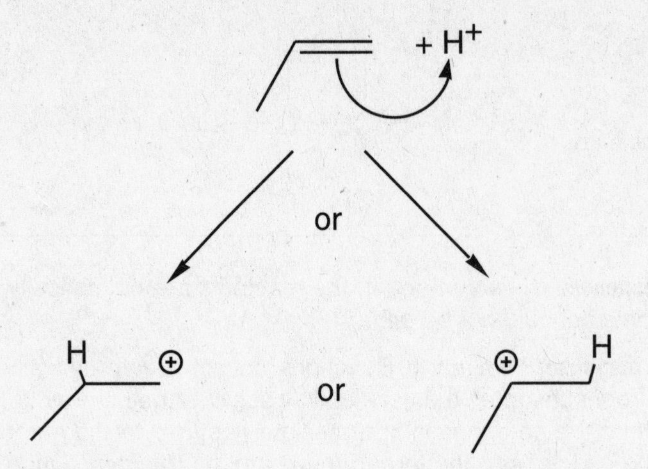

Next, the bromide ion swoops in to bond with the carbocation carbon and result in two different products, either 1-bromopropane or 2-bromopropane.

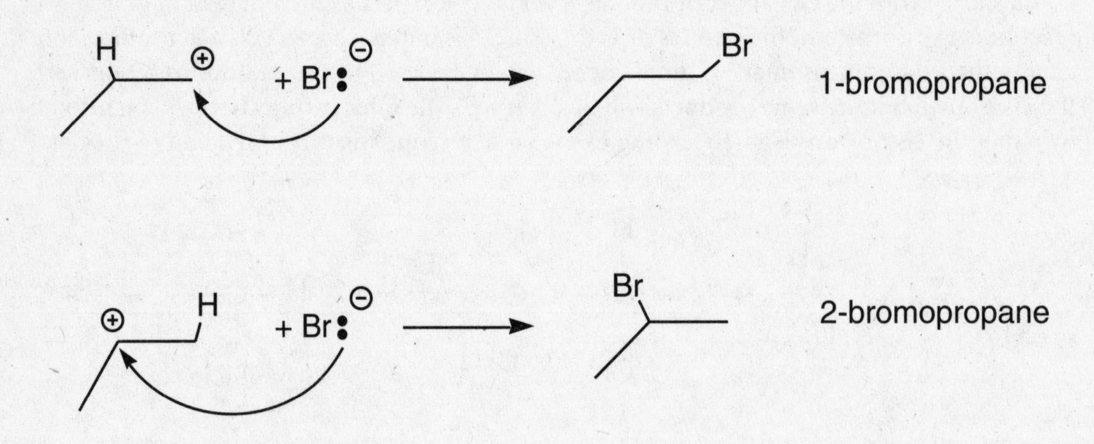

Instructor question: Any ideas about which of these two products might predominate in this reaction?

Student response: We know the formation of the carbocation is the rate-limiting step, and this is the key to determining which product is favored. Secondary carbocations are more stable, so the addition product of the secondary carbocation intermediate will be favored, the 2-bromopropane.

This observation follows **Markovnikov's rule**, which states that in an addition reaction involving the addition of a hydrogen halide like HBr to an alkene, the hydrogen will react with the less substituted carbon to create the more substituted and more stable carbocation intermediate.

Instructor question: Let's have a go at predicting the favored product for the reaction of HI with the following reactants. Remember, go for the carbocation intermediate that is more substituted and more stable.

a.

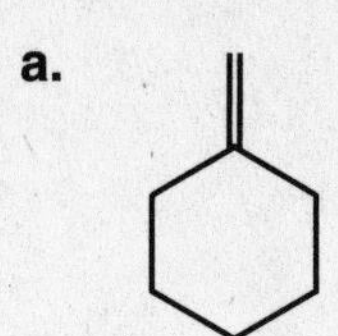

b.

Student response: In the first case, the favored product would result from the addition of hydrogen to the primary carbon on the end of the double bond to create a tertiary carbocation, which is nice and stable, and the tertiary iodine derivative. Similarly, the second reaction example will favor the creation of the tertiary carbocation and the tertiary iodine derivative.

Instructor response: By George, I think we've got it.

Electrophilic Aromatic Substitution Reactions

CHAPTER EIGHT

Benzene is not your ordinary garden variety organic molecule. Recall the structure of the benzene ring. The electrons in the pi bonds of benzene are delocalized in an orbital that spreads around the entire benzene ring. The unique delocalization of electrons in this circular pi bond provides benzene with an unusual degree of stability: the more delocalized electrons are, the lower the energy state and the more stable the molecule will be. The benzene ring does not like to change. Anything that disturbs the delocalized electrons in the aromatic ring would require the molecule to occupy a much higher energy state, and that's no good.

As you might expect, then, addition reactions that occur readily in alkenes are not observed in aromatic rings, since they would disrupt the stable aromatic structure. So what do aromatic rings do? They must do something if chemists love them so much. Noble gases tend not to be so popular in chemist circles, except perhaps for the challenge. Aromatic compounds do have a favorite activity and its name is **electrophilic aromatic substitution**. The electron-rich pi bonds of the aromatic ring attract electron-hungry electrophiles, like sugar attracts flies. What can the electrophile do once it reaches the aromatic ring? The electrophile manages to react with the ring and yet maintain the ring's aromaticity—not by undergoing addition, but rather by performing substitution. It simply boots off a hydrogen from the ring and merrily takes its place.

An example would be the reaction of benzene with bromine and a catalyst ($FeBr_3$). Catalysts in reaction equations are usually written above the arrow to indicate that they are neither reactants nor products. In the course of the reaction, the aromatic ring is preserved, with the replacement of one hydrogen by a bromine atom.

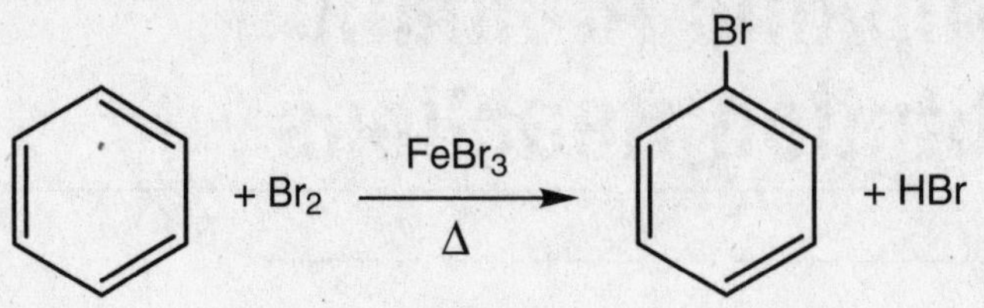

Br

FeBr$_3$

+ Br_2 $\xrightarrow[\Delta]{FeBr_3}$ + HBr

How does this all work? The electrophilic aromatic substitution reaction (EAS to its friends) takes place in two steps. When the mechanism is broken down, we see that beneath the guise of something brand-new are some reactions we have met before: first an addition, and then an elimination. In the first step, an electrophile attacks, adding itself on to one carbon in the aromatic ring. The electrophile draws electrons away from the ring in the process to form a sigma bond, and breaks the aromatic nature of the ring. The electrophile is represented by E^+ in the reaction diagram, with the arrow indicating the direction of electron movement to the electrophile to form a bond.

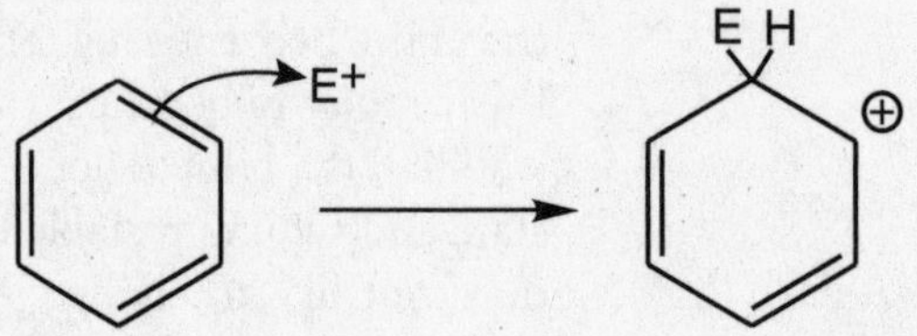

E^+

E H

⊕

What else can we deduce about the product of the first step in the EAS reaction? As in other addition reactions, the action of the electrophile leaves a positive charge on the adjacent carbon since it drew away electrons to form the new covalent bond.

> **Student question:** Hey, wait a second. Where did that cute little hydrogen come from?
>
> **Instructor response:** Actually that hydrogen in the product is not new – it was there all along in the reactant ring, but it wasn't drawn since that is the convention in the drawing of benzene rings.

This product of the first step is a reaction intermediate. Is it aromatic? Several resonance structures can be drawn and the positive charge is stabilized by delocalization provided by the resonance structures.

But we still have not answered the question—is the product of the first step, the reaction intermediate, aromatic? It turns out no. The carbon with the electrophile added on must be sp^3 hybridized to have four substituents, so the ring is not aromatic. If the product of the first step is not aromatic, what can we predict about the relative energies of the reactants and the reaction intermediate? The intermediate has lost the aromatic character of its electrons and must be higher energy, less stable. We can use this information to postulate the first section of a reaction diagram.

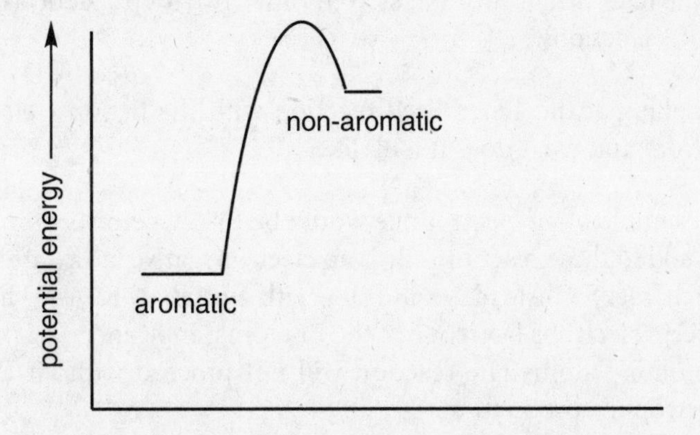

If the reaction stopped here, then it would pretty much be like your ordinary run-of-the-mill addition. The end-point of the first step is not a tremendously stable molecule, however. It wants to get back to its aromatic nature. To get back, it is willing and able to spit out the hydrogen bound next the electrophile. This is the second step of the EAS reaction, which resembles an elimination reaction. Electrons move from the covalent bond with the hydrogen back to the ring to reform the aromatic ring. With the second step, the EAS reaction is complete, the aromatic ring is restored and the people rejoice.

If the reaction is spontaneous, then the final product must be at a lower energy than the original reactants. The activation energy of the first step and the second step are labeled. The reaction profile of the overall reaction must look something like:

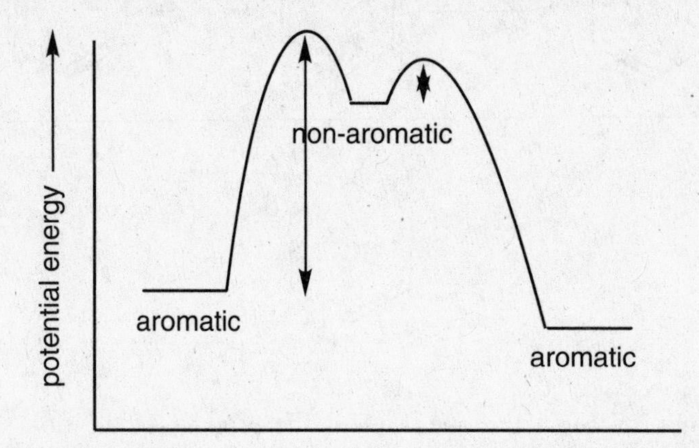

What will be the rate-limiting step in this reaction? One way to answer this question is by examining the reaction profile. The first step has a much higher activation energy than the second step so it will be much slower than the second step and therefore rate-limiting. Another way to answer the question is to remember that the first step involves the formation of a carbocation intermediate, a rate-limiting step in other two-step reactions that we have encountered: S_N1 and E1 reactions.

So far we have been looking at the generalized reaction with the fictional electrophile E^+. Where does E^+ come from and what does it look like?

Let's get specific. One example of an electrophile would be Br+. Chemists do not just have a bottle of Br+ that they add to these reactions. Br is an electronegative atom, after all; it would rather have a negative charge. So instead we add Br_2 with a catalyst, $FeBr_3$. The catalyst is a Lewis acid, acting to accept electrons from one of the Br atoms, and generate a partial positive charge on the other bromine atom. The reaction will not proceed without the catalyst to increase the rate of electrophile production.

$$\text{Br—Br:} + \text{FeBr}_3 \longrightarrow \overset{\delta+}{\text{Br}}\text{- - - -}\text{Br—}\overset{\delta-}{\text{Fe}}\text{Br}_3$$

Once the electrophile is generated (Br^+), then the first and second reaction steps can proceed. At the end of the second step, the released H^+ binds to one of the bromines attached to the catalyst (which has one too many bonds at this point) and thus recreates the original catalyst, $FeBr_3$. Recall that if the catalyst is not regenerated, it is not a true catalyst, since by definition catalysts are not consumed by the reaction.

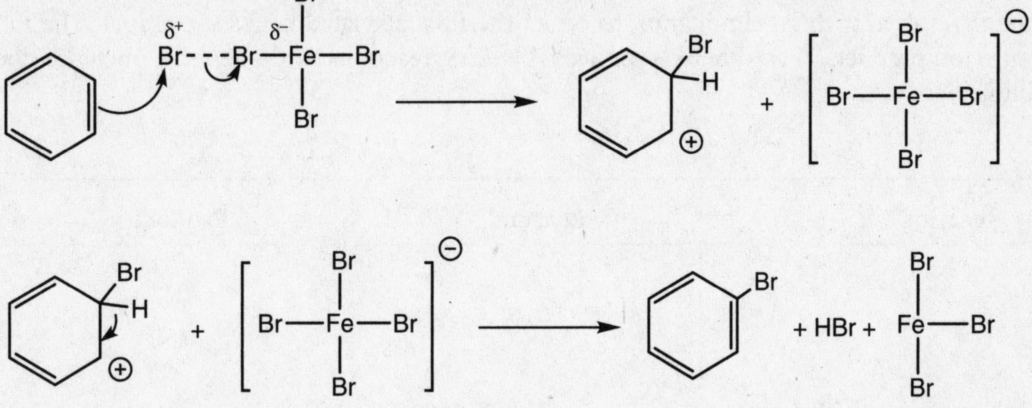

Adding a halogen to the benzene ring (halogenation) is only one type of EAS reaction. What else can you do with this nifty new reaction tool? What if we could substitute the benzene ring with an alkyl chain? An EAS reaction mechanism in which alkyl chains are substituted onto benzene rings is called the **Friedel-Crafts** reaction. Take for example the prototypical alkyl chain R that we want to add to a benzene ring. The reaction in this case could be written as:

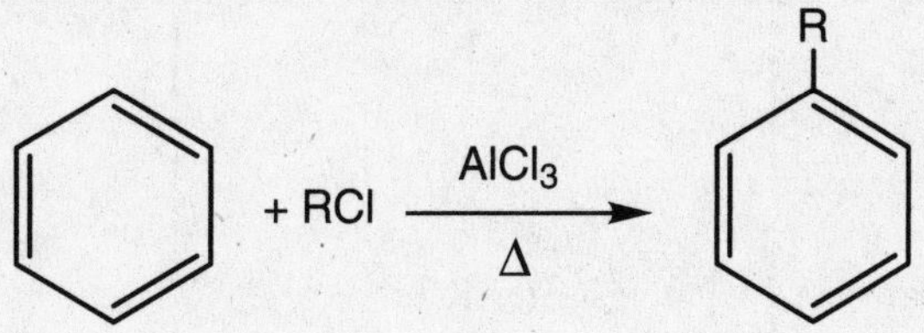

What would the mechanism be if the reactant used was 2-chloropropane? First, we must figure out the electrophile that is generated by the Lewis acid catalyst $AlCl_3$. The Lewis acid accepts electrons from the chlorine of the 2-chloropropane, to form $AlCl_4$- and a secondary propane carbocation. Where is the electrophile? The carbocation is the electrophile, and now it is ready to take on the task of attacking benzene to steal away some of the aromatic electrons.

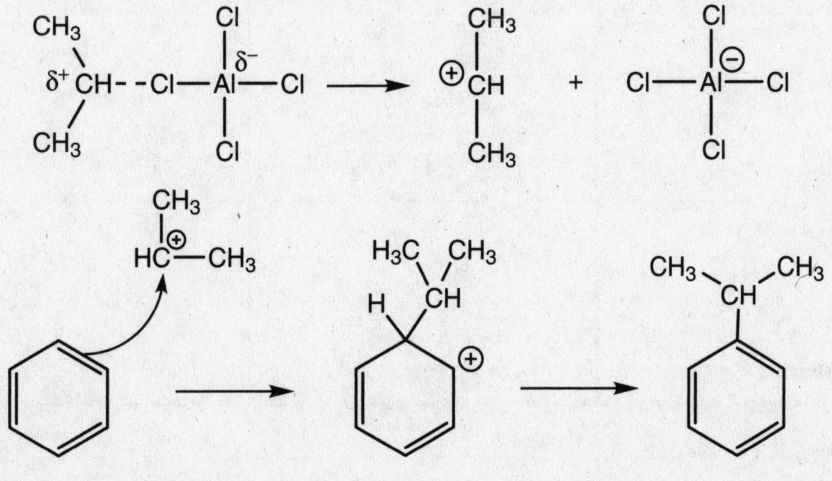

The carbocation adds to the benzene ring, creating a positively charged reaction intermediate that is resolved with an elimination to create the final alkylated benzene product. The full range of products that can be produced by EAS reactions with benzene includes the following:

Reaction	Reagent	Product
bromination	Br_2, $FeBr_3$	Br
chlorination	Cl_2, $FeCl_3$	Cl
Friedel-Crafts alkylation	RCl, $AlCl_3$	R
Friedel-Crafts acylation	O ‖C RC— l, $AlCl_3$	O C R
nitration	HNO_3, H_2SO_4	NO_2
sulfonation	SO_3, H_2SO_4	SO_3H

POLYSUBSTITUTED AROMATIC RINGS

With this handy-dandy EAS reaction tool you can brominate, you can chlorinate, and you can alkylate. But wait – there's more! Why stop with one group? Why not put some more groups on your aromatic rings? Friends, the EAS reaction is there for you once again, at your service to create polysubstituted aromatic rings.

All of the EAS reaction products presented so far have a single functional group substituted onto the benzene ring. The aromatic ring can and does however go on to add more substituents through more rounds of EAS reactions. First, a quick mention of the nomenclature of the positions of the substituents. The relative positions of aromatic rings with two substituents are indicated with the prefixes para, meta, and ortho. Two groups in the para position are on opposite sides of the ring from each other, two groups that are ortho are next to each other, and meta is in the middle between para and ortho.

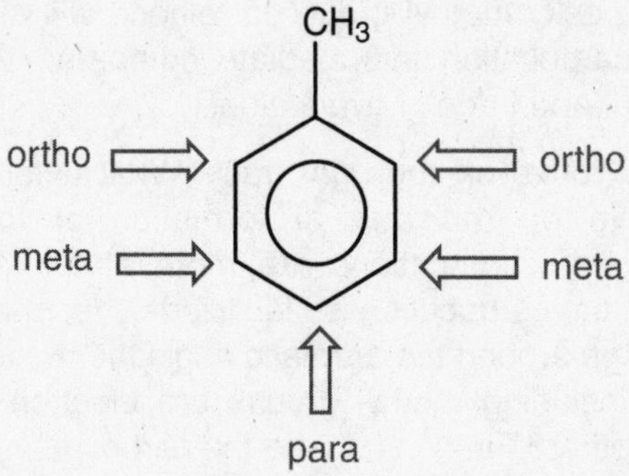

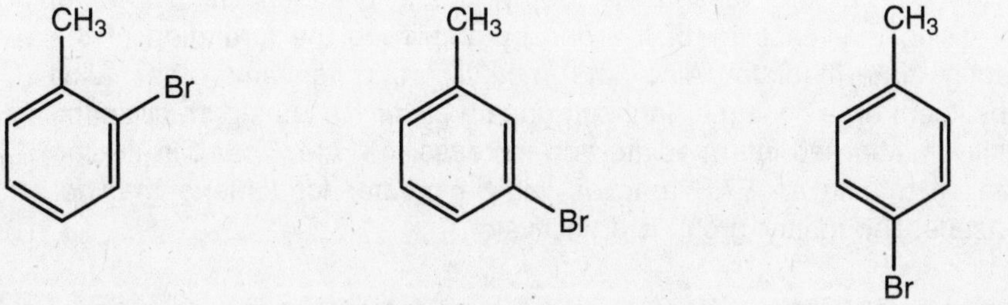

o-bromotoluene (1-bromo-2-methylbenzene)

m-bromotoluene (1-bromo-3-methylbenzene)

p-bromotoluene (1-bromo-4-methylbenzene)

Adding on second or third groups through more EAS reactions does indeed occur, but it does not occur randomly. Once a ring has one substituent, the position of this group determines how easily the ring will be able to add a second group at all, and in what position. The reason for this is the ability of the ring to form carbocations at other locations in the ring or in the ring as a whole. Groups that increase the ring's reaction rate are called **activating** and groups that decrease the reaction rate are called **deactivating**.

Instructor question: Let's look at how substituents affect the substitution of more groups on the ring. Toluene is the common name for benzene substituted with a methyl group. Would benzene or toluene react more quickly in a Friedel-Crafts alkylation reaction?

Student response: Great. Now I don't remember what the heck a Friedel-Crafts reaction is.

Instructor response: No problem. It's an EAS reaction to add an alkyl group onto the benzene ring.

Student response: Right. Knew that. Okay, if we want to know how the rates of reactions compare, then it seems to me we need to look at the rate-determining step, which is the reaction of the electrophile with the aromatic ring to form the carbocation intermediate. Anything that favors the formation of the carbocation will increase the reaction rate and anything that does not will slow the reaction rate. The question then boils down to how that pesky methyl group on toluene will affect the rate of formation of the carbocation intermediate compared to the rate this occurs in benzene...and I don't have a clue.

Instructor response: You're on the right track. What difference could the methyl group have on the rate of formation of the carbocation intermediate? Tertiary carbocations are more stable than secondary carbocations. Also, the carbocation is stabilized by resonance movement of the positive charge around the aromatic ring. One more piece of info is that alkyl groups including methyl groups are electron donating. How does this all fit together? First, in toluene the carbocation will be partially stabilized by the ability to form a tertiary carbocation, which benzene cannot do. For this reason the methyl group in toluene is *activating*, increasing the reactivity of the ring by stabilizing the formation of the reaction intermediate. Also, the methyl group, or any other alkyl substituent on a benzene ring, can donate electrons to help stabilize the positively charged intermediate and increase the rate. So, all in all, the reaction rate in an EAS reaction will be greater for toluene than for benzene. The methyl group is an activator.

Can a group influence the rate of EAS for a second substituent in a way that is specific to different sites around the ring (ortho, meta, para)? How would we figure this one out? Think resonance. The reaction intermediate in an EAS reaction is stabilized by resonance forms. Some resonance forms may be more stable than others depending on where the positive charge in the resonance form of the intermediate falls.

If toluene is reacted in EAS conditions with Br, will the ortho, meta, or para position be favored? The best and only way to see this is to sketch out the resonance forms that would be involved in the reaction of each of these, then figure out if any of the resonance forms are particularly good (stabilized), or particularly bad (de-stabilized).

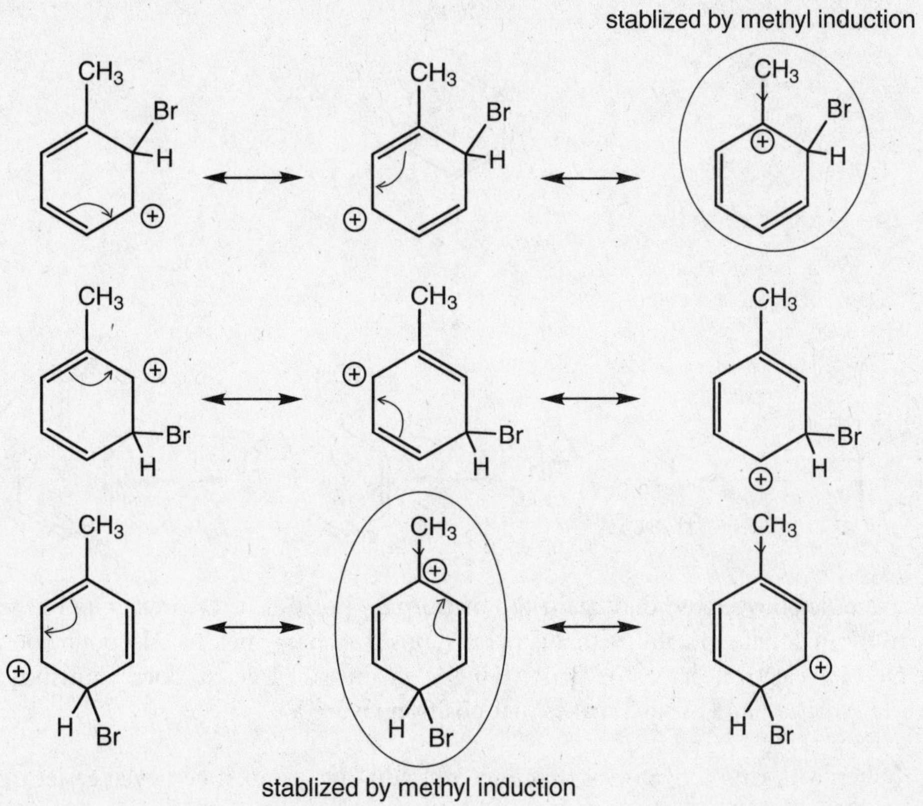

stablized by methyl induction

stablized by methyl induction

For the ortho and the para positions, at least one of the resonance forms of the reaction intermediate places the positive charge of the carbocation adjacent to the methyl group. This resonance form will be stabilized both because it is tertiary and by induction from the methyl group. This does not occur for the bromine in the meta position; the meta position isn't bad, per se, but it just doesn't result in any particularly favored resonance forms. The ortho and para substituted compounds have a particularly stable intermediate so an EAS reaction with toluene will produce more of the products in the ortho and para positions than in the meta position.

The way this is described is to say that the methyl group "directs" further substitution to the ortho and para positions. Since it is also an activator, putting it all together makes the methyl group an ortho/para-directing activator.

What other types of groups might also be ortho/para-directing activators? Hydroxyl groups on a benzene ring, such as in phenol, are strong ortho/para-directing groups. This might be surprising since oxygen is very electronegative. You might expect oxygen to withdraw electrons from the benzene ring and destabilize the carbocation intermediate. What happens instead is that the non-bonding lone electrons from oxygen contribute to the resonance stabilization of the intermediate. The resonance stabilization is drawn for the ortho substitution, but the para position would produce the same effect. A substituent at the meta position does not allow the same resonance contribution, so the hydroxyl group directs substitution to the ortho/para positions and not to the meta position.

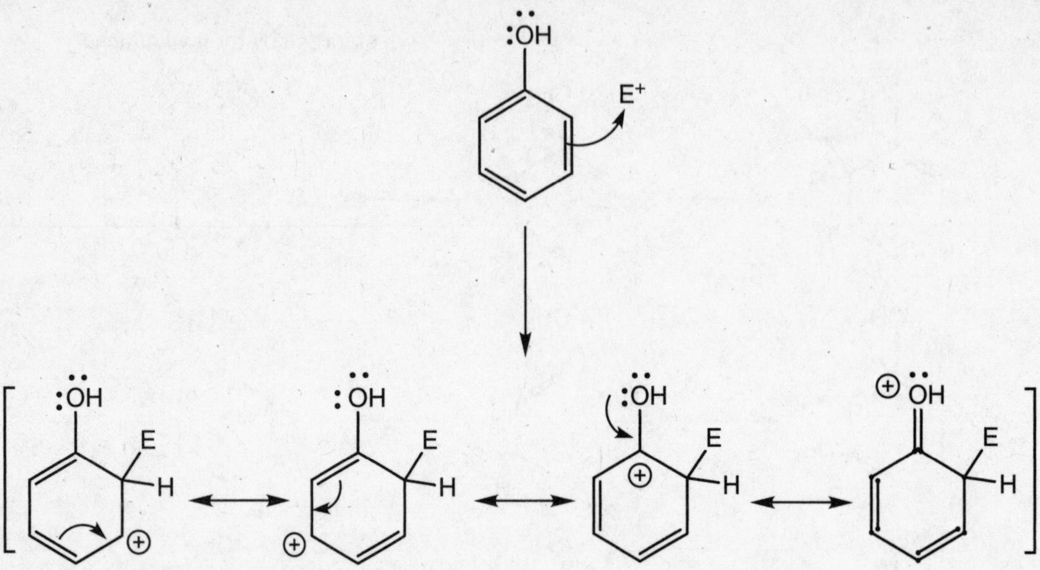

In the example above, note that again the ortho/para position is favored. The oxygen has electrons it can donate to stabilize that generated positive charge nearby. Although you might cringe at that electronegative oxygen having a positive charge, it does contribute as a resonance structure and as such makes that position favored.

One problem with ring activators is that once you add one group, it gets even easier to add a second group. (Bet you can't add just one.) If you are doing a Friedel-Crafts alkylation reaction starting with benzene as a reactant, once you add one alkyl group to the ring, that methyl activates the addition of more methyl groups on the ring. Rapidly you will end up with some of the rings having more than one methyl group. It is difficult to control the reaction in a way that limits addition to just one substituent. Reactions that are messy, with lots of different products, or reactions that are difficult to control, are not things that organic chemists like to deal with.

META-DIRECTING DEACTIVATORS

What other type of effects can groups have on the substitution of benzene derivatives? Let's have a look at the nitro group. One of the features of the nitro group is that it has its own resonance forms, and another feature is that it has a positive charge.

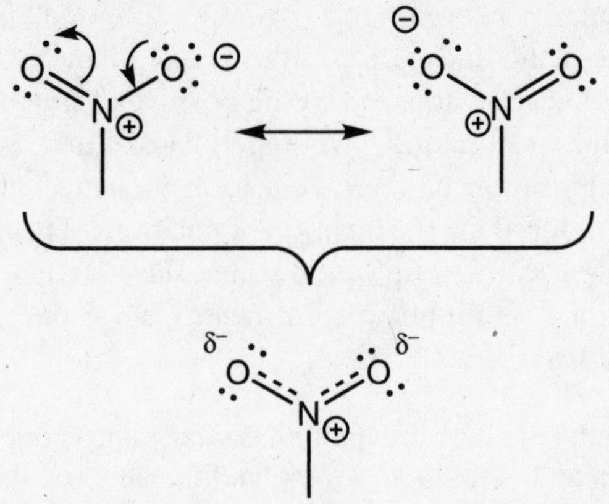

The positive charge on the attached nitro group means that the group will in general be deactivating for further substitution on the aromatic ring. How else might the nitro group affect further substitution through EAS reactions? Would a nitro group be a meta or ortho/para director? To answer this question, it is necessary once again to examine the possible resonance forms generated by substitution at different sites in the molecule:

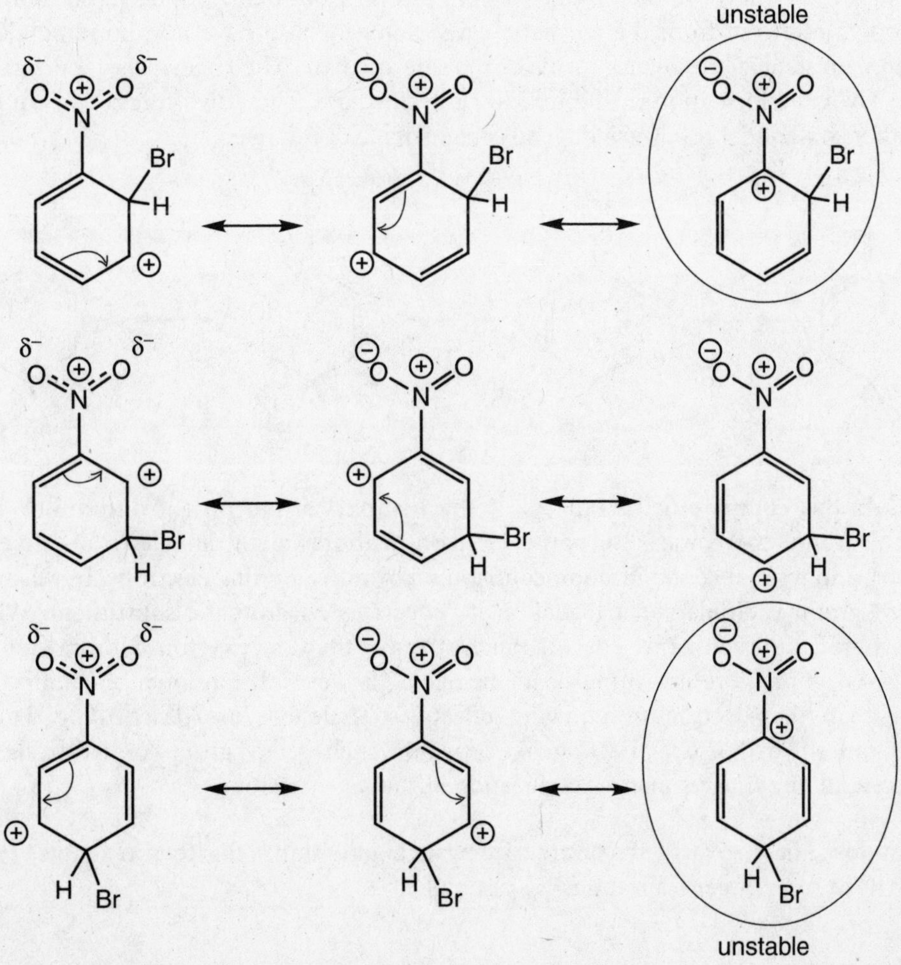

unstable

unstable

Here, we see that particularly bad resonance structures abound, as opposed to particularly good ones before. In the ortho and para positions, one of the resonance forms places the positive charge of the carbocation adjacent to the positive charge of the nitro group. This is not good. When it comes to charges, like repels like. This resonance form would be unfavorable, and the stabilization of the carbocation intermediate provided by resonance delocalization would be reduced for the ortho/para positions. The meta substituent does not share this trait, although the nitro group would still be deactivating. The end result is that the nitro group is deactivating overall, but meta-directing since the meta position is the least unfavorable, the lesser of two evils.

> **Instructor comment:** One nice thing about deactivating groups is that if you are doing a reaction to create an aromatic ring with a deactivating group, then once you add one group, it is harder to add a second. If one group is what you want, it should be possible to find conditions that provide this. A nice controllable reaction, fit to bring a smile to a chemist's face.

Warning! It is not the case that all deactivators are meta-directors. For example, there are certainly ortho/para-directing deactivators. What type of character might play this role? Halogens! What are the characteristics of halogens that might be relevant? Halogens are electronegative – they love their electrons, and pull them in tightly. In doing so they will be electron-withdrawing from the aromatic ring. Halogens also have non-bonding electrons that can help stabilize resonance forms of the intermediate. The resonance stabilization will tend to favor the formation of the reaction intermediate. These two forces run counter to each other so the final result will depend on which affect is larger.

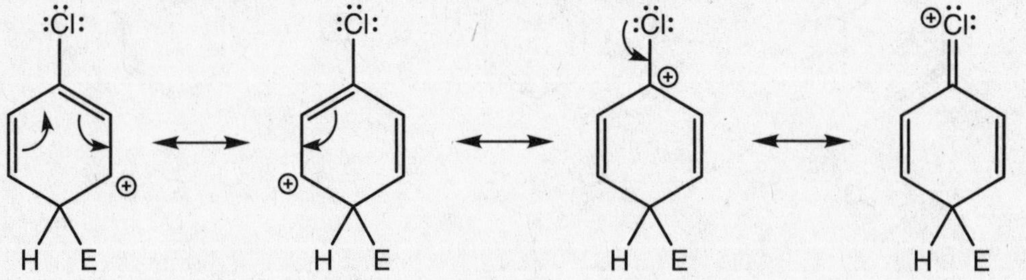

:Cl: ⟷ :Cl: ⟷ :Cl: ⟷ ⊕Cl:

H E H E H E H E

How does this compare to the affects of the hydroxyl group on substitution? With the hydroxyl group, there was also some electron withdrawal by an electronegative atom (oxygen), and resonance stabilization contributed by non-bonding electrons. In this way the hydroxyl group is similar to halogens in its effects as an aromatic substituent. With the hydroxyl group, the resonance effect is much stronger than the electron-withdrawing effect, so hydroxyl groups are activating. With halogens, however, the resonance stabilization is smaller than the electron-withdrawing effects, so halogens are *deactivating*. However, halogens are still ortho/para directing, like activators such as alkyl groups or hydroxyl groups, since they still provide resonance stabilization at the same positions.

So what do we have so far? Let's summarize by arranging things together in a table. Then we can see what sorts of generalizations we can make.

Ortho and para directors		Meta directors
Activators	Deactivators	Deactivators
$-NR_2$	-F	$-NO_2$
-NHC(=O)R	-Cl	$-CF_3$
-OH	-Br	-COOH
-OR	-I	-COOR
-R		-COR
$-C_6H_5$		-CHO
		$-SO_3H$
		-C≡N

The table includes some of the groups we have gone over here, as well as some additional groups. By understanding how each class works, it should be possible to predict which class a group will fall into *without* memorizing the table. Ortho/para directors are compounds that donate electrons to the aromatic ring, either by resonance or by induction. Some ortho/para directors are activating while others are deactivating. The difference depends on how strong the electron-withdrawing effects of the groups are compared to the resonance stabilization provided. Groups with stronger electron-withdrawing effects are deactivating, while groups with stronger resonance stabilization are activating.

The groups in the ortho/para-directing activators that we've discussed include alkyl groups and hydroxyl groups. We have not discussed amines so far, but they are included in this category as well. Can we figure out why? Amines have an electronegative nitrogen atom that would tend to be deactivating. On the other hand, this nitrogen also has a lone pair of electrons that contributes to resonance stabilization. As with the oxygen in the hydroxyl group, the resonance effects are stronger than the electron-withdrawing effects, making amines activating groups. The ortho or para positions would provide an additional resonance form not found with an amine in the meta position, as with the hydroxyl group, making amines ortho/para directors in addition to being activators. In fact, amines are especially strong activators; given the chance, they will add substituents to every possible position. This makes selectivity extremely difficult.

The only class of ortho/para directing deactivating groups is halogens, alone in this class due to their unique combination of resonance and induction effects.

In the class of meta-directors, all of the groups are deactivating. Meta-directors are always electron withdrawing. The groups in this class all have a positively charged atom directly bound to the aromatic ring or at the positive side of a polar covalent bond. The nitro group has a positively charged atom bound to the aromatic ring, and so it will draw electrons away from the ring and destabilize the carbocation intermediates. Most of the other groups in this class have a carbon as the atom connected immediately next to the aromatic ring. The carbon atom does not have a formal positive charge in any of these groups, but it does have a partial charge as a member of a polar covalent bond. For example, in -CF_3, the carbon is bound to three electronegative atoms that tend to draw charge away from the carbon, giving it a partial positive charge. Similarly, in groups with this carbon bound to oxygen in various forms, the oxygen withdraws electrons from the carbon and makes the group electron-withdrawing for aromatic rings.

Instructor question: It's problem time, so fasten your seatbelts. Predict for each of the following the major product resulting from EAS reaction:

CH_3 (toluene) $\xrightarrow{HNO_3,\ H_2SO_4}$

Cl (chlorobenzene) $\xrightarrow{CH_3\overset{O}{\overset{\|}{C}}Cl,\ AlCl_3}$

OCH_3, Cl (2-chloroanisole) $\xrightarrow{Br_2,\ FeBr_3}$

Student response: Okay, in reaction (a), we know the reaction will add a nitro group on to the aromatic ring, but where? That methyl group on the ring is activating, and like all activating groups it will be ortho/para directing. If the reaction is limited to the addition of one more group, then the major products will be o-nitrotoluene and p-nitrotoluene. The result of the reaction would be different if we started out with a nitro group on the ring which we then alkylated. Nitro groups are meta-directing, so in that case we would produce m-nitrotoluene instead.

Instructor response: Nicely done. As you noted, even though a nitro group on the ring would be deactivating, it still can add a group to the ring, which students sometimes find hard to grasp. Now, how about reaction (b)?

Student response: We start out in this case with chlorobenzene. Chlorine is a halogen, one of the small elect crew of deactivating ortho/para directors. The product, regardless of the identity of the reagent being added, will come out mostly as ortho- and para- substituted, known in this case as o-chloroacetophenone and p-chloroacetophenone.

Instructor response: You're smoking! Okay, there is time for one more, this time starting with a di-substituted molecule to kick things up a notch.

Student response: Who have we here? Ah, chlorine, an ortho/para-directing deactivator. Chlorine is accompanied by –OCH3, an ortho/para activating group. I'll try to sketch out the sites the two different groups will tend to direct substitution toward:

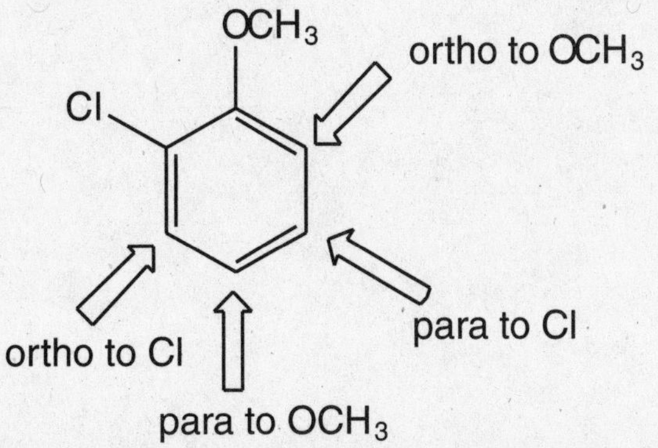

OCH_3 Cl ortho to OCH_3 para to Cl ortho to Cl para to OCH_3

Instructor response: Good sketch. As you can see, there is a bit of a conflict here, with the two different groups apparently working against each other. Who will win the struggle? How can we work this thing out? There is a way! The rate of reactions directed by the activating ether group will be FASTER than the reactions directed by the deactivating group. The reaction race goes to the swift, so the products will be mostly those directed by the activating ether group.

OCH_3 Cl Br OCH_3 Cl Br

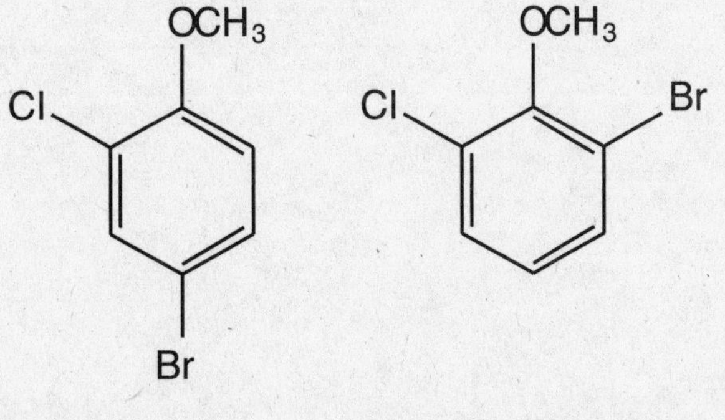

EAS reactions provide an important and versatile tool to manipulate the infamous aromatic ring, bending it to the chemist's will. They also round out the preliminary survey of some basic reaction mechanisms covered in the last three chapters.

The full reaction toolkit available to the modern organic chemist is far larger than the reaction mechanisms presented here. However, these mechanisms do provide a glimpse at common factors that are found in many other reactions.

Next, on to a few practical notes about how this all works in the laboratory.

Orgo Lab

CHAPTER NINE

All of the material that we have covered has a practical outcome. One outcome might be to do well on exams, but another outcome would be to learn the foundations required to enter the laboratory in the future and synthesize the compounds of your dreams: perhaps a new medicine, perhaps a new plastic, perhaps a superconducting nano-computing fullerene tube. So once you have studied all about atoms, bonds, and reactions, what's next? Time to hit the lab. One rite of passage is usually a laboratory class that accompanies lecture. This is a great opportunity to get your hands on some molecules and see how the material we have covered really acts. We are not making this stuff up (for the most part).

Unfortunately, as you may have noticed, this is just a book, so we won't actually be able to get our hands on anything but paper, but we can at least get a glimpse of how organic chemistry is conducted in the laboratory. And by the way, even if you don't plan on becoming a chemist, this stuff does appear on the MCAT so it's worth knowing.

SEPARATION AND PURIFICATION TECHNIQUES

Even the best chemist will seldom achieve 100% conversion of reactant to product or 100% of only the intended product. When a chemist is going through a synthesis to make a compound, she will often need to concentrate and purify a compound, to separate it from solvent, from reactants, and from by-products that creep into the reaction. What's a chemist to do? The techniques chemists use to separate molecules from each other include extraction, distillation, recrystallization, filtration and chromatography.

EXTRACTION

One thing that a chemist can do to separate substances is exploit their differences in physical properties. One such property is the solubility of molecules in different solvents. The physical properties of molecules affect how well a solvent can dissolve them in solution. What does it mean for a molecule to be in solution? A molecule in solution (a solute) is surrounded by and interacts with solvent molecules through intermolecular interactions. If a molecule is not in solution but in a crystalline lattice, it is surrounded by copies of itself that it interacts with. If a molecule is to leave the crystal and enter solution, then the interactions with solvent must be more favorable than interactions with its own kind.

Once a molecule is in solution, the next challenge is for it to stay there. If conditions in the solution begin to favor interactions between solute molecules, then they can begin to recrystallize, and come back out of solution.

The interactions of solutes with solvents follow a simple rule: like dissolves like. Why does sodium chloride dissolve well in water? Water is very polar, with a strong dipole moment, so it is a good solvent for both positive and negative ions, as well as many other polar molecules with their own dipole moments. The interactions of water with the ions are strong enough that they can favor the interaction with water rather than of sodium and chloride with each other in a crystal. Another example is the solubility of hydrophobic molecules like alkanes in water. Alkanes have almost no dipole moment, and interact with each other almost exclusively as a result of Van der Waals forces. There is not much in an alkane molecule that a water molecule can get a grip on since alkanes are so greasy. If you try mixing an alkane like hexane with water, it will not dissolve, but will instead form a separate layer. An alkane like octane mixed with hexane will readily dissolve, since the molecules can undergo non-polar interactions. Two liquids that are not soluble in each other, that do not mix together no matter how much you shake and stir, are called **immiscible**.

Let's say we are synthesizing something in the lab, and at the end of the reaction we have two products, one with a strong dipole moment and the other non-polar and hydrophobic. How can we separate them? What if we prepare two immiscible solvents, one polar and the other non-polar, along with the products? If there is enough of a difference in the solubility of the products, then after shaking up the mixture the products will each equilibrate into the solvent that dissolves it the best. The non-polar product will end up in the non-polar solvent and the polar product will end up in the polar solvent. This process is called an **extraction**.

The two immiscible solvents used in an extraction are called **phases**. The two phases will usually include an aqueous phase (water) and an organic phase composed of hydrophobic molecules. An extraction will often be conducted in a handy piece of glassware called a **separatory funnel**. The organic layer is usually the upper layer floating on top in these systems, since hydrocarbons are less dense than water. If a reaction is conducted in water, then when the reaction is complete, the reaction mixture can be poured into the top of a separatory funnel. The organic phase can be poured in the top as well, and then the flask can be sealed with a stopper. A valve at the bottom prevents liquids from leaking out. If the flask is shaken, then the hydrophobic molecules will partition into the organic phase. After the phases are allowed to separate, then the aqueous phase can be removed from the flask by opening the valve and letting it pour out the bottom, leaving the organic phase, and its solutes, behind. If the products are sufficiently hydrophobic, then during the extraction they will pass from the aqueous layer into the organic layer. At the end of the extraction, the organic layer can be removed and processed to further purify the desired compounds.

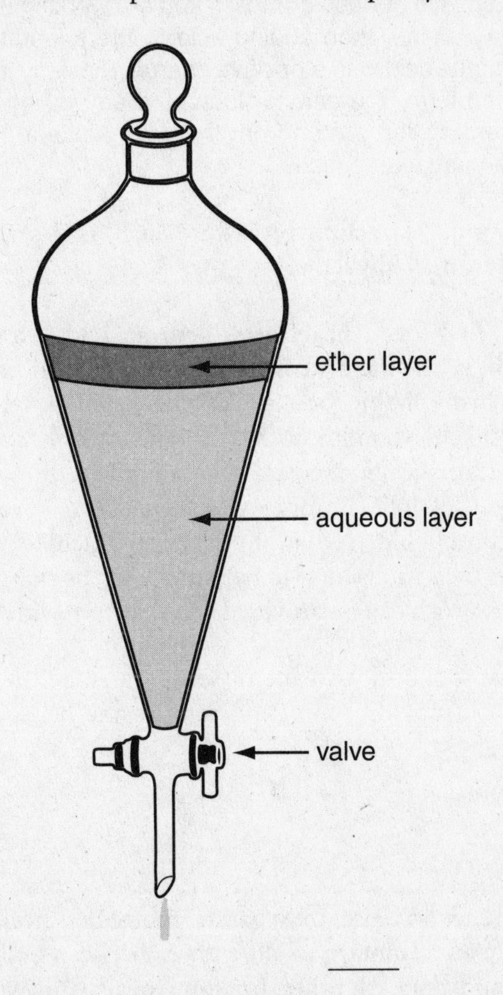

One useful trick to use in an extraction is to manipulate the charge on molecules to make them either more or less soluble in the organic phase.

> **Instructor question:** Will charged molecules be more or less soluble in the organic phase of an extraction?
>
> **Student response:** Charged molecules should be much less soluble in an organic phase, since the charge interacts well with water but not with hydrophobic groups. And the opposite should be true for uncharged molecules.

Instructor response: You're right. Another question: some molecules have charges that you can manipulate based on pH. Can you think of what type of molecules would change their charge based on pH?

Student response: Acids and bases?

Instructor response: Sure! The great contributors to charges in organic molecules are often acid or base groups. If you are working with a neutral base like a tertiary amine, then adding acid to the aqueous phase will produce the conjugate acid with a positive charge. Similarly if you starting with the protonated form of a carboxylic acid, then making the solution more basic will remove the proton from the acid, make it charged, and more soluble in an aqueous phase.

Practically speaking, using an extraction procedure such as described above, how could benzoic acid be extracted from diethyl ether?

To answer this question, first meet the players. Benzoic acid is an aromatic ring with a carboxylic acid side chain. When the acid is protonated, the acid group is neutral and the molecule will be fairly hydrophobic because of the aromatic ring. When the acid is deprotonated, such as in alkaline conditions, it will be negatively charged, and much more polar. Charged groups dramatically increase the solubility of molecules in water and dramatically decrease their solubility in organic solvents. The solvent in this case, diethyl ether, is non-polar and hydrophobic. If you want the benzoic acid to leave the organic solvent to enter an aqueous phase, then you want it to be more polar, or charged. This would happen best in alkaline conditions, which can be provided by extraction with basic aqueous solution.

The final answer: Dissolve sodium hydroxide in some water, add it to a separatory funnel along with the benzoic acid mixture, mix, and remove the benzoic acid in the aqueous mixture.

DISTILLATION

Another technique that chemists use to separate molecules from each other is called **distillation.** Distillation takes advantage of differences between boiling points of different molecules to separate them from each other. In a previous chapter the physical factors that affect the boiling point were discussed. If you have a solution with two different molecules and you start to heat up the solution, the molecule with the lower boiling point will start to boil at a lower temperature than the second molecule. If the vapor from the solution is collected as you heat the mixture, then your vapor will be enriched with the molecule with the lower boiling point.

An apparatus that is used to do a distillation might look like this:

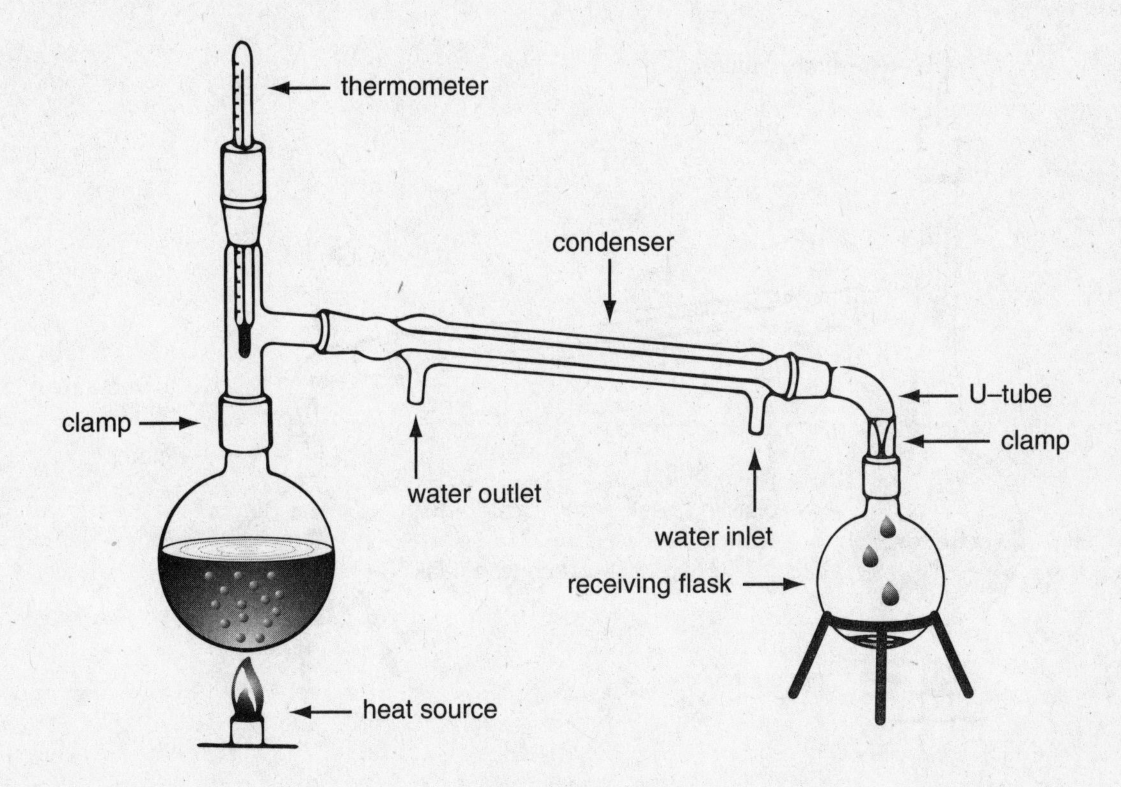

The bubbling liquid in the diagram is the original mixture of molecules in solution. Since you don't want to lose the molecules that are boiling off in the vapor phase, the vapor is forced to pass through a piece of glassware that has cold water running around the outside of the tube. As the vapor cools, it condenses back into liquid that drips down to collect in a second flask.

This procedure is a simple distillation, and works best for molecules that have a significant difference in boiling points. Even then, it is not a perfect separation, since you will always get at least a little of the other molecules in the receiving flask. The greater the difference in boiling points, the better the separation.

One variant on distillations can be used for molecules that have very high boiling points. If the boiling point is so high that you cannot get a molecule to boil, then a distillation will not work. The solution? If you lower the pressure, then molecules will boil at a lower temperature. Water boils at a lower temperature in the mountains, for example. This method is called a vacuum distillation, and the apparatus is shown below.

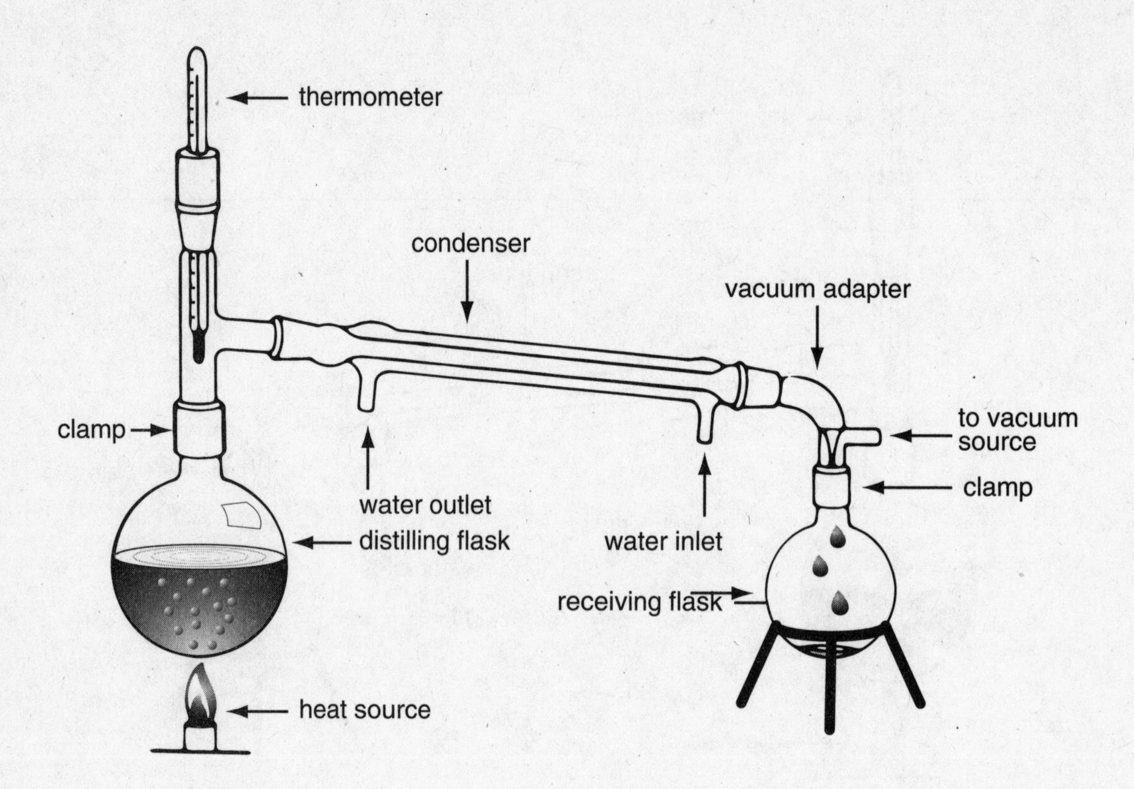

Fractional distillations are used in cases where boiling points are close together and a simple distillation does not achieve adequate separation. If we took a mixture and performed a simple distillation, going from a 50/50 mixture to a distillate that is enriched in the desired molecule 60% (from 50%), leaving 40% impurity, how could we separate the mixture further? Taking the 60/40 mixture and doing another round of distillation would be one way of doing it. If you do enough rounds of simple distillation, you will eventually improve the purity enough to get what you need.

In fractional distillations, many different fractionations occur in a single piece of equipment, allowing separation of molecules with relatively close boiling points. The material in the column allows the molecules to bind to the column then enter the vapor phase, then bind, vapor, bind, vapor, and so on, all the way up the column. The more rounds of switching back and forth, the more "distillations" to occur, the greater the separation will be between molecules as the leave the top of the column (see diagram below).

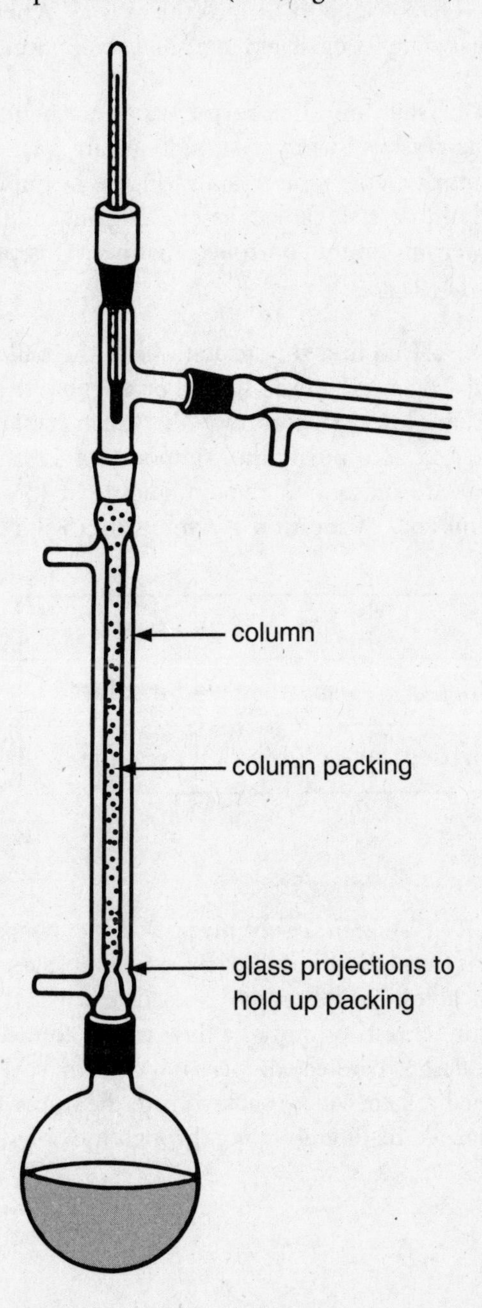

RECRYSTALLIZATION

Crystals are formed as a regular lattice matrix of a compound or chemical. Almost anything with a regular and fairly stable structure can form a crystal in the right conditions, ranging from sodium chloride to organic compounds, proteins and even viruses. Crystals are usually formed from only a single substance. The way that molecules pack into the crystal lattice will only allow that one type of molecule to fit into the crystal. When a crystal of a substance forms, therefore, it is usually highly enriched in a single compound.

When doing a synthesis, crystals are often formed, and are highly enriched in the desired substance. However, if the crystals form very rapidly or are formed from a solution with a high concentration of impurities, then the crystals will also be impure. To improve the purity of the compound desired, the crystals can be dissolved again, and the compound crystallized again, this time in a slower and more controlled manner. This process is used to generate high-quality, high-purity crystals.

The crystals are usually dissolved in a solvent with heat, then allowed to recrystallize via a controlled rate of cooling. The solvent used should be one that dissolves the compound at a high temperature, but not at a low temperature so that it can crystallize out of solution. Since the goal with recrystallization is to purify the compound, then the conditions used should allow any impurities that are present to remain soluble at low temperatures. When the compound of interest comes out of solution, the impurities will remain behind in solution.

	hot	**cold**
product	soluble	insoluble
impurities	soluble	soluble

FILTRATION

When you have bunch of crystals in a solvent, how do you separate the solids from the liquids? One way to do it is through filtration. A filtration apparatus has a funnel on top, with a flat bottom with small holes. A filter paper is used to cover the bottom of this funnel, covering the holes. The funnel rests on top of a flask with a connection to a vacuum line on the side. When the lower flask is connected, there is a vacuum seal with the upper flask, and when the vacuum is turned on, air will be sucked from the upper flask down into the lower flask, drawing the substance to be filtered down through the holes.

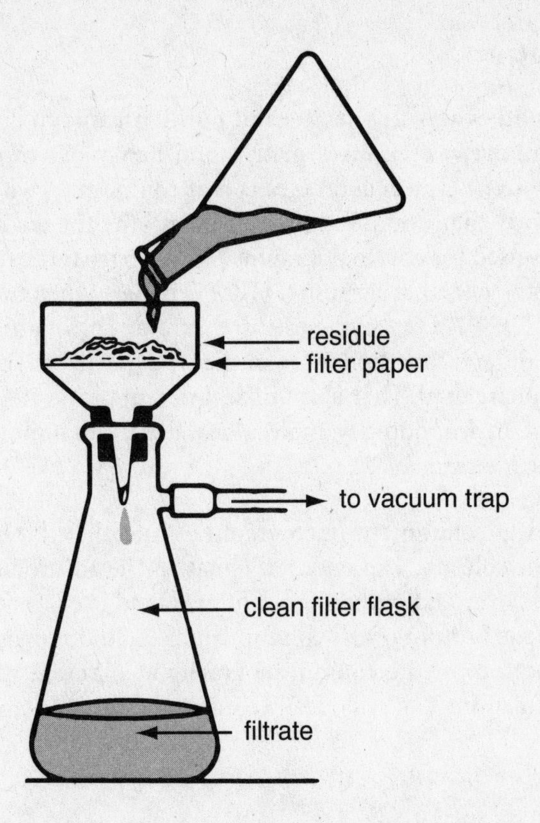

To do a filtration, the mixture of crystals and solvent are poured into the funnel, and the liquid is sucked by vacuum down through the filter paper and holes into the lower flask, leaving the crystals on the filter paper.

> **Instructor question:** Can you think of an instance when vacuum filtration might not be necessary?
>
> **Student response:** I guess for filtrations that are easy, like sand from water. That should be pretty simple to filter.
>
> **Instructor response:** Exactly. In cases like that one, a vacuum line is not necessary, and gravity alone can be used to drain the solvent from the crystals. We call this simply gravity filtration. Though this can be effective, letting gravity take its course is slower and sometimes provides less efficient drying of the crystals involved.

Usually the solid that is collected would be the desired product, leaving impurities in solution to discard. However, it is possible to crystallize impurities, leaving the product in solution, filter out the impurities, and then do another crystallization to collect the product.

CHROMATOGRAPHY

Chromatography is another way to separate and purify organic molecules from each other. Compounds are passed over a stationary matrix, and they move through the matrix along with a mobile phase solvent. The general idea is that compounds with more affinity for the matrix move more slowly than compounds with less affinity for the matrix. There is a great variety of technologies used for chromatography. The primary types of chromatography that we use include gas chromatography (GC), paper chromatography, thin-layer chromatography (TLC), HPLC (high-pressure liquid chromatography) and column chromatography. The different methods vary in the type of solvent (mobile phase) used and the type of stationary phase used. They also utilize different properties including electrostatic charge, dipole moment, hydrophobicity, molecular weight, and boiling point. However, they all operate under the same principle.

The figure below depicts column chromatography. Material with the solute compound is poured in the top of the column. The stationary phase is the adsorbent, or the solid material that the solute will pass over and perhaps bind to. It is held in place in the column, often by sand. The stopcock at the bottom of the column controls fluid movement. The liquids drip, or **elute**, out of the bottom of the column by gravity at different rates. They can then be collected in different fractions that will ideally contain different compounds.

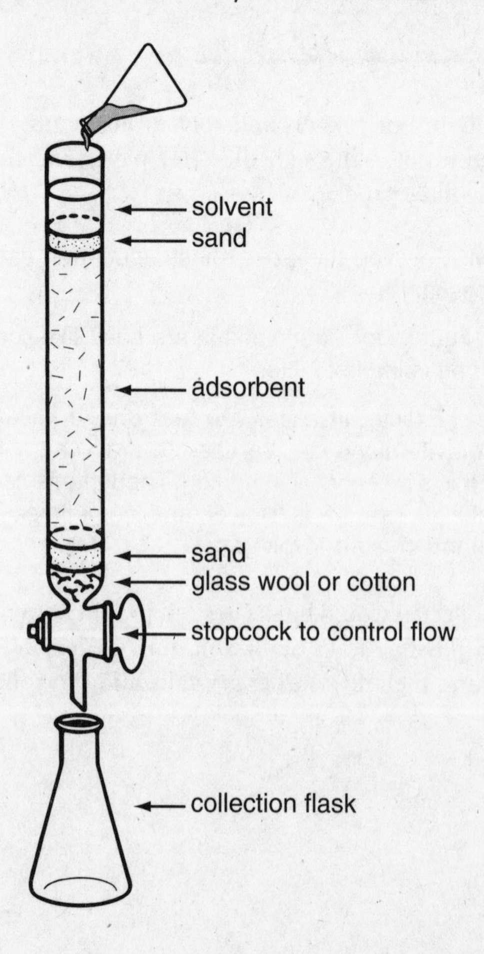

As the solvent moves through the stationary phase, a compound will at times bind to the stationary phase and at times remain in the mobile phase. The more a molecule favors binding to the stationary phase, the less time it will spend in solution, and the less it will move with the mobile phase. The slow-moving compounds in chromatography therefore have higher affinity for the stationary phase, and the fast moving compounds that come out of the column quicker have more affinity for the solvent. The fastest that a compound can pass through the column is the amount of time it takes for solvent to pass through, unimpeded. The longest a compound can take is, well, forever. Sometimes material will have so much affinity for the stationary phase that you basically can't get it out. That is not usually the goal, since this will mess up your column.

Instructor question: Polar materials like silica are often used as the stationary phase in column chromatography. If you have the following three compounds in solution in ether and load them onto a silica column chromatography system using diethyl ether as the mobile phase, which compound will elute first from the column?

a. 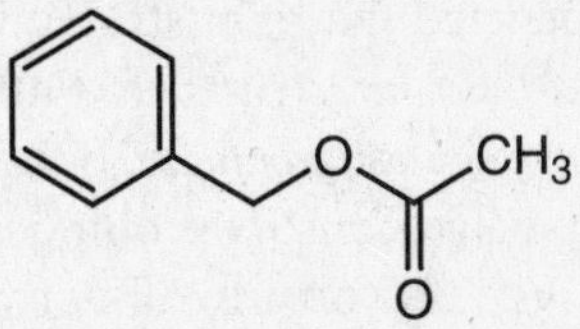 O, CH_3, O

benzyl acetate

b. 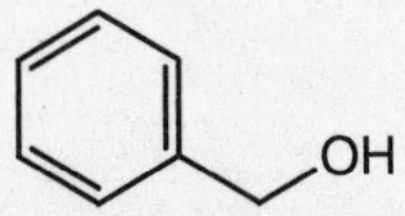 OH

benzyl alcohol
(phenylmethanol)

c.

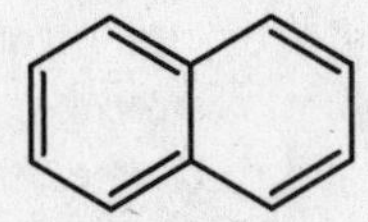

naphthalene

Student response: Well, the mobile phase is ether, which we said is fairly non-polar and hydrophobic. The stationary phase is silica, a polar solid material. Compounds will be separated in this system on the basis of their polarity. Things with a more hydrophobic nature will favor solution in the non-polar mobile phase solvent, ether, and will move quickly through the column. Hydrophilic compounds will bind more to the matrix and will move more slowly through the column, taking longer to elute. Compounds a and b are both relatively polar compared to compound c. Compound c, naphthalene, is composed of two hydrophobic aromatic rings and will take the cake for being the most hydrophobic member of the trio. Therefore it will probably spend its time associated with the solvent, and will pass most rapidly through the column compared to the other compounds.

SPECTROSCOPY TECHNIQUES

Once a chemist has slaved away in the lab for weeks to make the compound of her dreams, how does she know if she has concocted the right structure or not? She needs a way to look at the compound, to probe its molecular structure. Putting on super-power magnifying goggles here is not going to work. What she can do, however, is to shine different types of electromagnetic energy on a sample, like sensors probing molecular space. The way that the compounds absorb and emit energy tells you a lot about what the molecule is made of and how it is put together.

We will touch on two ways of analyzing molecular structure. The first, infrared spectroscopy, is useful for looking at functional groups, indicating what types of chemical bonds are present in a molecule. The second, proton NMR, is useful in seeing what atoms are present and how these atoms are connected.

INFRARED SPECTROSCOPY

Infrared energy involves photons that are just below the wavelength of visible light. Infrared energy is usually associated with heat. To perform infrared spectroscopy (called IR), infrared light is passed through a sample containing a chemical of interest, and the energy coming out the other side is measured. The wavelength of the infrared light can be varied across a spectrum described as the wavenumber. The wavenumber is the number of light waves that fit into one centimeter. The wavenumber used in IR spectroscopy can vary from 4000 to 400 cm-1. What end of this spectrum is higher in frequency? The higher the wavenumber (more waves/cm), the higher the frequency (waves/second).

As infrared energy passes through the sample, some energy will be absorbed by the compound in solution. Why does the sample absorb energy? The frequencies used in infrared spectroscopy interact with chemical bonds to induce stretching and bending of chemical bonds. How does this tell you anything about a specific molecule? Different types of chemical bonds absorb infrared light at different wavelengths.

Functional group/Bond		Frequency range (cm^{-1})
Alcohols (O–H stretch)		3200–3650
Carbonyl compounds (C=O stretch)		1650–1750
Alkenes	(C=C stretch)	1620–1680
	(C–H stretch)	3020–3150
Alkynes	(C≡C stretch)	2100–2260
	(C–H stretch)	3260–3330
Ethers (C–O stretch)		1000–1260
Nitriles (C≡N stretch)		2220–2260
Amines (N–H stretch)		3250–3500
Aromatics (C–H stretch)		2900–3100

In scanning a chemical sample across a range of frequencies, a spectrum is observed. The data is expressed in the spectrum as % transmittance, the inverse of absorbance. 100% transmittance means that no light was absorbed, and 0% transmittance indicates that all of the light was absorbed by the sample. Since the absorbance is highly characteristic to specific functional groups from one molecule to the next, chemists can use a table like the one above to check their spectrum and determine the most likely groups that corresponds to each peak (see example below).

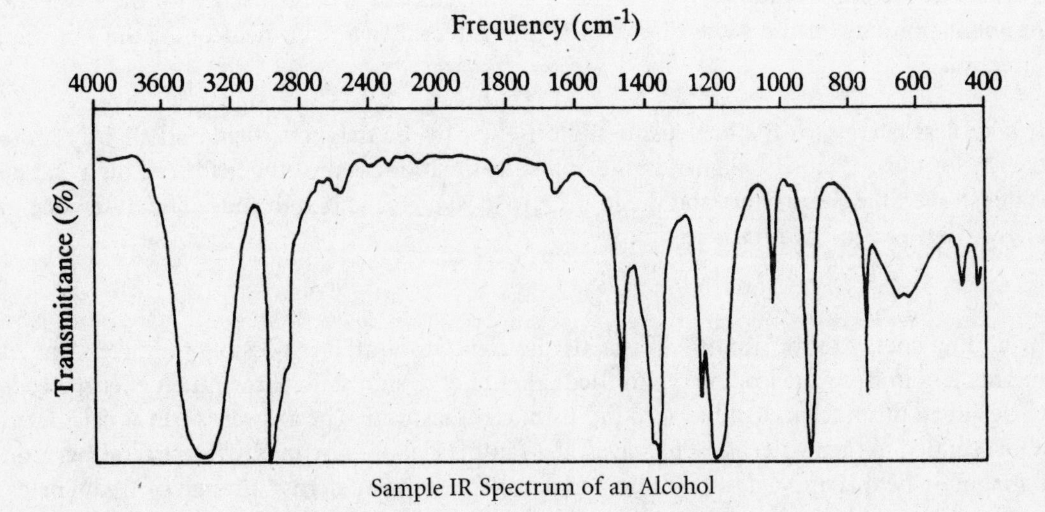

Sample IR Spectrum of an Alcohol

From looking at an IR spectrum, one might learn what types of bonds, and therefore what functional groups, are present. But the spectrum does not tell you how a molecule is put together. For this, other techniques such as NMR are required.

PROTON NUCLEAR MAGNETIC RESONANCE (NMR) SPECTROSCOPY

Now for a look at a very different type of spectroscopy, NMR. It all starts with spinning nuclei. It may not matter most of the time, but nuclei of some atoms spin. The nucleus, chock full of protons, is a charged object, and when a charged object moves, (like when spinning) it generates a magnetic field. Not all nuclei spin, but fortunately some very useful nuclei in organic chemistry do have spin including the isotopes ^{1}H, the most abundant hydrogen isotope and ^{13}C. We will in this book only discuss ^{1}H-NMR, but chemists do use other NMR isotopes for additional information about structures.

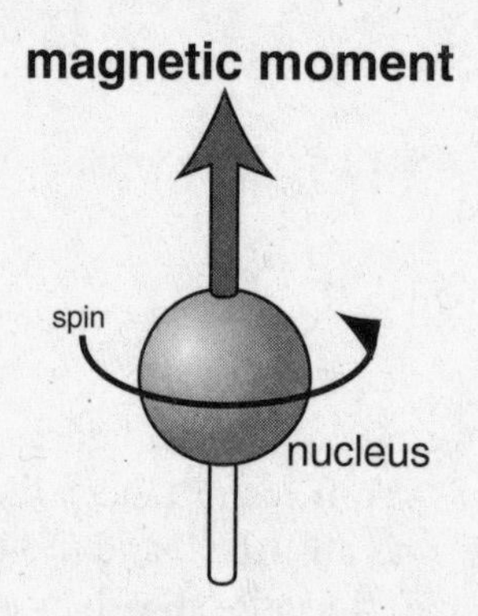

Since the ^{1}H nucleus, a proton, is a tiny spinning magnet, it has a magnetic moment that points in a specific direction. Normally nuclei are oriented randomly in space. If the ^{1}H nucleus is placed in a strong magnetic field, the field will influence the arrangement of the nucleus with respect to the magnetic field. The nucleus can arrange itself with the magnetic moment pointing in the same direction as the external magnetic field or in the opposite direction.

It takes less energy for the nucleus to align itself with the magnetic field and slightly more energy for the nuclei to be aligned in the opposite direction, against the field. The high energy state is called the β spin state and the lower energy state is the α spin state. The difference in energy between the two states is:

$$\Delta E = E_{\beta} - E_{\alpha}$$

By adding energy to the mixture in exactly the right amount, it is possible to make some of the nuclei flip from the low energy to the higher energy spin state. How much energy needs to be added to a nucleus to allow it to flip its nucleus around? The answer is ΔE, as calculated above, added to the nuclei as radio waves. If a sample is placed in an NMR test chamber, and a magnetic field is applied, then radio waves can be supplied across a range of frequencies, until energy is absorbed by the sample at the right radio frequency for nuclei to flip from low to high energy.

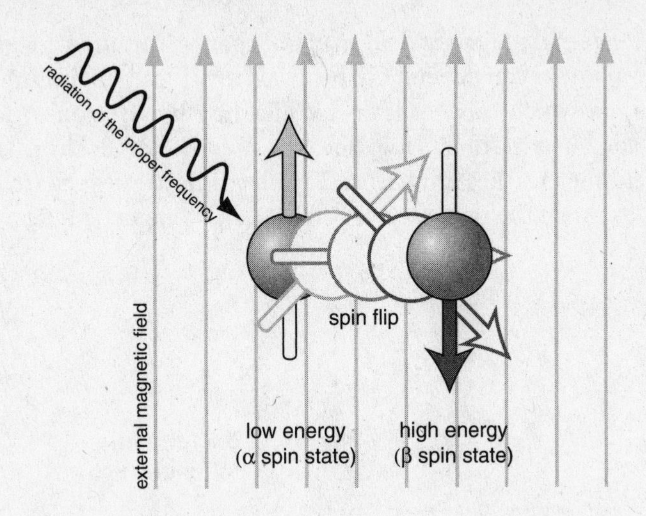

The amount of energy that is required to make nuclei flip depends on how strong the magnetic field is. The stronger the magnetic field, the greater the amount of energy that it takes for nuclei to flip from the low energy to the high energy orientation.

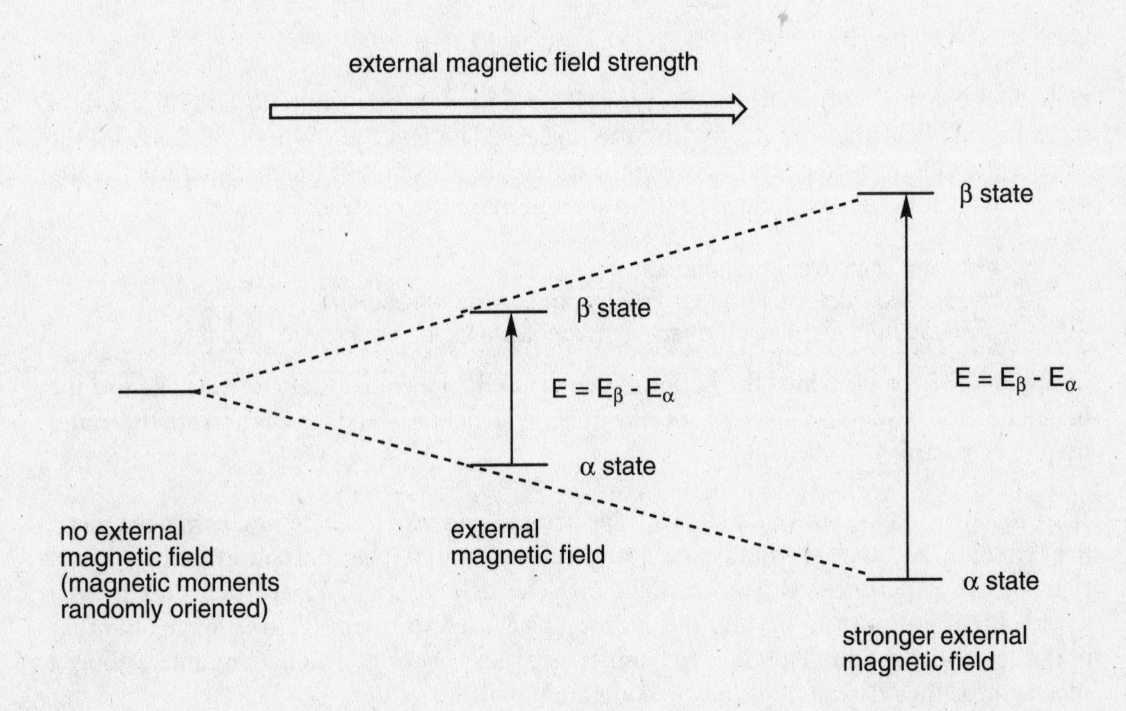

The strength of the external magnetic field may not change, but the strength of the field that is observed by a given nuclei can depend on the local (very local) environment of the nucleus. One important way in which this occurs is by shielding by electrons. Most nuclei are not naked, but surrounded by electrons. How does the presence of electrons affect the strength of the magnetic field that reaches a nucleus? The electrons are also charged, and they move in the field, acting to shield the nucleus from the external magnetic field.

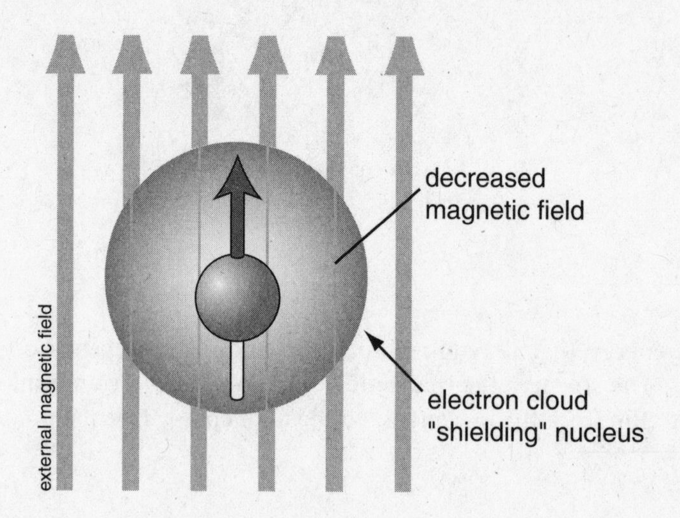

How does the shield of electrons affect ^{1}H NMR? The electrons will reduce the strength of the magnetic field that is perceived by the nucleus. The less the strength of the field, the smaller the difference in energy between the different spin states of the nucleus, and the less the energy that needs to be added to the system to make the nucleus resonate.

> **Instructor comment**: In a nutshell...
> *deshielded nucleus:* stronger field, more energy to resonate
> *shielded nucleus:* weaker effective magnetic field, less energy to resonate

In doing NMR experiments, the radio frequency used to stimulate is held constant and the magnetic field strength is varied, or the magnetic field is held constant and the radio frequency is varied.

Time to have a look at our very first NMR spectrum. The x-axis indicates increasing magnetic field strength, and the y-axis indicates absorbance of energy. The compound farthest to the right in this experiment is TMS, a standard. The protons in TMS are the most heavily shielded that protons can be, and the TMS point is used to mark "0" to compare all other peaks to in the graph. This helps chemists get their bearings when they are reading a spectrum, so they have an idea how to compare signals.

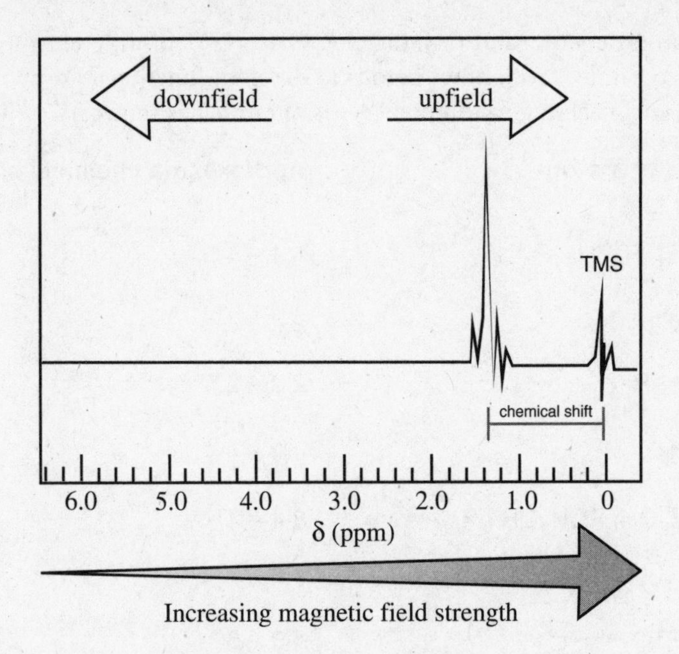

The distance from the TMS peak to any other signal that is observed is called the **chemical shift**. The peak for ethane in this figure has a chemical shift of 1.25 ppm (parts per million). Different types of protons give signals in different places, depending on the shielding of each different type of nuclei in a molecule. Nuclei with signals on the right side of the spectra, close to TMS, are said to be **upfield**, and signals on the left side are **downfield**. The more shielded the nucleus is, the more upfield the signal will be, and the less shielded a nuclei is, the more downfield the signal will be. You can think of it as going alphabetically; the "d" (downfield) is to the left and "u" (upfield) is to the right.

Instructor question: In the molecule shown below, will proton a or proton b appear more upfield?

```
     Ha    Hb
     |     |
  H—C————C—Cl
     |     |
     H     H
```

Student response: Um, not sure. How do you tell them apart? How does shielding come into play?

Instructor response: Think about whether the two protons are the same, or in other words, if they have the same environment. In this case, these protons are not equivalent. Proton b is adjacent to a chlorine, while proton a is one carbon removed from the chlorine. Will the chlorine affect the chemical shift of proton b? The chlorine is electronegative, sucking electrons toward itself from the neighborhood. In so doing, it will draw electrons away from proton b and leave proton b less shielded. Proton a may have a slight inductive affect caused by chlorine as well, but proton b is closer so the effect will be larger. Proton b, being more unshielded, will be more downfield and proton a, being more shielded, will be further upfield.

Through time and experience, and running NMR on a LOT of different molecules, chemists have become pretty familiar with how protons in different functional groups appear in NMR. The approximate chemical shift for different types of protons is summarized in the table below.

Type of proton	Approximate chemical shift (δ)
RCH_3, R_2CH_2, R_3CH	~ 1 – 1.5
RC≡CH	2–3
$ArCH_2R$	2.2 – 2.5
RCH_2X	~ 3 – 4
RCH_2OH, RCH_2OR	3.4 – 4
C=C–H (vinylic H)	4.6 – 6
ArH	6 – 9.5
–C(=O)–H	9.5 – 9.9

Looking at an NMR spectrum with this table in hand, we can tell what types of functional groups are in a molecule. The same trick can be accomplished with IR spectra though. What's so special about NMR? There is more to come still. NMR spectra can also tell you how many different types of protons there are, and how many of each type of proton are present. The chemical shifts in the table are approximate, and the specific chemical shift of each proton depends on the local neighborhood.

Looking at chloroethane again, how many different peaks will be observed in the spectrum of chloroethane and what will they look like?

To know how many peaks will be observed in an NMR spectrum, we need to observe how many different types of protons are present in a molecule. In chloroethane, the three protons further from the chlorine are all the same. With bond rotation between the two carbons, there is no way that these three protons can be distinguished – they are equivalent. Since they are equivalent, they will each give the same chemical shift and their signals will pile on top of each other to make one large signal. Similarly, the two hydrogen nuclei next to the chlorine are equivalent to each other, and distinct from the other protons since they are next to the chlorine. They will also sum up to give one large peak since they are both the same. In total, there will be two different signals from chloroethane.

What will the two different signals look like? This depends on how many of each different proton contribute to each signal. The proton signals pile on top of each other, so there is a mathematical relationship between the signal size and the number of equivalent protons that contribute to that signal. If there are three equivalent protons then the signal will be three times the size of a signal from one proton. The way that this is measured is to measure the area underneath the peak that forms an NMR signal by integrating the area under the peak. In the case of chloroethane, there are three protons contributing to one signal and two contributing to the other, so the size of the peaks will be in the ratio 3:2 between the two peaks. No need for calculus here; we usually just eyeball the peaks.

SPLITTING

One more factor that makes NMR a rich source of information about the structure of molecules is called **splitting.** Sometimes, a larger NMR peak is split into several smaller peaks that are located very close to each other in the spectrum. Splitting of the signal occurs when non-equivalent protons are located nearby on an adjacent carbon. The cluster of smaller peaks is a **multiplet**. If there is no splitting, then a peak remains as one peak and can be called a **singlet**. If the large peak is split into two peaks, it's called a **doublet**, three peaks makes a **triplet** and so on. Splitting occurs because the flipping of one proton in the magnetic field can affect the local magnetic field of other nearby nuclei. The number of peaks caused by splitting into a multiplet is equal to the number of hydrogens on the nearby carbon plus one. Are you thoroughly confused? Good.

Some examples:

In the first example, let's look at the signal from three protons where there is no adjacent hydrogen. In this case, the signal will be 3X the size of a single proton, and will be in one peak. There is no splitting because there are no adjacent, non-equivalent protons to affect them.

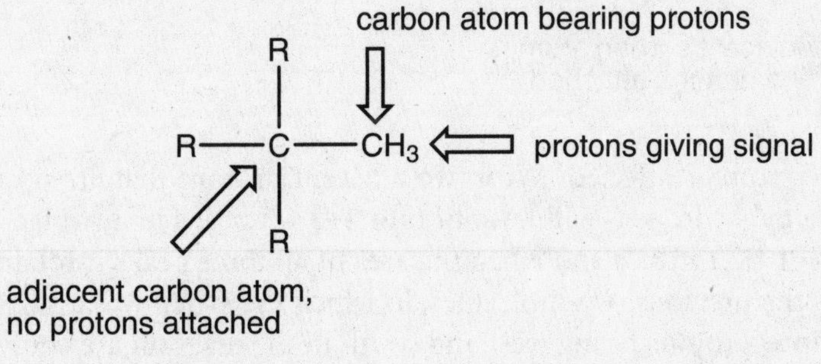

carbon atom bearing protons

$R_3C{-}CH_3$

protons giving signal

adjacent carbon atom, no protons attached

Now, what happens if we change the molecule so that there are only two R groups on the left hand carbon rather than three?

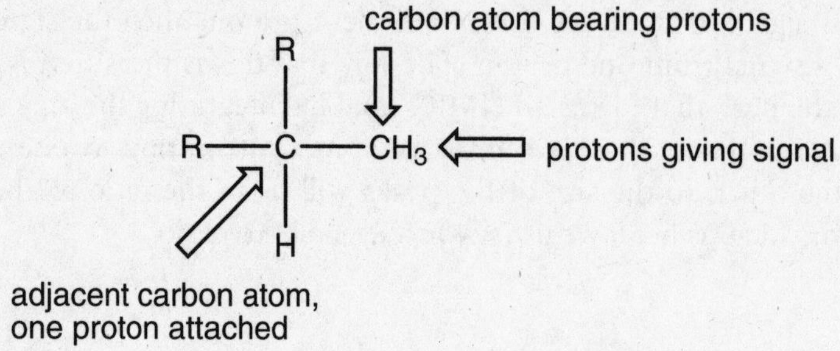

carbon atom bearing protons

R—C(R)(H)—CH_3 ⟸ protons giving signal

adjacent carbon atom, one proton attached

Now there is a hydrogen on the carbon adjacent to the methyl group. What will happen to the spectrum? The signal from the three methyl protons will still be three protons in size. With one hydrogen adjacent, the signal will be split however, from one peak into two peaks (# peaks = # adjacent protons + 1). The two peaks (doublet) will be equal in size, and if the area under the two peaks is totaled it will equal the area under the peak for the first molecule in which no splitting occurred. Similarly, the peak for the lone hydrogen will actually be split into four peaks, because that hydrogen is next to three hydrogen (the methyl).

Finally, how about taking away one more R group so that now there are two adjacent, non-equivalent protons.

carbon atom bearing protons

H—C(R)(H)—CH_3 ⟸ protons giving signal

adjacent carbon atom, two protons attached

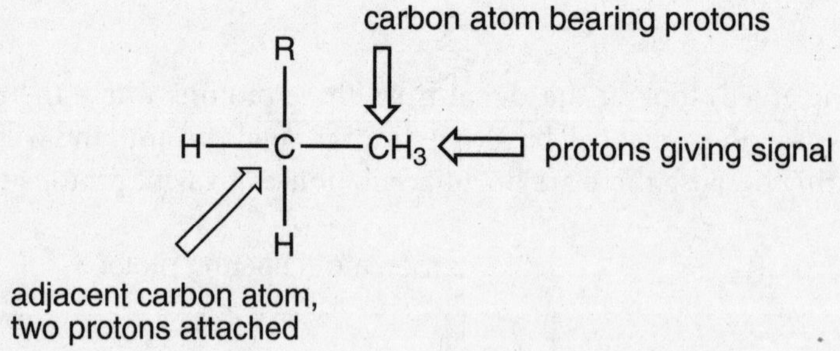

How will the spectrum be affected? With two adjacent protons that are non-equivalent, the signal for the methyl hydrogens will be split into 2+1= 3 peaks in a triplet. The three peaks will be in a ratio of 1:2:1 in size and when the area of all three peaks is totaled, it will sum to the same area as the previous two molecules, in which the signal was a singlet or a doublet. Meanwhile, the lone protons facing west and south in the diagram are equivalent; they each have three neighbors (the methyl) and so will be represented as four peaks.

All together now, let's go back to chloroethane. What have we predicted so far about the spectrum of this molecule? It should have two different signals, from each of the two different types of non-equivalent protons, and that the ratio of area in the two different signals should be 3:2. The signal from the two protons next to the chlorine is more downfield than the signal from the other three protons. Let's see how we did:

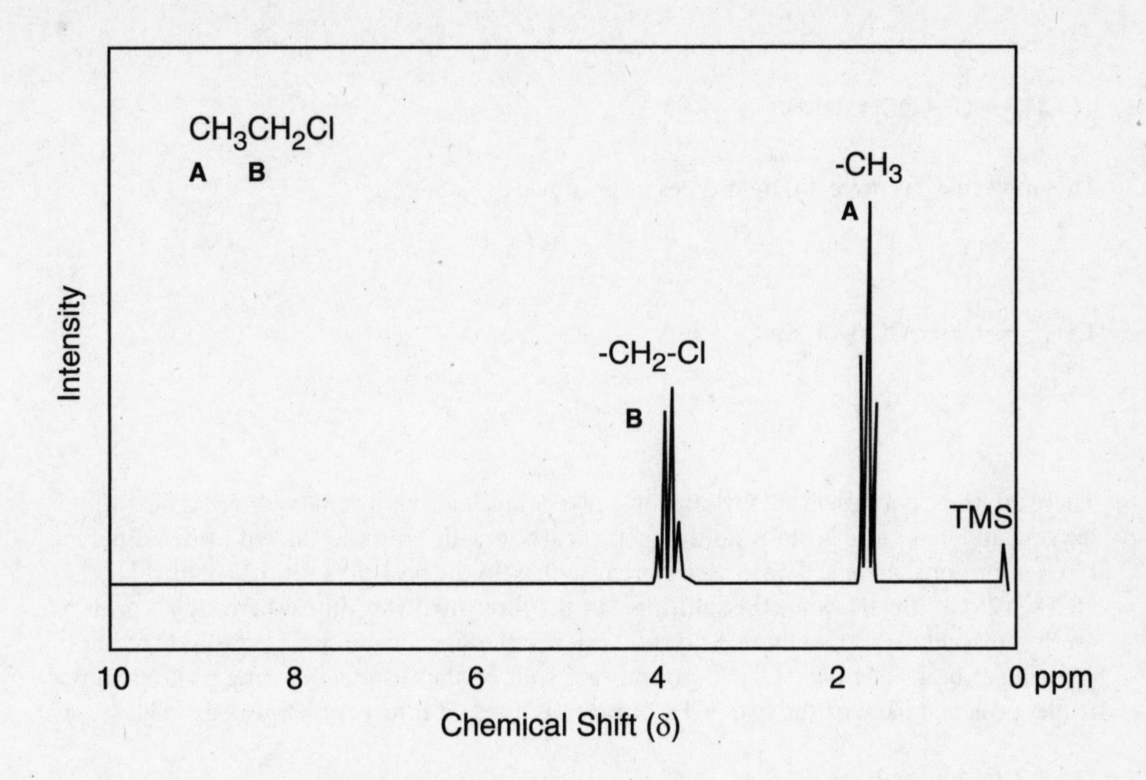

As we noted above, what we observe is two different signals split into multiplets. The downfield proton signal is split into 4 peaks, a quartet, by the three neighboring hydrogens, and the upfield signal is split into three peaks by the two neighboring hydrogens.

Got that? We knew you could do it. Let's try some more.

a) cyclopropane

We recommend drawing these molecules out to get a good look at how many and what type of equivalent protons there are in each of these examples. Cyclopropane has six hydrogens, and they are all equivalent to each other. There is nothing different about any of the protons that makes one stand out from the others. There will not be any splitting, since there are adjacent hydrogens, but all of the hydrogens are the same. For splitting to occur, the adjacent hydrogens must be non-equivalent. There will be a singlet peak and the peak size will be six protons in area.

b) 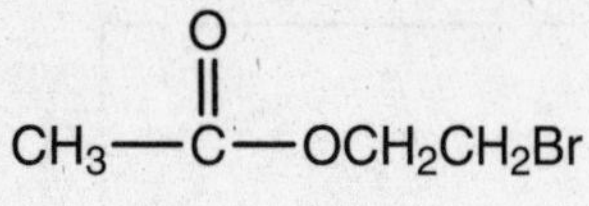 $CH_3-\overset{O}{\overset{\|}{C}}-OCH_2CH_2Br$

This molecule has three different types of protons.

$CH_3-\overset{O}{\overset{\|}{C}}-OCH_2CH_2Br$

a (under CH_3), b (under OCH_2), c (under CH_2Br)

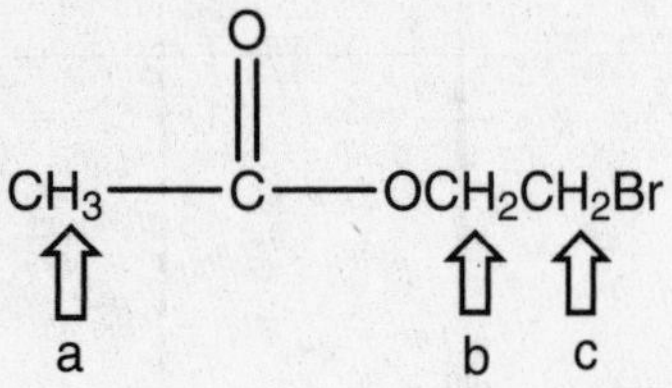

There are three equivalent methyl protons, two equivalent methylene hydrogens next to the oxygen, and two more protons on the methyl carbon with bromine. So, with three different types of protons, there will be three different signals, in the peak area ratio 3 (methyl H): 2 (-OCH_2): 2 (CH_2Br). Now for the splitting. For the three methyl protons there are no protons on the adjacent carbon, so there is no splitting, just the signal from three protons together in one singlet peak. The two –OCH_2 protons are split by the two neighboring protons into a triplet peak and likewise the two –CH_2Br protons are split into a triplet peak as well.

c) 1,1,2-trichloroethane

First, draw the structure:

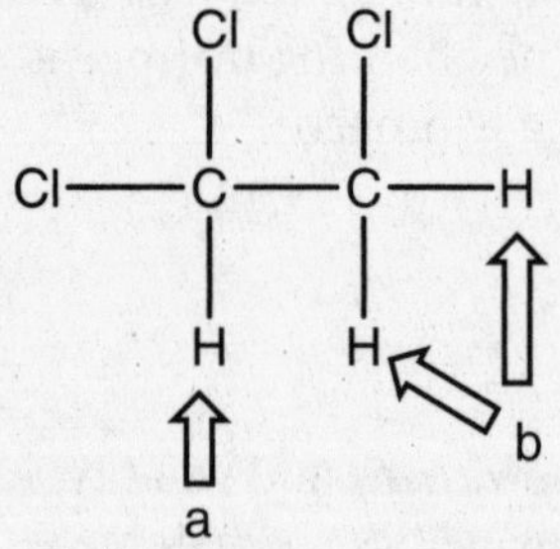

$Cl-\underset{H}{\overset{Cl}{C}}-\underset{H}{\overset{Cl}{C}}-H$

a (H on the left carbon), b (the two H on the right carbon)

There are two protons of one type on the right and one proton of another type of the left, distinguished by the different number of chlorines on the carbons they share. The two chlorines that proton a shares its carbon with make this proton less shielded than the other two protons that only need to share their carbon with one chlorine. The signal from proton a will be less shielded, or more downfield. It will be split by the two adjacent protons into a triplet peak. The signal from the two b protons will be more upfield and will be split into a doublet by the single neighbor proton.

d) 2-iodopropane

If you draw this molecule and label the different types of equivalent protons you should see:

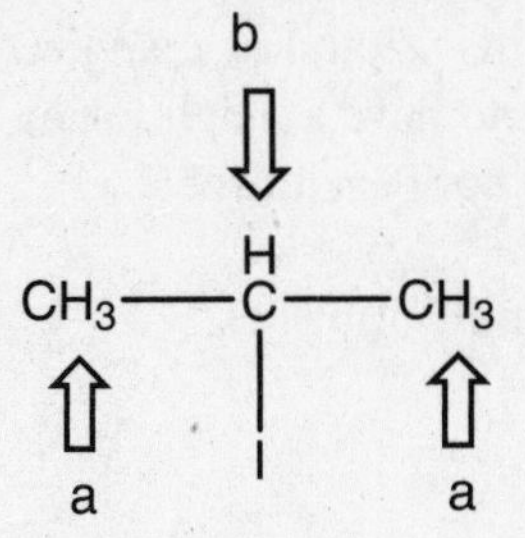

There are two different types of protons, the six methyl protons on the ends, labeled a, and the single proton on the middle carbon, labeled b. The signal from the a protons will be split into a doublet peak by the single b proton. The signal from the single b proton will be split into seven peaks by the six a protons. The overall total area of each multiplet will be 6:1 for the total a multiplet: b multiplet areas.

d) 1,4-dimethylbenzene

Draw it:

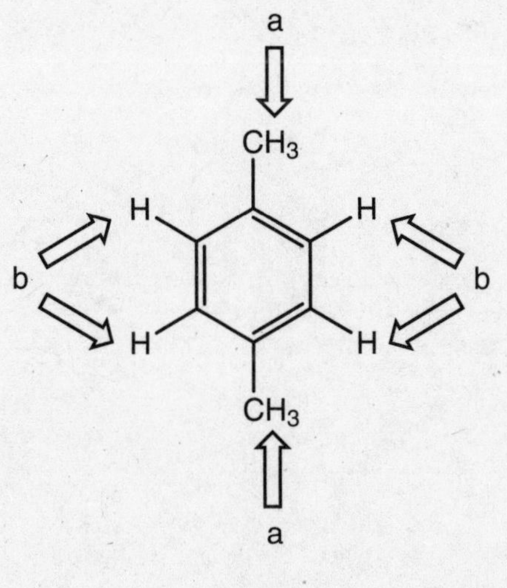

There are two different types of protons, the a protons on the two methyl groups and the b protons on the aromatic ring. The molecule is symmetric so the a protons on both ends of the molecule are equivalent and the b protons are all equivalent as well. Since there are 6 a protons and 4 b protons, the signal strengths will have the ratio 6:4 (or 3:2). The a protons, on the methyl groups, do not have any protons on the adjacent carbon, so their signal is a singlet. Likewise, the signal from the b protons is also a singlet. The b protons do have a neighboring proton, but this proton is just like them and equivalent protons do not split each other.

Don't discouraged if splitting is difficult to grasp at first. It just takes a bit of practice.

Carbonyl Chemistry and Biological Chemistry

CHAPTER TEN

The carbonyl group is found in many different places, including some of the most popular functional groups - ketones, aldehydes, carboxylic acids, amides, esters – need we go on? In addition, the carbonyl is found in many biological molecules and their reactions. Plus, the MCAT loves carbonyls. This group and its chemistry deserve a closer look – shall we? No, please, we insist, after you.

The carbonyl group is a carbon double-bonded to an oxygen. What properties does this group have? Oxygen is more electronegative than the carbon it is bound to, so it will draw electrons inward to give itself a partial negative charge, leaving its carbon partner a partial positive charge. Will any part of the carbonyl be prone to attack by nucleophiles?

Put yourself in the place of a nucleophile, out on the streets. What are you looking for? You have a non-bonding pair of electrons and you want a willing electron acceptor. The positive charge on the carbonyl carbon makes it a likely candidate to attract a nucleophile and to act as an electron acceptor.

Now, how exactly does that nucleophile react with the carbonyl?

To answer this question, we need to think about the shape of the group that the nucleophile is attacking. The carbonyl carbon must be sp^2 hybridized, since it has three groups bound to it. As an sp^2 carbon, what geometry will it have with the groups it is bound to? Sp^2 hybrid orbitals are planar, and at 120 degrees from each other. As a planar molecule, a nucleophile can attack the carbonyl from either the top or bottom side of the molecule. What about steric hindrance of the substrate to block nucleophile access? Not to worry. The substrate is planar, so there is no reason why a nucleophile cannot get access to the carbon.

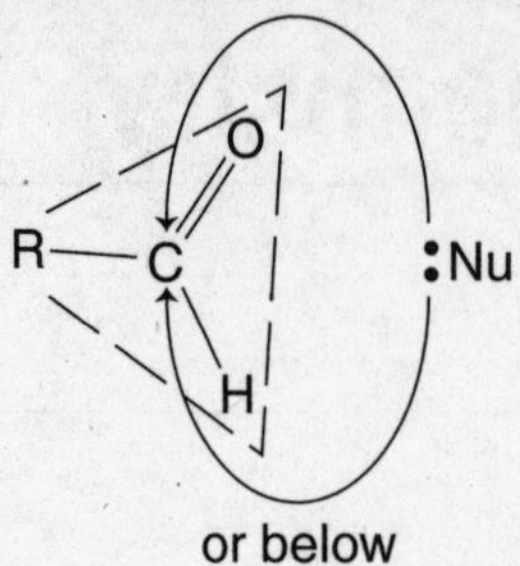

This is just what happens. A nucleophile enters the carbonyl scene with its electrons, and one way or the other donates them to the carbon. This can occur as either an addition reaction or a substitution reaction. Both of these we have seen before in the context of the alkene double bond, but now we will investigate how they apply to carbonyl groups, a different type of double bond.

NUCLEOPHILIC ADDITION REACTIONS ON CARBONYLS

One reaction that can occur with the carbonyl is an addition reaction initiated by nucleophile attack on the carbonyl. If the nucleophile is a hydroxide ion, then the reaction becomes a base-catalyzed hydration, creating an alcohol with two hydroxyl groups.

As you can see, the intermediate in the reaction is a negatively charged oxygen species. How stable will this negatively charged species be? Not very. It will have a strong urge to donates electrons to a hydrogen in a water molecule to form an alcohol and regenerate a new hydroxide ion as the leaving group. The product will be sp^3 hybridized with tetrahedral geometry, instead of the sp^2 orbitals of the reactant carbonyl.

> **Instructor question:** Can you see why this reaction is called base-catalyzed?
>
> **Student response:** Yup, because each molecule of base that goes into the reaction will be regenerated at the end of the reaction.

A similar reaction can occur in acidic conditions. If conditions are acidic, then a positively charged hydronium ion can accept electrons from the carbonyl oxygen, creating a positively charged intermediate. How stable will this intermediate be? This species, like the negatively charged species in the base-catalyzed reaction, is reactive and a target for nucleophilic attack and benefits from resonance stabilization to increase the stability of the intermediate. The protonation of the oxygen by acid helps in this way to make the nucleophilic attack more favorable. The positively charged species is attacked by a nucleophile, as in the base-catalyzed reaction, but this time the nucleophile is the weak nucleophile, water. After water acts as the nucleophile, donation of a proton back to water creates the same product as the base-catalyzed reaction, arrived at by a different route. There is more than one way to attack a carbonyl.

O
H–O–H ⊕
H
⊕ OH
+ H_2O

⊕ OH
OH ⊕
OH
H_2O⊕
$H_2\ddot{O}$:

OH
H_2O⊕
+ H_2O
OH
HO
+ $H_3O^{\oplus}$

Hemiacetals are structures related to the alcohol product of acid or base-catalyzed addition of –OH to a carbonyl. In the hemiacetal though, a carbon is bound to an –OH group and an –OR group, not to two –OH groups. To create a hemiacetal starting with a carbonyl once again, an acid (or base) catalyzed reaction can add an alcohol group to the carbonyl. In the acid-catalyzed reaction, protonation of the carbonyl oxygen creates a complex that initiates the reaction, followed by nucleophilic attack of the alcohol on the carbon, and the loss of a proton once again to complete the product.

Hemiacetal

Taking this one step further, we can create an **acetal** type of molecule. While the hemiacetal contains a carbon bound to one –OH group and one –OR group, an acetal contains two –OR groups. An acetal can be created from a hemiacetal in an acid-catalyzed reaction similar to the one that created the hemiacetal in the first place. The creation of the acetal is a bit more complicated though, with four steps required to go from the hemiacetal to the acetal. The overall reaction looks like:

elimination | addition

hemiacetal A B C acetal

Step A: Protonate the hydroxyl group. What does this accomplish? It makes the hydroxyl into a water, a much better leaving group. If we are to add another R group, we need to get rid of the –OH first.

Step B: Time for the water, as the designated leaving group, to take off. What will happen to the rest of the molecule if the water leaves? Why, a carbocation is formed, of course, and a pretty good one at that. The carbocation is stabilized by resonance forms, with electrons moving between the oxygen and the C-O bond.

Step B continued: Time for the attack. Nucleophilic attack, that is. An alcohol molecule will take on this role; unbonded electrons from the oxygen will donate electrons to the carbocation, to form a new C-O bond.

Step C: A proton is lost (water steals it). Loss of a hydrogen from C-O-R creates the final product. Along the way, the process includes an elimination reaction. Put it all together and have you got? An acetal.

Now for something completely different. Or at least it appears that way on the surface. If we reduce a carbonyl group, that is to say we add hydrogen on to it, then it is going through a type of addition reaction, adding a hydride at the carbonyl double bond. When something is reduced, the product of the reaction has more hydrogens than the parent reactant. Another way to look at redox reactions, like most reactions, is to follow the electrons. Protons are a convenient way to look at redox reactions for the moment, though.

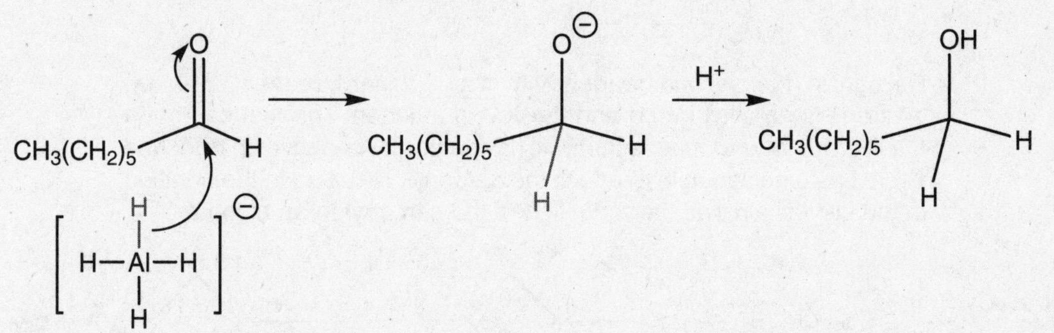

In this example, a carbonyl is reduced by lithium aluminum hydride ($LiAlH_4$) to an alcohol. Notice that alcohols are more reduced than carbonyls, since they result from the addition of a hydrogen to both the oxygen and the carbon of the carbonyl group. Another famous reducing agent is $NaBH_4$.

The order of decreasing reduced state/increasing oxidation is:

alkanes>alcohols>carbonyl>carboxylic acid

The Grignard reaction is a very popular fixture in organic chemistry. This reaction involves the addition of an alkyl chain to the carbonyl at the carbon end. In the process the carbonyl is reduced once again to an alcohol. Grignard reagents that are involved in these reactions are usually created by reacting an alkyl halide with magnesium metal to make R-MgX. Magnesium is very electropositive (wants to give up its electrons), so a bond between carbon and magnesium is a polar bond, in which carbon is partially negative and magnesium is positive. Note that this is the opposite of what occurs when carbon is bound to a more electronegative atom like oxygen. The carbon in the carbonyl bond has a partial positive charge, so that the Grignard carbon can act as a nucleophile with the carbonyl carbon, donating electrons to form the new covalent bond.

Does this reaction look familiar? What about a mechanism for addition at alkene double bonds? The Friedel-Crafts alkylation used to add alkyl groups is comparable in its result, if not the mechanism.

Instructor question: Let's practice. What are the products of the following reactions?

a. (cyclobutyl)$C(=O)CH_3$ + CH_3CH_2MgCl $\xrightarrow{\text{diethyl ether}}$

b. (3-methylbutanal, CHO with H) $\xrightarrow{NaBH_4}$

a. Recognize the second reagent? It is a Grignard reagent, R-MgX, where R is an alkyl group and the X is a halogen. The first carbon in the R group next to magnesium will have a partial negative charge, and will act as a nucleophile to attack the carbonyl, resulting in the addition of the alkyl chain and the reduction of the carbonyl to an alcohol.

(cyclobutyl)C(=O)CH_3 + $CH_3CH_2^{\delta-}$—$MgCl^{\delta+}$ ⟶ (cyclobutyl)C(O^-Mg^+)(CH_2CH_3)—CH_3 $\xrightarrow{H^+}$ (cyclobutyl)C(OH)(CH_2CH_3)—CH_3

b. Recognize $NaBH_4$? It is a reducing agent, and will add hydrogen to the carbonyl bond to create a reduced version of the reactant, an alcohol.

(aldehyde, H) + $[BH_4]^{\ominus}$ ⟶ (O^-, H, H) $\xrightarrow{H^+}$ (OH, H, H)

Reductions and Grignard reactions are examples of additions to the carbonyl, transforming the double bond to a single bond. But there are also substitution reactions that we must examine.

NUCLEOPHILIC SUBSTITUTIONS ON CARBONYLS

If a nucleophile attacks a carbonyl, and there is a good leaving group on hand, then indeed the reaction will be one of nucleophilic substitution instead of addition.

CH3 Cl :O: :OH⊖ → CH3 Cl :O:⊖ OH → CH3 OH O + Cl⁻ → rapid deprotonation (irreversible) → CH3 O O⊖

This story starts out with an attack of a killer nucleophile once again, on the carbonyl carbon. The electrons can in turn push up to the oxygen to create a transient negatively charged species. The next step depends on the nature of the leaving group in the molecule. If there is a good leaving group available, then it will exit stage left, and the carbonyl will be recreated. What makes a leaving group good? Just as in the nucleophilic substitutions we have come to know and love, a good leaving group is one that is a weaker base, and a weaker nucleophile, than the nucleophile that initiated the reaction. One of those nucleophiles has got to leave, and it is going to be the weaker one. In the example shown, the strong nucleophile that attacks is a hydroxyl, and the weak nucleophile/weak base that is the leaving group is chloride. Another feature of this reaction is that it creates a carboxylic acid, and can deprotonate to stabilize itself through resonance. The basic conditions of the reaction, with hydroxide ions as one of the reactants, will help the acid to dissociate itself from the unsightly proton as rapidly as possible.

Instructor question: For each of the following, predict if the reaction will proceed as written by a nucleophilic substitution mechanism. Remember, a good place to start is by examining the reactants. Do we have a good strong nucleophile to go on the attack? Do we have a carbonyl carbon that can be attacked? And how about a good leaving group?

a. $CH_3CH_2O^-$ + $CH_3CH_2C(=O)Cl$ →

b. $Cl^- + CH_3CH_2C(=O)OH \longrightarrow$

c. $CH_3CH_2C(=O)O^{\ominus} + CH_3CH_2C(=O)OCH_3 \longrightarrow$

d. $Br^- + CH_3C(=O)NH_2 \longrightarrow$

e. $OH^- + CH_3C(=O)OC(=O)CH_3 \longrightarrow$

Student response: Okay, reaction (a). What does it have? I'll start with the nucleophile. The ethoxide ion should be a strong base/strong nucleophile since it is the conjugate base of a very poor acid. It will have a strong tendency to donate electrons. There is a carbonyl there, and to top things off there is a chloride ready and willing to act as the ideal leaving group.

Instructor response: Exactly. This reaction will go like a dream, and the product will be:

$$CH_3CH_2O^- + CH_3CH_2C(=O)Cl \longrightarrow CH_3CH_2C(=O)OCH_2CH_3$$

Student response: Okay, on to (b). The nucleophile, Cl-, is not a great nucleophile. It is the conjugate base of a strong acid. There is a carbonyl, which is okay, but the leaving group, -OH, is not a good one. It is a strong base, much stronger than the chloride ion that is trying to replace.

Instructor response: You got it. Thermodynamics would have this reaction running in the opposite direction from how it's written. This one is a no-go.

Student response: Now for (c). I guess that O- is the nucleophile, and it's negatively charged, so it's a good one. It—

Instructor response: Wait a minute – what kind of a nucleophile is that? Look closer. This thing is the conjugate base of a carboxylic acid, so it's not a very strong base, nor a good nucleophile. There is a carbonyl, but it is part of a carboxylate group as well. The leaving group would have to be a methoxylate, a good nucleophile and poor leaving group by any account. This reaction has two strikes against it and is going nowhere.

Student response: Oops. Moving on to problem (d). I know that Br^- is just an outright bad nucleophile. And $-NH_2$ is a poor leaving group. This reaction ain't happening.

Instructor response: Well done. Now problem (e) is a new animal, known as an anhydride. It is created through the union of two carboxylic acids. The nucleophile is a good one. We are off to a good start. And the carbonyl that is being attacked? There seem to be two carbonyls to chose from, an embarrassment of riches. Which one to chose? Either of them will do just fine. It does not matter at all, since the result will be the same either way. The leaving group will be the conjugate base of a strong acid, a carboxylic acid, a good leaving group. All systems are go – this reaction is going places, forward to be exact. The product will be two carboxylic acid molecules, created from the two halves of the original symmetric anhydride reactant.

OH^- + $CH_3C(=O)OC(=O)OCH_3$ → $CH_3C(=O)OH$ + $^-OC(=O)CH_3$ → 2 $CH_3C(=O)O^-$

Now for a look at a nucleophilic acyl substitution. The example we will discuss forms an ester group, a variant of carboxylates in which an R group is added to the acid to form –COOR instead of –COOH. This reaction takes place in the presence of an acid catalyst.

O, OH ⇌ (H^+) $^+$OH, OH ↔ OH, +, OH (A) ⇌ (ROH) OH, OH, O, R, +, H (B)

B ⇌ ($-H^+$) C ⇌ (H^+) D

D ⇌ E ($^+$OH, OR) + H_2O ⇌ ($-H^+$) O, OR (ester)

How does the mechanism work? Can we just stick the R on there and be done with it? It turns out to be slightly more complicated than this, curiously. Let's break it down one step at a time. The initial reactant is a carboxylic acid. In the first reaction step, the acid is protonated on the carbonyl oxygen to form a positively charged intermediate. What then? In comes a nucleophile, R-OH, an alcohol. Perhaps not the strongest nucleophile, but the carbocation is reactive enough that even a weak nucleophile like the oxygen of an alcohol will be enough to get the job done. The oxygen attacks the carbon of the carbonyl in this step. The product of this step is rather awkward-looking, with a positively charged oxygen. What to do? Get rid of a proton from the alcohol nucleophile, of course, to lose the positive charge and get a neutral species. Another proton from the solvent adds to one of the hydroxyls, creating a water as a leaving group. We know water is pretty stable, and makes one heck of a leaving group. When the water leaves, the carbonyl of the carboxylate is reformed, bringing things back around to the last step, the deprotonation of the carbonyl oxygen. With this, the reaction is complete and the ester product is ready for the world.

What would happen if you reverse an esterification reaction? In this case, water would add to the ester, and the leaving group would be the alcohol, recreating a carboxylic acid in the process.

BIOLOGICAL CHEMICALS

Living organisms are excellent chemists. Even the simplest bacterium is capable of a wide range of reactions that would be the envy of any human scientist. The full range of reactions that biological organisms conduct will be left to a biochemistry course. For the moment, we will introduce some of the basic molecules of life, perhaps forming a bridge to future explorations in this area.

The main building molecules of life are carbohydrates (sugars), lipids (fats), proteins, and nucleic acids (DNA and RNA). Each of these molecules plays its own unique role in maintaining and propagating life. Carbohydrates mainly supply the energy that life needs to do just about anything. Fats are also an important energy source, and form the membranes that separate the inside and outside of cells and structures inside cells. Nucleic acids are informational molecules that store the codes with instructions for how to build and maintain a living organism. Proteins are the main actors that carry out the orders given by nucleic acids.

And now, let's dig a bit deeper...

CARBOHYDRATES

Carbohydrates have the general formula $C_nH_{2n}O_n$. This can also be written as $C_n(H_2O)_n$. These molecules originally got their name when chemists found that when combusted, carbohydrates released a certain quantity of water related to the number of carbons present. There is no water in carbohydrates, however. Carbohydrates include small simple sugars like glucose, disaccharides like sucrose, table sugar (two sugar molecules joined together), and large carbohydrate polymers like starch and cellulose.

The central player in energy metabolism in organisms from bacteria to trees to humans is the sugar glucose. Glucose is a carbohydrate with six carbons and has the molecular formula $C_6H_{12}O_6$. Glucose has a hydroxyl group on every carbon but one, the carbon at the end, which instead has a carbonyl, specifically an aldehyde group.

Are there any chiral carbons in glucose? There are in fact not one, not two, but four chiral carbons in glucose. That is a lot of stereochemistry. One bizarre thing about the chemistry of living organisms is that life conducts very selective and very complex stereospecific reactions. How many stereoisomers of glucose are there? With four chiral centers, there are a maximum of 2^4 possible stereoisomers = 16. Some of these stereoisomers are also important biological molecules in their own right, making it important that life can carry out its chemistry in a very stereospecific way.

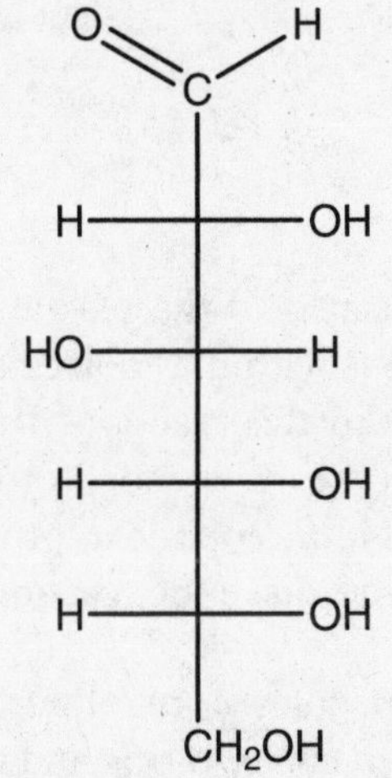

O H
C
H OH
HO H
H OH
H OH
CH_2OH

In the above Fischer projection, glucose is drawn as a straight chain of carbons. Glucose in solution can react with itself, one end of the molecule with the other, to go from a straight chain to a ring-shaped structure. Glucose mixed in water does this readily, and is in fact mostly found in the ring form rather than the straight chain. The sugars that form starch, glycogen and cellulose are all mostly in the ring form. How exactly does this intramolecular reaction occur?

It is easier to visualize this reaction if the molecule is drawn as a ring (see below). The ring that is formed by glucose has six members, one of them an oxygen and the rest carbons, and is similar in shape to the cyclohexane ring we talked about earlier. In fact, the glucose ring can go through some of the same conformational changes to produce the chair and boat configurations, and has the same rules for the placement of substituents in the axial or equatorial positions.

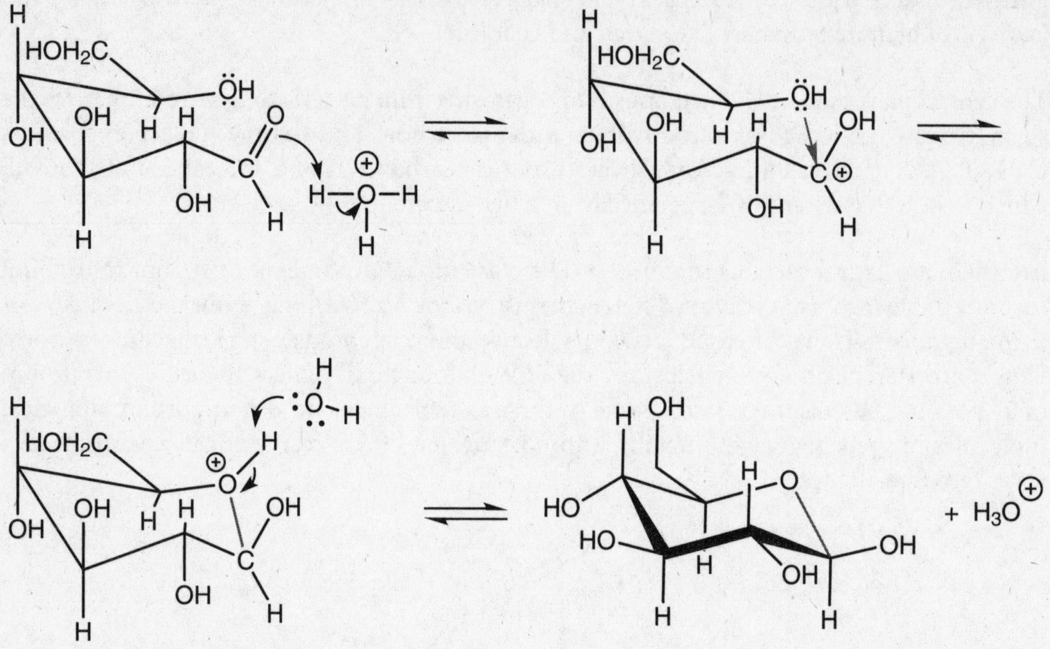

This reaction to form the ring is related to the reaction presented earlier for hemiacetals, except that in this case the reaction is within a molecule rather than between two different molecules. What is the nucleophile in this reaction? It is a hydroxyl from one end of the molecule that attacks the carbonyl at the other end. Protonation of the carbonyl oxygen helps things along, allowing the weak –OH nucleophile to proceed with the reaction. After attack, a water will act as a proton acceptor to make the reaction complete.

The formation of the ring form of glucose, or other sugars, is a reversible process. The reaction goes in both directions, from chain to ring and at the same time from ring to chain. The ring actually coexists with the chain in equilibrium in solution.

In the process of forming the ring, an additional result is that a new chiral carbon is formed in the ring that is not chiral in the straight chain form. The original four chiral carbons are conserved in their stereochemistry. The carbonyl carbon that is attacked in the reaction becomes a new chiral carbon, indicated below with a star. This chiral carbon in the ring form is called the **anomeric** carbon.

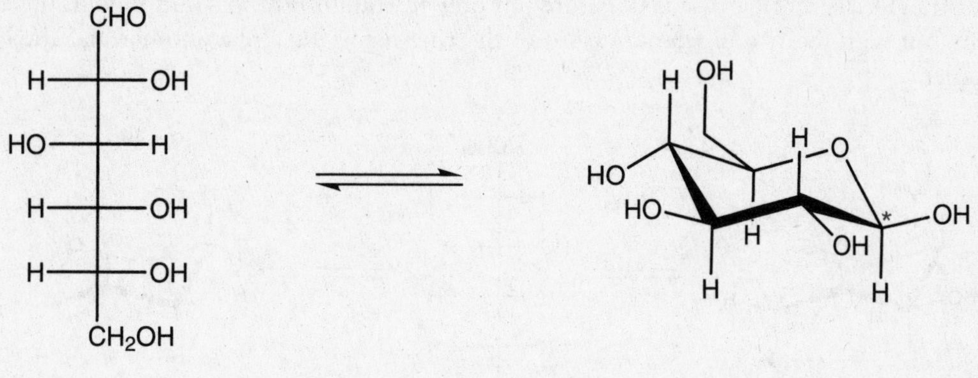

Depending on the orientation of the carbonyl when it undergoes nucleophilic attack, the hydroxyl group of the anomeric group can end up in one of two orientations, either axial or equatorial. The two different stereoisomers of the ring form are called alpha and beta anomers.

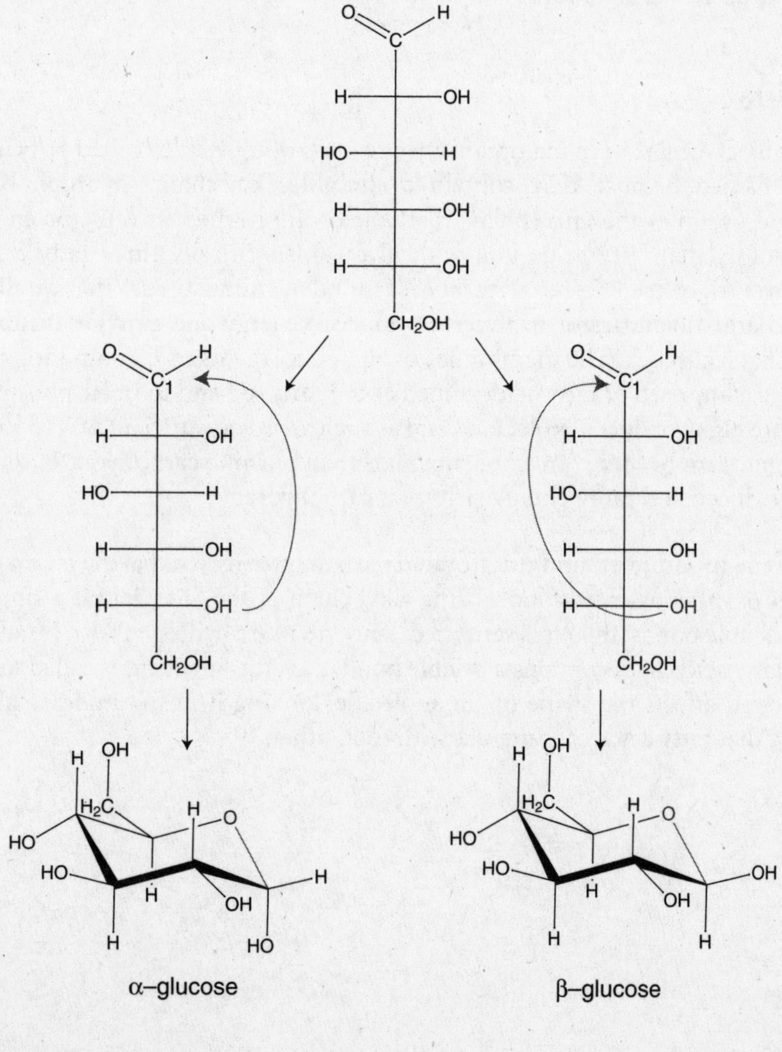

α–glucose

β–glucose

The straight chain of glucose is therefore not only in equilibrium in solution with the ring form, but with the two different version of the ring form, the alpha anomer and the beta anomer.

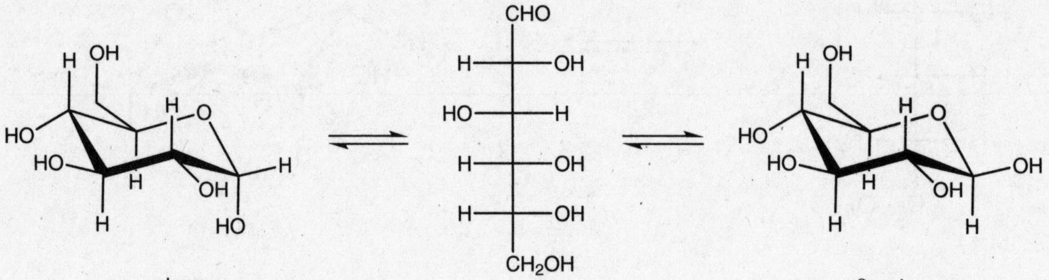

α—glucose ⇌ open chain ⇌ β—glucose

Student comment: I think of it as "Alpha = Axial." In the alpha conformation, the OH is either axially up or down. Beta is always equatorial, either slightly up or slightly down.

FATTY ACIDS

The building block of fats in living organisms is a class of molecules called fatty acids. Fatty acids are so-named because they contain a straight alkyl chain (hydrophobic), and a carboxylic acid group at the end (polar). Fatty acids are burned in cells for energy. To be stored energy for later, fatty acids join with glycerol, specifically three fatty acids to one glycerol, to form triglycerides. The white fat cells found in animal tissue (that we all know and love) contain large quantities of triglycerides to store energy and provide insulation from cold and mechanical shock. The membranes of all living organisms contain molecules called phospholipids, comprised of fatty acids joined at the carboxyl end to polar phosphate-based groups. Fats are highly reduced molecules, and as such are a very efficient way to store energy. Animals get much more energy from burning fats than burning carbohydrates, since fats are more highly reduced and more energy rich than carbohydrates.

Fatty acids come in saturated and unsaturated varieties, terms you have seen on your jar of peanut butter or other assorted foods. If the alkyl chain of the fatty acid is a simple alkane, without any double bonds, then it is saturated, since no more hydrogens can be added to this chain. If the fatty acid chain contains a double bond, then the fatty acid is called unsaturated. The double bond affects the shape of the molecule, kinking it in the middle, and this will affect the way that fatty acids can interact with each other.

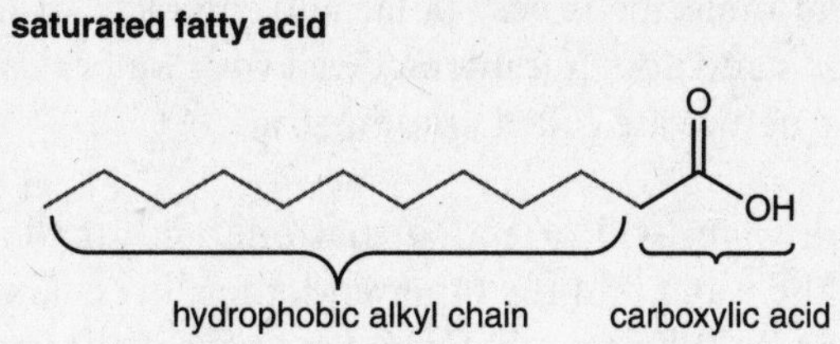

saturated fatty acid

hydrophobic alkyl chain carboxylic acid

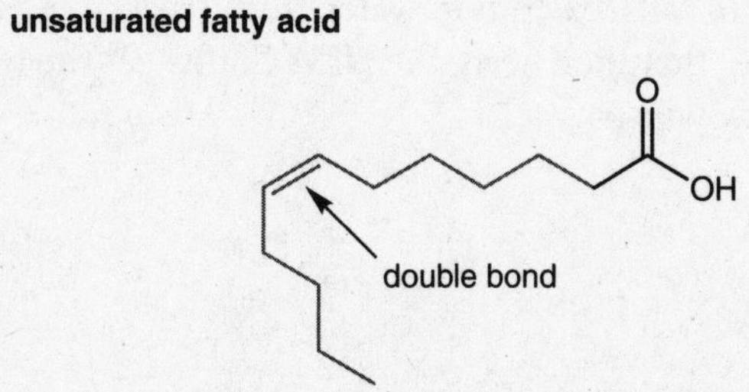

unsaturated fatty acid

double bond

The main thing we will discuss about fatty acids is how they interact with each other and with solvent and what this means for us. The way that fatty acids interact with each other is a function of their physical characteristics. What are the physical characteristics of fatty acids? As we noted earlier, at one end is the carboxylic acid, fairly acidic; when deprotonated, as it often is, it will be charged. When the fatty acid is linked to another molecule through the carboxylate by an ester linkage, as it is in a tryglyceride, then it will not be charged. When the fatty acid is free on its own and charged, the carboxylate end will favor interaction with water as a solvent, since water is a good solvent for polar and charged groups.

What about the alkyl end of the fatty acid? How do alkanes interact with water? Not very well, thank you. Alkanes are composed entirely of non-polar bonds, and are hydrophobic. What do alkanes interact with? Other hydrophobic molecules, of course. What sort of interactions occur between alkanes? Van der Waals transient dipoles, a weak interaction that occurs between hydrophobic molecules. If you try to mix alkanes with water, the alkane molecules will tend to cluster together, staying away from water. Think salad dressing, with oil and water. Triglycerides, which are very non-polar, form water insoluble clumps inside of cells that create the white layer of insulating fat in tissues, keeping water out. Why might alkanes do this?

Water has structure, with lots of dipole-dipole interactions between water molecules. If an alkane molecule inserts itself into the midst of the solvent, it will break up the structure of water and prevent the stabilizing bonds between water molecules, forcing the water to assume a higher energy state around the alkanes. The mixing of alkanes with water is also disfavored in terms of entropy. The more favorable state for alkanes is for the alkanes to cluster together, and interact with each other rather than with water. This way alkanes have the most favored arrangement, interacting with each other, and water molecules have their most energetically favored state as well.

So one end of a fatty acid molecule is very polar and interacts strongly with water and the other end is very non-polar and does not interact well with water at all. Molecules with a split hydrophilic/hydrophobic nature are called amphipathic.

Fatty acids if mixed with water will orient so that one end of the molecule, the carboxyl group, faces out toward the water and the other end, the alkyl chain, faces the alkyl chain of other neighboring fatty acids. Who says that you can't have it all? Fatty acids accomplish this feat by forming a structure called a lipid micelle. A micelle is a round organized cluster of fatty acids, with the charged carboxyl groups on the outside, facing the water, and the inside packed with the alkyl chain tails, excluding water from the center. The carboxyl groups get the interactions with water that they need, the alkyl chains get the hydrophobic interactions that they need – everybody wins.

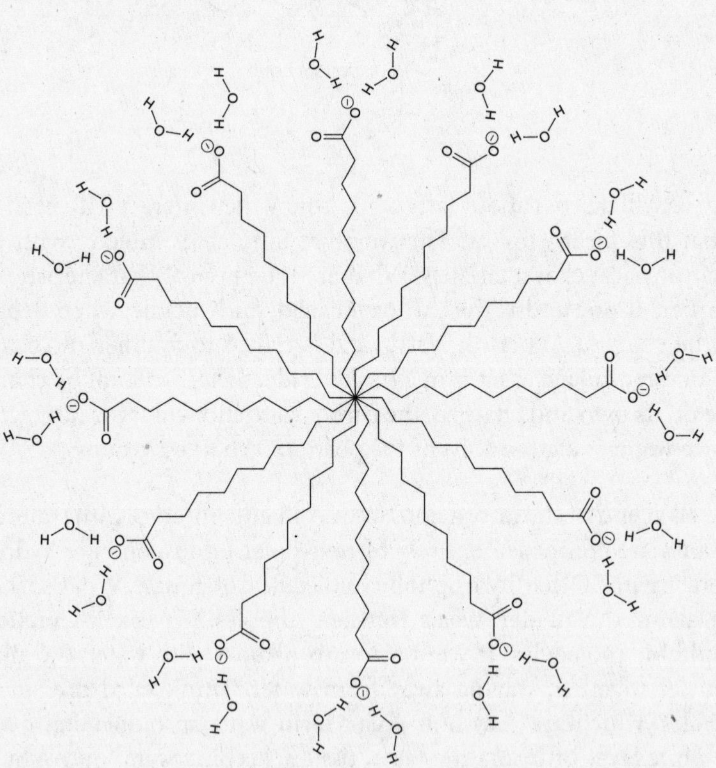

Instructor question: If your hands get covered in motor grease or your clothes get greasy food stains, how can you get it off? Soap, obviously. But how does this work?

Student response: Grease and oil are rich in alkanes, very hydrophobic and not soluble in water. The reason we use soap or detergent to get the grease out is that fatty acids are a type of detergent and can be created from animal fat by hydrolyzing the ester linkages in triglycerides. Oil alone has the same insolubility as the fatty acid alkyl chains in water. If the oil can hide itself with the alkyl chains inside the micelle, however, then it be carried away inside the micelle, to get washed down the drain by the polar water which interacts with the outer polar ends of the micelle.

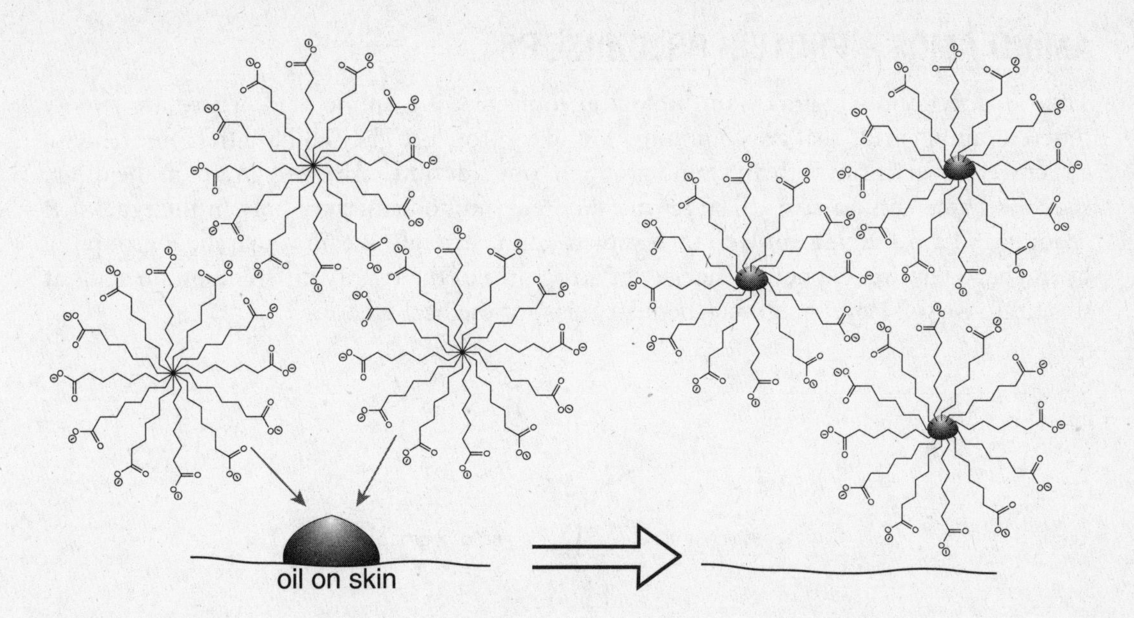

Saturated fatty acids have no double bonds while unsaturated fatty acids do. How does this affect the behavior of fats? The double bonds in unsaturated fatty acids change the shape of these molecules, making them kinked. How would this affect the melting point of fatty acids? For long chain fatty acids, the main interactions between molecules that hold them in a solid are the Van der Waals interactions between the fatty acid tails. If the fatty acid chains are saturated, they are relatively linear, although flexible. These straight chain alkyl groups can pack together very evenly and densely, maximizing their contact, and maximizing the interaction between molecules. Unsaturated fatty acids, however, will be bent and kinked, so they pack together unevenly and loosely with fewer interactions between molecules. The more interactions there are, the more energy it takes to break the interactions, and the higher the melting point will be. Saturated fatty acids therefore have a higher melting point that unsaturated fatty acids. This is why butter, with lots of saturated fatty acids, is a solid in the refrigerator while vegetable oil, with lots of unsaturated fatty acids, is a liquid.

> **Instructor comment:** As you might guess, it's the harder saturated fats that sit on your arteries and cause health problems.

Phospholipids are biological molecules containing fatty acids linked to a polar phosphate group. Phospholipids act to minimize the contact of fatty acid alkyl chains when mixed with water, but in a different way. Rather than forming micelles when mixed with water, phospholipids spontaneously assemble themselves into sandwich-like sheets, with polar phosphate groups sticking out both sides of the sheet and alkyl chains sandwiched in the middle. If the sheet rounds up and forms a ball, then you are well on your way to building the membrane of a cell. One of the features of the cell that allows life to exist, and probably one of the requirements that allowed life to begin on earth, is that lipid membranes create a barrier separating the insides from the outside of cells.

AMINO ACIDS – PROTEIN PRECURSORS

Time to meet another important biological molecule, the amino acid. There are twenty different amino acids that are commonly found in proteins. They all have the same general structure shown below, with an amino group on one side and a carboxyl group on the other, thus the name "amino acid." The twenty different amino acids vary only in the type of R group they have, starting out with the simplest amino acid glycine, in which the R group is a hydrogen. Sadly, we will not memorize the structures of the twenty different amino acids at this time. We will leave that to biochemistry class or medical school.

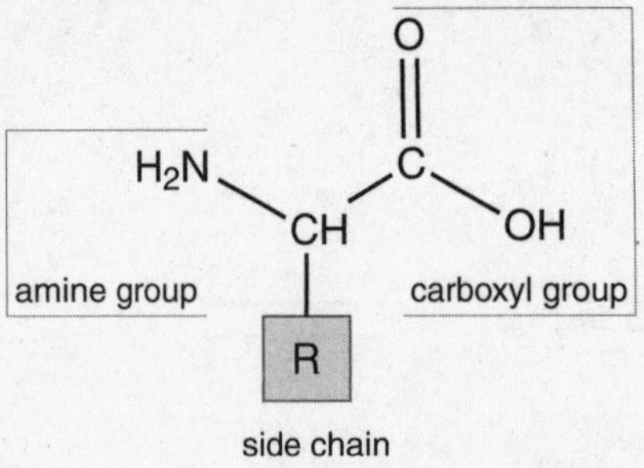

O
H_2N C
CH OH
amine group
carboxyl group
R
side chain

If the R group = H, is the carbon that R is bound to a chiral carbon? No, clearly it is not, since in this case the carbon would have two hydrogens bound to it, rather than four unique substituents. Will this carbon be chiral if the R is anything *but* H? Sure! If the R group is one of the other 19 possible amino acid side chains, then the amino acid will be chiral. Almost all naturally occurring amino acids that are used in living organisms are of the same chirality, known as L-amino acids for their optical rotation (L=levarotatory, rotates to the left).

One important feature of amino acids is that they contain both an acidic group and a basic group in the same molecule. At acidic pHs, both the amino group and the acid will be protonated, and the molecule will have a positive charge. At basic pH conditions, both the amino and carboxyl groups will be deprotonated, and the molecule as a whole will have a negative charge. What will happen to the charge on the molecule in pH conditions that are in the middle range? The middle pH range includes pH conditions around 7 such as those found in the human body. Under these conditions, the acid loses its proton to have a negative charge and the amine group remains protonated, to have a positive charge. The molecule thus has both a positive and a negative charge, but since these charges cancel each other out, the molecule as a whole is neutral. Molecules such as this that contain both a positive and negative charge in the same molecule are called **amphoteric**. These molecules are also referred to as zwitterions, meaning that they possess positive and negative charges not adjacent to each other.

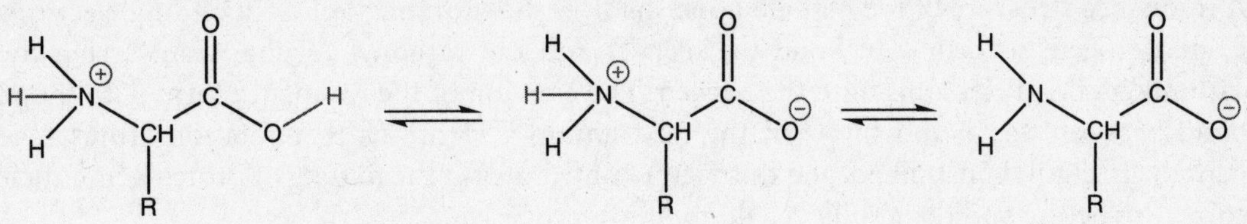

To form proteins, amino acids are joined together in long chains. The linkage between amino acids is called the peptide bond, which forms through an amide linkage between the carboxyl group of one amino acid and the amino group of the next. The reaction that creates the amide linkage is a nucleophilic substitution reaction, where the nucleophile is the amine group which attacks the carbonyl, and the leaving group is the hydroxyl.

The longer the protein is, the more amino acids there are, joined together with peptide bonds. Peptides in the body range in size, from hormones that are just a few amino acids long to large proteins that have a molecular weight over a million grams per mole. Proteins are made by the cell to do work, such as proteins called enzymes that act as catalysts of biological reactions. The work that proteins do depends on the sequence of amino acids that the protein is made of and how the polypeptide chain folds together in space. The example shown below is a dipeptide between alanine (a type of amino acid), with a methyl R side chain, and serine, which has a methyl alcohol side chain.

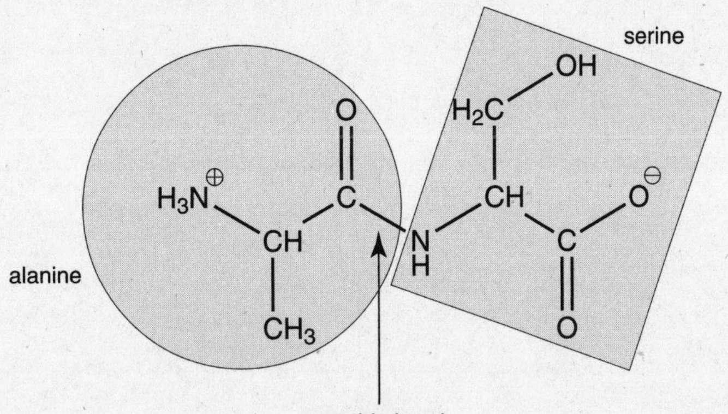

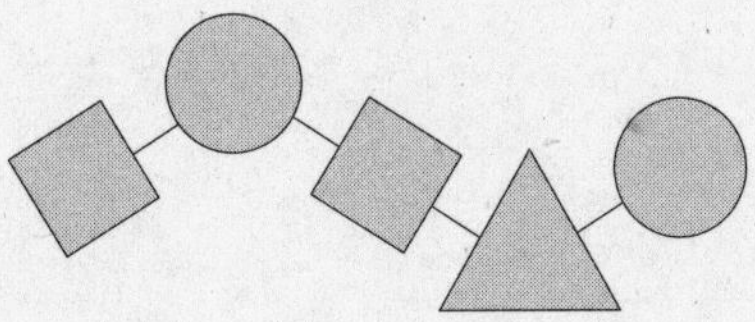

A resonance structure of the peptide bond involves the movement of nonbonding electrons from nitrogen to form a pi bond between N and the carbonyl carbon, transferring the carbonyl pi bond electrons onto the oxygen. The creation of the resonance form with the pi bond between the N and the C of the next amino acid means that rotation around the peptide linkage is constrained, and this factor contributes to the folding of proteins into their correct structure and their ability to do their job properly.

Protein folding is extremely complex and probably will be covered in your biology class. At this point, we will stop, and let you try your hand at some review and practice problems in the next section of this book. We hope you have enjoyed this introductory view of some of the basic concepts in organic chemistry. Y'all come back now, hear?

PART II

ORGANIC CHEMISTRY REVIEW & PRACTICE

Test your skills. This section contains a review of the content covered in an organic chemistry course. It also contains realistic problems, complete with detailed explanations, that you can use to build your organic chemistry skills. The explanations will tell you—not only what the right answer is and why—but also why the other choices are wrong. You can use these practice problems to develop effective strategies to help you approach even the most difficult organic chemistry questions. Or use this section to review for an organic chemistry test—whether it's the final in your college course or a standardized test like the MCAT or DAT.

Nomenclature

CHAPTER ELEVEN

Nomenclature, the set of accepted conventions for naming compounds, is crucial to a discussion of organic chemistry. The rules of nomenclature presented in this chapter are for general cases only. More specific examples will be discussed in the chapters dealing with particular types of compounds.

You may see specific nomenclature questions on an organic chemistry test, such as "Name the following compound," or "Which structure represents the following named compound?" But more importantly, nomenclature represents the basic language of organic chemistry. If you don't know it, you may feel like you're taking a test in a foreign language—which, in a way, you would be!

ALKANES

Alkanes are the simplest organic molecules, consisting only of carbon and hydrogen atoms held together by single bonds.

A. STRAIGHT-CHAIN ALKANES

The names of the four simplest alkanes are:

CH_4	CH_3CH_3	$CH_3CH_2CH_3$	$CH_3CH_2CH_2CH_3$
methane	ethane	propane	butane

The names of the longer-chain alkanes consist of prefixes derived from the Greek root for the number of carbon atoms, with the ending **-ane**.

C_5H_{12} = **pent**ane
C_6H_{14} = **hex**ane
C_7H_{16} = **hept**ane
C_8H_{18} = **oct**ane
C_9H_{20} = **non**ane
$C_{10}H_{22}$ = **dec**ane
$C_{11}H_{24}$ = **undec**ane
$C_{12}H_{26}$ = **dodec**ane

These prefixes are applicable to more complex organic molecules and should be memorized.

Note:

You must memorize the names of the 4 simplest alkanes:

- *Meth-*
- *Eth-*
- *Prop-*
- *But-*

Note:

All straight chain alkanes have the general formula C_nH_{2n+2} (n is an integer).

B. BRANCHED-CHAIN ALKANES

The International Union of Pure and Applied Chemistry (IUPAC) has proposed a set of simple rules for naming complex molecules. This basic system can be used to name all classes of organic compounds. Throughout these notes, the IUPAC names will be listed as the primary name, and common names will appear in parentheses.

In a Nutshell:

1. Identify the longest backbone.
2. Number it, keeping numbers for the substituents as low as possible.
3. Name substituents.
4. Assign numbers.
5. Put the whole name together, remembering to alphabetize substituents.

1. Find the longest chain in the compound.

The longest continuous carbon chain within the compound is taken as the backbone. If there are two or more chains of equal length, the most highly substituted chain takes precedence. The longest chain may not be obvious from the structural formula as it is drawn. For example, the backbone shown below is an octane (it contains eight carbon atoms).

```
                C   C
            C    \ /
            |     C
            |     |
   [C—C—C—C—C—C—C—C]
            |   |
            C   C
                 \
                  C
```

2. Number the chain.

Number the chain from one end in such a way that the lowest set of numbers is obtained for the substituents.

```
          C   C                        C   C
      C    \ /                     C    \ /
     3|  4  C                     6|  5  C
C—C—C—C—C—C—C—C              C—C—C—C—C—C—C—C
1 2   |  | 5 6 7 8           8 7   |  | 4 3 2 1
      C  C               NOT       C  C
          \                            \
           C                            C
```

3. Name the substituents.

Substituents are named according to their appropriate prefix with the ending **-yl**. More complex substituents are named as derivatives of the longest chain in the group.

CH_3–	CH_3CH_2–	$CH_3CH_2CH_2$–
methyl	ethyl	*n*-propyl

The prefix *n*- in the above example indicates an unbranched ("normal") compound. There are special names for some common branched alkanes, and these are usually used in the naming of substituents.

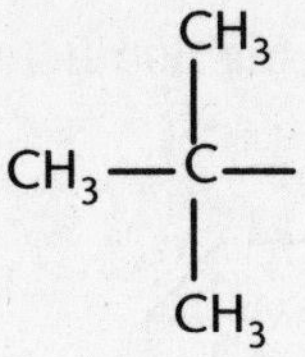

$CH_3-C(CH_3)_2-$

t-butyl

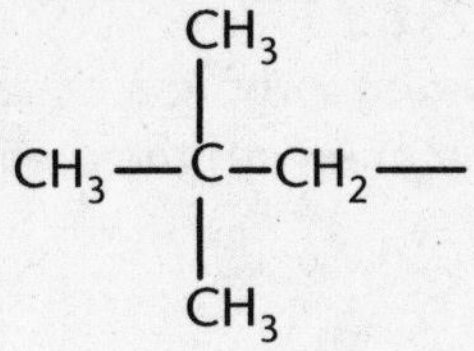

$CH_3-C(CH_3)_2-CH_2-$

neopentyl

$(CH_3)_2HC-$

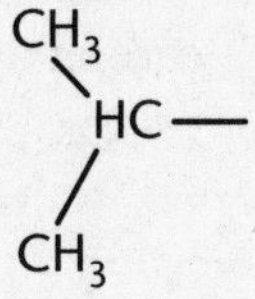

isopropyl

Note:

You should memorize these common structures.

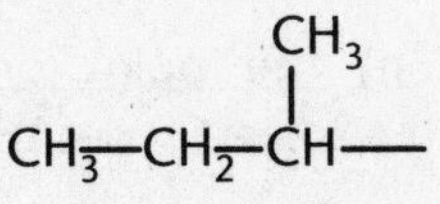

$CH_3-CH_2-CH(CH_3)-$

sec-butyl

$CH_3-CH(CH_3)-CH_2-$

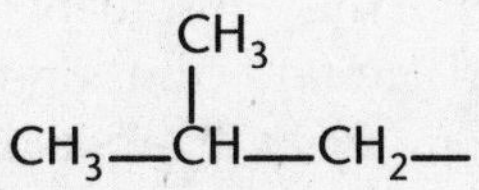

isobutyl

If there are two or more equivalent groups, the prefixes **di-**, **tri-**, **tetra-**, etc. are used.

4. **Assign a number to each substituent.**
   Each substituent is assigned a number to identify its point of attachment to the principal chain. If the prefixes **di-**, **tri-**, **tetra-**, etc., are used, a number is still necessary for each individual group.

5. **Complete the name.**
   List the substituents in alphabetical order with their corresponding numbers. Prefixes such as di–, tri–, etc., as well as the hyphenated prefixes (*tert-* [or *t-*], *sec-*, *n-*) are ignored in alphabetizing. In contrast, **cyclo-**, **iso-**, and **neo-** are considered part of the group name and are alphabetized. Commas should be placed between numbers, and dashes should be placed between numbers and words. For example:

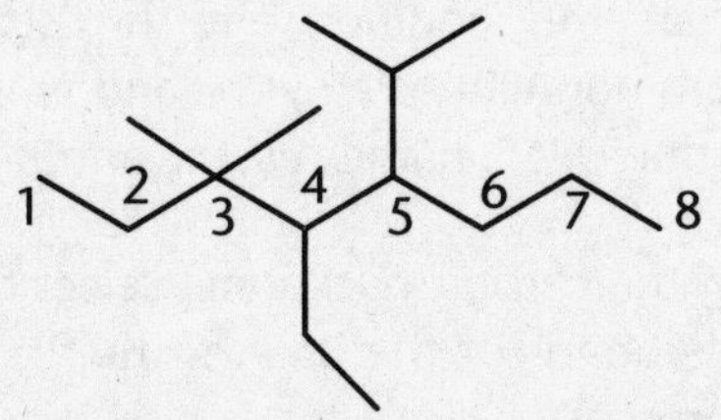

1 2 3 4 5 6 7 8

4-ethyl-5-isopropyl-3,3-dimethyl octane

You may also need to indicate the isomer you are describing—e.g., *cis* or *trans*, R or S, etc. Isomers will be discussed in detail in Chapter 12.

C. CYCLOALKANES

Alkanes can form rings. These are named according to the number of carbon atoms in the ring with the prefix **cyclo-**.

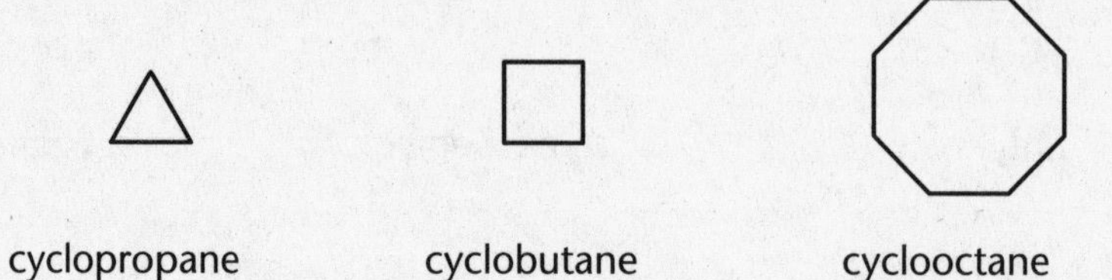

cyclopropane cyclobutane cyclooctane

Substituted cycloalkanes are named as derivatives of the parent cycloalkane. The substituents are named, and the carbon atoms are numbered around the ring starting from the point of greatest substitution. Again, the goal is to provide the lowest series of numbers as in rule number 2 above.

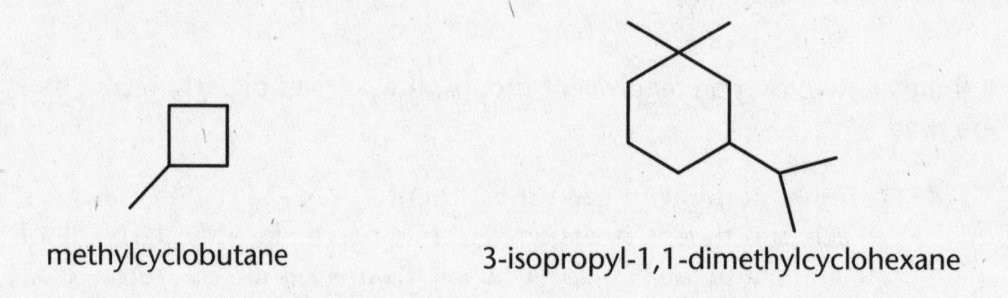

methylcyclobutane 3-isopropyl-1,1-dimethylcyclohexane

MORE COMPLICATED MOLECULES

Organic molecules that are more complicated than simple alkanes can also be named using this 5-step process, with a few additional considerations.

MULTIPLE BONDS

A. ALKENES

Alkenes (or **olefins**) are compounds containing carbon-carbon double bonds. The nomenclature rules are essentially the same as for alkanes, except that the ending **-ene** is used rather than **-ane**. (Exceptions: the common names *ethylene* and *propylene* which are used preferentially over the IUPAC names *ethene* and *propene*).

When identifying the carbon backbone, select the longest chain that contains the double bond (or the greatest number of double bonds, if more than one is present).

Note:

Noncyclic alkenes with one double bond have the general formula C_nH_{2n} (n is an integer).

Note:

For alkenes (and alkynes), the carbon backbone should include the double (or triple) bond. Double bonds, however, take priority.

1 4 2 5 3 NOT 5 3 1 6 4 2

Number the backbone so that the double bond receives the lowest number possible. Remember that multiple double bonds must be named using the prefixes di-, tri-, etc. and that each must receive a number. Also, you may need to name the configurational isomer (*cis*/*trans*, Z/E). This topic will be discussed further in Chapter 12.

Substituents are named as they are for alkanes, and their positions are specified by the number of the backbone carbon atom to which they are attached.

Frequently, an alkene group must be named as a substituent. In these cases, the systematic names may be used, but common names are more popular. **Vinyl-** derivatives are monosubstituted ethylenes (**ethenyl-**), and **allyl-** derivatives are propylenes substituted at the C3 position (**2-propenyl-**). **Methylene-** refers to the $-CH_2$ group.

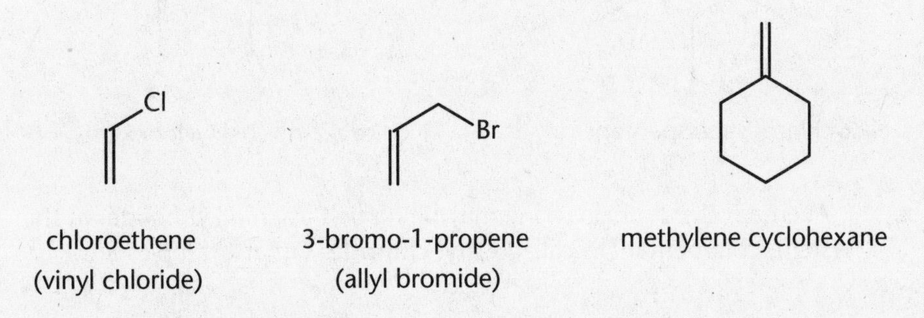

chloroethene (vinyl chloride) 3-bromo-1-propene (allyl bromide) methylene cyclohexane

B. CYCLOALKENES

Cycloalkenes are named like cycloalkanes but with the suffix **-ene** rather than **-ane**. If there is only one double bond and no other substituents, a number is not necessary.

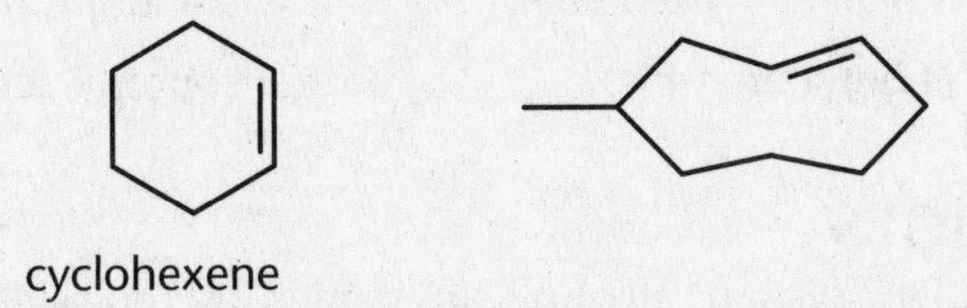
cyclohexene

C. *ALKYNES*

Alkynes are compounds that possess carbon-carbon triple bonds. The suffix **-yne** replaces *-ane* in the parent alkane. The position of the triple bond is indicated by a number when necessary. The common name for ethyne is **acetylene,** and this name is used almost exclusively.

Note:

Noncyclic alkynes with one triple bond have the general formula C_nH_{2n-2} *(n is an integer).*

Note:

Regardless of how they are drawn, triple bonds are linear.

HC≡CH
ethyne
(acetylene)

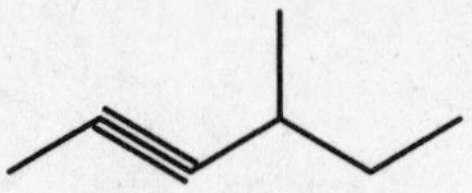

4-methyl-2-hexyne

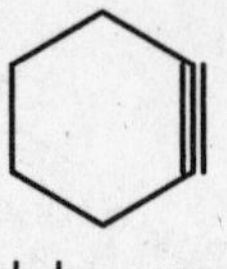

cyclohexyne

SUBSTITUTED ALKANES

A. HALOALKANES

Compounds that contain a halogen substituent are named as **haloalkanes**. The appendages are numbered and alphabetized as alkyl groups are treated. Notice that the presence of the halide does not dramatically affect the numbering of the chain—you should still proceed so that substituents receive the lowest possible numbers. For example:

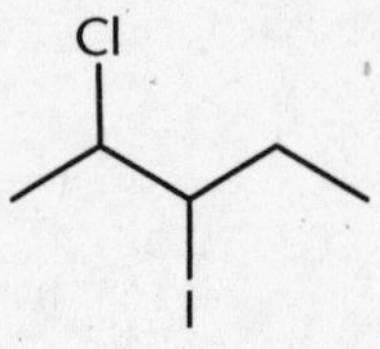

Cl

I

2-chloro-3-iodopentane

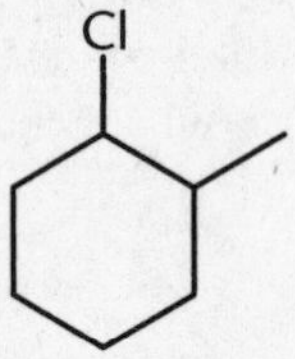

Cl

1-chloro-2-methylcyclohexane

Alternatively, the haloalkane may be named as an **alkyl halide**. In this system, chloroethane is called **ethyl chloride**. Other examples are:

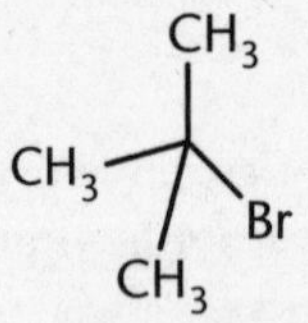

CH_3
CH_3
Br
CH_3

2-bromo-2-methylpropane
(*t*-butyl bromide)

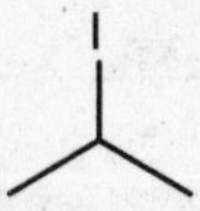

I

2-iodopropane
(isopropyl iodide)

B. ALCOHOLS

Note:

–OH has priority over a multiple bond when numbering the chain.

In the IUPAC system, **alcohols** are named by replacing the *-e* of the corresponding alkane with **-ol**. The chain is numbered so that the carbon attached to the hydroxyl group (–OH) receives the lowest number possible.

In compounds that possess a multiple bond and a hydroxyl group, numerical priority is given to the carbon attached to the –OH.

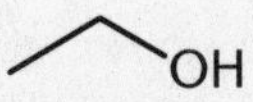

OH

ethanol

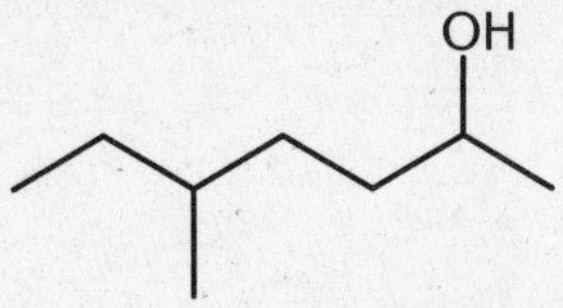

5-methyl-2-heptanol

hept-6-en-1-ol

A common system of nomenclature exists for alcohols in which the name of the alkyl group is combined with the word *alcohol*. These common names are used for simple alcohols. For example, methanol may be named "methyl alcohol," while 2-propanol may also be named "isopropyl alcohol."

Molecules with two hydroxyl groups are called **diols** (or **glycols**) and are named with the suffix **-diol**. Two numbers are necessary to locate the two functional groups. Diols with hydroxyl groups on adjacent carbons are referred to as **vicinal**, and diols with hydroxyl groups on the same carbon are **geminal**. Geminal diols (also called **hydrates**) are not commonly observed because they spontaneously lose water (**dehydrate**) to produce carbonyl compounds (containing C=O; see Chapter 18).

C. ETHERS

In the IUPAC system, **ethers** are named as derivatives of alkanes, and the larger alkyl group is chosen as the backbone. The ether functionality is specified as an **alkoxy-** prefix, indicating the presence of an ether (-oxy-), and the corresponding smaller alkyl group (alk-). The chain is numbered to give the ether the lowest position. Common names for ethers are frequently used. They are derived by naming the two alkyl groups in alphabetical order and adding the word *ether*. The generic term "ether" refers to diethyl ether, a commonly used solvent.

For **cyclic ethers**, numbering of the ring begins at the oxygen and proceeds to provide the lowest numbers for the substituents. Three-membered rings are termed **oxiranes** by IUPAC, although they are commonly called **epoxides**.

Note:

Ethers may have the prefix alkoxy- or be given a common name (e.g., ethyl methyl ether).

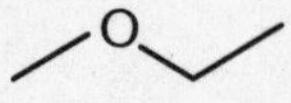

methoxyethane
(ethyl methyl ether)

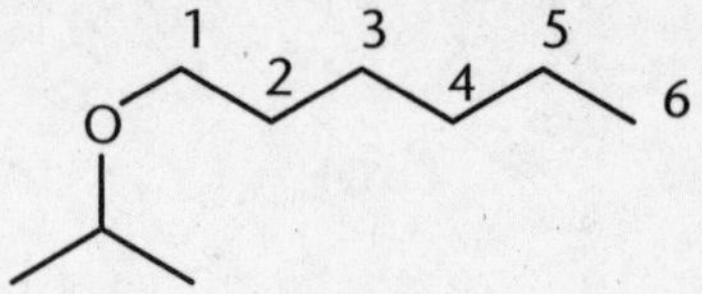

1 2 3 4 5 6

1-isopropoxyhexane
(n-hexyl isopropyl ether)

Note:

Cyclic ethers with 3 members are epoxides.

oxirane
(ethylene oxide)

2-methyloxirane
(propylene oxide)

tetrahydrofuran
(THF)

D. ALDEHYDES AND KETONES

Aldehydes are named according to the longest chain containing the aldehyde functional group. The suffix **-al** replaces the *-e* of the corresponding alkane. The carbonyl carbon receives the lowest number, although numbers are not always necessary since by definition an aldehyde is terminal and receives the number 1.

Note:

An aldehyde is a terminal functional group: it defines the C-1 of the backbone.

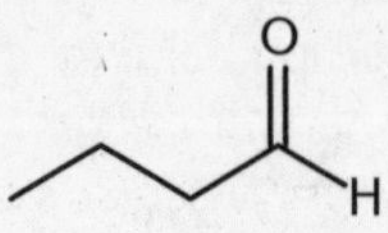

n-butanal

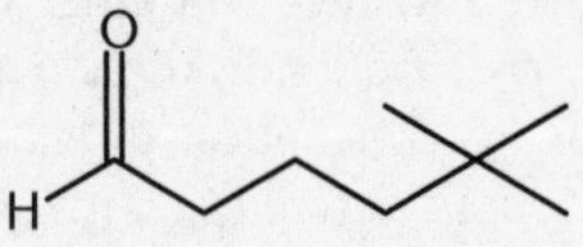

5,5-dimethylhexanal

The common names *formaldehyde, acetaldehyde,* and *propionaldehyde* are used almost exclusively instead of the IUPAC names *methanal, ethanal,* and *propanal*, respectively.

methanal
(formaldehyde)

ethanal
(acetaldehyde)

propanal
(propionaldehyde)

Ketones are named analogously, with **-one** as a suffix. The carbonyl group has to be assigned the lowest possible number. In complex molecules, the carbonyl group can be named as a prefix with the term **oxo-**. Alternatively, the individual alkyl groups may be listed in alphabetical order, followed by the word **ketone**.

Note:

The carbonyl in a ketone should receive the lowest number possible unless there is a higher priority group.

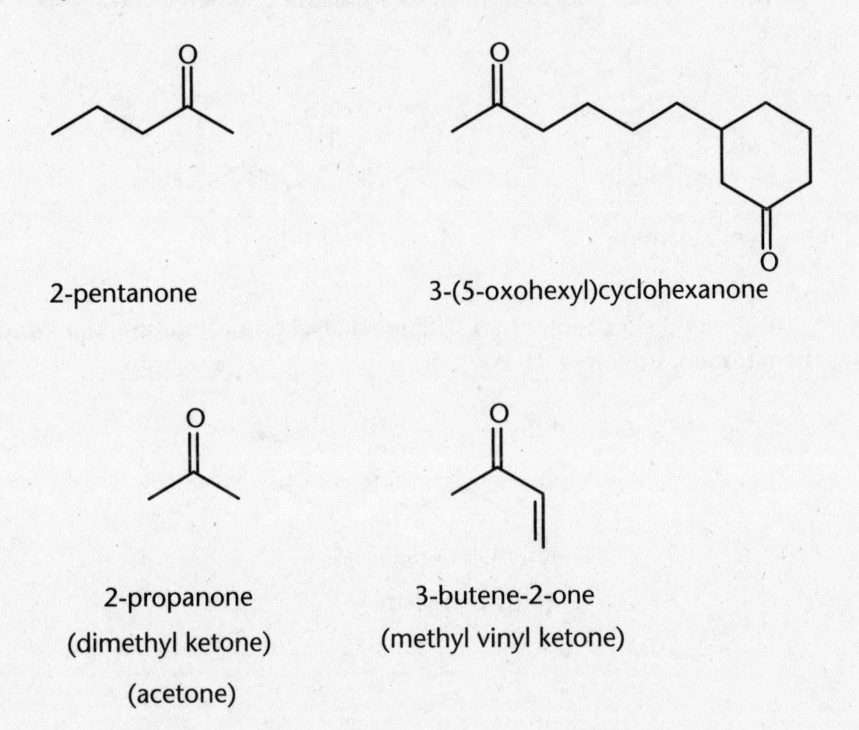

2-pentanone

3-(5-oxohexyl)cyclohexanone

2-propanone
(dimethyl ketone)
(acetone)

3-butene-2-one
(methyl vinyl ketone)

A commonly used alternative to the numerical designation of substituents is to term the carbon atom adjacent to the carbonyl carbon as α and the carbon atoms successively along the chain as β, γ, δ, etc. This system is encountered with dicarbonyl compounds and halocarbonyl compounds.

Note:

Carbons near a carbonyl can also be designed as α (adjacent) or β, γ, δ, etc. (moving farther away from the carbonyl.)

E. CARBOXYLIC ACIDS

Carboxylic acids are named with the ending **-oic** and the word **acid** replacing the *-e* ending of the corresponding alkane. Carboxylic acids are terminal functional groups and, like aldehydes, are numbered one (1). The common names formic acid (methanoic acid), acetic acid (ethanoic acid), and propionic acid (propanoic acid) are used almost exclusively.

Note:

Like aldehydes, the carboxylic acid group is a terminal functional group.

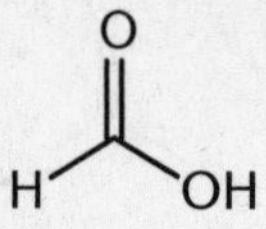

methanoic acid (formic acid)

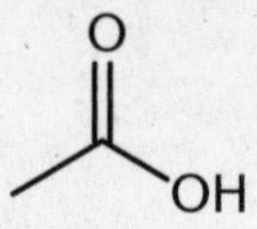

ethanoic acid (acetic acid)

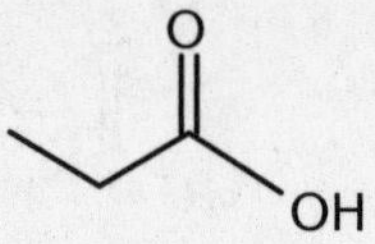

propanoic acid (propionic acid)

F. AMINES

The longest chain attached to the nitrogen atom is taken as the backbone. For simple compounds, name the alkane and replace the final "e" with "**amine.**" More complex molecules are named using the prefix amino-.

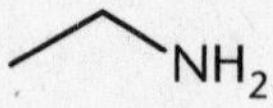

ethanamine

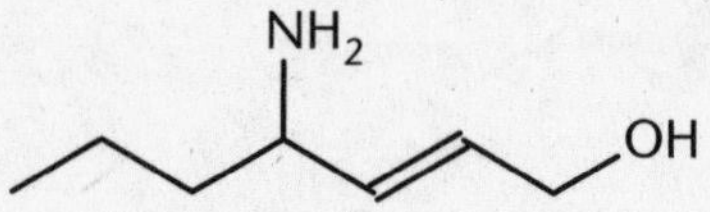

4-aminohept-2-en-1-ol

Note:

When additional alkyl groups are attached to the nitrogen, use the prefix N-.

To specify the location of an additional alkyl group that is attached to the nitrogen, the prefix ***N***- is used:

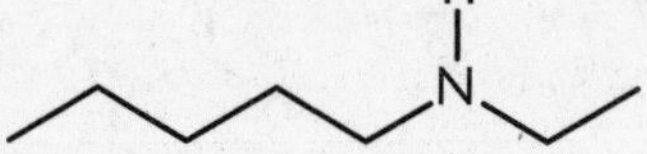

N-ethylpentanamine
(ethylpentylamine)

SUMMARY OF FUNCTIONAL GROUPS

Functional Group	Structure	IUPAC Prefix	IUPAC Suffix
Carboxylic acid	R–C(=O)–OH	carboxy-	-oic acid
Ester	R–C(=O)–OR	alkoxycarbonyl-	-oate
Acyl halide	R–C(=O)–X	halocarbonyl-	-oyl halide
Amide	R–C(=O)–NH_2	amido-	-amide
Nitrile/Cyanide	RC≡N	cyano-	-nitrile
Aldehyde	R–C(=O)–H	oxo-	-al
Ketone	R–C(=O)–R	oxo-	-one
Alcohol	ROH	hydroxy-	-ol
Thiol	RSH	sulfhydryl-	-thiol
Amine	RNH_2	amino-	-amine
Imine	$R_2C=NR'$	imino-	-imine
Ether	ROR	alkoxy-	-ether
Sulfide	R_2S	alkylthio-	
Halide	-I, -Br, -Cl, -F	halo-	
Nitro	RNO_2	nitro-	
Azide	RN_3	azido-	
Diazo	RN_2^+	diazo-	

In a Nutshell:

More complex molecules can also be named with the same 5 steps, with a few additional considerations:

1. *Multiple bonds should be on the main carbon backbone whenever possible.*
2. *–OH is a high priority functional group, placed above multiple bonds in numbering.*
3. *Haloalkanes, ethers, and ketones are often given common names (e.g., methyl chloride, ethyl methyl ether, diethyl ketone).*
4. *Aldehydes and carboxylic acids are terminal functional groups. If present, they define C–1 of the carbon chain (taking precedence over hydroxy, –OH, or multiple bonds.*
5. *Remember to specify the isomer, if relevant (such as cis or trans, R or S, etc.).*

REVIEW PROBLEMS

1. What is the IUPAC name of the following compound?

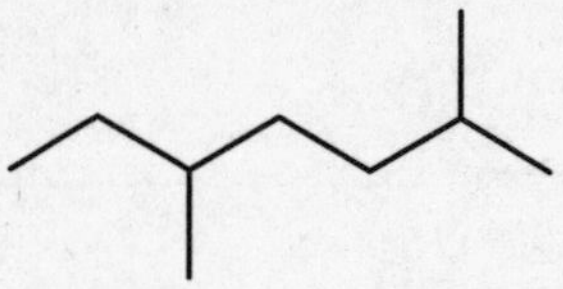

A. 2,5-dimethylheptane
B. 2-ethyl-5-methylhexane
C. 3,6-dimethylheptane
D. 5-ethyl-2-methylhexane

2. What is the structure of 5-ethyl-2,2-dimethyloctane?

3. What is the name of the following compound?

CH_3 CH_3 CH_3

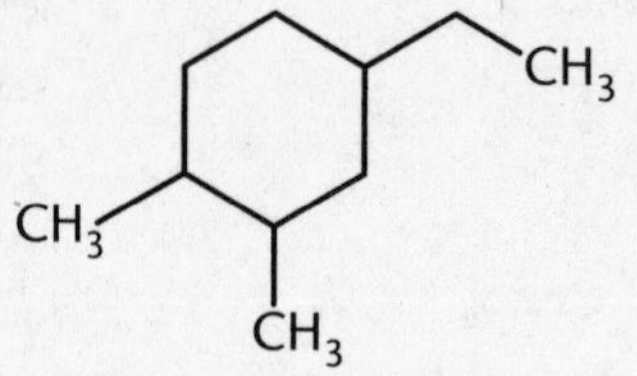

A. 1-ethyl-3,4-dimethylcycloheptane
B. 2-ethyl-4,5-dimethylcyclohexane
C. 1-ethyl-3,4-dimethylcyclohexane
D. 4-ethyl-1,2-dimethylcyclohexane

4. What is the name of the following compound?

I Br Cl Cl

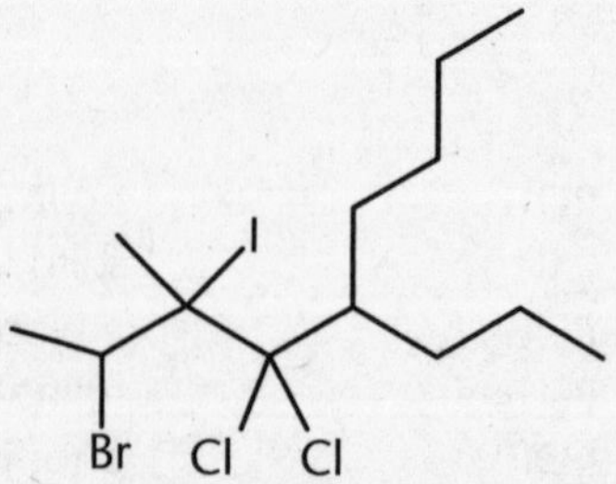

A. 2-bromo-5-butyl-4,4-dichloro-3-iodo-3-methyloctane
B. 7-bromo-4-butyl-5,5-dichloro-6-iodo-6-methyloctane
C. 2-bromo-4,4-dichloro-3-iodo-3-methyl-5-propylnonane
D. 2-bromo-5-butyl-4,4-dichloro-3-iodo-3-methylnonane

5. What is the name of the following compound?

$$\begin{array}{c} OH \\ | \\ CHCH_3 \\ | \\ CHCH_2CH_3 \end{array}$$
H / H_3C — C=C — H (trans)

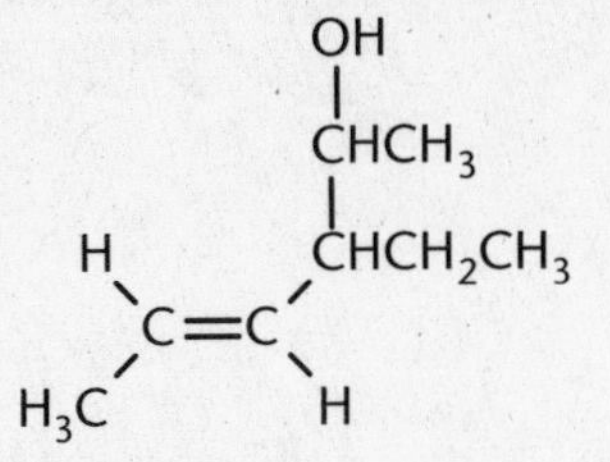

A. *trans*-3-ethyl-4-hexen-2-ol
B. *trans*-4-ethyl-2-hexen-5-ol
C. *trans*-3-ethanol-2-hexene
D. *trans*-4-ethanol-2-hexene

6. Indicate the α, β, γ, and δ carbons in the following compound.

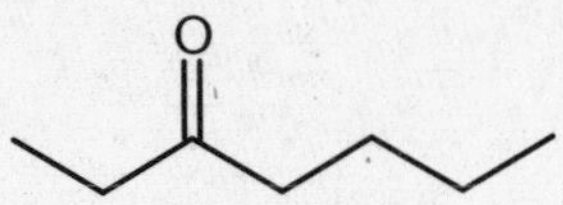

7. An alkane can be synthesized from its corresponding alkene by a reaction with hydrogen in the presence of a platinum catalyst. If 4-methyl-2-propyl-1-hexene (shown below) were treated with hydrogen in the presence of a platinum catalyst, what would be the name and structure of the alkane that would be produced?

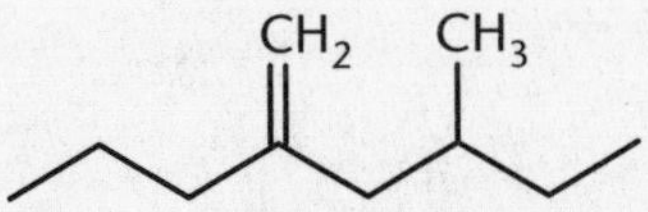

CH_2 CH_3

8. What is the correct structure for *cis*-1-ethoxy-2-methoxycyclopentane?

A. O O

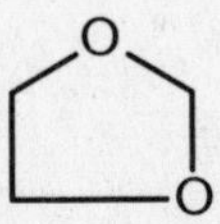

B. OCH_2CH_3 OCH_3

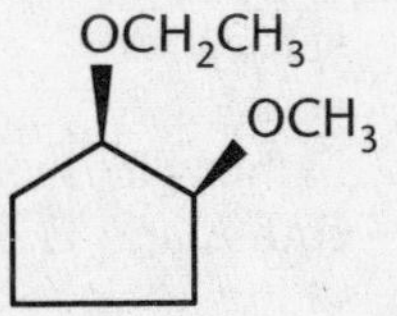

C. OCH_2CH_3 OCH_3

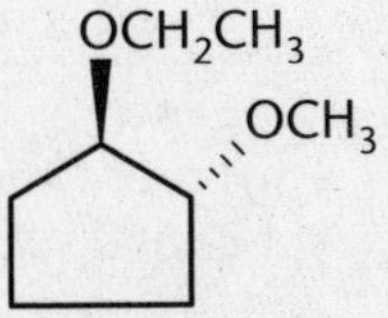

D. OCH_3 O

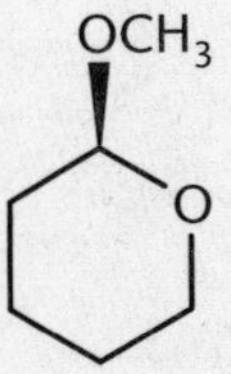

9. Do the following structures show the same compound or different compounds? Give a name for each structure.

A. 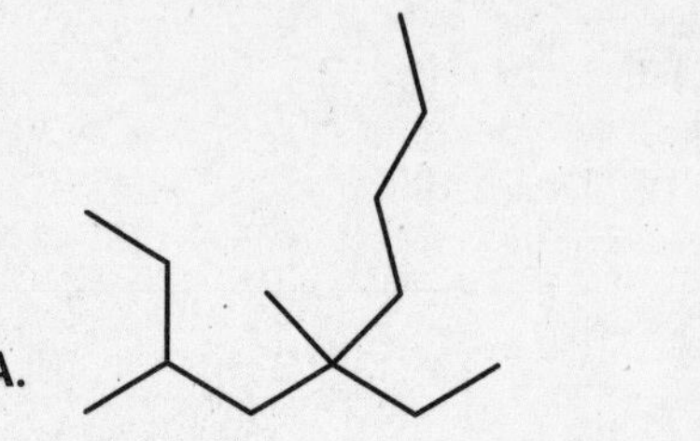

B. 

10. Match the name with the correct structure.

A. *t*-butyl
B. diene
C. β-keto acid
D. cyclohexanol
E. *sec*-butyl

1. 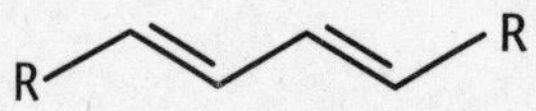 R ... R

2. 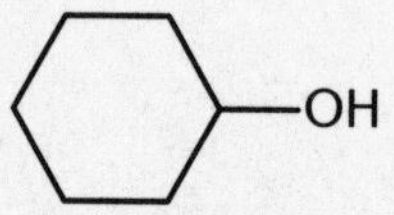 —OH

3. CH_3—CH— with C_2H_5 below

4. $(CH_3)_3C$—

5. 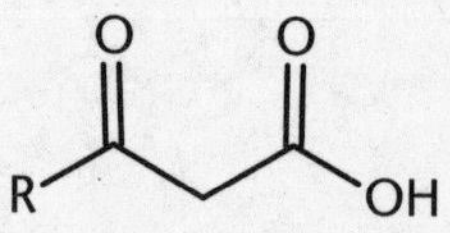 R, O, O, OH

11. Which of the following are considered terminal functional groups?

A. Aldehydes
B. Ketones
C. Carboxylic acids
D. Both A and C

SOLUTIONS TO REVIEW PROBLEMS BEGIN ON THE FOLLOWING PAGE

SOLUTIONS TO REVIEW PROBLEMS

1. **A** The first task in naming alkanes is to identify the longest chain. In this case, the longest chain has seven carbons, so the parent alkane is heptane. Choices B and D can be eliminated. The next step is to identify the substituents on the alkane chain. This compound has two methyl groups at carbons 2 and 5, so the correct IUPAC name is 2,5-dimethylheptane. Choice C is wrong because the position numbers of the substituents are not minimized.

2.

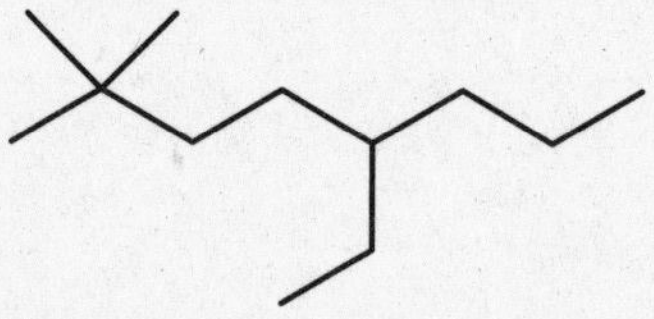

3. **D** Substituted cycloalkanes are named as derivatives of their parent cycloalkane, which in this case is cyclohexane. Thus, choice A can be ruled out immediately. Then the substituents are listed in alphabetical order and the carbons are numbered so as to give the lowest sum of substituent numbers. This cyclohexane has an ethyl and two methyl substituents; it is therefore an ethyl dimethyl cyclohexane. All of the remaining answer choices recognize this; they only differ in the numbers assigned. In order to give the lowest sum of substituent numbers, the two methyl substituents must be numbered 1 and 2, and the ethyl substituent must be numbered 4. The correct name for this compound is thus 4-ethyl-1,2-dimethylcyclohexane.

4. **C** This question requires the application of the same set of rules that was laid out in Question 1. The longest backbone has nine carbons, so the compound is a nonane. Thus, choices A and B can be ruled out immediately. The substituent groups are, in alphabetical order: bromo, chloro, iodo, methyl, and propyl; these substituents must be given the lowest possible number on the hydrocarbon backbone. The resulting name is 2-bromo-4,4-dichloro-3-iodo-3-methyl-5-propylnonane.

5. **A** The first step is to locate the longest carbon chain containing the functional groups (C=C and OH). The backbone has six carbons (hex). Since the alcohol group has higher priority than the double bond, it dictates the ending (ol) and is given the lower position (2). The alkene is named according to the position of the double bond on the backbone followed by ene or en if it's in the middle (4-hexene). Thus the chain is called 4–hexene–2–ol. The substituents on the backbone are an ethyl group and an alkene group an C–3 and C–4 respectively. Since these constituents lie on opposite sides of the double bond, the molecule is *trans*. Therefore, this compound is called *trans*–3–ethyl–4–hexen–2–ol.

6. Numbering the carbons on the backbone from left to right, C–2 and C–4 are α carbons, C–1 and C–5 are β carbons, C–6 is a γ carbon, and C–7 is a δ carbon. This nomenclature is used to specify how far a given carbon in the backbone is from a reactive center, usually a carbonyl carbon.

7. Treating this alkene with hydrogen in the presence of a platinum catalyst will cause hydrogen atoms to be added across the double bond; the product will be 3,5-dimethyloctane, shown below.

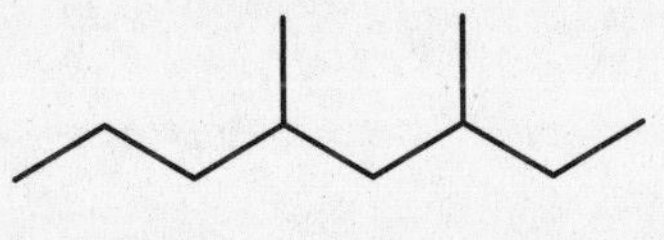

8. **B** A cyclopentane is a cyclic alkane with five carbons. A *cis* cyclic compound has both of its substituents on the same side of the ring. Only choices B and C have two substituents, so A and D can be ruled out. In fact, choice C is a *trans* compound, so the correct answer must be B. Ethoxy and methoxy represent ether substituents, and they must be on adjacent carbons on the same side of the molecule. Thus, the structure of *cis*-1-ethoxy-2-methoxycyclopentane is given by choice B.

9. These structures both represent the same compound, 5-ethyl-3,5-dimethylnonane. The best way to see this is to name each one; this forces you to figure out the backbone and the substituents.

10. A — **4**

B — **1**

C — **5**

D — **2**

E — **3**

11. **D** Both aldehyde and carboxylic acid functional groups are located on the terminal ends of carbon backbones. As a result, the carbon to which they are attached is named C–1, and choice D is correct. Ketones are always internal to the carbon chain.

Isomers

CHAPTER TWELVE

Isomers are chemical compounds that have the same molecular formula but differ in structure—that is, in their atomic connectivity, rotational orientation, or the 3-dimensional position of their atoms. Isomers may be extremely similar, sharing most or all of their physical and chemical properties, or they may be very different.

In a Nutshell:

Isomers have the same molecular formula but <u>differ</u> <u>in structure</u>. They may be extremely similar or extremely different.

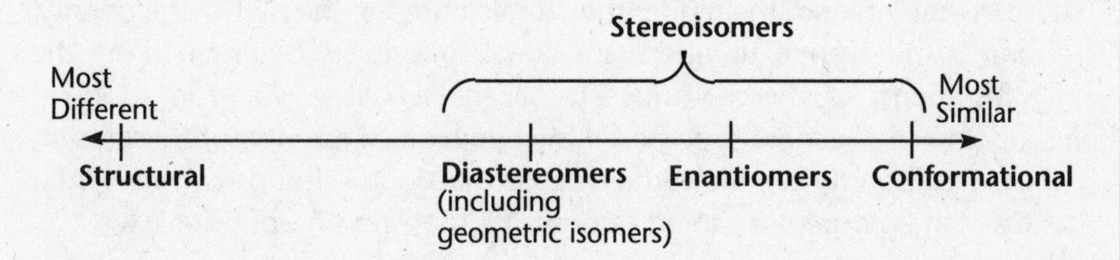

Structural isomers are the most unlike each other, while conformational isomers are the most similar.

STRUCTURAL ISOMERISM

Structural isomers, also known as constitutional isomers, are compounds that share only their molecular formula. Because their atomic connections may be completely different, they often have very different chemical and physical properties (such as melting point, boiling point, solubility, etc.). For example, five different structures exist for compounds with the formula C_6H_{14}.

In a Nutshell:
Structural isomers share only their molecular formula; their atomic connectivity is different. Therefore, they may have very different chemical and physical properties.

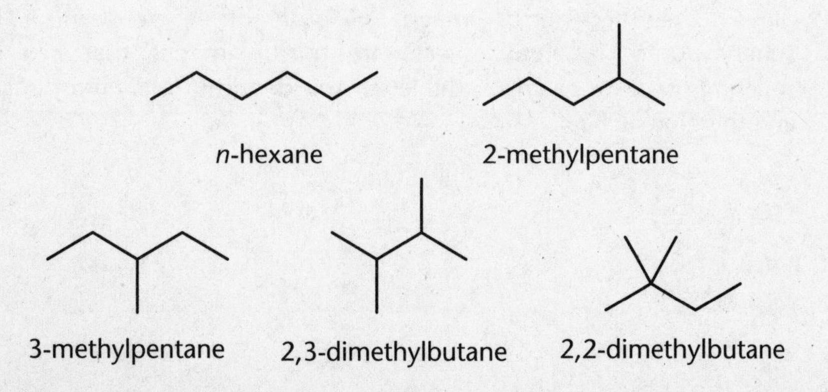

All have the same formula, but they differ in their carbon framework and in the number and type of atoms bonded to each other.

In a Nutshell:

Stereoisomers have the same atomic connections, but the atoms are arranged differently in space.

STEREOISOMERISM

Stereoisomers are compounds that differ from each other only in the way that their atoms are oriented in space. Geometric isomers, enantiomers, diastereomers, *meso* compounds, and conformational isomers all fall under this heading.

In a Nutshell:

Geometric isomers differ in their arrangement of atoms around a double bond. They may have different physical properties, such as melting and boiling points. However, they tend to have quite similar chemical reactivity.

A. GEOMETRIC ISOMERS

Geometric isomers are compounds that differ in the position of substituents attached to a double bond. If two substituents are on the same side, the double bond is called ***cis***. If they are on opposite sides, it is a ***trans*** double bond.

For compounds with polysubstituted double bonds, the situation can be confusing, and an alternative method of naming is employed. The highest priority substituent attached to each double bonded carbon has to be determined: the higher the atomic number, the higher the priority, and if the atomic numbers are equal, priority is determined by the substituents of these atoms. The alkene is called (*Z*) (from German *zusammen*, meaning together) if the two highest priority substituents on each carbon lie on the same side of the double bond, and (*E*) (from German *entgegen*, meaning opposite) if they are on opposite sides.

Mnemonic:

Z = zame zide (same side)
E = epposite (opposite sides)

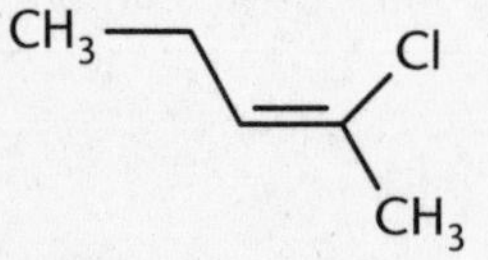

CH_3 Cl CH_3

(Z)-2-chloro-2-pentene

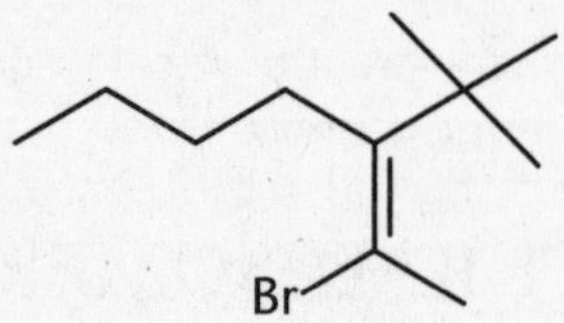

Br

(E)-2-bromo-3- *t*-butyl-2-heptene

Note:

Chirality = handedness. Non-superimposable mirror images exist—like your right and left hands.

B. CHIRALITY

An object that is not superimposable upon its mirror image is called **chiral**. Familiar chiral objects are your right and left hands. Although essentially identical, they differ in their ability to fit into a right-handed glove. They are mirror images of each other, yet cannot be superimposed. **Achiral** objects are mirror images that can be superimposed; for example, the letter A is identical to its mirror image and therefore achiral.

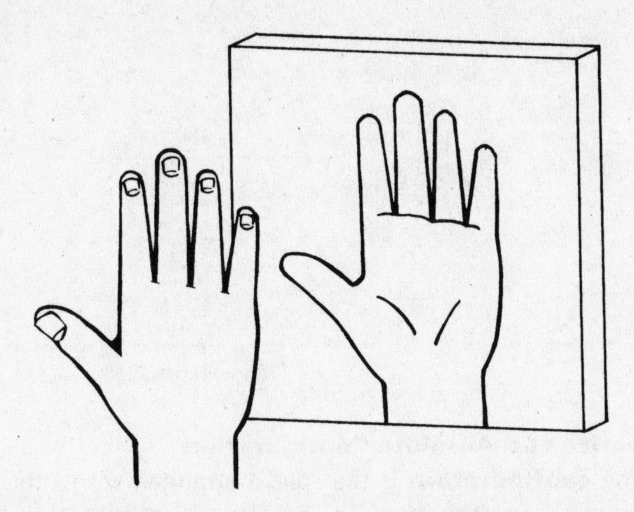

In organic chemistry, chirality is most frequently encountered when carbon atoms have four different substituents. Such a carbon atom is called *asymmetric* because it lacks a plane or point of symmetry. For example, the C1 carbon atom in 1-bromo-1-chloroethane has four different substituents. The molecule is chiral because it is not superimposable on its mirror image. Chiral objects that are non-superimposable mirror images are called **enantiomers** and are a specific type of stereoisomer.

Note:

A carbon must have four different substituents to be a chiral center.

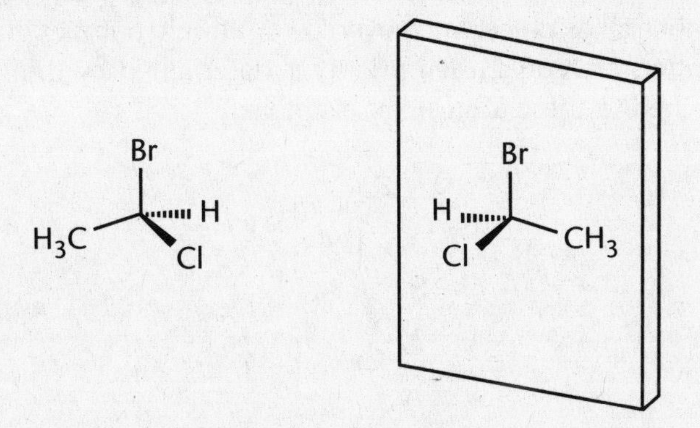

A carbon atom with only three different substituents, such as 1,1-dibromoethane, has a plane of symmetry and is therefore achiral. A simple 180° rotation along the *y*-axis allows the compound to be superimposed upon its mirror image.

Note:

Rotating a molecule in space does not change its chirality.

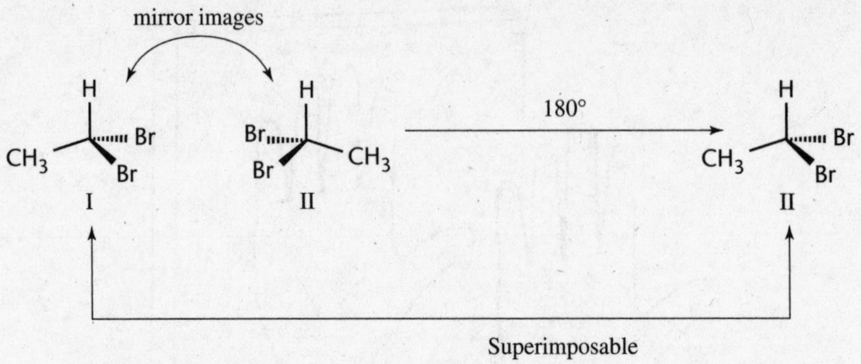
mirror images
H
Br
CH3
Br
I
H
Br
CH3
Br
II
180°
H
Br
CH3
Br
II
Superimposable

1. Relative and Absolute Configuration

The **configuration** is the spatial arrangement of the atoms or groups of a stereoisomer. The **relative configuration** of a chiral molecule is its configuration in relation to another chiral molecule. The **absolute configuration** of a chiral molecule describes the spatial arrangement of these atoms or groups. There is a set sequence to determine the absolute configuration of a molecule at a single chiral center:

Step 1:

Assign priority to the four substituents, looking only at the first atom that is directly attached to the chiral center. Higher atomic number takes precedence over lower atomic number. If the atomic numbers are equal, then priority is determined by the substituents attached to these atoms. For example:

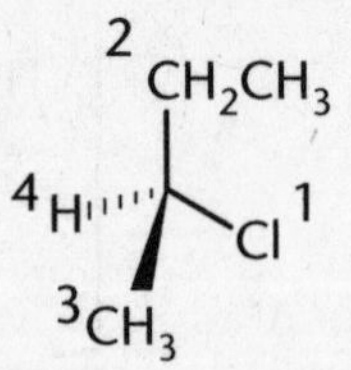
2
CH2CH3
4
H
1
Cl
3
CH3

Note:

When assigning priority, look only at the first atom attached to the chiral carbon, not at the group as a whole! The higher the atomic number, the higher the priority—this same system is used when determining Z and E.

Step 2:

Orient the molecule in space so that the line of sight proceeds down the bond from the asymmetric carbon atom (the chiral center) to the substituent with lowest priority. The three substituents with highest priority should radiate from the asymmetric atom like the spokes of a wheel.

Step 3:

Proceeding from highest priority (#1) on down, determine the order of substituents around the wheel as either clockwise or counterclockwise. If the order is clockwise, the asymmetric atom is called ***R*** (from Latin *rectus,* meaning right). If it is counterclockwise, it is called ***S*** (from Latin *sinister,* meaning left).

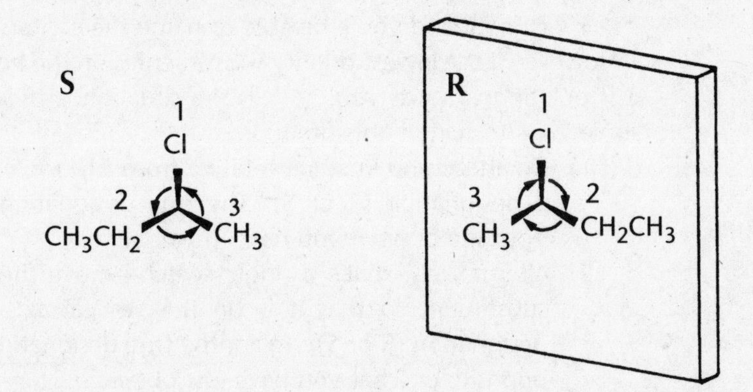

Step 4:

Provide a full name for the compound. The terms *R* and *S* are put in parentheses and separated from the rest of the name by a dash. If there is more than one asymmetric carbon, location is specified by a number preceding the *R* or *S* within the parentheses, without a dash.

2. Fischer Projections

A three-dimensional molecule can be conveniently represented in two dimensions in a **Fischer projection**. In this system, horizontal lines indicate bonds that project out from the plane of the page, while vertical lines indicate bonds behind the plane of the page. The

Note:

R and S are conventions of notation. They do not predict the direction of light rotation.

Mnemonic:

Think of a steering wheel. The lowest priority substituent (#4) should point away from you, down the column, while #1, #2, and #3 lie on the wheel itself.

Mnemonic:

Clockwise is like turning the steering wheel clockwise, which makes the car turn Right—so the chirality at that center is R. Or, think of the way you write an R and an S. An R is drawn with a clockwise movement, while an S is drawn with an anti-clockwise movement.

In a Nutshell:

To determine the absolute configuration at a single chiral center:

1. *Assign priority by atomic number.*
2. *Orient the molecule with the lowest priority substituent in the back.*
3. *Move around the molecule from highest to lowest priority (1 → 2 → 3).*
4. *Clockwise = R Counterclockwise = S.*

point of intersection of the lines represents a carbon atom. They can be interconverted by interchanging any two pairs of substituents, or by rotating the projection in the plane of the page by 180°. If only one pair of substituents is interchanged, or if the molecule is rotated by 90°, the mirror image of the original compound is obtained.

Mnemonic:

Think of a man with a bow-tie on. This will remind you that horizontal lines come <u>out</u> of the page.

Cl
H—Br ≡ H—Br
CH_3

Cl, H—Br, CH_3 ≡ Br, CH_3—Cl, H (180°) ≡ H, Cl—CH_3, Br

In a Nutshell:

Differences in chirality (handedness) can produce isomers such as enantiomers (nonsuperimposable mirror images) as well as diastereomers (isomers that differ in their chirality but are not mirror images). Meso compounds contain chiral centers but are not optically active due to an internal mirror plane of symmetry.

This provides another way to determine the chirality at a chiral center. If the lowest priority substituent is on the vertical axis, it is already pointing away from you. Simply picture moving from #1 → #2 → #3, and you'll be able to name the center.

However, if the lowest priority substituent is on the horizontal axis, it is pointing towards you, and so the situation is trickier. Here are some ways to handle this situation:

1) Go ahead and imagine rotating from #1 → #2 → #3. Obtain a designation (R or S). The <u>true</u> designation will be the opposite of what you have just obtained.
2) Alternatively, make a single switch—move the low priority substituent so that it is on the vertical axis. Obtain the designation (R or S). Again, the <u>true</u> designation will be the opposite of what you have just obtained.
3) Another approach is to make two switches or interconversions—that is, move the low-priority atom to the vertical axis and "trade" some other pair of atoms at the same time. This new molecule has the same configuration as the molecule you started with. So you can go ahead and determine the correct designation right away.

3. Optical Activity

Enantiomers have identical chemical and physical properties with one exception: **optical activity**. A compound is optically active if it has the ability to rotate plane-polarized light. Ordinary light is unpolarized. It consists of waves vibrating in all possible planes

Note:

Enantiomers have <u>identical</u> properties, except for optical activity.

perpendicular to its direction of motion. A polarizer allows light waves oscillating only in a particular direction to pass, producing plane-polarized light.

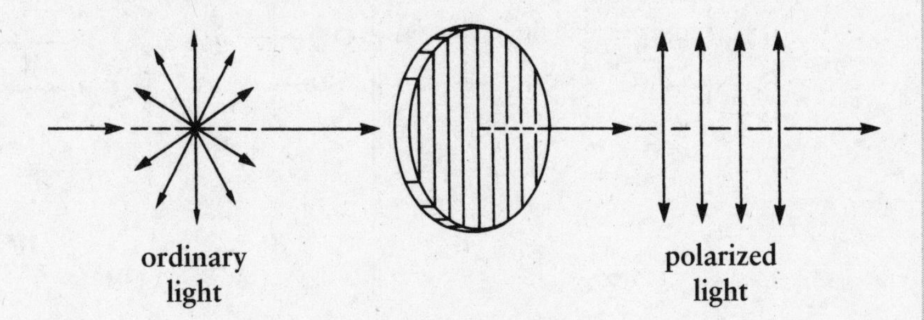

If plane-polarized light is passed through an optically active compound, the orientation of the plane is rotated by an angle α. The enantiomer of this compound will rotate light by the same amount, but in the opposite direction. A compound that rotates the plane of polarized light to the right, or clockwise (from the point of view of an observer seeing the light approach), is **dextrorotatory** and is indicated by (+). A compound that rotates light toward the left, or counterclockwise, is **levorotatory** and is labeled (–). The direction of rotation cannot be determined from the structure of a molecule and must be determined experimentally.

MCAT Favorite:

The direction of rotation can only be determined experimentally. S or R says nothing about the direction of rotation.

The amount of rotation depends on the number of molecules that a light wave encounters. This depends on two factors: the concentration of the optically active compound and the length of the tube through which the light passes. Chemists have set standard conditions of 1 g/ml for concentration and 1 dm for length in order to compare the optical activities of different compounds. Rotations measured at different concentrations and tube lengths can be converted to a standardized **specific rotation** (α) using the following equation:

$$\text{specific rotation } ([\alpha]) = \frac{\text{observed rotation}(\alpha)}{\text{concentration (g/ml)} \times \text{length (dm)}}$$

Bridge:

This equation can be rewritten as:
$\alpha = [\alpha] \bullet \text{conc} \bullet \text{length}$

A **racemic mixture,** or **racemic modification,** is a mixture of equal concentrations of both the (+) and (–) enantiomers. The rotations cancel each other and no optical activity is observed.

Note:

A racemic mixture displays no optical activity.

C. OTHER CHIRAL COMPOUNDS

1. Diastereomers

For any molecule with *n* chiral centers, there are 2^n possible stereoisomers. Thus, if a compound has two chiral carbon atoms, it has four possible stereoisomers (see figure below).

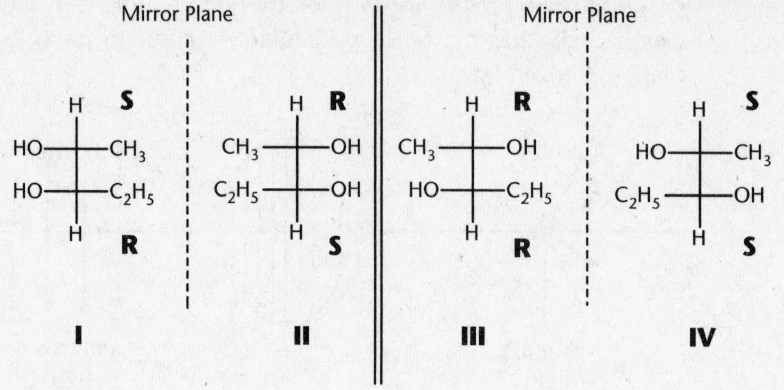

Mirror Plane | Mirror Plane

I: H, S; HO—C—CH_3; HO—C—C_2H_5; H, R
II: H, R; CH_3—C—OH; C_2H_5—C—OH; H, S
III: H, R; CH_3—C—OH; HO—C—C_2H_5; H, R
IV: H, S; HO—C—CH_3; C_2H_5—C—OH; H, S

Note:

Diastereomers may have different physical properties (such as solubility) and can therefore be separated by physical means.

I and II are mirror images of each other and are therefore enantiomers. Similarly, III and IV are enantiomers. However, I and III are not. They are stereoisomers that are not mirror images, and so they are called **diastereomers**. Notice that other combinations of nonmirror image stereoisomers are also diastereomeric. Hence I and IV, II and III, I and III, and II and IV are all pairs of diastereomers.

Note:

Meso compounds have a mirror image that is superimposable. Thus, they are not optically active.

Mnemonic:

MeSo compounds have a Mirror plane of Symmetry.

2. Meso Compounds

The criterion for optical activity of a molecule containing a single chiral center is that it has no plane of symmetry. The same applies to a molecule with two or more chiral centers. If a plane of symmetry exists, the molecule is not optically active, even though it possesses chiral centers. Such a molecule is called a *meso* compound. For example:

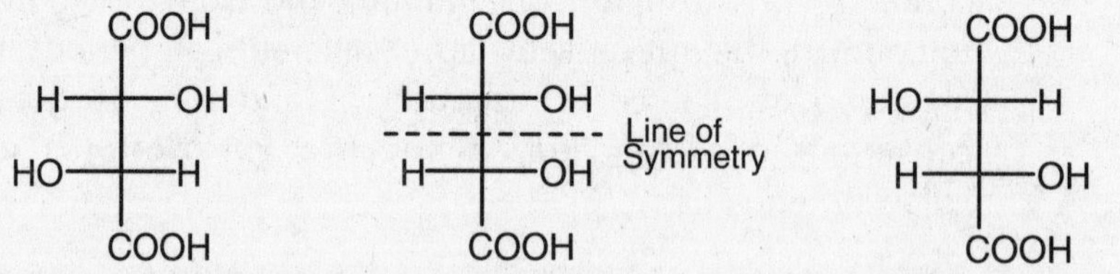

L-tartaric acid: COOH; H—C—OH; HO—C—H; COOH

Meso-tartaric acid: COOH; H—C—OH; ---- Line of Symmetry ----; H—C—OH; COOH

D-tartaric acid: COOH; HO—C—H; H—C—OH; COOH

D- and L-tartaric acid are both optically active, but meso-tartaric acid has a plane of symmetry and is not optically active. Although meso-tartaric acid has two chiral carbon atoms, the lack of optical activity is a function of the molecule as a whole.

D. CONFORMATIONAL ISOMERISM

Conformational isomers are compounds that differ only by rotation about one or more single bonds. Essentially, these isomers represent the same compound in a slightly different position—analogous to a person who may be either standing up or sitting down. These different conformations can be seen when the molecule is depicted in a **Newman projection,** in which the line of sight extends along a carbon-carbon bond axis. The conformations are encountered as the molecule is rotated about this axis. The classic example for demonstrating conformational isomerism in a straight chain is *n*-butane. In a Newman projection, the line of sight extends through the C2-C3 bond axis.

90°

CH_3 H H H H CH_3

CH_3 H H H H CH_3

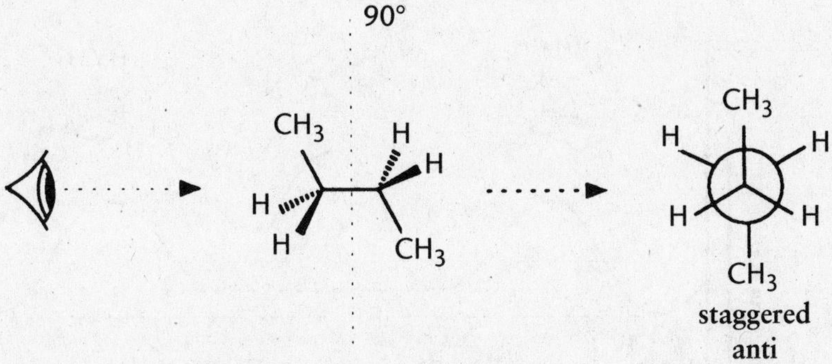

staggered
anti

1. Straight-chain Conformations

The most stable conformation is when the two methyl groups (C1 and C4) are oriented 180° from each other. There is no overlap of atoms along the line of sight (besides C2 and C3), so the molecule is said to be in a **staggered** conformation. Specifically, it is called the ***anti*** conformation, because the two methyl groups are antiperiplanar to each other. This particular orientation is very stable and thus represents an energy minimum because all atoms are far apart, minimizing repulsive steric interactions.

The other type of staggered conformation, called ***gauche,*** occurs when the two methyl groups are 60° apart. In order to convert from the anti to the gauche conformation, the molecule must pass through an **eclipsed** conformation, in which the two methyl groups are 120° apart and overlap with the H atoms on the adjacent carbon. When the two methyl groups overlap with each other, the molecule is said to be **totally eclipsed** and is in its highest energy state.

Note:

Think of conformational isomers as being different positions (conformations) of a compound—like a person who may be standing up or sitting down. Thus, they are often easily interconverted.

Mnemonic:

It's "gauche" (or inappropriate) for one methyl group to stand too close to another group!

Mnemonic:

Groups are "eclipsed" when they are completely in line with one another—think of a solar eclipse!

Note:

Notice that the anti isomer has the lowest energy, while the totally eclipsed isomer has the highest energy.

60° gauche

120° eclipsed

0° totally eclipsed

A plot of potential energy versus the degree of rotation about the C2-C3 bond shows the relative minima and maxima the molecule encounters throughout its various conformations.

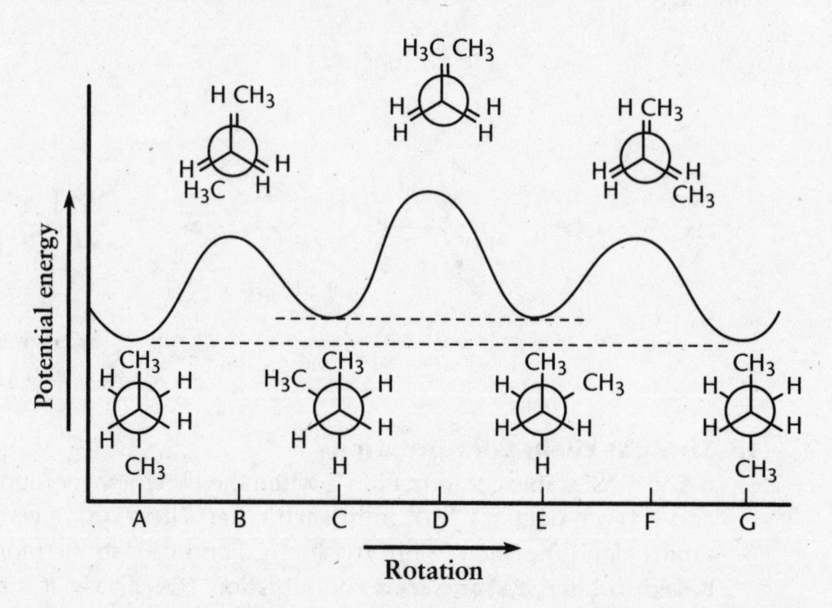

It is important to note that these barriers are rather small (3–4 kcal/mol) and are easily overcome at room temperature. Very low temperatures will slow conformational interconversion. If the molecules do not possess sufficient energy to cross the energy barrier, they may not rotate at all.

Note:

At room temperature, these forms easily interconvert—so if someone comes up and offers you a jar of all anti-butane, be suspicious! All forms are present, to some degree.

2. Cyclic Conformations

a. Strain Energies

In cycloalkanes, ring strain arises from three factors: angle strain, torsional strain, and nonbonded strain. Angle strain results when bond angles deviate from their ideal values; torsional strain results when cyclic molecules must assume conformations that have eclipsed interactions; and nonbonded strain (Van der Waals repulsion) results when atoms or groups compete for the same space. In order to alleviate these three types of strain, cycloalkanes attempt to adopt nonplanar conformations. Cyclobutane puckers into a slight V shape, cyclopentane adopts what is called the **envelope** conformation, and cyclohexane exists mainly in three conformations called the **chair,** the **boat,** and the **twist** or **skew-boat**.

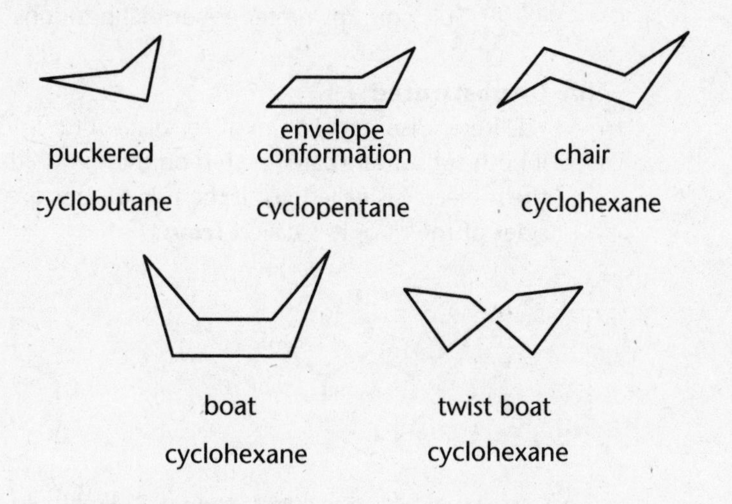

Conformations of cyclic hydrocarbons

Note:

Chair and boat are the most important forms of cyclohexane.

b. Cyclohexane

i. Unsubstituted

The most stable conformation of cyclohexane is the chair conformation. In this conformation, all three types of strain are eliminated. The hydrogen atoms that are perpendicular to the plane of the ring are called axial, and those parallel are called equatorial. The axial-equatorial orientations alternate around the ring.

The boat conformation is adopted when the chair "flips" and converts to another chair. In such a process, hydrogen atoms that were equatorial become axial, and vice versa, in the new chair. In the boat conformation, all of the atoms are eclipsed, creating a high-energy state. To avoid this strain, the boat can twist into a slightly more stable form called the twist or skew-boat conformation.

Mnemonic:

When you have low energy, you sit down in a chair to rest. Boats can be tippy, so they are less stable.

Mnemonic:

Axial substituents are on a vertical axis, like your axial skeleton (skull and spine). Equatorial substituents go around the middle, like the earth's equator.

ii. Monosubstituted

The interconversion between the two chairs can be slowed or even prevented if a sterically bulky group is attached to the ring. The equatorial position is favored over the axial position because of steric repulsion with other axial substituents. Hence, a large group such as *t*-butyl can lock the molecule in one conformation.

Note:

A bulky substituent can prevent the ring from adapting certain conformations.

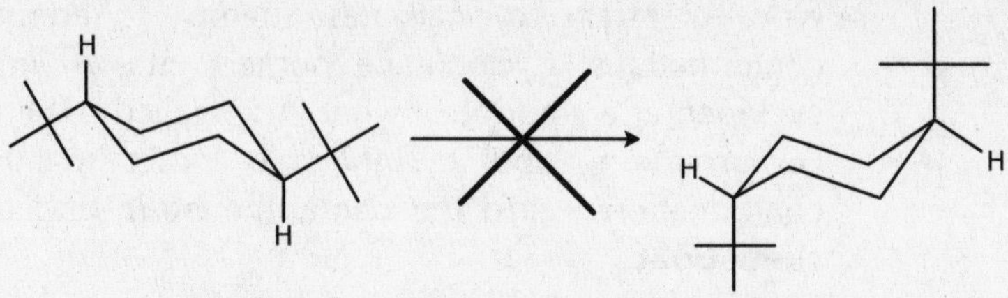

Bulky groups prefer equatorial positions

iii. Disubstituted

Different isomers can exist for disubstituted cycloalkanes. If both substituents are located on the same side of the ring, the molecule is called ***cis;*** if the two groups are on opposite sides of the ring, it is called ***trans.***

Note:

Cis *and* trans *apply to cycloalkanes too!*

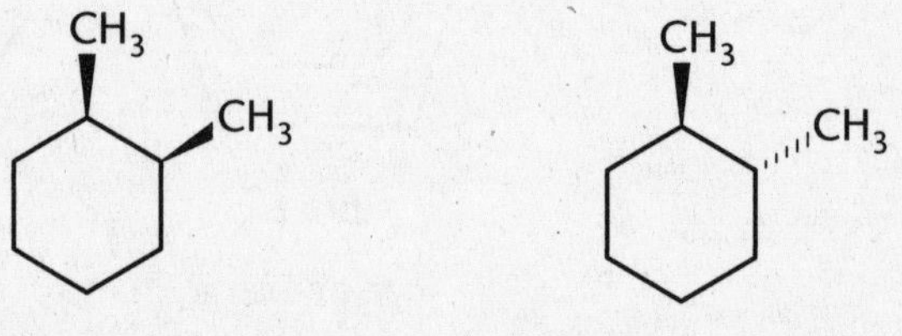

cis-1,2-dimethylcyclohexane *trans*-1,2-dimethylcyclohexane

In *trans*-1,4-dimethylcyclohexane, both of the methyl groups are equatorial in one chair conformation and axial in the other, but in either case they point in opposite directions relative to the plane of the ring.

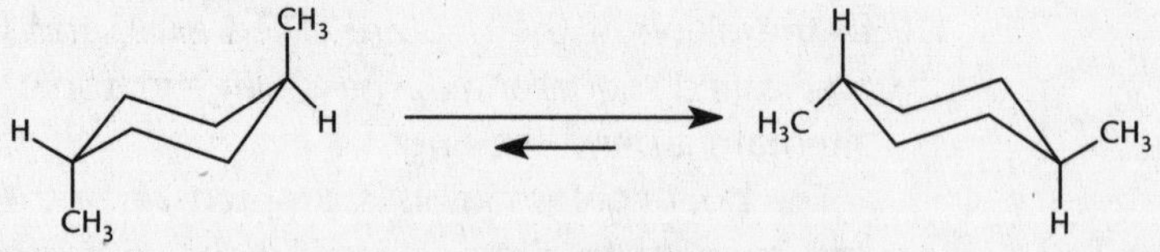

trans-1 ,4-dimethylcyclohexane

REVIEW PROBLEMS

1. Categorize the following pairs as enantiomers, diastereomers, structural isomers, molecules of the same compound, or different compounds.

a. dimethyl ether and ethanol

b.

```
      CH3                    CH3
       |                      |
HO ----+---- Br   and   Br ---+--- OH
       |                      |
      CH3                    CH3
```

c.

```
      CH3                CH3
       |                  |
H -----+---- Br   and  H -+--- Cl
       |                  |
       Cl                 Br
```

d.

```
       C2H5                     Cl
        |                        |
HO -----+---- Br        HO ------+---- C2H5
        |          and           |
HO -----+---- Cl        HO ------+---- Br
        |                        |
       C2H5                     C2H5
```

2. Which of the following do not show optical activity?

A. (*R*)-2-butanol
B. (*S*)-2-butanol
C. A solution containing 1 M (*R*)-2-butanol and 2 M (*S*)-2-butanol
D. A solution containing 2 M (*R*)-2-butanol and 2 M (*S*)-2-butanol

3. How many stereoisomers exist for the following aldehyde?

```
         O
         ‖
         C—H
         |
   HO—C—H
         |
   HO—C—H
         |
   HO—C—H
         |
   HO—C—H
         |
         H
```

A. 2
B. 4
C. 8
D. 16

4. Which of the following compounds is optically inactive?

A. 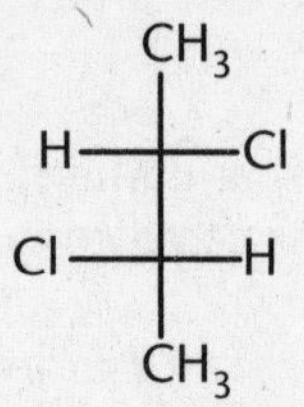

B. 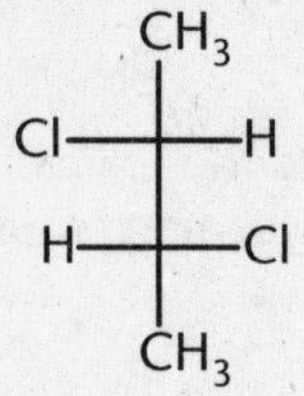

C. 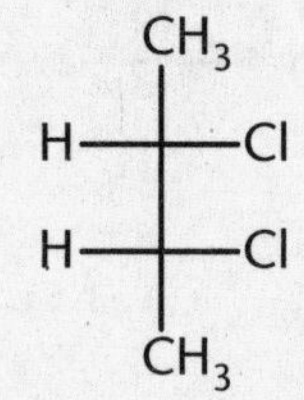

D. 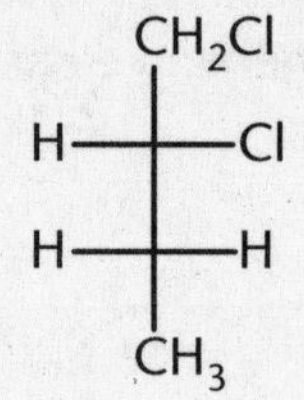

5. Which isomer of 2-pentene is more stable, the *cis*-isomer or the *trans*-isomer?

6 Assign (*R*) and (*S*) designations to the following compounds:

a. 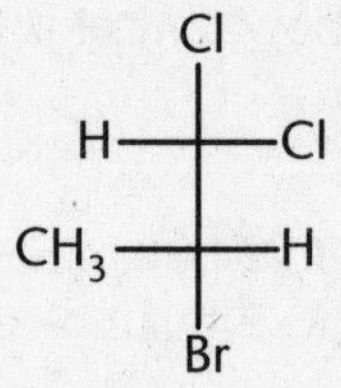

b. 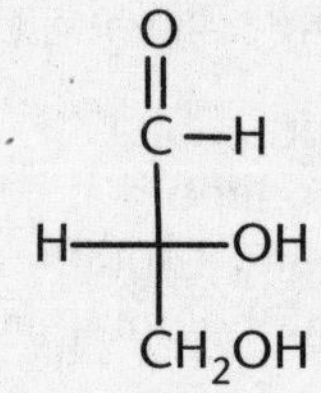

7. Cholesterol, shown below, contains how many chiral centers?

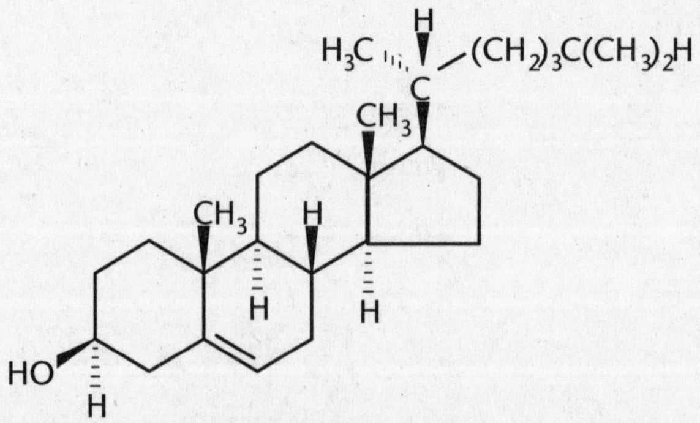

CHOLESTEROL

A. 5
B. 7
C. 8
D. 9

8. Which isomer of the following compound is the most stable?

A. 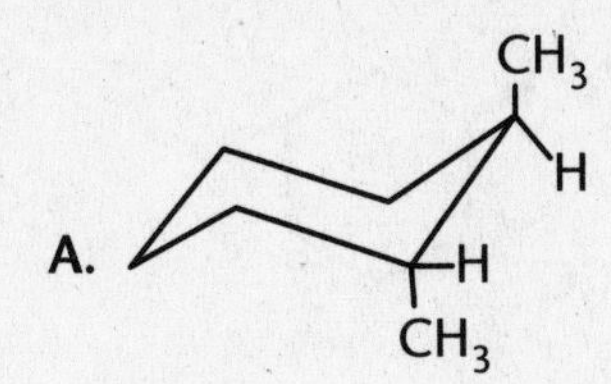

B. 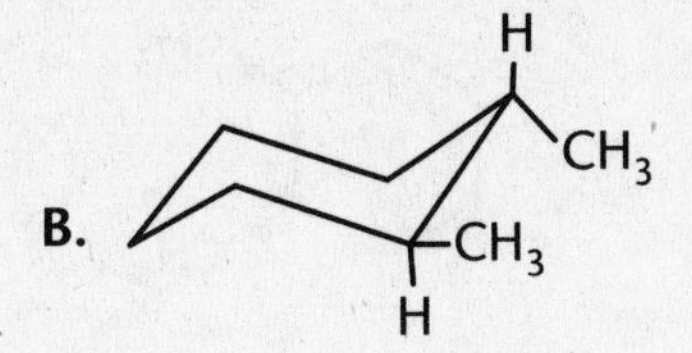

C. 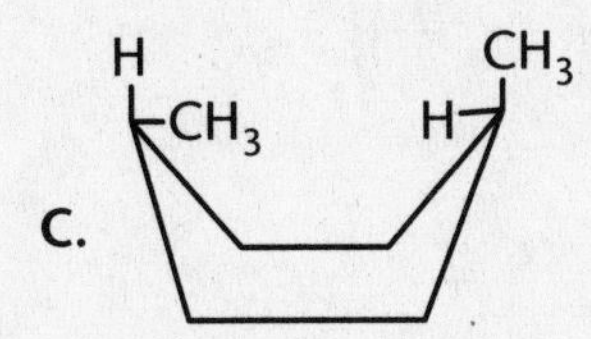

D. They are all equally stable.

9. Designate the following compounds as (*R*) or (*S*):

a. 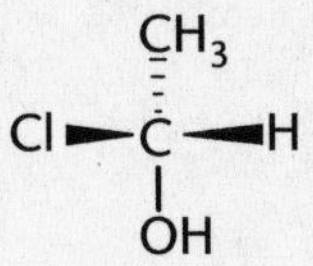 CH_3 (top); Cl—C—H; OH (bottom)

b. H (top); HO_2C—C—CH_2CH_3; CH_3 (bottom)

c. Br (top); HO—C—Cl; F (bottom)

d. $HC{=}CH_2$ (top); H—C—CH_2CH_3; $C(CH_3)_3$ (bottom)

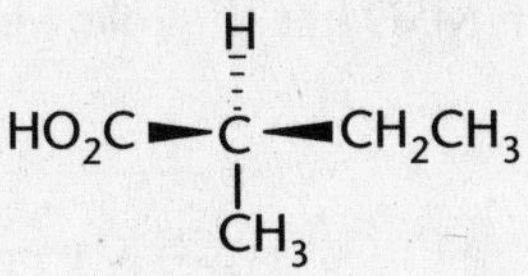

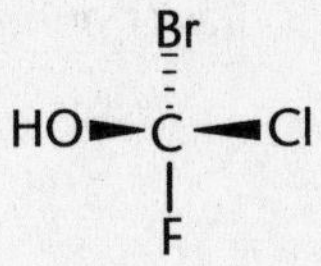

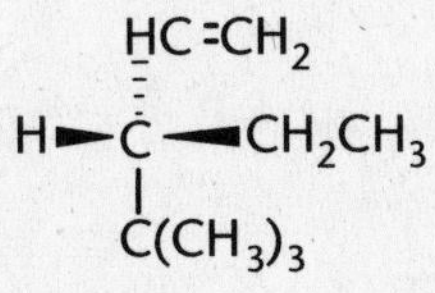

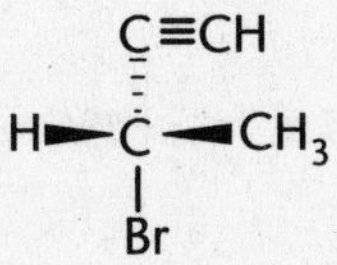

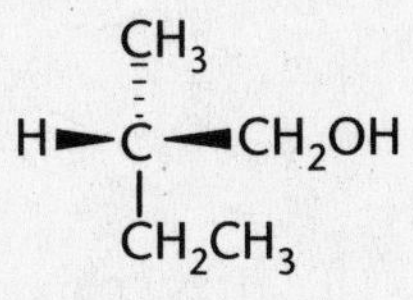

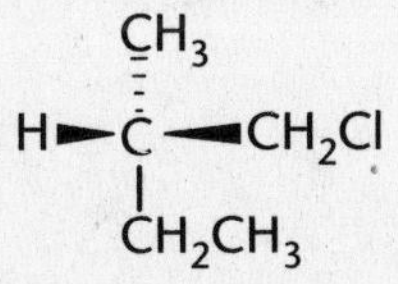

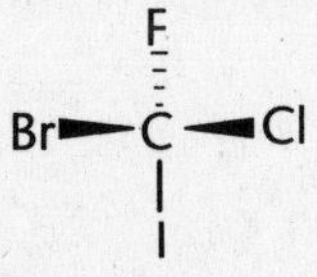

e. C≡CH (top); H—C—CH_3; Br (bottom)

f. CH_3 (top); H—C—CH_2OH; CH_2CH_3 (bottom)

g. CH_3 (top); H—C—CH_2Cl; CH_2CH_3 (bottom)

h. F (top); Br—C—Cl; I (bottom)

i. NH_2 (top); H_3C—C—Br; OH (bottom)

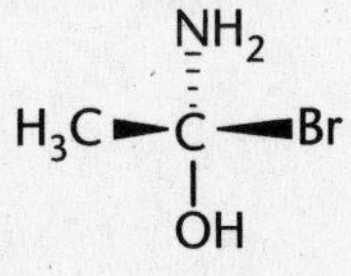

10. The following reaction results in:

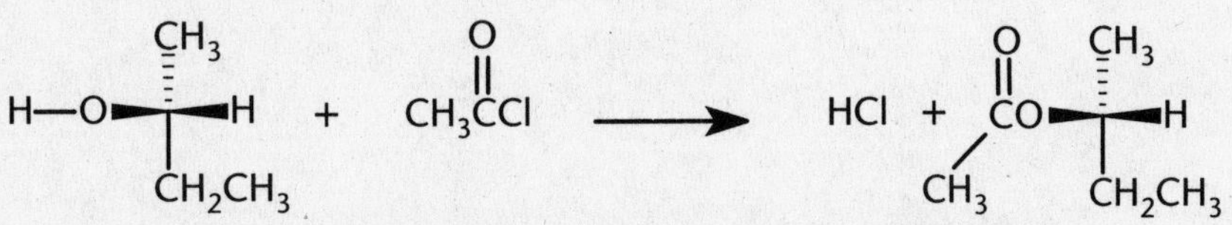

H—O—C(CH_3)(H)(CH_2CH_3) + CH_3CCl (C=O) ⟶ HCl + CH_3C(=O)O—C(CH_3)(H)(CH_2CH_3)

A. retention of relative configuration and a change in the absolute configuration.
B. a change in the relative and absolute configurations.
C. retention of the relative and absolute configurations.
D. retention of the absolute configuration and a change in the relative configuration.

11. The following pair of structures are:

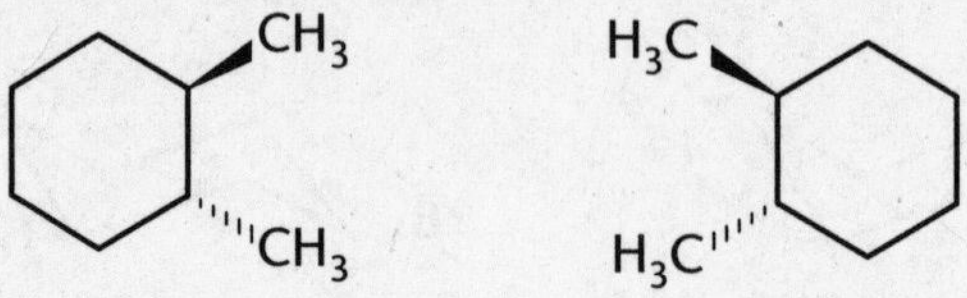

CH_3 CH_3 H_3C H_3C

I II

A. enantiomers.
B. diastereomers.
C. meso compounds.
D. structural isomers.

12. MSG ([*S*]-monosodium glutamate) is a compound widely used as a flavor enhancer. MSG has the following structure:

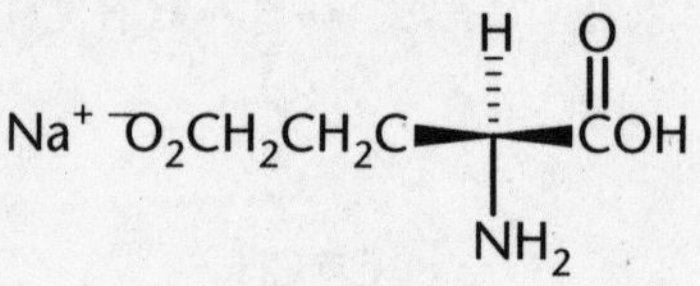

$Na^+ \ ^-O_2CH_2CH_2C$ — H, NH_2 — COOH

(*S*)-MSG

This enantiomer of MSG has a specific rotation of +24°. In a racemic mixture of MSG, what would be the specific rotation? What is the specific rotation of (*R*)-monosodium glutamate?

SOLUTIONS TO REVIEW PROBLEMS BEGIN ON THE FOLLOWING PAGE

SOLUTIONS TO REVIEW PROBLEMS

1. a. **Structural isomers**

The two compounds, CH_3CH_2OH and CH_3OCH_3, have the same molecular formula, C_2H_6O, but differ in how the atoms are connected to each other. For instance, the oxygen atom of ethanol is bonded to a carbon atom on one side and to a hydrogen atom on the other, whereas the oxygen atom of dimethyl ether is bonded to carbon atoms on both sides.

b. **Molecules of the same compound**

The two structures are achiral. Rotating one gives the other.

Another way to look at this is to recall that, in order for a compound to be chiral, at least one of its carbons must be bonded to four different substituents. In both of these compounds, all the carbons are attached to two identical groups, so the compounds are achiral.

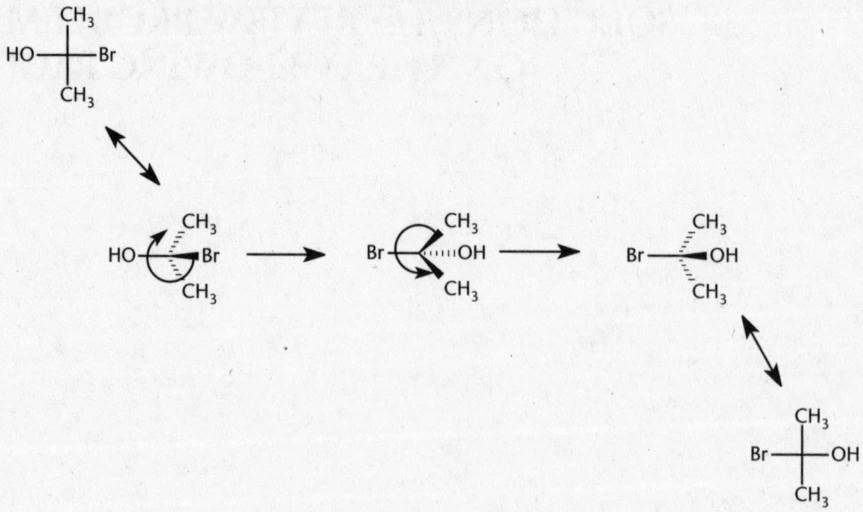

CH_3
HO—Br
CH_3

CH_3 HO Br CH_3 ⟶ CH_3 Br OH CH_3 ⟶ CH_3 Br OH CH_3

CH_3
Br—OH
CH_3

c. **Enantiomers**

These two compounds look alike, except the Cl and the Br have been interchanged. You may remember that one "trade" produces the mirror image of the original compound. Thus, these two molecules are enantiomers.

Another way to solve this problem is to remember that according to the rules governing Fischer projections, interchanging two pairs of substituents in the first compound, as shown below, will give the same compound. The two compounds can now be easily seen as mirror images of each other; hence, they are enantiomers.

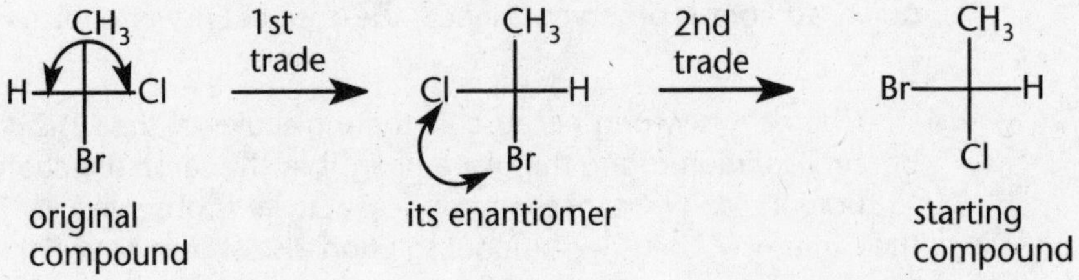

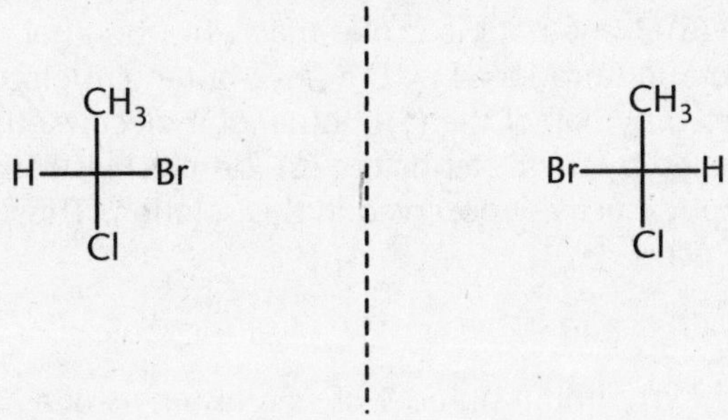

original compound (CH$_3$ top, H left, Cl right, Br bottom) —1st trade→ its enantiomer (CH$_3$ top, Cl left, H right, Br bottom) —2nd trade→ starting compound (CH$_3$ top, Br left, H right, Cl bottom)

CH$_3$ top, H left, Br right, Cl bottom	CH$_3$ top, Br left, H right, Cl bottom

d. **Diastereomers**

Rotate the second compound in the plane of the paper by 180°. The two compounds now are:

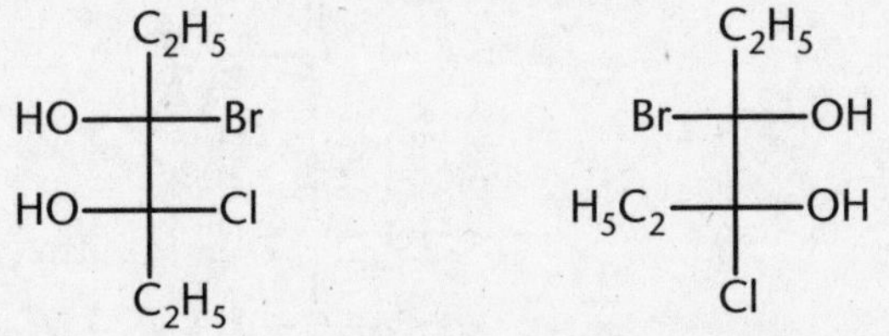

C$_2$H$_5$ top; HO—C—Br; HO—C—Cl; C$_2$H$_5$ bottom	C$_2$H$_5$ top; Br—C—OH; H$_5$C$_2$—C—OH; Cl bottom

Assign (*R*) and (*S*) designations to the compounds (discussed in notes).

C$_2$H$_5$ top; HO—C(*S*)—Br; HO—C(*R*)—Cl; C$_2$H$_5$ bottom	C$_2$H$_5$ top; Br—C(*R*)—OH; H$_5$C$_2$—C(*R*)—OH; Cl bottom

It is clear from the above figure that one compound is an *S, R* stereoisomer and the other is an *R, R* stereoisomer; hence, they are diastereomers.

2. **D** This is a racemic mixture of 2-butanol since it consists of equimolar amounts of (*R*)-2-butanol and (*S*)-2-butanol. The (*R*)-2-butanol molecule rotates the plane of polarized light in one direction, and the (*S*)-2-butanol molecule rotates it by the same angle but in the opposite direction. For every (*R*)-2-butanol molecule there is an (*S*)-2 butanol molecule; as a result exact cancellation of all rotation occurs, so no net rotation of polarized light is observed. Hence, the correct answer choice is D.

Choice A is wrong because all the molecules of the (*R*)-2-butanol solution rotate the plane of light in the same direction, so rotations do not cancel and optical activity is observed. In the same way, the (*S*)-2-butanol solution also shows optical activity. Thus, choices A and B are incorrect. Choice C has more (*S*)-2-butanol molecules than (*R*)-2-butanol molecules. All the rotation produced by the (*R*)-2-butanol molecules is canceled by half of the (*S*)-2-butanol molecules; the rotation produced by the other half of (*S*)-2-molecules contributes to the optical activity observed in this solution. Thus, choice C is incorrect.

3. **C** The maximum number of stereoisomers of a compound equals 2^n, where n is the number of chiral carbons in the compound. Here, there are three chiral carbon atoms (n=3) marked by asterisks in the following figure:

```
        O
        ‖
        C—H
        |
   HO—C—H
       *|
   HO—C—H
       *|
   HO—C—H
       *|
   HO—C—H
        |
        H
```

So the number of stereoisomers it can form is $2^n = 2^3 = 8$. Hence, the correct choice is C.

4. **C** The answer choice is an example of a *meso* compound: a compound that contains chiral centers but is superimposable on its mirror image. A *meso* compound can also be recognized by the fact that one half of the compound is the mirror image of the other half:

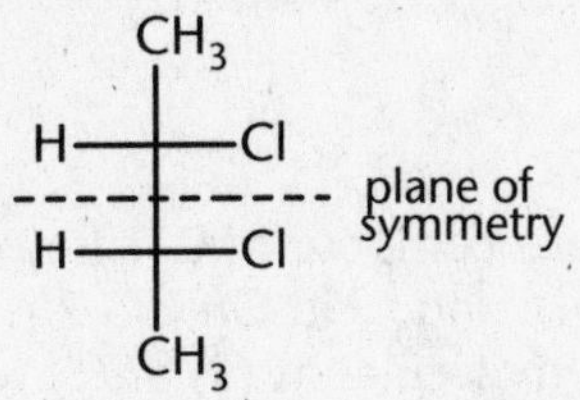

$$\begin{array}{c} CH_3 \\ H - C - Cl \\ \text{- - - - - - - - -} \\ H - C - Cl \\ CH_3 \end{array} \quad \text{plane of symmetry}$$

As a result of this internal plane of symmetry, the molecule is achiral and hence optically inactive. Choices A and B are enantiomers of each other and will certainly show optical activity on their own. Choice D, since it contains a chiral carbon, is optically active as well.

5. The *trans* isomer of 2-pentene is the most stable. If you draw out this isomer, you can see that on one side of the double bond there is a methyl group and a hydrogen, and on the other side, an ethyl group and a hydrogen. In this configuration, the van der Waals repulsion (nonbonding interaction) between the methyl and the ethyl groups is minimized. If these groups were oriented *cis* to each other (on the same side of the double bond), the van der Waals repulsion would be maximized.

6. a. **(R)-1,1-dichloro-2-bromopropane**

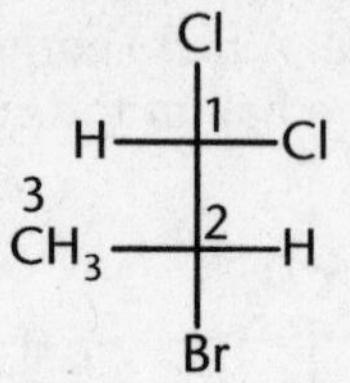

Cl
H—1—Cl
3 CH$_3$—2—H
Br

In the above compound, C–1 is not chiral because it has two Cl atoms bound to it. Only C–2 is chiral, so we have to designate the molecules as (*R*) or (*S*) with respect to C–2. Assign priority numbers to the atoms connected directly to C–2. The atom with the highest atomic number is given priority **1,** followed by the atom with next highest atomic number, which is given the priority number **2,** and so on. In this compound, Br has the highest atomic number and so is assigned priority **1**. Next come two carbon atoms connected to the chiral carbon. Since these have the same atomic number, we consider their substituents. C–1 has two Cl atoms and one hydrogen, whereas C–3 has three hydrogens. Since chlorine has a higher atomic number than hydrogen, we assign a higher priority to C–1. Thus, C–1 is assigned **2** and C–3 is assigned **3**. The hydrogen is assigned the lowest priority, **4**.

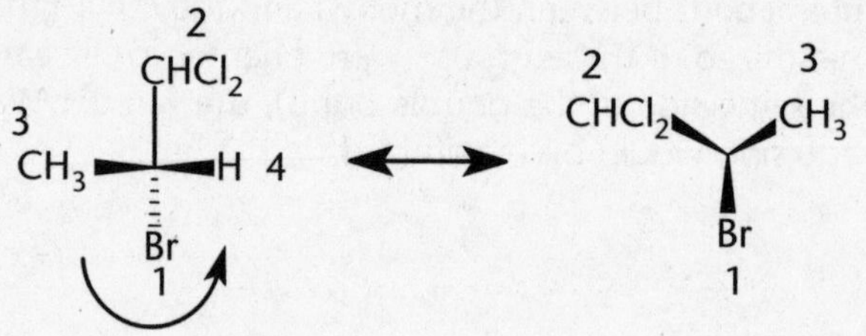

2 $CHCl_2$; 3 CH_3; H 4; Br 1 ⟷ 2 $CHCl_2$; 3 CH_3; Br 1

Now, rotate the molecule such that the group with lowest priority (H) is pointing away from us. Draw a curved arrow from **1** → **2** → **3**. If the direction of the arrow is clockwise, the molecule is designated (*R*), and if the direction is counterclockwise, it is designated (*S*). In this example the direction is clockwise, so the molecule is designated (*R*).

b. ***(R)***

Double- or triple-bonded atoms are considered to have the appropriate number of single bonds to the same atom to which they are multiply-bonded. For instance, in –C=O, C would be considered to have two single bonds with O, and O would be considered to have two single bonds with carbon. In this question, –OH is assigned the highest priority **1;** –C=O (here considered to be two [–C–O)'s] is assigned the next priority **2;** –CH_2OH is assigned priority **3;** and hydrogen, with the lowest atomic number, is assigned the lowest priority **4**. The direction of the curved arrow from **1** → **2** → **3** is clockwise and the molecule is designated (*R*).

7. **C** Remember, to be a chiral center, a carbon must have four different substituents. There are eight stereocenters in this molecule; these are marked below with asterisks.

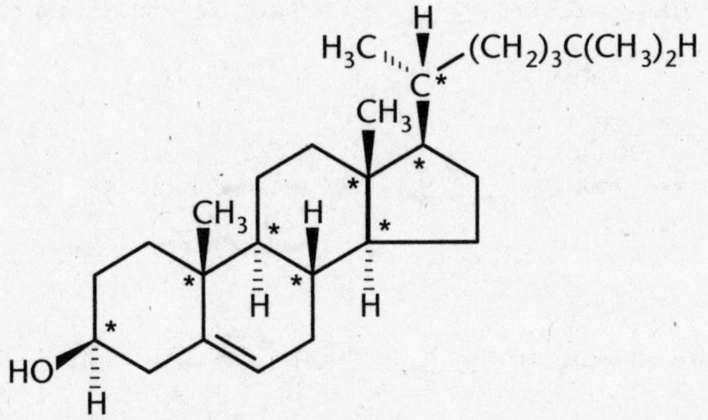

CHOLESTEROL

Other carbons are not chiral, for various reasons. Many are bonded to two hydrogens; others participate in double bonds, which count as two bonds to the same thing (another C atom).

8. **B** This is a chair conformation where the two equatorial methyl groups are *trans* to each other. Since the methyl hydrogens do not compete for the same space as the hydrogens attached to the ring, this conformation ensures the least amount of steric strain. Choice A would be more unstable than choice B since the diaxial methyl group hydrogens are closer to the hydrogens on the ring, causing greater steric strain. Choice C is wrong because it is in the more unstable boat conformation. Choice D is incorrect because these are all different structures with different stabilities.

9.

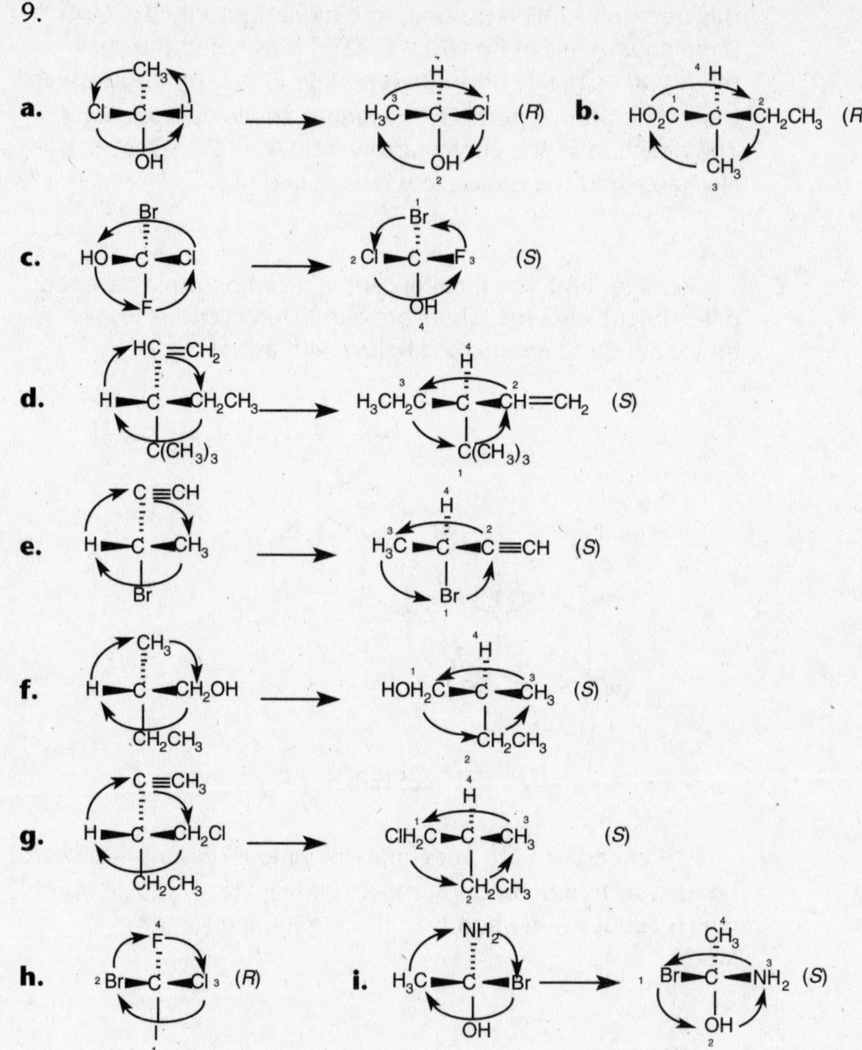

10. **C** The relative configuration is retained because the bonds of the stereocenter are not broken. The bond that is cleaved is one between a substituent of the stereocenter (the O atom) and another atom attached to the substituent (the H attached to the O). The absolute configuration is also retained, because the (*R*)/(*S*) designation is the same for the reactant and the product.

11. **A** The correct answer is choice A, enantiomers. If you look at the two structures you can see that they are mirror images of each other. Also, if you turn one of the structures around 180°, (structure III) you will be able to see that structures I and III are nonsuperimposable.

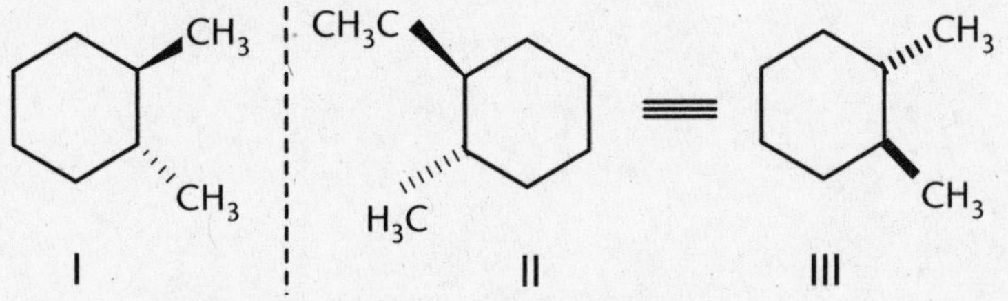

CH_3 CH_3 I | CH_3C H_3C II ≡ CH_3 CH_3 III

Choice B is incorrect because diastereomers are stereoisomers which are not mirror images of each other. Choice C is incorrect because in order for a compound to be designated as a *meso* compound, it must have a plane of symmetry, which neither of these structures contains. Choice D is wrong because structural isomers are compounds with the same molecular formula but different atomic connections. These compounds do have the same atomic connections. The only difference is that they differ in their spatial arrangement of atoms. As a result, they are in the class of stereoisomers, not structural isomers.

12. The specific rotation of a racemic mixture of MSG, or any other racemic mixture, is zero. A racemic mixture, by definition, is a mixture that contains equal amounts of the (+) and (–) enantiomers, which cancel each other's optical rotations. The specific rotation of (*R*)–monosodium glutamate is –24°. An enantiomer of a compound rotates plane polarized light by the same amount but in the opposite direction.

Bonding

CHAPTER THIRTEEN

As discussed in general chemistry, there are two types of chemical bonds: **ionic**, in which an electron is transferred from one atom to another, and **covalent**, in which pairs of electrons are shared between two atoms. In organic chemistry, it is important to understand the details of covalent bonding, as these play a crucial role in determining the properties and reactions of organic compounds.

ATOMIC ORBITALS

The first three quantum numbers, *n*, *l*, and *m*, describe the size, shape, and number of the atomic orbitals an element possesses. The number *n*, which can equal 1, 2, 3, . . . , corresponds to the energy levels in an atom and is essentially a measure of size. Within each electron shell, there can be several types of orbitals (s, p, d, f, g,. . . corresponding to the quantum numbers *l* = 0, 1, 2, 3, 4,...). Each type of atomic orbital has a specific shape. An s orbital is spherical and symmetrical, centered around the nucleus. A p orbital is composed of two lobes located symmetrically about the nucleus and contains a **node** (an area where the probability of finding an electron is zero). A d orbital is composed of four symmetrical lobes and contains two nodes. Both d and f orbitals are complex in shape and are rarely encountered in organic chemistry.

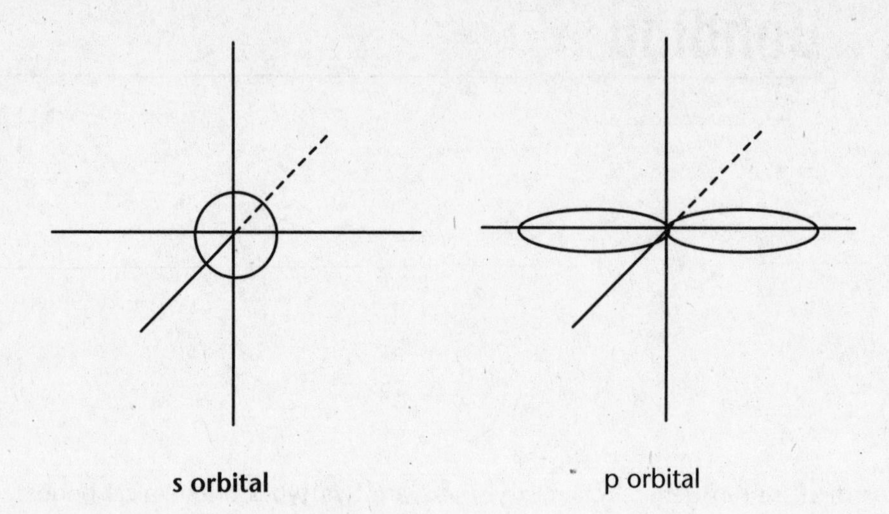

MOLECULAR ORBITALS

A. SINGLE BONDS

Two atomic orbitals can be combined to form what is called a **molecular orbital (MO)**. Molecular orbitals are obtained mathematically by adding the wave functions of the atomic orbitals. If the signs of the wave functions are the same, a lower-energy **bonding orbital** is produced. If the signs are different, a higher-energy **antibonding orbital** is produced. This is represented schematically by the addition of two s orbitals. Two p orbitals or one p and one s orbital can also be combined in a similar fashion.

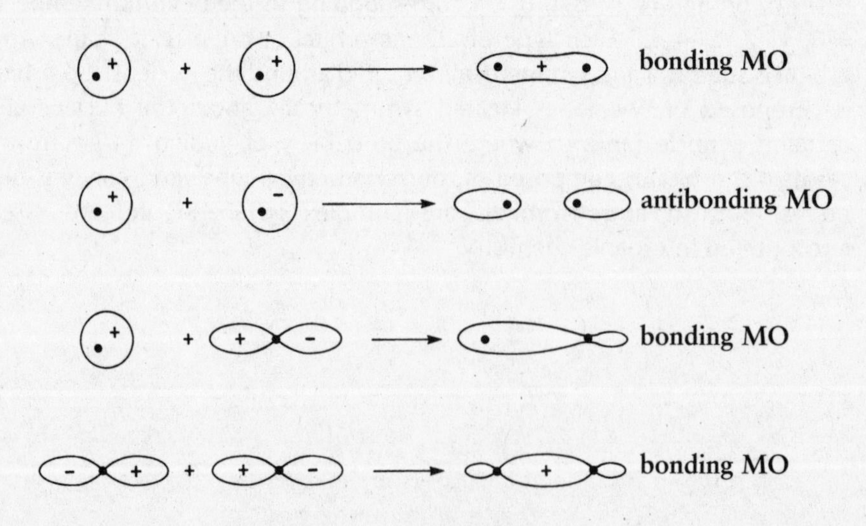

When a molecular orbital is formed by head-to-head overlap as in the figure above, the resulting bond is called a **sigma (σ) bond**. All single bonds are sigma bonds, accommodating two electrons. Shorter single bonds are stronger than longer single bonds.

B. DOUBLE AND TRIPLE BONDS

When two p orbitals overlap in a parallel fashion, a bonding MO is formed, called a **pi (π) bond**. When both a sigma and a pi bond exist between two atoms, a **double bond** is formed. When a sigma bond and two pi bonds exist, a **triple bond** is formed. As can be seen in the figure below, the overlap of the p orbitals involved in a p bond hinder rotation about double and triple bonds.

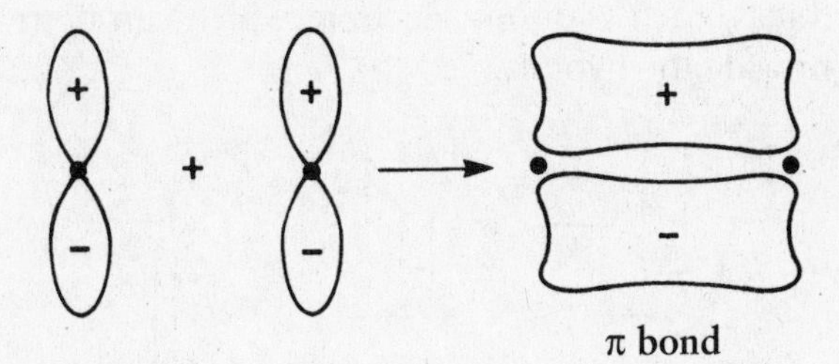

π bond

A pi bond cannot exist independently of a sigma bond. Only after the formation of a sigma bond will the p orbitals of adjacent carbons be parallel, because without the bond the three p orbitals are orthogonal to one another.

In general, pi bonds are weaker than sigma bonds; it is possible to break one bond of a double bond, leaving a single bond intact.

HYBRIDIZATION

The carbon atom has the electron configuration $1s^2 2s^2 2p^2$ and therefore needs four electrons to complete its octet. A typical molecule formed by carbon is methane, CH_4. Experimentation shows that the four sigma bonds in methane are equal. This is inconsistent with the unsymmetrical distribution of valence electrons: two electrons in the 2s orbital, one in the p_x orbital, one in the p_y orbital, and none in the p_z orbital.

A. sp^3

The theory of **orbital hybridization** was developed to account for this discrepancy. Hybrid orbitals are formed by mixing different types of atomic orbitals. If one s orbital and three p orbitals are mathematically combined, the result is four sp^3 hybrid orbitals that have a new shape.

s atomic orbital + p_x + p_y + p_z → $4sp^3$ hybrid MOs

These four orbitals will point toward the vertices of a tetrahedron, minimizing repulsion. This explains the preferred tetrahedral geometry adopted by carbon.

The hybridization is accomplished by promoting one of the 2s electrons into the $2p_z$ orbital (see figure below). This produces four valence orbitals, each with one electron, which can be mathematically mixed to provide the hybrids.

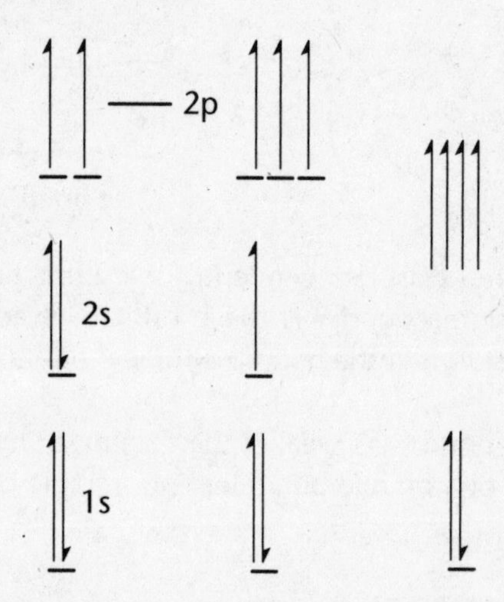

unhybridized ground state | unhybridized excited state | hybridized ground state

B. sp^2

Although carbon is most often found with sp^3 hybridization, there are other possibilities. If one s orbital and two p orbitals are mixed, three sp^2 hybrid orbitals are obtained.

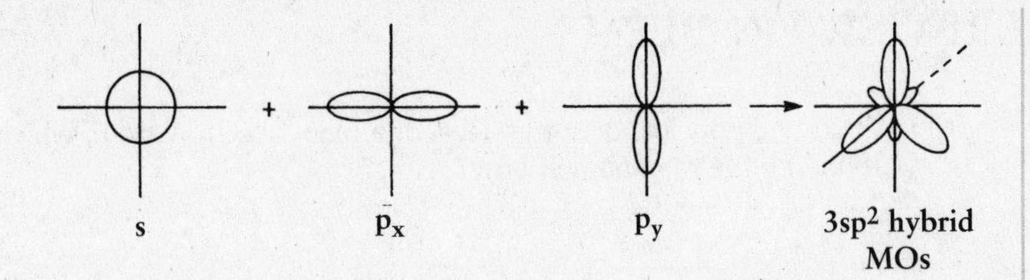

This occurs, for example, in ethylene. The third p orbital of each carbon atom is left unhybridized and participates in the pi bond. The three sp^2 orbitals are 120° apart, allowing maximum separation. These orbitals participate in the formation of the C=C and C–H single bonds.

C. sp

If two p orbitals are used to form a triple bond, and the remaining p orbital is mixed with an s orbital, two sp hybrid orbitals are obtained. They are oriented 180° apart, explaining the linear structure of molecules like acetylene.

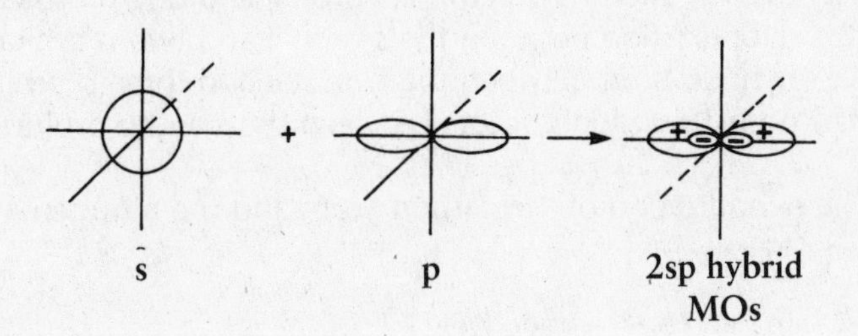

BONDING SUMMARY

The following table summarizes the major features of bonding in organic molecules.

Bond Order	Component Bonds	Hybridization	Angles	Examples
single	sigma	sp^3	109.5°	C–C; C–H
double	sigma	sp^2	120°	C=C; C=O
	pi			
triple	sigma	sp	180°	C≡C; C≡N
	pi			
	pi			

REVIEW PROBLEMS

1. Within one principal quantum level of a many-electron atom, which orbital has the minimum energy?

   A. s
   B. p
   C. d
   D. f

2. Which of the following compounds possesses at least one σ bond?

   A. CH_4
   B. C_2H_2
   C. C_2H_4
   D. All of the above

3. In a double-bonded carbon atom:

   A. hybridization between the s orbital and one p orbital occurs.
   B. hybridization between the s orbital and two p orbitals occurs.
   C. hybridization between the s orbital and three p orbitals occurs.
   D. no hybridization occurs between the s and p orbitals.

4. The hybridization of the carbon atom and the nitrogen atom in the ion CN^- are

   A. sp^3 and sp^3, respectively.
   B. sp^3 and sp, respectively.
   C. sp and sp^3, respectively.
   D. sp and sp, respectively.

5. Which of the following hybridizations does the Be atom in BeH_2 assume?

   A. sp
   B. sp^2
   C. sp^3
   D. None of the above

6. Two atomic orbitals combine to form:

   I. a bonding molecular orbital.
   II. an antibonding molecular orbital.
   III. new atomic orbitals.

   A. I only
   B. I, II and III
   C. III only
   D. I and II only

7. Molecular orbitals can contain a maximum of:

   A. 1 electron.
   B. 2 electrons.
   C. 4 electrons.
   D. $2n^2$ electrons, where n is the principal quantum number of the combining atomic orbitals.

8. π bonds are formed by which of the following orbitals?

   A. two s orbitals
   B. two p orbitals
   C. one s and one p orbital
   D. All of the above

9. How many σ bonds and π bonds are present in the following compound?

   H–C(H)(H)–C(=O)–H

   acetaldehyde

   A. 6 σ bonds and 1 π bond
   B. 6 σ bonds and 2 π bonds
   C. 7 σ bonds and 1 π bond
   D. 7 σ bonds and 2 π bonds

10. The four C–H bonds of CH_4 point toward the vertices of a tetrahedron. This indicates that the hybridization of a carbon atom is

   A. sp.
   B. sp^2.
   C. sp^3.
   D. None of the above.

SOLUTIONS TO REVIEW PROBLEMS

1. **A** In a many-electron atom, the energy of the orbitals within the principal quantum numbers is as follows: 2s < 2p; 3s < 3p < 3d; 4s < 4p < 4d < 4f. From this ranking it is clear that the correct choice is A.

2. **D** σ bonds are formed when orbitals overlap end-to-end. All the single bonds are σ bonds; double and triple bonds contain one σ bond each. The compounds CH_4, C_2H_2, and C_2H_4 (choices A, B, and C, respectively) all contain at least one single bond, so the correct choice is D.

3. **B** In a double-bonded carbon, sp^2 hybridization occurs, i.e., one s orbital hybridizes with two p orbitals to form three sp^2 hybrid orbitals. Therefore, the correct choice is B. Note that the sp^2 orbitals take part in σ bond formation, making choice D incorrect. The third p orbital of the carbon atom remains unhybridized and takes part in the formation of the π bond of the double bond.

4. **D** The carbon atom and the nitrogen atoms are connected by a triple bond in CN^-.

$$:N \equiv C:^-$$

A triple-bonded atom is sp hybridized; one s orbital hybridizes with one p orbital to form two sp hybridized orbitals. The two remaining unhybridized p orbitals take part in the formation of two π bonds. The correct choice, therefore, is D.

5. **A** BeH_2 is a linear molecule, which means that the angle between the two Be–H bonds is 180°. Since the sp orbitals are oriented at an angle of 180°, the Be atom is sp hybridized. Therefore, the correct choice is A. Note that sp^2 orbitals are oriented at an angle of 120°, and sp^3 orbitals are oriented at an angle of 109.5°.

6. **D** When atomic orbitals combine to form molecular orbitals, the number of molecular orbitals obtained equals the number of atomic orbitals that take part in the process. Half of the molecular orbitals formed are bonding molecular orbitals and the other half are antibonding molecular orbitals. In this case, therefore, two atomic orbitals combine to form one low-energy bonding molecular orbital and one high-energy antibonding molecular orbital. New atomic orbitals do not form, so answer choices B and C can be eliminated. Finally, A is incorrect since the bonding molecular orbital will have a corresponding antibonding molecular orbital. The correct choice is D.

7. **B** Each molecular orbital, like an atomic orbital, can contain a maximum of two electrons with opposite spins.

8. **B** π bonds are formed by the parallel overlap of unhybridized p orbitals. The electron density is concentrated above and below the bonding axis. A σ bond can be formed by the head-to-head overlap of two s orbitals, two p orbitals, or one s and one p orbital. Here, the density of the electrons is concentrated between the two nuclei of the bonding atoms.

9. **A** Each single bond has 1 σ bond, and each double bond has one σ and one π bond. In this question, there are five single bonds (5 σ bonds) and one double bond (one σ bond and one π bond), which gives a total of six σ bonds and one π bond. Thus, the correct choice is A.

10. **C** The four bonds point to the vertices of a tetrahedron, which means that the angle between two bonds is 109.5°. sp^3 orbitals have an angle of 109.5° between them. Hence, the carbon atom of CH_4 is sp^3-hybridized. The correct choice, therefore, is C.

Alkanes

CHAPTER FOURTEEN

Alkanes are fully saturated hydrocarbons, compounds consisting only of hydrogen and carbon atoms joined by single bonds. Their general formula is C_nH_{2n+2}, which means they have the maximum possible number of hydrogen atoms attached to each carbon atom.

Flashback:

Refer to Chapter 11 for the general rules of nomenclature for the alkanes.

NOMENCLATURE

Once again, be sure that you are familiar with the common, frequently encountered names. These include:

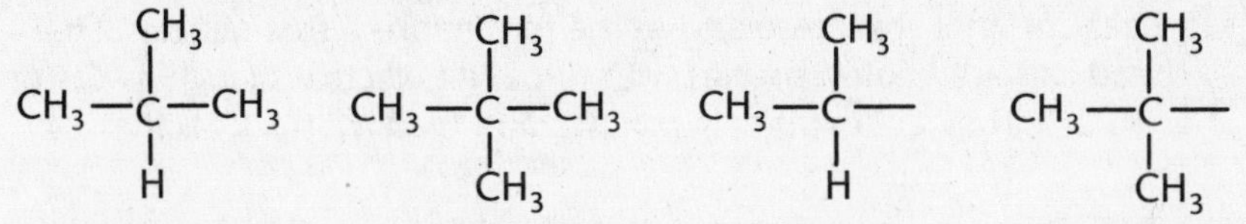

CH_3—$C(CH_3)(H)$—CH_3	CH_3—$C(CH_3)_2$—CH_3	CH_3—$C(CH_3)(H)$—	CH_3—$C(CH_3)_2$—
isobutane	neopentane	isopropyl	*t*-butyl

Carbon atoms can be characterized by the number of other carbon atoms to which they are directly bonded. A **primary** carbon atom (written as **1°**) is bonded to only one other carbon atom. A **secondary (2°)** carbon is bonded to two; a **tertiary (3°)** to three, and a **quaternary (4°)** to four other carbon atoms. In addition, hydrogen atoms attached to 1°, 2°, or 3° carbon atoms are referred to as 1°, 2°, or 3°, respectively.

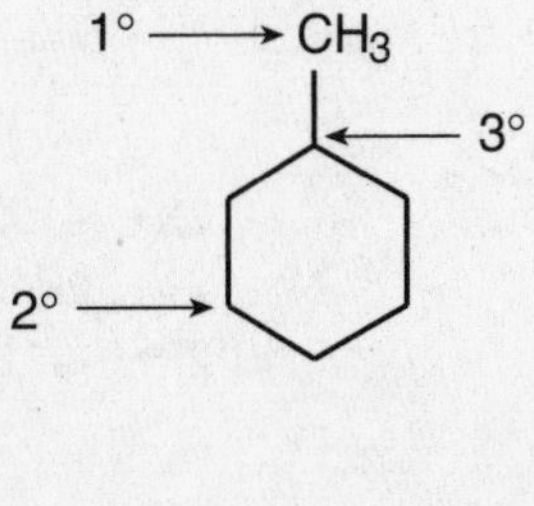

1° → CH_3

3°

2° →

In a Nutshell:

↑ chain length → ↑ b.p., ↑ m.p., ↑ density

↑ branching → ↓ b.p., ↓ density

PHYSICAL PROPERTIES

The physical properties of alkanes vary in a regular manner. In general, as the molecular weight increases, the melting point, boiling point, and density also increase. At room temperature, the straight-chain compounds C_1 through C_4 are gases, C_5 through C_{16} are liquids, and the longer-chain compounds are waxes and harder solids. Branched molecules have slightly lower boiling and melting points than their straight-chain isomers. Greater branching reduces the surface area of a molecule, decreasing the weak intermolecular attractive forces (Van der Waals forces). Hence, the molecules are held together less tightly, effectively lowering the boiling point. In addition, branched molecules are usually more difficult to pack into a tight, three-dimensional structure. This difficulty is often reflected in the lower melting points of branched alkanes.

REACTIONS

A. FREE RADICAL HALOGENATION

One frequently encountered reaction of alkanes are **halogenations**, in which one or more hydrogen atoms are replaced by halogen atoms (Cl, Br, or I) via a **free-radical substitution** mechanism. These reactions involve three steps:

1. **Initiation**—Diatomic halogens are homolytically cleaved by either heat or light (hv), resulting in the formation of free radicals. Free radicals are neutral species with unpaired electrons (such as Cl• or R_3C•). They are extremely reactive and readily attack alkanes.

$$\text{Initiation: } X_2 \xrightarrow[\text{or } \Delta]{h\nu} 2X\bullet$$

2. **Propagation**—A propagation step is one in which a radical produces another radical that can continue the reaction. A free radical reacts with an alkane, removing a hydrogen atom to form HX, and creating an alkyl radical. The alkyl radical can then react with X_2 to form an alkyl halide, generating X•.

$$\text{Propagation: } X\bullet + RH \rightarrow HX + R\bullet$$
$$R\bullet + X_2 \rightarrow RX + X\bullet$$

3. **Termination**—Two free radicals combine with one another to form a stable molecule.

$$\text{Termination: } 2X\bullet \rightarrow X_2$$
$$X\bullet + R\bullet \rightarrow RX$$
$$2R\bullet \rightarrow R_2$$

A single free radical can initiate many reactions before the reaction chain is terminated.

Larger alkanes have many hydrogens that the free radical can attack. Bromine radicals react fairly slowly, and primarily attack the hydrogens on the carbon atom that can form the most stable free radical, i.e., the most substituted carbon atom.

$$\bullet CR_3 > \bullet CR_2H > \bullet CRH_2 > \bullet CH_3$$
$$3° > 2° > 1° > \text{methyl}$$

Thus, a tertiary radical is the most likely to be formed in a free-radical bromination reaction.

CH_3 CH_3 Br
hν
Br_2

Free-radical chlorination is a more rapid process and thus depends not only on the stability of the intermediate, but on the number of hydrogens present. Free-radical chlorination reactions are likely to replace primary hydrogens because of their abundance, despite the relative instability of primary radicals. Unfortunately, free-radical chlorination reactions produce mixtures of products, and are preparatively useful only when just one type of hydrogen is present.

Note:

Free radicals, as well as carbocations, follow the same patterns of stability: 3° > 2° > 1° > methyl

B. COMBUSTION

The reaction of alkanes with molecular oxygen, to form carbon dioxide, water, and heat, is a process of great practical importance. It is an unusual reaction because heat, not a chemical species, is generally the desired product. The reaction mechanism is very complex and is believed to proceed through a radical process. The equation for the complete **combustion** of propane is:

$$C_3H_8 + 5O_2 \rightarrow 3CO_2 + 4H_2O + \text{heat}$$

Combustion is often incomplete, producing significant quantities of carbon monoxide instead of carbon dioxide. This frequently occurs, for example, in the burning of gasoline in an internal combustion engine.

Real World Analogy:

Nitrogen in the air is also often oxidized in the internal combustion engine. Various nitrogen oxides are major contributors to air pollution.

C. PYROLYSIS

Pyrolysis occurs when a molecule is broken down by heat. Pyrolysis, also called **cracking**, is most commonly used to reduce the average molecular weight of heavy oils and to increase the production of the more desirable volatile compounds. In the pyrolysis of alkanes, the C–C bonds are cleaved, producing smaller-chain alkyl radicals. These radicals can recombine to form a variety of alkanes:

$$CH_3CH_2CH_3 \xrightarrow[\Delta]{} CH_3\bullet + \bullet CH_2CH_3$$

$$2\ CH_3\bullet \longrightarrow CH_3CH_3$$

$$2\ \bullet CH_2CH_3 \longrightarrow CH_3CH_2CH_2CH_3$$

Alternatively, in a process called **disproportionation**, a radical transfers a hydrogen atom to another radical, producing an alkane and an alkene:

$$CH_3\bullet + \bullet CH_2CH_3 \rightarrow CH_4 + CH_2 = CH_2$$

SUBSTITUTION REACTIONS OF ALKYL HALIDES

Alkyl halides and indeed other substituted carbon atoms can take part in reactions known as *nucleophilic substitutions*. **Nucleophiles** ("nucleus lovers") are electron-rich species that are attracted to positively polarized atoms.

A. NUCLEOPHILES

1. BASICITY

If the nucleophiles have the same attacking atom (for example, oxygen) then nucleophilicity is roughly correlated to basicity. In other words, the stronger the base, the stronger the nucleophile. For example, nucleophilic strength decreases in the order:

$$RO^- > HO^- > RCO_2^- > ROH > H_2O$$

2. SIZE AND POLARIZABILITY

If the attacking atoms differ, nucleophilic ability doesn't necessarily correlate to basicity. In a protic solvent, large atoms tend to be better nucleophiles as they can shed their solvent molecules and are more polarizable. Hence, nucleophilic strength decreases in the order:

In a Nutshell:

In protic solvents (solvents capable of hydrogen bonding), larger atoms are better nucleophiles. In aprotic solvents, more basic atoms are better nucleophiles.

$$CN^- > I^- > RO^- > HO^- > Br^- > Cl^- > F^- > H_2O$$

In aprotic solvents however, the nucleophiles are "naked"; they are not solvated. In this situation, nucleophilic strength is related to basicity. For example in DMSO, the order of nucleophilic strength is the same as base strength:

$$F^- > Cl^- > Br^- > I^-$$

Note that this is the opposite of what happens in polar solvents.

B. LEAVING GROUPS

The ease with which nucleophilic substitution takes place is also dependent on the leaving group. The best leaving groups are those that are weak bases, as these can accept an electron pair and dissociate to form a stable species. In the case of the halogens, therefore, this is the opposite of base strength:

$$I^- > Br^- > Cl^- > F^-$$

In a Nutshell:

Weak bases make good leaving groups.

C. S_N1 REACTIONS

S_N1 is the designation for **unimolecular nucleophilic substitution** reaction. It is called unimolecular because the rate of the reaction is dependent upon only one species. Generally, the rate-determining step is the dissociation of this species to form a stable, positively-charged ion called a **carbocation** or **carbonium ion**.

1. Mechanism of S_N1 Reactions

S_N1 reactions involve two steps: the dissociation of a molecule into a carbocation and a good leaving group, followed by the combination of the carbocation with a strong nucleophile.

$$(CH_3)_3C{-}Br \rightleftharpoons (CH_3)_3C^+ + Br^-$$

$$(CH_3)_3C^+ + Nu^- \rightleftharpoons (CH_3)_3C{-}Nu$$

In the first step, a carbocation is formed. Carbocations are stabilized by polar solvents that have lone electron pairs to donate (e.g., water, acetone). Carbocations are also stabilized by charge delocalization. More highly substituted cations are therefore more stable. The order of stability for carbocations is:

tertiary > secondary > primary > methyl

To get the desired product, the original substituent should be a better leaving group than the nucleophile, so that at equilibrium, RNu is the main product. Conditions are usually chosen so that the second step of the reaction is essentially irreversible.

Note:

The kinetics of unimolecular reactions are first order.

2. Rate of S_N1 Reactions

The rate at which a reaction occurs can never be greater than the rate of its slowest step. Such a step is termed the **rate-limiting** or **rate-determining step** of the reaction, because it limits the speed of the reaction. In an S_N1 reaction, the slowest step is the dissociation of the molecule to form a carbocation, a step that is energetically unfavorable. The formation of a carbocation is therefore the rate-limiting step of an S_N1 reaction. The only reactant in this step is the original molecule, and so the rate of the entire reaction, under a given set of conditions, depends only on the concentration of this original molecule (a so-called *first-order reaction*). The rate is *not* dependent on the concentration or the nature of the nucleophile, because it plays no part in the rate-limiting step.

The rate of an S_N1 reaction can be increased by anything that accelerates the formation of the carbocation. The most important factors are as follows:

a. Structural factors: Highly substituted alkyl halides allow for distribution of the positive charge over a greater number of carbon atoms, and thus form the most stable carbocations.

b. Solvent effects: Highly polar solvents are better at surrounding and isolating ions than are less polar solvents. Polar protic solvents such as water work best since solvation stabilizes the intermediate state.

c. Nature of the leaving group: Weak bases dissociate more easily from the alkyl chain and thus make better leaving groups, increasing the rate of carbocation formation.

D. S_N2 REACTIONS

The formation of a carbocation is not always favorable. Under certain conditions, substitution can proceed by a different mechanism, which does not involve a carbocation. An S_N2 (**bimolecular nucleophilic substitution**) reaction involves a nucleophile pushing its way into a compound while simultaneously displacing the leaving group. Its rate-determining, and only, step involves two molecules: the **substrate** and the nucleophile.

$$CH_3CH_2Cl + {}^{-}OH \longrightarrow \left[Cl^{\delta-} \cdots CH_3CH_2 \cdots OH^{\delta-} \right] \longrightarrow CH_3CH_2OH + Cl^{-}$$

1. Mechanism of S_N2 reactions

In S_N2 reactions, the nucleophile actively displaces the leaving group. For this to occur, the nucleophile must be strong, and the reactant cannot be sterically hindered. The nucleophile attacks the reactant from the backside of the leaving group, forming a trigonal bipyramidal **transition state**. As the reaction progresses, the bond to the nucleophile strengthens while the bond to the leaving group weakens. The leaving group is displaced as the bond to the nucleophile becomes complete.

Note:

An intermediate is distinct from a transition state. An intermediate is a well-defined species with a finite lifetime. On the other hand, a transition state is a theoretical structure used to define a mechanism.

2. Rate of S_N2 Reactions

The single step of an S_N2 reaction involves *two* reacting species: the substrate (the molecule with a leaving group, usually an alkyl halide), and the nucleophile. The concentrations of both therefore play a role in determining the rate of an S_N2 reaction; the two species must "meet" in solution, and raising the concentration of either will make such a meeting more likely. Since the rate of the S_N2 reaction depends on the concentration of two reactants, it follows **second-order kinetics**.

S_N1 vs. S_N2

Certain reaction conditions favor one substitution mechanism over the other. It is also possible for both to occur in the same flask. Sterics, nucleophilic strength, leaving group ability, reaction conditions, and solvent effects are all important in determining which reaction will occur.

STEREOCHEMISTRY OF SUBSTITUTION REACTIONS

A. S_N1 STEREOCHEMISTRY

S_N1 reactions involve carbocation intermediates, which are approximately planar and therefore achiral.

$CH_3CH_2C(CH_3)(CH(CH_3)_2)Br \longrightarrow CH_3CH_2C^+(CH_3)CH(CH_3)_2 \xrightarrow{^-OH} CH_3CH_2C(CH_3)(OH)CH(CH_3)_2$ (both enantiomers)

If the original compound is optically active because of the reacting chiral center, then a racemic mixture will be produced. S_N1 reactions result in a loss of optical activity.

Flashback:

Refer to Chapter 12 for further discussion of optical activity.

B. S_N2 STEREOCHEMISTRY

The single step of an S_N2 reaction involves a chiral transition state. Since the nucleophile attacks from one side of the central carbon and the leaving group departs from the opposite side, the reaction "flips" the bonds attached to the carbon.

$CH_3CO_2^- + C(H)(R')(R'')-I \longrightarrow CH_3CO_2-C(H)(R')(R'') + I^-$

If the reactant is chiral, optical activity is usually retained; however, in the case of S_N2 reactions, an inversion of configuration occurs.

In a Nutshell:

S_N1	*S_N2*
• *2 steps*	• *1 step*
• *Favored in polar protic solvents*	• *Favored in polar aprotic solvents*
• *3° > 2° > 1° > methyl*	• *1° > 2° > 3°*
• *rate = k[RX]*	• *rate = k[Nu][RX]*
• *racemic products*	• *optically active/inverted products*
• *favored with the use of bulky nucleophiles*	

REVIEW PROBLEMS

1. Under the following conditions,

$$\text{Br-}CH_2CH_2CH_2CH(CH_3)_2 \xrightarrow{CN^-} \text{NC-}CH_2CH_2CH_2CH(CH_3)_2 + Br^-$$

1-bromo-4-methylpentane will most probably react via

A. S_N1.
B. S_N2.
C. both S_N1 and S_N2.
D. neither S_N1 nor S_N2.

2. The following molecule can be classified as having:

(cyclohexane ring bearing adjacent $CH_2C(CH_3)_3$ and CH_2CH_3 substituents)

A. 4 primary, 2 secondary, 4 tertiary, and 3 quaternary carbon atoms.
B. 3 methyl groups, 2 ethyl groups, and 4 secondary carbon atoms.
C. 4 primary, 6 secondary, 2 tertiary, and 1 quaternary carbon atoms.
D. 3 primary, 3 secondary, 4 tertiary, and 3 quaternary carbon atoms.

3. The following reactions are part of a free-radical halogenation sequence:

	Reaction	ΔH kcal/mol
a.	$Cl_2 \xrightarrow{h\nu} 2\,Cl\bullet$	+ 58
b.	$Cl\bullet + CH_4 \rightarrow \bullet CH_3 + HCl$	+ 1
c.	$\bullet CH_3 + Cl_2 \rightarrow CH_3Cl + Cl\bullet$	– 26
d.	$\bullet CH_3 + Cl\bullet \rightarrow CH_3\,Cl$	– 84

Identify initiation, propagation, and termination steps.

4. S_N1 reactions show first-order kinetics because:

A. the rate-limiting step is the first step to occur in the reaction.
B. the rate-limiting step involves only one molecule.
C. there is only one rate-limiting step.
D. the reaction involves only one molecule.

5. The following reaction sequence is typical of S_N1 reactions. Which is the rate-limiting step(s)?

Step 1 $(CH_3)_3C{-}Cl \longrightarrow (CH_3)_3C^+ + Cl^-$

Step 2 $(CH_3)_3C^+ \xrightarrow{CH_3CH_2OH} (CH_3)_3C{-}\overset{+}{O}(H){-}CH_2CH_3$

Step 3 $(CH_3)_3C{-}\overset{+}{O}(H){-}CH_2CH_3 \longrightarrow (CH_3)_3C{-}O{-}CH_2CH_3 + H^+$

A. Step 1
B. Step 2
C. Step 3
D. Steps 1 and 2

6. Which of the following would be the best solvent for an S_N2 reaction?

A. H_2O
B. CH_3CH_2OH
C. CH_3SOCH_3
D. $CH_3CH_2CH_2CH_2CH_2CH_3$

7. What choices for X and Y would most favor the following reaction?

$$R_3C-X \xrightarrow{-X^-} R_3C^+ \xrightarrow{+Y^-} R_3C-Y$$

A. $X = I^-$, $Y = Cl^-$
B. $X = EtO^-$, $Y =$ tosylate
C. $X =$ tosylate, $Y = CN^-$
D. $X = OH^-$, $Y = H_2O$

8. What would be the major product of the following reaction?

$$CH_3CH_2CH_3 + Br_2 \xrightarrow{h\upsilon}$$

A. $CH_3CH_2CH_2Br$
B. $CH_3CH_2CH_2CH_2CH_2CH_3$
C. $(CH_3)_2CHBr$
D. CH_3CH_2Br

9. Treatment of (*S*)–2-bromobutane with sodium hydroxide results in the production of a compound with an (*R*) configuration. The reaction has most likely taken place through:

A. an S_N1 mechanism.
B. an S_N2 mechanism.
C. both an S_N1 and S_N2 reaction.
D. cannot be determined.

10. What is the correct order of the boiling points of the following compounds?

I. *n*-hexane
II. 2-methylpentane
III. 2,2-dimethylbutane
IV. *n*-heptane

A. I > IV > II > III
B. IV > III > II > I
C. IV > I > II > III
D. I > II > III > IV

11. The reaction of isobutane with an unknown halogen is catalyzed by light. The two major products obtained are:

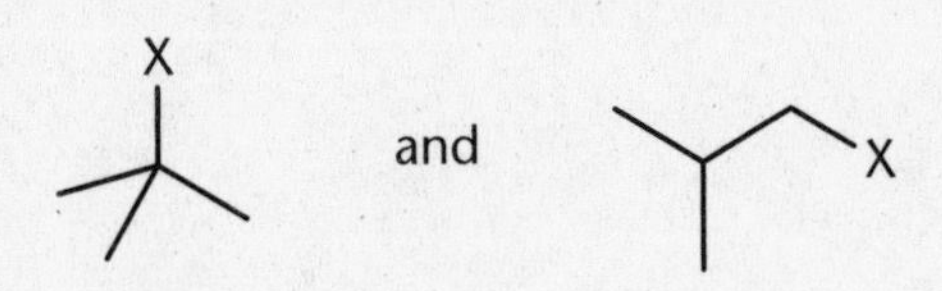

What is the unknown halogen?

A. Cl_2
B. Br_2
C. I_2
D. F_2

12. Place the following species in order of increasing stability.

a. $CH_3CH_2CH_2^{\bullet}$ $\quad (CH_3)_2C^{\bullet}H$ (with H bonded below C) $\quad CH_3^{\bullet}$ $\quad (CH_3)_3C^{\bullet}$

b. R_2C^+H (with H bonded below C) $\quad R_3C^+$ $\quad H_3C^+$ $\quad H_2C^+R$ (with R bonded below C)

SOLUTIONS TO REVIEW PROBLEMS BEGIN ON THE FOLLOWING PAGE.

SOLUTIONS TO REVIEW PROBLEMS

1. **B** In this question, a primary alkyl halide is treated with cyanide, which is a good nucleophile. Primary alkyl halides are not sterically hindered and are readily displaced in S_N2 reactions. Cyanide will displace the bromide to produce 5-methylheptanenitrile in good yield. The correct answer is therefore choice B. An S_N1 reaction will probably not occur because formation of a primary carbocation, which would result from the loss of bromide, is highly unfavorable.

2. **C** The molecule shown in question 2 is 1-(2,2-dimethylpropyl)-2-ethylcyclohexane. A primary carbon atom is one that is bonded to only one other carbon atom, a secondary carbon atom is bonded to two, a tertiary to three, and a quaternary to four. Thus, this molecule has four primary, six secondary, two tertiary, and one quaternary carbon atoms, so choice C is correct. Choice A is incorrect because there are six, not two, secondary carbon atoms. These include four on the ring and one on each substituent. Choice B is incorrect because there are six, not four, secondary carbon atoms, and choice D is incorrect because there are four primary carbon atoms, not three.

3. Step a is an initiation step because a nonradical species forms two radicals. Both steps b and c are propagation steps in which one radical species reacts to form another radical species. Step d is a termination step in which two radicals form a nonradical species.

4. **B** An S_N1 reaction is a first-order nucleophilic substitution reaction. It is called first-order because the rate-limiting step involves only one molecule; thus the correct answer is choice B. Choice A is incorrect because the rate-limiting step is not necessarily the first step to occur in a reaction. It is simply the slowest step. Choice C is a true statement, but is incorrect because it is irrelevant to the term "first-order." Finally, choice D is incorrect because it is the rate-limiting step, not the reaction, that involves only one molecule.

5. A The formation of a carbocation is the rate-limiting step. This step is the slowest to occur and its rate determines the rate of the reaction. Step 2 is a nucleophilic attack on the carbocation by ethanol. In step 3 a proton is lost from the protonated ether. Both steps 2 and 3 occur very rapidly in solution and are not rate-limiting steps. Answer choice D is incorrect because there can be only one rate-limiting step.

6. C The correct answer is choice C, dimethyl sulfoxide. S_N2 reactions give the best results if a polar aprotic solvent is used. S_N2 reactions occur via a one-step mechanism in which a nucleophile attacks a substrate. Polar aprotic solvents accelerate this reaction by allowing the nucleophile to be "naked," i.e., not surrounded by hydrogen-bonded solvation spheres. The nucleophile therefore has easy access to the substrate. In addition, the solvent should be polar to dissolve the reactants. Choice A, water, and choice B, ethanol, are both incorrect because although these are polar, they are also protic and would diminish the power of the nucleophile. Choice D is hexane and is incorrect because it is a nonpolar solvent.

7. C This reaction is S_N1 and in order to occur, there must be a good leaving group and a strong incoming nucleophile. In choice C, X is a tosylate ion, an excellent leaving group and Y is a cyanide, an excellent nucleophile. Therefore, choice C is the correct answer. Choice A is incorrect because although iodide is a better nucleophile than chloride, it is also a better leaving group. Choice B is incorrect because ethoxide is a poor leaving group and tosylate is a weak nucleophile. Finally, choice D is incorrect because again, hydroxide is a poor leaving group and water is a poor nucleophile.

8. C This question concerns the free-radical bromination of an alkane. The bromination reaction occurs in such a way as to produce the most stable alkyl radical. This is because bromine radicals are very selective, as opposed to chlorine radicals, which react indiscriminately. In this question, the most stable radical is a secondary radical, which further reacts to form 2-bromopropane, choice C, the correct answer. Choice A is incorrect because 1-bromopropane results from reaction of a primary radical, and although this may occur to an extent, the major product will be 2-bromopropane. Choice B would result from the combination of two primary radicals, and is expected to be a very minor product. Finally, choice D is incorrect because a carbon atom is lost in the reaction, and this does not occur in free-radical bromination.

9. **B** Inversion of configuration is a trademark of the S_N2 reaction, whereas racemization is typical of S_N1 reactions. When (*S*)-2-bromobutane is treated with hydroxide, a compound with an *R* configuration is obtained. The most likely occurrence is a substitution reaction, and the fact that the absolute configuration has changed suggests an S_N2 reaction. If the reaction proceeded by S_N1, the products would have both *R* and *S* configurations because the hydroxide ion could attack the planar carbocation from either side. There is only one configuration in this case, and therefore the correct answer is S_N2.

10. **C** The correct answer is choice C, IV > I > II > III, corresponding to: *n*-heptane > *n*-hexane > 2-methylpentane > 2,2-dimethylbutane. As the chain length of a straight-chain alkane is increased, the boiling point also increases, approximately 25–30°C for each additional carbon atom. Therefore, *n*-heptane is expected to boil at a higher temperature than *n*-hexane. Isomeric alkanes follow a typical trend: as branching increases, boiling point decreases. Compounds I, II, and III are isomeric hexanes, listed in increasing order of branching. Therefore, *n*-hexane boils at a higher temperature than 2-methylpentane, which boils at a higher temperature than 2,2-dimethylbutane.

11. **A** Free-radical halogenation reactions are practical only for bromine and chlorine; iodine and fluorine do not react efficiently, and can therefore be eliminated. Bromine radicals react slowly in comparison to chlorine radicals, and are therefore more likely to react in a manner that forms the most stable alkyl radical, i.e., the most substituted radical. This leads to the production of one major bromination product. Chlorine radicals, on the other hand, react so quickly that they become rather indiscriminate, and generally produce several different products. In this particular reaction, two different products are isolated in comparable yields.

12. a. The order of stability of free radicals follows this sequence: 3° > 2° > 1° > CH_3. Stability is enhanced by an increase in the number of alkyl substituents bonded directly to the radical carbon atom. These substituents allow the extra electron density to be spread out or delocalized throughout the molecule.

b. The stability of carbocations is increased when the positive charge can be distributed over more than one carbon atom. This means that tertiary carbocations are more stable than secondary, and so on.

Alkenes and Alkynes

CHAPTER FIFTEEN

ALKENES

Alkenes are hydrocarbons that contain carbon-carbon double bonds. The general formula for a straight-chain alkene with one double bond is C_nH_{2n}. The degree of unsaturation (the number N of double bonds or rings) of a compound of molecular formula C_nH_m can be determined according to the equation:

$$N = \frac{1}{2}(2n + 2 - m)$$

Double bonds are considered functional groups, and alkenes are more reactive than the corresponding alkanes.

NOMENCLATURE

Alkenes, also called **olefins**, may be described by the terms *cis, trans,* E, and Z. The common names *ethylene, propylene,* and *isobutylene* are often used over the IUPAC names.

$CH_2{=}CH_2$

ethene
(ethylene)

$CH_3CH{=}CH_2$

propene
(propylene)

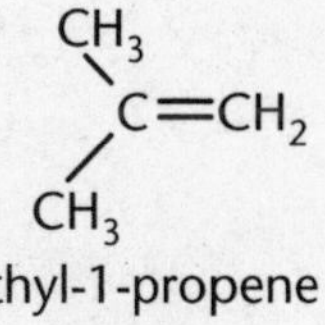

$(CH_3)_2C{=}CH_2$

2-methyl-1-propene
(isobutylene)

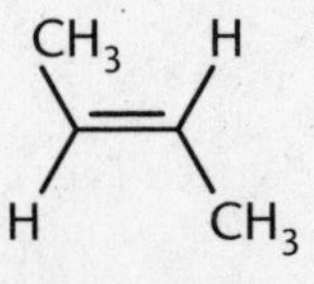

CH₃ H / H CH₃

trans-2-butene

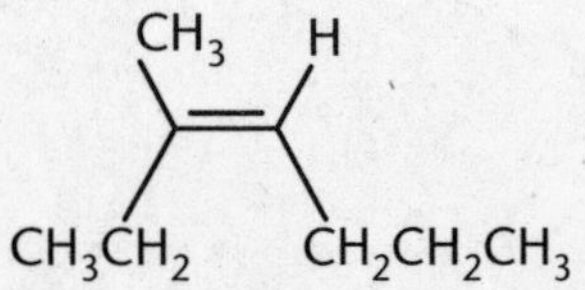

(Z)-3-methyl-3-heptene

Flashback:

Refer to Chapter 11 for the general rules of nomenclature for alkenes.

PHYSICAL PROPERTIES

The physical properties of alkenes are similar to those of alkanes. For example, the melting and boiling points increase with increasing molecular weight and are similar in value to those of the corresponding alkanes. *Trans*-alkenes generally have higher melting points than *cis*-alkenes because their higher symmetry allows better packing in the solid state. They also tend to have lower boiling points than *cis*-alkenes because they are less **polar**.

In a Nutshell:

Trans-alkenes have higher melting points than cis-alkenes due to higher symmetry.
Cis-alkenes have higher boiling points than trans-alkenes due to polarity.

Polarity is a property that results from the asymmetrical distribution of electrons in a particular molecule. In alkenes, this distribution creates dipole moments that are oriented from the electropositive alkyl groups toward the electronegative alkene. In *trans*-2-butene, the two dipole moments are oriented in opposite directions and cancel each other. The compound possesses no net dipole moment and is not polar. On the other hand, *cis*-2-butene has a net dipole moment, resulting from addition of the two smaller dipoles. The compound is polar, and the additional intermolecular forces tend to raise the boiling point.

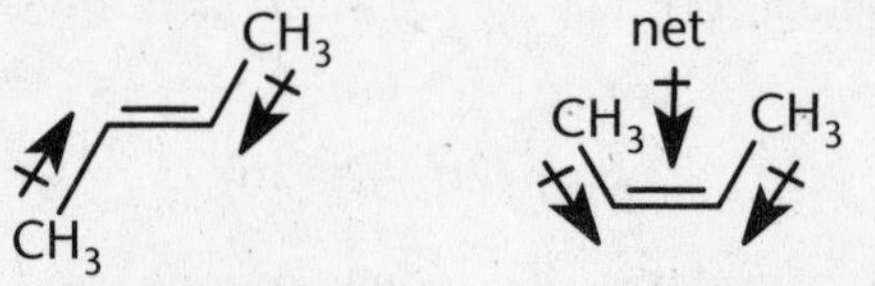

CH_3 CH_3 (non-polar)   net CH_3 CH_3 (polar)

SYNTHESIS

Alkenes can be synthesized in a number of different ways. The most common method involves **elimination reactions** of either alcohols or alkyl halides. In these reactions the carbon skeleton loses HX (where X is a halide), or a molecule of water, to form a double bond:

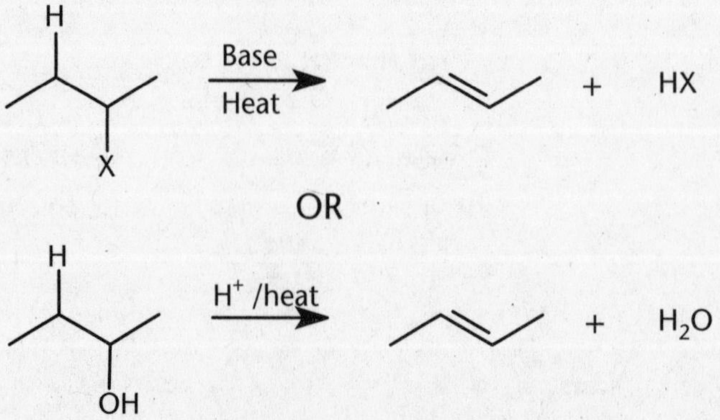

H, X — Base / Heat → + HX

OR

H, OH — H^+ /heat → + H_2O

Elimination occurs by two distinct mechanisms, unimolecular and bimolecular, which are referred to as **E1** and **E2**, respectively.

A. UNIMOLECULAR ELIMINATION

Unimolecular elimination, abbreviated E1, is a two-step process proceeding through a carbocation intermediate. The rate of reaction is dependent on the concentration of only one species, namely the substrate. The elimination of a leaving group and a proton results in the production of a double bond. In the first step, the leaving group departs, producing a carbocation. In the second step, a proton is removed by a base.

E1 is favored by the same factors that favor S_N1: highly polar solvents, highly branched carbon chains, good leaving groups, and weak nucleophiles in low concentration. These mechanisms are therefore competitive, and directing a reaction toward either E1 or S_N1 alone is difficult, although high temperatures tend to favor E1.

B. BIMOLECULAR ELIMINATION

Bimolecular elimination, termed E2, occurs in one step. Its rate is dependent on the concentration of two species, the substrate and the base. A strong base such as the ethoxide ion ($C_2H_5O^-$) removes a proton, while a halide ion *anti* to the proton leaves, resulting in the formation of a double bond.

$$\text{(H)(H)(CH}_3\text{)C–C(H)(CH}_3\text{)(Cl)} \xrightarrow{C_2H_5O^-} \text{(H)(CH}_3\text{)C=C(CH}_3\text{)(H)} + C_2H_5OH + Cl^-$$

Often there are two possible products. In such cases, the more substituted double bond is formed preferentially.

Controlling E2 vs. S_N2 is easier than controlling E1 vs. S_N1.

1. Steric hindrance does not greatly affect E2 reactions. Therefore, highly substituted carbon chains, which form the most stable alkenes, undergo E2 most easily and S_N2 rarely.

2. A strong base favors E2 over S_N2. S_N2 is favored over E2 by weak Lewis bases (strong nucleophiles).

Other factors, such as the polarity of the solvent and branching of the carbon chain, can be modified in order to reduce the competition between E1 and S_N1 reactions.

REACTIONS

A. REDUCTION

Catalytic hydrogenation is the reductive process of adding molecular hydrogen to a double bond with the aid of a metal catalyst. Typical catalysts are platinum, palladium, and nickel (usually Raney nickel, a special powdered form), but occasionally rhodium, iridium, or ruthenium are used.

The reaction takes place on the surface of the metal. One face of the double bond is coordinated to the metal surface, and thus the two hydrogen atoms are added to the same face of the double bond. This type of addition is called *syn* addition.

Note:

Reactions where one stereoisomer is favored are termed stereospecific reactions.

$$CH_3(Cl)C{=}C(Cl)CH_3 \xrightarrow[Pd]{H_2} CH_3CH(Cl){-}CH(Cl)CH_3 \quad (syn)$$

B. ELECTROPHILIC ADDITIONS

The π bond is somewhat weaker than the σ bond, and can therefore be broken without breaking the σ bond. As a result, one can *add* compounds to double bonds while leaving the carbon skeleton intact. Though many different **addition reactions** exist, most operate via the same essential mechanism.

The electrons of the π bond are particularly exposed and are thus easily attacked by molecules that seek to accept an electron pair (Lewis acids). Because these groups are electron-seeking, they are more often termed **electrophiles** (literally, "lovers of electrons").

1. Addition of HX

The electrons of the double bond act as a Lewis base and react with electrophilic HX molecules. The first step yields a carbocation intermediate after the double bond reacts with a proton. In the second step, the halide ion combines with the carbocation to give an alkyl halide. In cases where the alkene is asymmetrical, the initial protonation proceeds to produce the *most stable carbocation.* The proton will add to the less substituted carbon atom (the carbon atom with the most protons), since alkyl substituents stabilize carbocations. This phenomenon is called **Markovnikov's rule**. An example is:

Note:

Markovnikov's rule refers to the addition of something (e.g., halide, hydroxyl group) to the most substituted carbon in the double bond.

$$CH_2{=}CHCH_3 \xrightarrow{+H^+} CH_3\overset{+}{C}HCH_3 \xrightarrow{Br^-} CH_3CHBrCH_3$$

2. Addition of X_2

The addition of halogens to a double bond is a rapid process. It is frequently used as a diagnostic tool to test for the presence of double bonds. The double bond acts as a nucleophile and attacks an X_2 molecule, displacing X^-. The intermediate carbocation forms a **cyclic halonium ion,** which is then attacked by X^-, giving the dihalo compound. Note that this addition is *anti*, because the X^- attacks the cyclic halonium ion in a standard S_N2 displacement.

Br—Br
H H C=C CH_3 CH_3 → H Br+ H CH_3 CH_3 Br— → CH_3 H C—C Br Br H CH_3

Anti-addition

If the reaction is carried out in a nucleophilic solvent, the solvent molecules can compete in the displacement step, producing, for example, a **halo alcohol** (rather than the **dihalo** compound).

3. Addition of H_2O

Water can be added to alkenes under acidic conditions. The double bond is protonated according to Markovnikov's rule, forming the most stable carbocation. This carbocation reacts with water, forming a protonated alcohol, which then loses a proton to yield the alcohol. The reaction is performed at low temperature because the reverse reaction is an acid-catalyzed **dehydration** favored by high temperatures.

$\xrightarrow[H_2O]{H^+}$ + $\xrightarrow{H_2O}$ H_2O^+ $\xrightarrow{-H^+}$ OH

Direct addition of water is generally not useful in the laboratory because yields vary greatly with reaction conditions; therefore, this reaction is generally carried out indirectly using mercuric acetate, $Hg(CH_3COO)_2$.

C. FREE RADICAL ADDITIONS

An alternate mechanism exists for the addition of HX to alkenes, which proceeds through **free-radical intermediates**, and occurs when peroxides, oxygen, or other impurities are present. Free-radical additions disobey the Markovnikov rule because X• adds first to the double bond, producing the most stable free radical, whereas H^+ adds first in standard electrophilic additions, producing the most stable carbocation. The reaction is useful for HBr, but is not practical for HCl or HI, because the energetics are unfavorable.

Pavlov's Dog:

When peroxides are present, expect free radical reactions that do not follow Markovnikov's rule.

Br• → (radical intermediate) Br; HBr → Br + Br•

most stable radical

D. HYDROBORATION

Diborane (B_2H_6) adds readily to double bonds. The boron atom is a Lewis acid and attaches to the less sterically hindered carbon atom. The second step is an oxidation-hydrolysis with peroxide and aqueous base, producing the alcohol with overall anti-Markovnikov, *syn* orientation.

3 (1-methylcyclohexene, CH_3) → BH_3 → (CH_3, H, BR_2) → H_2O_2, OH^- → 3 (CH_3, H, OH)

E. OXIDATION

1. Potassium Permanganate

Alkenes can be oxidized with $KMnO_4$ to provide different types of products, depending upon the reaction conditions. Cold, dilute, aqueous $KMnO_4$ reacts to produce 1,2 diols (vicinal diols), which are also called glycols, with *syn* orientation:

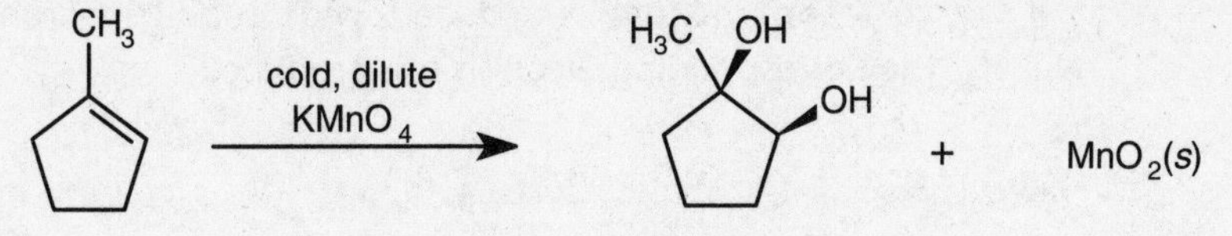
CH3
cold, dilute
KMnO4
H3C
OH
OH
+
MnO2(s)

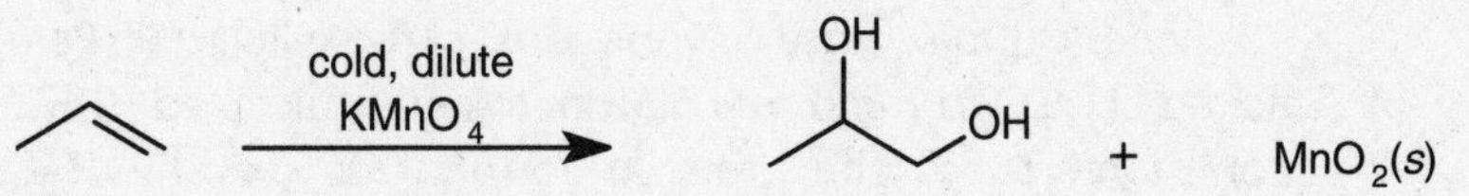
cold, dilute
KMnO4
OH
OH
+
MnO2(s)

If a hot, basic solution of potassium permanganate is added to the alkene and then acidified, nonterminal alkenes are cleaved to form 2 molar equivalents of carboxylic acid, and terminal alkenes are cleaved to form a carboxylic acid and carbon dioxide. If the nonterminal double bonded carbon is disubstituted, however, a ketone will be formed:

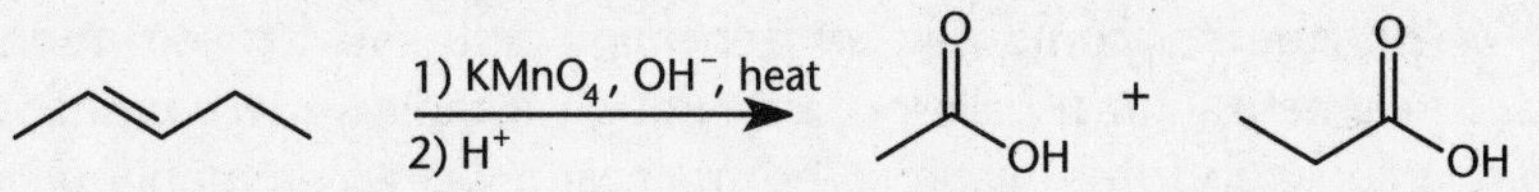
1) KMnO4, OH−, heat
2) H+
O
OH
+
O
OH

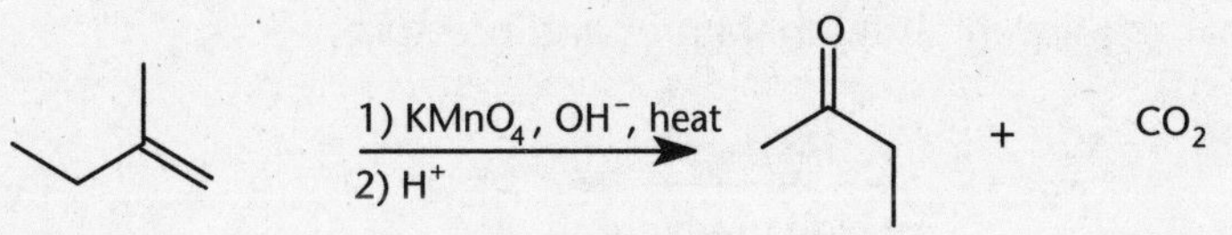
1) KMnO4, OH−, heat
2) H+
O
+
CO2

2. **Ozonolysis**

Treatment of alkenes with ozone followed by reduction with zinc and water results in cleavage of the double bond in the following manner:

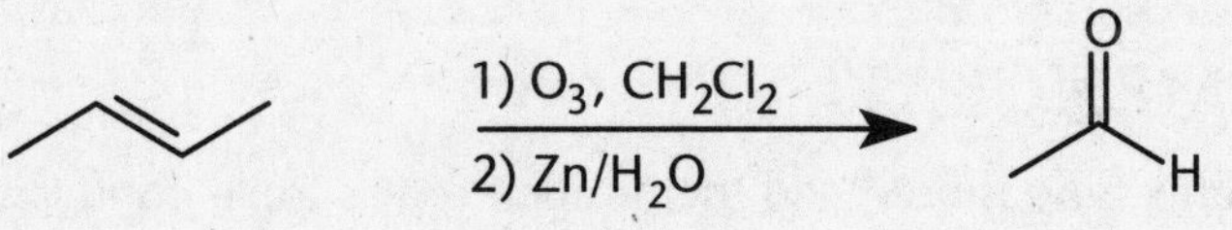
1) O3, CH2Cl2
2) Zn/H2O
O
H

If the reaction mixture is reduced with sodium borohydride, $NaBH_4$, the corresponding alcohols are produced:

$$\text{(hex-3-ene)} \xrightarrow[\text{2) } NaBH_4,\ CH_3OH]{\text{1) } O_3,\ CH_2Cl_2} \text{(butanol) OH} + \text{(propanol) OH}$$

3. **Peroxycarboxylic Acids**

Alkenes can be oxidized with peroxycarboxylic acids. Peroxyacetic acid (CH_3CO_3H) and *m*-chloroperoxybenzoic acid (MCPBA) are commonly used. The products formed are **oxiranes** (also called **epoxides**):

$$\text{(cyclopentene)} \xrightarrow{\text{MCPBA}} \text{(cyclopentene oxide)}$$

F. POLYMERIZATION

Polymerization is the creation of long, high molecular weight chains (**polymers**), composed of repeating subunits (called **monomers**). Polymerization usually occurs through a radical mechanism, although anionic and even cationic polymerizations are commonly observed. A typical example is the formation of polyethylene from ethylene (ethene) that requires high temperatures and pressures:

$$CH_2{=}CH_2 \xrightarrow[\text{high pressure}]{R\bullet,\ \text{heat}} RCH_2CH_2(CH_2CH_2)_nCH_2CH_2R$$

ALKYNES

Alkynes are hydrocarbon compounds that possess one or more carbon-carbon triple bonds.

Flashback:

Refer to Chapter 11 for the general rules of nomenclature of alkynes.

NOMENCLATURE

The suffix **-yne** is used and the position of the triple bond is specified when necessary. A common exception to the IUPAC rules is ethyne, which is called *acetylene*. Frequently, compounds are named as derivatives of acetylene.

$CH_3CH_2CH_2CHC{\equiv}CCH_3$ (Cl on C-4)
4-chloro-2-heptyne

$CH{\equiv}CH$
ethyne
(acetylene)

$CH_3C{\equiv}CH$
propyne
(methylacetylene)

PHYSICAL PROPERTIES

The physical properties of the alkynes are similar to those of the analogous alkenes and alkanes. In general, the shorter-chain compounds are gases, boiling at somewhat higher temperatures than the corresponding alkenes. Internal alkynes, like alkenes, boil at higher temperatures than terminal alkynes.

Asymmetrical distribution of electron density causes alkynes to have dipole moments which are larger than those of alkenes, but still small in magnitude. Thus, solutions of alkynes can be slightly polar.

Terminal alkynes are fairly acidic, having pKa's of approximately 25. This property is exploited in some of the reactions of alkynes, which will be discussed later.

SYNTHESIS

Triple bonds can be made by the elimination of two molecules of HX from a geminal or vicinal dihalide:

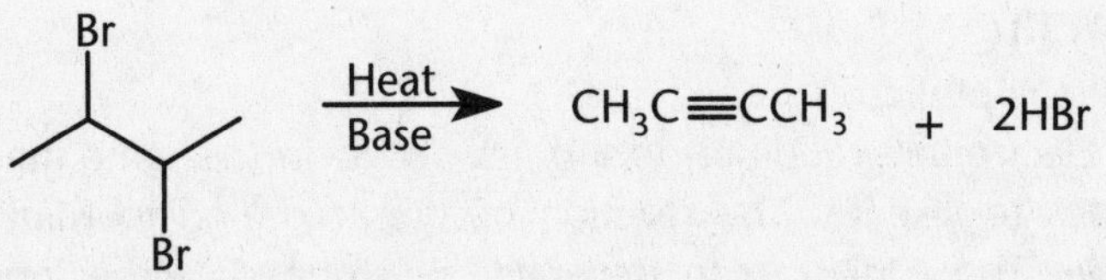

$$\text{2,3-dibromobutane} \xrightarrow[\text{Base}]{\text{Heat}} CH_3C{\equiv}CCH_3 + 2HBr$$

This reaction is not always practical and requires high temperatures and a strong base. A more useful method adds an already existing triple bond into a particular carbon skeleton. A terminal triple bond is converted to a nucleophile by removing the acidic proton with strong base, producing an *acetylide ion*. This ion will perform nucleophilic displacements on alkyl halides at room temperature:

$$CH{\equiv}CH \xrightarrow{n\text{-BuLi}} CH{\equiv}C^{-}\,Li^{+} \xrightarrow{CH_3Cl} CH{\equiv}CCH_3$$

REACTIONS

A. REDUCTION

Alkynes, just like alkenes, can be hydrogenated with a catalyst to produce alkanes. A more useful reaction stops the reduction after addition of just one equivalent of H_2, producing alkenes. This partial hydrogenation can take place in two different ways. The first uses **Lindlar's catalyst**, which is palladium on barium sulfate ($BaSO_4$) with quinoline, a poison which stops the reaction at the alkene stage. Because the reaction occurs on a metal surface, the product alkene is the *cis* isomer. The other method uses sodium in liquid ammonia below –33°C (the boiling point of ammonia), and produces the *trans* isomer of the alkene via a free radical mechanism:

$$CH_3C{\equiv}CCH_3 \xrightarrow[\text{Quinoline (Lindlar's catalyst)}]{H_2,\ Pd/BaSO_4} (CH_3)(H)C{=}C(CH_3)(H)$$

2-butyne → *cis*-2-butene

$$CH_3C{\equiv}CCH_3 \xrightarrow{Na,\ NH_3(liq)} (CH_3)(H)C{=}C(H)(CH_3)$$

2-butyne → *trans*-2-butene

B. ADDITION

1. Electrophilic

Electrophilic addition to alkynes occurs in the same manner as it does to alkenes. The reaction occurs according to Markovnikov's rule. The addition can generally be stopped at the intermediate alkene stage, or carried further. The following examples are illustrative:

$$CH_3C{\equiv}CH \xrightarrow{Br_2} (CH_3)(Br)C{=}C(Br)(H)$$

$$CH_3C{\equiv}CH \xrightarrow{2Br_2} CH_3CBr_2CBr_2H$$

2. Free Radical

Radicals add to triple bonds as they do to double bonds—with anti-Markovnikov orientation. The reaction product is usually the *trans* isomer, because the intermediate vinyl radical can isomerize to its more stable form.

$$CH_3CH_2C{\equiv}CH + X\cdot \longrightarrow (CH_3CH_2)\dot{C}{=}C(X)(H) \rightleftharpoons (CH_3CH_2)\dot{C}{=}C(X)(H) \longrightarrow (H)(CH_3CH_2)C{=}C(X)(H)$$

C. HYDROBORATION

Addition of boron to triple bonds occurs by the same method as addition of boron to double bonds. Addition is *syn,* and the boron atom adds first. The boron atom can be replaced with a proton from acetic acid, to produce a *cis* alkene:

$$3\ H_3CC{\equiv}CCH_3 + {}^1/_2\ B_2H_6 \longrightarrow (CH_3)(H)C{=}C(CH_3)\text{–}B[C(CH_3){=}CHCH_3]_2$$

$$\xrightarrow{CH_3COOH} 3\ (CH_3)(H)C{=}C(CH_3)(H)$$

With terminal alkynes, a disubstituted borane is used to prevent further boration of the vinylic intermediate to an alkane. The vinylic borane intermediate can be oxidatively cleaved with hydrogen peroxide (H_2O_2), creating an intermediate vinyl alcohol, which rearranges to the more stable carbonyl compound (via **keto-enol tautomerism**).

$$CH_3C{\equiv}CH \xrightarrow[H_2O_2,\ OH^-]{R_2BH} \underset{H}{\overset{H_3C}{}}C{=}C\underset{OH}{\overset{H}{}} \longrightarrow CH_3CH_2\underset{\underset{O}{\|}}{C}H$$

D. OXIDATION

Alkynes can be oxidatively cleaved with either basic potassium permanganate (followed by acidification) or ozone.

$$\text{2-hexyne} \xrightarrow[2)\ H^+]{1)\ KMnO_4,\ OH^-} CH_3COOH + CH_3CH_2CH_2COOH$$

$$\text{1-hexyne} \xrightarrow[2)\ H_2O]{1)\ O_3,\ CCl_4} CH_3CH_2CH_2CH_2COOH + HCOOH$$

REVIEW PROBLEMS

1. The major product of the reaction below is

   (3-methyl-2-butanol) $\xrightarrow[\Delta]{H^+}$?

   A. 3-methyl-1-butene.
   B. 2-methyl-3-butene.
   C. 3-methyl-2-butene.
   D. 2-methyl-2-butene.

2. (tert-butyl chloride) $\xrightarrow[-HCl]{CH_3O^-, CH_3OH}$ (2-methylpropene)

   The above reaction takes place mostly by which of the following mechanisms?

   A. S_N1
   B. S_N2
   C. E1
   D. E2

3. Which of the following products would be formed if 2-methyl-2-butene was reacted with hot basic $KMnO_4$?

   A. 1 mole of acetic acid and 1 mole of propanoic acid.
   B. 2 moles of pentanoic acid.
   C. 1 mole of acetic acid and 1 mole of acetone.
   D. 2 moles of acetic acid and 1 mole of CO_2.

4. Which of the following are true about E2 reactions?

   A. They are greatly affected by steric hindrance.
   B. They need a strong base to abstract the proton.
   C. They are favored over S_N2 by weak Lewis bases.
   D. They are favored over S_N1 and E1 by polar solvents.

5.

(2-methylpropene) $\xrightarrow[\text{2. Zn, } H_2O]{\text{1. } O_3/CH_2Cl_2}$?

A. (isopropanol) OH + (formaldehyde) H, H

B. (acetone) + CH_3OH

C. (acetone) + CO_2

D. (formaldehyde) H, H + (acetone)

6.

$CH_3-C\equiv C-CH(CH_3)_2$ $\xrightarrow[\text{2) } H^+]{\text{1) hot } KMnO_4\text{, } OH^-}$?

A. (acetaldehyde) H + (2-methylpropanal) H

B. (acetaldehyde) H + HO (2-methylpropanoic acid)

C. (acetic acid) OH + HO (2-methylpropanoic acid)

D. (enediol) OH, OH

7.

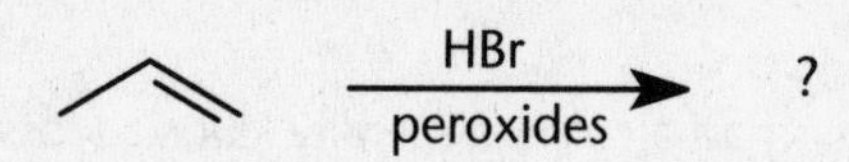

Which of the following represents the product obtained in the above reaction?

A.

B.

C. Br

D. Br

SOLUTIONS TO REVIEW PROBLEMS

1. **D** Heating an alcohol generally leads to loss of a water molecule. Two products can be obtained depending on which H atom is used by the OH group to form water; either

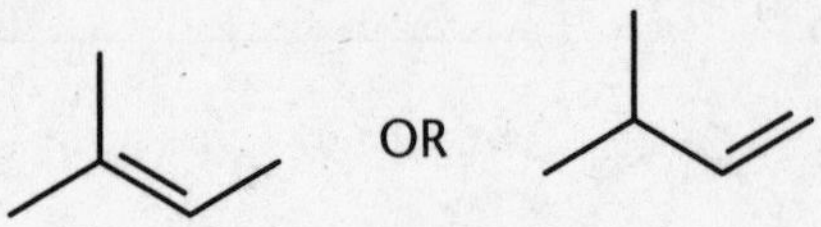

OR

The most stable alkene, which is the most substituted one, is formed. Of the two alkenes, 2-methyl-2-butene is more stable, and the correct choice is D.

2. **D** Since a tertiary haloalkane has been converted to an alkene, this is an elimination reaction. This particular compound,2-chloro-2-methylpropane, can react through either E1 or E2 depending on the conditions: Reaction with strong bases leads to E2, and reaction with weak bases leads to E1. Since methoxide is a strong base, elimination occurs by the E2 mechanism. Choice D is the correct response.

3. **C** The double bond of 2-methyl-2-butene is cleaved by hot, basic potassium permangenate to form acetone and acetic acid. If the double-bonded carbon is a monosubstituted carbon, a carboxylic acid is obtained, but if it is a disubstituted carbon, a ketone is obtained.

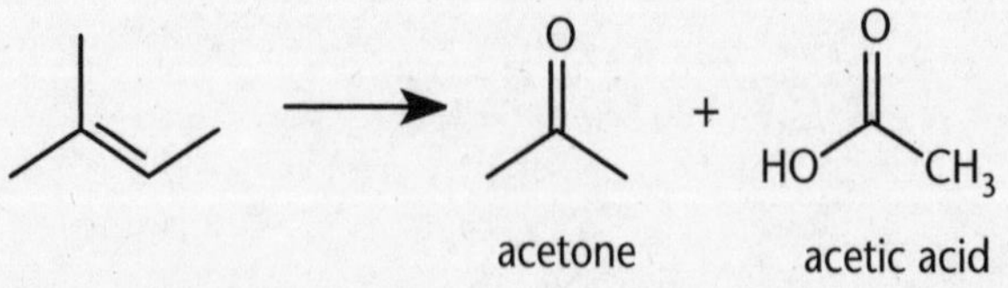

O O
HO CH$_3$
acetone acetic acid

The C–2 is a disubstituted carbon, so a ketone, $(CH_3)_2$ C=O, is obtained; the C–3 is a monosubstituted carbon, so an acid, CH_3CO_2H is obtained on reacting it with $KMnO_4$. Thus, the correct choice is C.

4. **B** Discussed on page 279.

5. **D** Ozonolysis of an alkene, and a subsequent treatment with zinc and water, produces carbonyl compounds. The double bond is broken and a disubstituted double-bonded carbon is converted to a ketone, whereas a monosubstituted double-bonded carbon is converted to an aldehyde. In this reaction, C–1 is a monosubstituted carbon and C–2 is a disubstituted carbon, and thus the products obtained are a ketone and an aldehyde.

6. **C** Treating alkynes with hot basic $KMnO_4$ leads to the cleavage of the triple bond and the formation of carboxylic acids.

7. **A** In the presence of peroxides, the addition of HBr to the double bond takes place in an anti-Markovnikov manner in a series of free-radical reactions initiated by peroxides.

1. $ROOR \xrightarrow{h\nu} 2RO\bullet$
2. $HBr + RO\bullet \longrightarrow ROH + Br\bullet$
3. $CH_3CH = CH_2 + Br\bullet \longrightarrow CH_3\text{–}\dot{C}H\text{–}CH_2Br$
4. $CH_3\text{–}\dot{C}H\text{–}CH_2Br + HBr \longrightarrow CH_3\text{–}CH_2\text{–}CH_2Br + Br\bullet$

In step 3, $CH_3\text{–}\dot{C}H_2\text{–}CH_2Br$ is formed instead of $CH_3\text{–}CHBr\text{–}\dot{C}H_2$, since the more substituted free-radical is more stable than the less substituted one. Thus, the correct choice is A. Note that in the absence of peroxides, HBr adds to the double bond in a Markovnikov manner.

Aromatic Compounds

CHAPTER SIXTEEN

The terms **aromatic** and **aliphatic**, meaning "fragrant" and "fatty," respectively, were used originally to distinguish types of organic compounds. The terms persist with new definitions. "Aromatic" now describes any unusually stable ring system. These compounds are cyclic, conjugated polyenes that possess $4n + 2$ pi electrons and adopt planar conformations to allow maximum overlap of the conjugated pi orbitals. "Aliphatic" describes all compounds that are not aromatic.

Note:

n can be any nonnegative integer; thus, 4n + 2 can be 2, 6, 10, 14, 18, etc.

The criterion of $4n + 2$ pi electrons is known as **Hückel's rule**, and is an important indicator of aromaticity. In general, if a cyclic conjugated polyene follows Hückel's rule, then it is an aromatic compound. Neutral compounds, anions, and cations may all be aromatic. Some typical aromatic compounds and ions are:

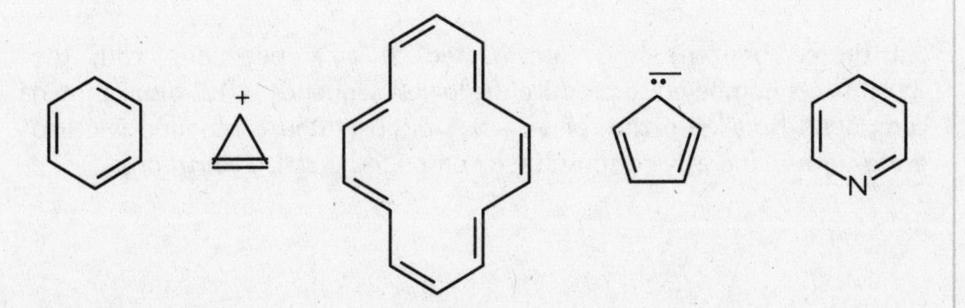

A cyclic, conjugated polyene that possesses $4n$ electrons is said to be **antiaromatic** (a cyclic, conjugated polyene that is destabilized). Some typical antiaromatic compounds are:

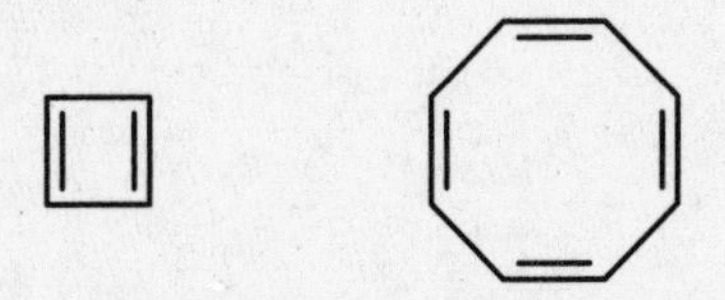

NOMENCLATURE

Aromatic compounds are referred to as **aryl** compounds, or **arenes,** and are represented by the symbol **Ar**. Aliphatic compounds are called **alkyl** and are represented by the symbol **R**. Common names exist for many mono- and di-substituted aromatic compounds.

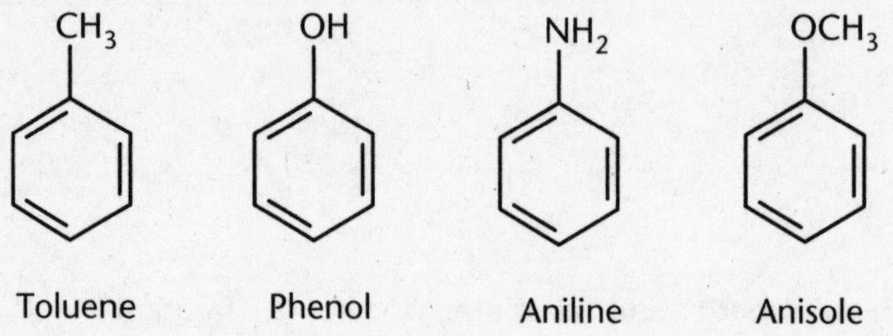

CH_3 OH NH_2 OCH_3

Toluene Phenol Aniline Anisole

The benzene group is called a **phenyl** group **(Ph)** when named as a substituent. The term **benzyl** refers to a toluene molecule substituted at the methyl position.

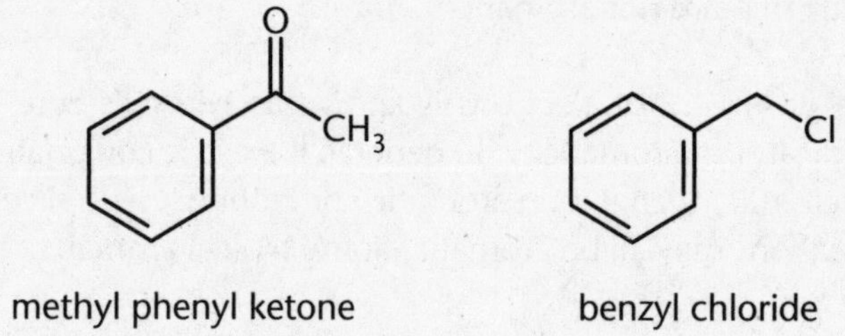

O CH_3 Cl

methyl phenyl ketone benzyl chloride

Substituted benzene rings are named as alkyl benzenes, with the substituents numbered to produce the lowest sequence. A 1,2-disubstituted compound is called ***ortho***- or ***o***-; a 1,3 disubstituted compound is called ***meta***- or ***m***-; and a 1,4 disubstituted compound is called ***para***- or ***p***-.

2,4,6-trinitrotoluene (TNT) *o*-nitrotoluene *m*-dichlorobenzene

p-methylbenzoic acid

There are many polycyclic and heterocyclic aromatic compounds.

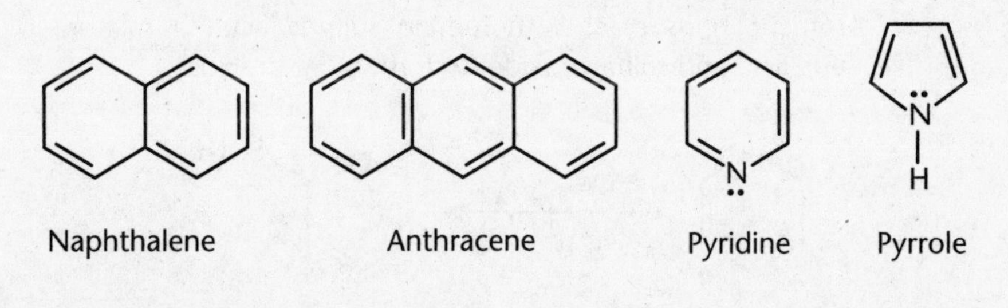

Naphthalene Anthracene Pyridine Pyrrole

PROPERTIES

The physical properties of aromatic compounds are generally similar to those of other hydrocarbons. By contrast, chemical properties are significantly affected by aromaticity. The characteristic planar shape of benzene permits the ring's six pi orbitals to overlap, delocalizing the electron density. All six carbon atoms are sp^2 hybridized, and each of the six orbitals overlaps equally with its two neighbors. As a result, the delocalized electrons form two "pi electron clouds," one above and one below the plane of the ring. This delocalization stabilizes the molecule, making it fairly unreactive: in particular, benzene does not undergo addition reactions as do alkenes. The same holds true for other aromatic compounds, since the definition of an aromatic compound includes the condition that it have a delocalized pi electron system.

Note:

The nonbonding electron pair in pyridine is in a nitrogen sp^2 orbital. This orbital is perpendicular to the p orbitals around the ring and therefore is not involved in the conjugated pi system. On the other hand, the nonbonding pair in pyrrole is in a nitrogen sp^3 orbital parallel to the ring p orbitals and therefore can participate in the delocalized pi system.

REACTIONS

A. ELECTROPHILIC AROMATIC SUBSTITUTION

The most important reaction of aromatic compounds is electrophilic aromatic substitution. In this reaction an electrophile replaces a proton on an aromatic ring, producing a substituted aromatic compound. The most common examples are halogenation, sulfonation, nitration, and acylation.

1. Halogenation

Aromatic rings react with bromine or chlorine in the presence of a Lewis acid, such as $FeCl_3$, $FeBr_3$ or $AlCl_3$, to produce monosubstituted products in good yield. Reaction of fluorine and iodine with aromatic rings is less useful, as fluorine tends to produce multisubstituted products, and iodine's lack of reactivity requires special conditions for the reaction to proceed.

benzene + Br_2 $\xrightarrow[\text{heat}]{FeBr_3}$ bromobenzene (Br)

2. Sulfonation

Aromatic rings react with fuming sulfuric acid (a mixture of sulfuric acid and sulfur trioxide) to form sulfonic acids.

benzene $\xrightarrow[\text{heat}]{SO_3/\ H_2SO_4}$ benzenesulfonic acid (SO_3H)

3. Nitration

The nitration of aromatic rings is another synthetically useful reaction. A mixture of nitric and sulfuric acids is used to create the nitronium ion, NO_2^+, a strong electrophile. This reacts with aromatic rings to produce nitro compounds.

benzene $\xrightarrow[HNO_3]{H_2SO_4}$ [H, O_2N arenium ion, +] $\xrightarrow{-H^+}$ nitrobenzene (O_2N)

Figure 6.9

4. Acylation (Friedel-Crafts Reactions)

In Friedel-Crafts acylation reaction, a carbocation electrophile, usually an acyl group, is incorporated into the aromatic ring. These reactions are usually catalyzed by Lewis acids such as $AlCl_3$.

$$C_6H_6 + CH_3\overset{O}{\overset{\|}{C}}Cl \xrightarrow{AlCl_3} C_6H_5\overset{O}{\overset{\|}{C}}CH_3 + HCl$$

5. Substituent Effects

Substituents on an aromatic ring strongly influence the susceptibility of the ring to electrophilic aromatic substitution, and also strongly affect what position on the ring an incoming electrophile is most likely to attack. Substituents can be grouped into three different classes according to whether substitution is enhanced (activating) or inhibited (deactivating), and where the reaction is likely to take place with respect to the group already present. These effects depend on whether the group tends to donate or withdraw electron density, and how it does so; the specifics of these mechanisms will not be discussed here. Arranged in order of decreasing strength of the substituent effect, the three classes are listed below:

a. Activating, *ortho/para*-directing substituents (electron-donating): NH_2, NR_2, OH, NHCOR, OR, OCOR, and R.
b. Deactivating, *ortho/para*-directing substituents (weakly electron-withdrawing): F, Cl, Br, and I.
c. Deactivating, *meta*-directing substituents (electron-withdrawing): NO_2, SO_3H, and carbonyl compounds, including COOH, COOR, COR, and CHO.

Example: when toluene undergoes electrophilic aromatic substitution, the methyl group directs substitution to occur at the *ortho* and *para* positions:

CH_3 $\xrightarrow[H_2SO_4]{HNO_3}$ CH_3 / NO_2 + CH_3 / NO_2 + CH_3 / NO_2

63% 34% 3%

B. REDUCTION

1. Catalytic Reduction

Benzene rings can be reduced by catalytic hydrogenation under vigorous conditions (elevated temperature and pressure) to yield cyclohexane. Ruthenium or rhodium on carbon are the most common catalysts; platinum or palladium may also be used.

OCH_3 $\xrightarrow[100°C]{H_2,\ Rh/C}$ OCH_3

REVIEW PROBLEMS

1. Predict aromatic, antiaromatic, or nonaromatic behavior for each of the following compounds.

a. 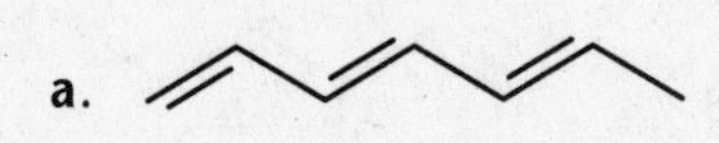

b. 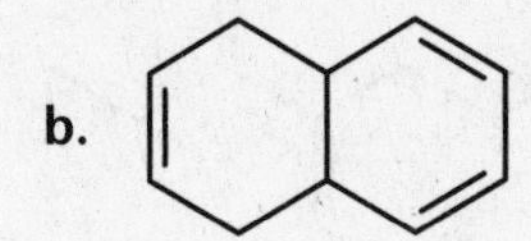

c.

d.

e. H, N^+—R, H

2. Which of the following represents the correct structure for *para*-nitrotoluene?

A. CH_2NH_2

B. NO_2, CH_3

C. NO_2, H_3C

D. NH_2, H_3C

3. What would be the major product of the following reaction?

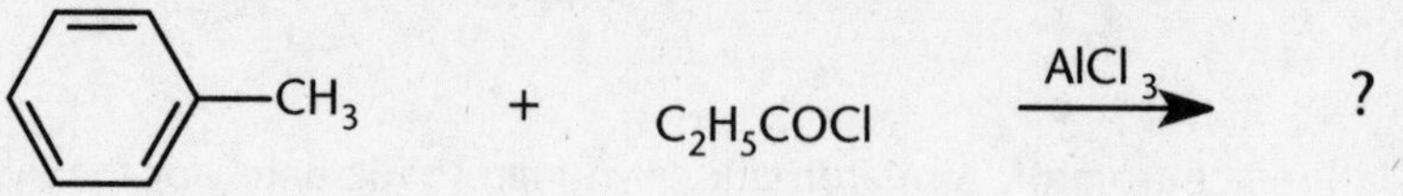

C_6H_5–CH_3 + C_2H_5COCl $\xrightarrow{AlCl_3}$?

A. (meta-methylpropiophenone)

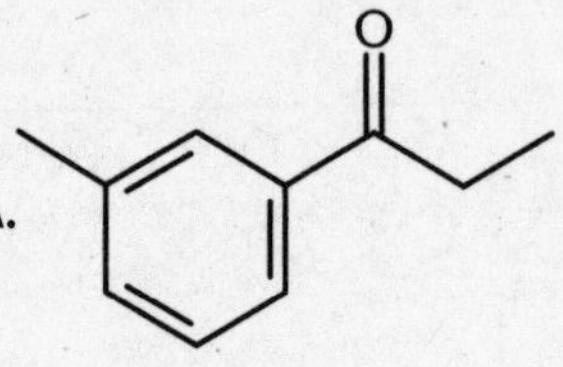

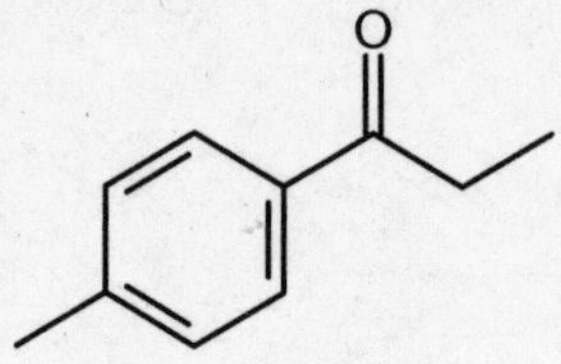

B. (para-methylpropiophenone)

C. Cl (meta-chlorotoluene)

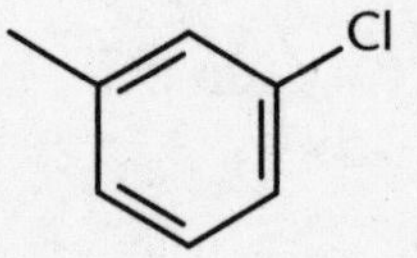

D. Cl (para-chlorotoluene)

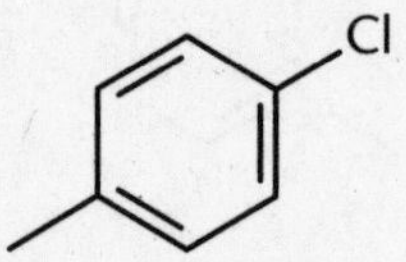

4. Nitration of benzene at 30°C leads to a 95 percent yield of nitrobenzene. When the temperature is increased to 100°C, dinitrobenzene is produced. Which of the following is the predominant product?

A. NO_2, NO_2 (ortho)

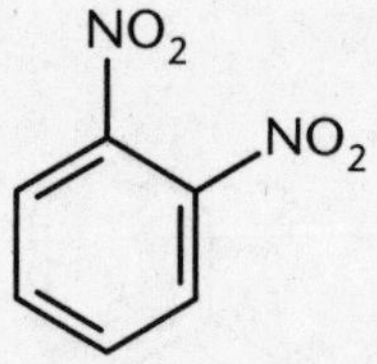

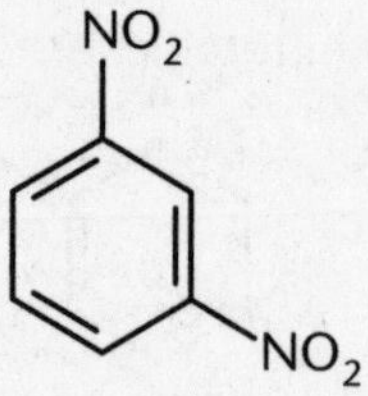

B. NO_2, NO_2 (para)

C. NO_2, NO_2 (meta)

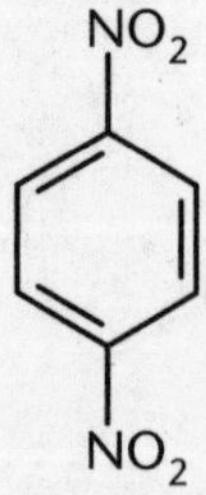

D. NO_2

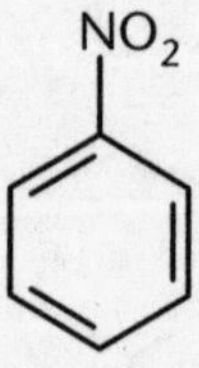

5. What is the major product of the nitration reaction below?

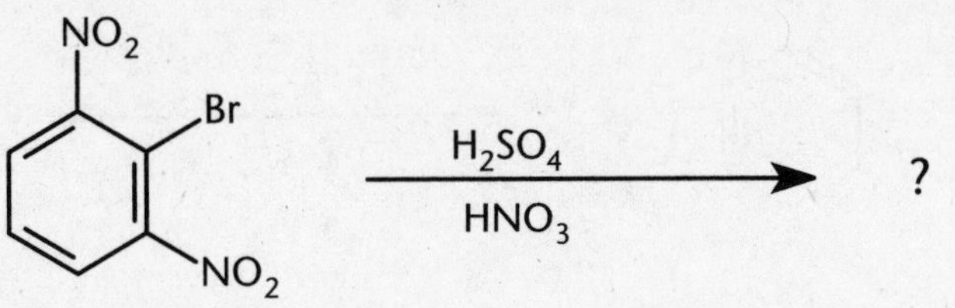

NO_2, Br, NO_2 $\xrightarrow[HNO_3]{H_2SO_4}$?

A. NO_2, Br, NO_2, O_2N

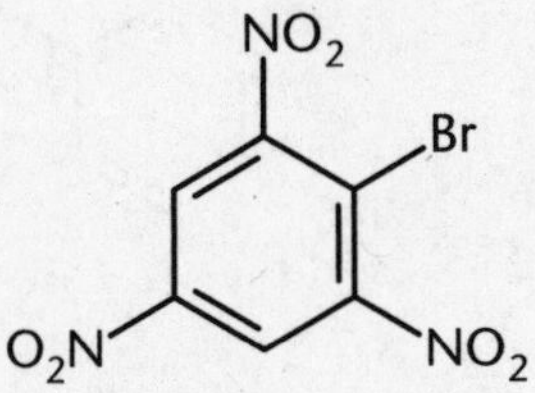

C. NO_2, O_2N, Br, NO_2

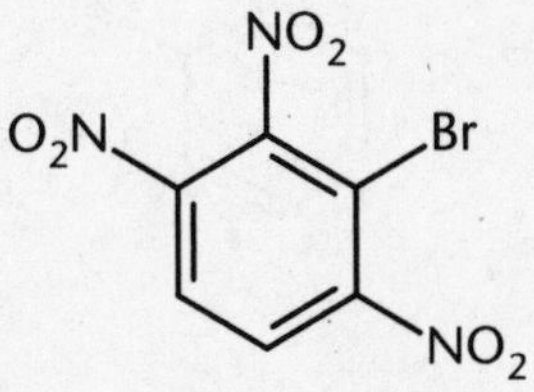

B. NO_2, Br, NO_2, Br

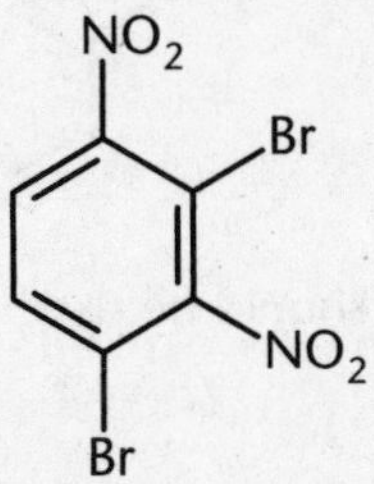

D. Br, NO_2, Br, O_2N

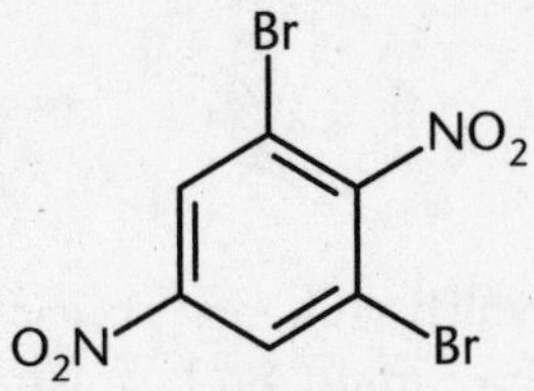

6. Give the predominant product(s) of the reaction below.

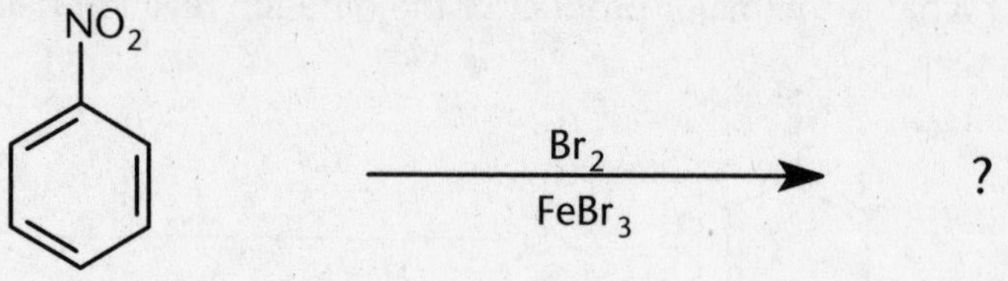

Nitrobenzene (NO_2) $\xrightarrow[FeBr_3]{Br_2}$?

A. 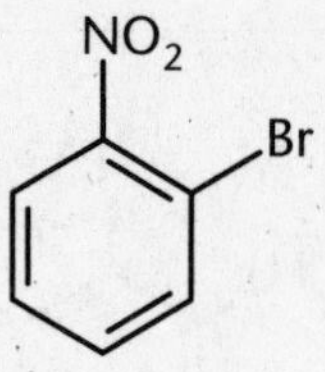 (NO_2, Br ortho)

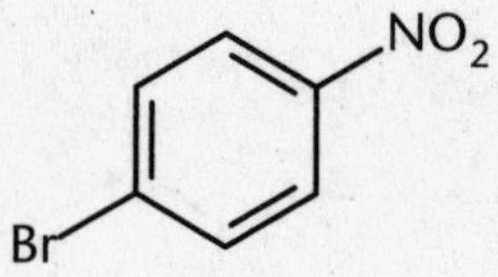

B. 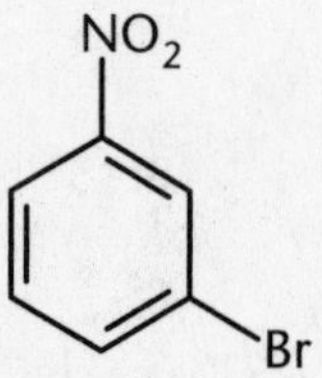 (NO_2, Br meta)

C. (NO_2, Br para)

D. 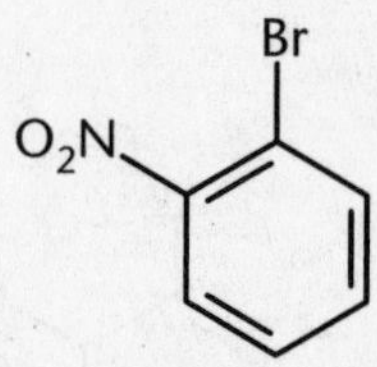 (O_2N, Br ortho)

7. Which sequence of reaction conditions should be used to produce the compound below from benzene?

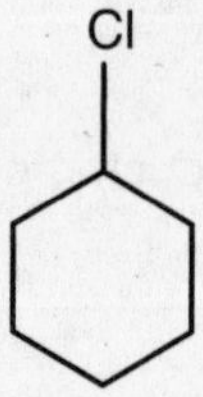

Cl (chlorocyclohexane)

A. $AlCl_3/Cl_2$; H_2/Pt
B. Cl_2/UV light; H_2/Pt
C. H_2/Pt; $AlCl_3/Cl_2$
D. HCl; H_2/Pt

SOLUTIONS TO REVIEW PROBLEMS BEGIN ON THE FOLLOWING PAGE.

SOLUTIONS TO REVIEW PROBLEMS

1. a. Nonaromatic. This compound has $4n + 2$ pi electrons ($n = 1$), and is a conjugated system; however, it is not cyclic.

   b. Nonaromatic. This compound has $4n + 2$ pi electrons ($n = 1$) and is cyclic. However, there is no conjugation of the double bonds.

   c. Aromatic. This compound (naphthalene) is cyclic, has $4n + 2$ pi electrons ($n = 2$), and has a conjugated system of double bonds.

   d. Antiaromatic. This compound is cyclic and conjugated. However, it has only $4n$ pi electrons ($n = 1$).

   e. Aromatic. This class of compounds (alkyl anilinium ions) have $4n + 2$ pi electrons ($n = 1$), are cyclic, and have conjugated double bond systems.

2. C Toluene is the common name for methylbenzene: a methyl group attached to a benzene ring. In *para*-nitrotoluene, the nitro group (NO_2) is attached to the ring directly across from the methyl group. Choice C is the correct response. Choice B is wrong because the nitro group is meta, not *para* substituted. Choices A and D can be eliminated since nitro groups are not present in these compounds.

3. B The reaction shown, a Friedel-Crafts acylation of toluene, will yield a product containing a $–CH_3$ substituent and a $–C = O$ substituent. Chlorination of the ring also uses $AlCl_3$ as a reagent, but there the second reagent would have to be Cl_2, not CH_2COCl; thus, choices C and D (both chlorotoluenes) can be eliminated. Since CH_3 is an activating, *ortho/para*-directing group, the *meta* isomer, choice A, would be a minor product at best. Choice B, the *para* isomer, would be favored by the substituent effect of the methyl group, and along with the *ortho* isomer (which is not among these choices) would be the major product.

4. C *meta*-Dinitrobenzene is the predominant product, since $–NO_2$ is a *meta*-directing group.

5. **A** All three substituents of 2-bromo-1,3-dinitrobenzene direct reaction to C–5, which is *meta* to both nitro groups and *para* to the bromine atom, so choice A will be the major product. Since all three groups are deactiving, the reaction will be slow and required elevated temperatures.

6. **B** This reaction shows bromination of nitrobenzene. This is a trick question, because choices A and D are different views of the same compound. Since aromatic rings are planar, rings that are mirror images are identical (although alkyl substituents of benzene rings may still have chiral centers, and molecules containing such groups may be chiral). B shows *meta*-bromonitrobenzene, which is the favored product of this reaction since the nitro substituent is *meta*-directing. Choices A, C, and D show the *ortho* and *para* isomers, which are less favored in this reaction.

7. **A** In order to produce chlorocyclohexane, two different procedures must be carried out: the benzene ring must be chlorinated and then hydrogenated. A suitable way to chlorinate the ring is to use Cl_2 and the Lewis acid $AlCl_3$, which is choice A. Using chlorine in the presence of UV light will not be effective, so choice B is wrong. Choice D is wrong because HCl will not chlorinate the ring. Now for the second step: hydrogenation. Hydrogenation of the benzene ring can be accomplished by using hydrogen in the presence of a platinum catalyst, so choice A is the correct answer. If the procedure were carried out according to choice C, reduction would occur, forming cyclohexane, but chlorination would not, since cyclohexane is unreactive towards chlorine and the Lewis acid catalyst.

Alcohols and Ethers

CHAPTER SEVENTEEN

ALCOHOLS

Alcohols are compounds with the general formula **ROH**. The functional group **–OH** is called the **hydroxyl** group. An alcohol can be thought of as a substituted water molecule, with an alkyl group R replacing one H atom.

NOMENCLATURE

Alcohols are named in the IUPAC system by replacing the **-e** ending of the root alkane with the ending **-ol**. The carbon atom attached to the hydroxyl group must be included in the longest chain and receives the lowest possible number. Some examples are:

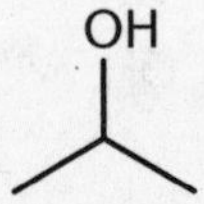

OH

OH

2-propanol 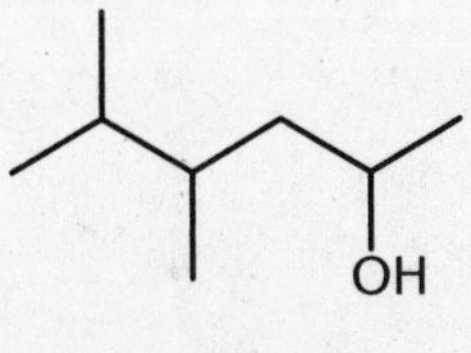 4, 5-dimethyl-2-hexanol

Alternatively, the alkyl group can be named as a derivative, followed by the word *alcohol*.

OH OH

ethyl alcohol isobutyl alcohol

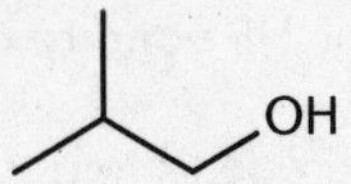

Compounds of the general formula ArOH, with a hydroxyl group attached to an aromatic ring, are called **phenols**.

Note:

The –OH group has high priority, so its C must be in the carbon backbone, with the lowest number possible.

Note:

Aromatic alcohols (ArOH) are called phenols. Resonance (through the ring) gives them special properties.

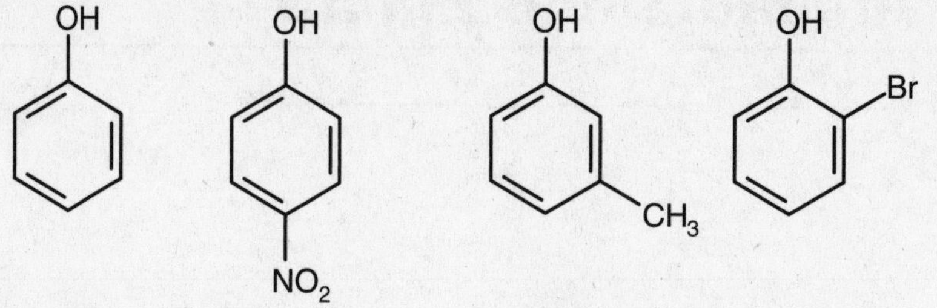

phenol | *p*-nitrophenol | *m*-cresol (*m*-methylphenol) | *o*-bromophenol

In a Nutshell:

Hydrogen bonding results in both elevated boiling points and better solubility in water.

PHYSICAL PROPERTIES

The boiling points of alcohols are significantly higher than those of the analogous hydrocarbons, due to **hydrogen bonding**.

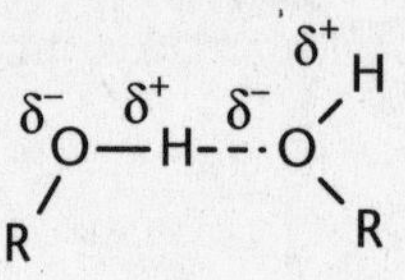

$$R-\overset{\delta^-}{O}-\overset{\delta^+}{H}\cdots\overset{\delta^-}{O}(R)-\overset{\delta^+}{H}$$

Molecules with more than one hydroxyl group show greater degrees of hydrogen bonding, as is evident from the following boiling points.

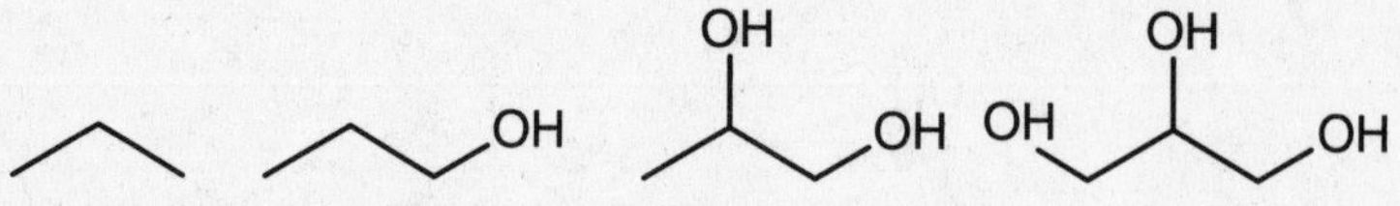

Boiling Point (°C)	–42.1	97.4	189.0	290.0

Mnemonic:

Hydrogen bonds form on the "phone"—the "FON"! (Fluorine, Oxygen, Nitrogen)

Hydrogen bonding can also occur when hydrogen atoms are attached to other highly electronegative atoms, such as nitrogen and fluorine. HF has particularly strong hydrogen bonds because the high electronegativity of fluorine causes the HF bond to be highly polarized.

Note:

Alcohols are weakly acidic:
$ROH \rightleftharpoons RO^- + H^+$

The hydroxyl hydrogen atom is weakly acidic, and alcohols can dissociate into protons and alkoxy ions just as water dissociates into protons and hydroxide ions. pK_a values of several compounds are listed below.

Dissociation		pK$_a$
H_2O ⇌	$HO^- + H^+$	15.7
CH_3OH ⇌	$CH_3O^- + H^+$	15.5
C_2H_5OH ⇌	$C_2H_5O^- + H^+$	15.9
i-PrOH ⇌	*i*-PrO$^-$ + H$^+$	17.1
t-BuOH ⇌	*t*-BuO$^-$ + H$^+$	18.0
CF_3CH_2OH ⇌	$CF_3CH_2O^- + H^+$	12.4
PhOH ⇌	PhO$^-$ + H+	≈10.0

The hydroxyl hydrogens of phenols are more acidic than those of alcohols, due to resonance structures that distribute the negative charge throughout the ring, thus stabilizing the anion. As a result, these compounds form intermolecular hydrogen bonds and have relatively high melting and boiling points. Phenol is slightly soluble in water (presumably due to hydrogen bonding), as are some of its derivatives. Phenols are much more acidic than aliphatic alcohols and can form salts with inorganic bases such as NaOH.

The presence of other substituents on the ring has significant effects on the acidity, boiling points, and melting points of phenols. As with other aromatic compounds, electron-withdrawing substituents increase acidity, and electron-donating groups decrease acidity.

Bridge to Chemistry:

$pK_a = -logK_a$
Strong acids have high K_a's and small pK_a's. Thus, phenol, which has the smallest pKa, is the most acidic.

Note:

Acidity decreases as more alkyl groups are attached because the electron-donating alkyl groups destabilize the alkoxide anion. Electron-withdrawing groups stabilize the alkoxy anion, making the alcohol more acidic.

Note:

Charge likes to be spread out as widely as possible!

↑acidity — *resonance, e⁻withdrawing*

↓acidity — *e⁻ donating*

KEY REACTION MECHANISMS FOR ALCOHOLS AND ETHERS

As you read about synthesis of (and from) alcohols and ethers, you'll see the same basic reaction mechanisms recurring over and over. Rather than memorizing each reaction individually, try to think of them in broad categories. Focus on how the basic mechanism works and on how this particular reaction exemplifies it. The "Big Three" mechanisms for alcohols and ethers are:

1) S_N1, S_N2: nucleophilic substitution

   e.g., $CH_3Br + OH^- \longrightarrow CH_3OH + Br^-$

   See Chapter 14 for review.

2) Electrophilic addition to a double bond,
   This and other reactions adding H_2O to double bonds are covered in Chapter 15.

3) Nucleophilic addition to a carbonyl,
   This mechanism is discussed further in Chapters 18-20.

Also, when thinking about alcohols, you should keep in mind their place on the oxidation-reduction continuum:

Note:

More bonds to oxygen means more oxidized.

OXIDATION →

1° alcohols ⟷ aldehydes ⟷ carboxylic acids
2° alcohols ⟷ ketones

← REDUCTION

As you read about the individual reactions in which alcohols participate, try to fit them into this framework (possible for most reactions, though not all).

SYNTHESIS

Alcohols can be prepared from a variety of different types of compounds. Methanol, also called wood alcohol, is obtained from the destructive distillation of wood. It is toxic and can cause blindness if ingested. Ethanol, or grain alcohol, is produced from the fermentation of sugars and can be metabolized by the body; however, in large enough quantities, it too is toxic.

A. ADDITION REACTIONS

Alcohols can be prepared via several reactions which involve addition of water to double bonds (discussed in Chapter 15). Alcohols can also be prepared from the addition of organometallic compounds to carbonyl groups (discussed in Chapter 20).

B. SUBSTITUTION REACTIONS

Both S_N1 and S_N2 reactions can be used to produce alcohols under the proper conditions (discussed in Chapter 14).

C. REDUCTION REACTIONS

Alcohols can be prepared from the reduction of aldehydes, ketones, carboxylic acids, or esters. Lithium aluminum hydride ($LiAlH_4$, or LAH) and sodium borohydride ($NaBH_4$) are the two most frequently used reducing reagents. LAH is more powerful and more difficult to work with, whereas $NaBH_4$ is more selective and easier to handle. For example, LAH will reduce carboxylic acids and esters, while $NaBH_4$ will not.

O ‖ OH —— 1) $LiAlH_4$ / 2) H_3O^+ ——> OH

O ‖ H —— 1) $NaBH_4$ / 2) H_3O^+ ——> OH H

D. PHENOL SYNTHESIS

Phenols may be synthesized from arylsulfonic acids with hot NaOH, as described in Chapter 16. However, this reaction is useful only for phenol or its alkylated derivatives, as most functional groups are destroyed by the harsh reaction conditions.

A more versatile method of synthesizing phenols is via hydrolysis of diazonium salts.

$$\text{4-BrC}_6\text{H}_4\text{NH}_2 \xrightarrow[H_2SO_4]{HNO_2} \text{4-BrC}_6\text{H}_4\text{N}_2^+\,HSO_4^- \xrightarrow{H_3O^+} \text{4-BrC}_6\text{H}_4\text{OH}$$

REACTIONS

A. ELIMINATION REACTIONS

Note:

Think of electrophilic addition of H_2O to a double bond, only this time it's in reverse—elimination.

Alcohols can be **dehydrated** in a strongly acidic solution (usually H_2SO_4) to produce alkenes. The mechanism of this dehydration reaction is E1, and proceeds via the protonated alcohol.

OH $\xrightarrow[\Delta]{H_2SO_4}$ + → minor

H ↓

+ → major

Note:

In E1, you want to form the most stable (that is, most substituted) carbocation.

Notice that two products are obtained, with the more stable alkene being the major product. This occurs via movement of a proton to produce the more stable 2° carbocation. This type of rearrangement is commonly encountered with carbocations.

B. SUBSTITUTION REACTIONS

Note:

Alcohols can participate in S_N1/S_N2 reactions, but only if you turn the –OH into a better leaving group by one of the following methods:
- *—protonate it;*
- *—convert to a tosylate;*
- *—form an inorganic ester.*

The displacement of hydroxyl groups in substitution reactions is rare because the hydroxide ion is a poor leaving group. If such a transformation is desired, the hydroxyl group must be made into a good leaving group. Protonating the alcohol makes water the leaving group, which is good for S_N1 reactions; even better, the alcohol can be converted into a tosylate (*p*-toluenesulfonate) group, which is an excellent leaving group for S_N2 reactions (see below).

$$-OH + HBr \longrightarrow -OH_2^+ + Br^-$$

$$Br^- + -OH_2^+ \longrightarrow -Br + H_2O$$

$$OH + ClSO_2-C_6H_4-CH_3 \longrightarrow O{=}S{=}O \xrightarrow{NaI} I + {}^-SO_3-C_6H_4-CH_3$$

tosyl chloride

A common method of converting alcohols into alkyl halides involves the formation of inorganic esters, which readily undergo S_N2 reactions. Alcohols react with thionyl chloride to produce an intermediate inorganic ester (a chlorosulfite) and HCl. The chloride ion of HCl displaces SO_2 and regenerates Cl^-, forming the desired alkyl chloride.

$$CH_3OH + SOCl_2 \longrightarrow CH_3OSOCl + HCl$$

$$Cl^- \quad CH_3{-}O{-}\overset{O}{\overset{\|}{S}}{-}Cl \longrightarrow ClCH_3 + SO_2 + Cl^-$$

An analogous reaction, where the alcohol is treated with PBr_3 instead of thionyl chloride, produces alkyl bromides.

Phenols readily undergo electrophilic aromatic substitution reactions; because it has lone pairs that it can donate to the ring, the –OH group is a strongly activating, *ortho/para*-directing ring substituent (see Chapter 16).

Note:

Phenols are good substrates for electrophilic aromatic substitution; the -OH is activating and ortho/para-directing.

C. OXIDATION REACTIONS

The oxidation of alcohols generally involves some form of chromium (VI) as the oxidizing agent, which is reduced to chromium (III) during the reaction. PCC (pyridinium chlorochromate, $C_5H_6NCrO_3Cl$) is commonly used as a mild oxidant. It converts primary alcohols to aldehydes without overoxidation to the acid. (In contrast, $KMnO_4$ is a very strong oxidizing agent that will take the alcohol all the way to the carboxylic acid.) It can also be used to form ketones from 2° alcohols. Tertiary alcohols cannot be oxidized for valence reasons.

OH —PCC→ O

Another reagent used to oxidize secondary alcohols is alkali (either sodium or potassium) dichromate salt. This will also oxidize 1° alcohols to carboxylic acids.

OH —$Na_2Cr_2O_7$, H_2SO_4→ O

A stronger oxidant is chromium trioxide, CrO_3. This is often dissolved with dilute sulfuric acid in acetone; the mixture is called Jones' reagent. It oxidizes primary alcohols to carboxylic acids and secondary alcohols to ketones.

CH_2OH —CrO_3, H_2SO_4, acetone→ CO_2H

Treatment of phenols with oxidizing reagents produces compounds called quinones (2,5-cyclohexadiene-1,4-diones).

OH, OH —$Na_2Cr_2O_7$, H_2SO_4→ O, O

1,4-benzenediol p-benzoquinone

Note:

When you see a transition metal (such as Cr or Mn) with lots of oxygen (Cr_2O_7, CrO_3, MnO_4.), think OXIDATION.

In a Nutshell:

Alcohols can be formed by and can participate in many of the key reaction mechanisms you've already seen:

1) *S_N1, S_N2 (although –OH must be made into a better leaving group);*
2) *addition of H_2O to a double bond (including electrophilic addition), and reverse reactions such as E1 (see Chapter 15);*
3) *nucleophilic addition to a carbonyl;*

Nu^- + O → OH, Nu

4) *oxidation/reduction reactions.*

ETHERS

An ether is a compound with two alkyl (or aryl) groups bonded to an oxygen atom. The general formula for an ether is **ROR**. Ethers can be thought of as disubstituted water molecules. The most familiar ether is diethyl ether, once used as a medical anesthetic, and still often used that way in the laboratory.

NOMENCLATURE

Ethers are named according to IUPAC rules as **alkoxyalkanes,** with the smaller chain as the prefix and the larger chain as the suffix. There is a common system of nomenclature in which ethers are named as alkyl alkyl ethers. In this system, methoxyethane would be named ethyl methyl ether. The alkyl substituents are alphabetized.

methoxyethane (ethyl methyl ether)

ethoxybenzene (ethyl phenyl ether)

Exceptions to these rules occur for cyclic ethers, for which many common names also exist.

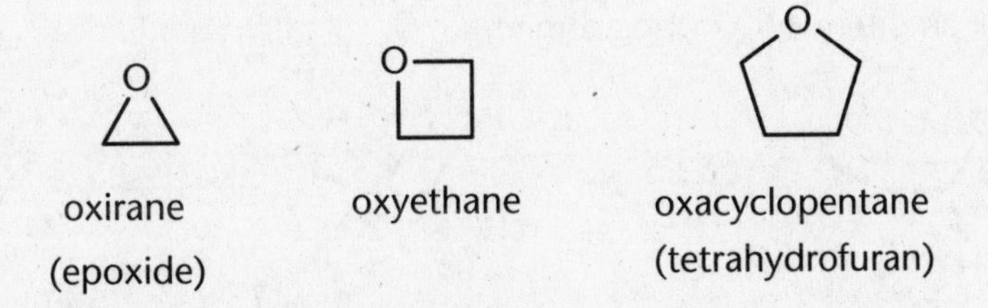

oxirane (epoxide)

oxyethane

oxacyclopentane (tetrahydrofuran)

Note:

Remember that smaller rings have more angle strain, making them less stable.

PHYSICAL PROPERTIES

Ethers do not undergo hydrogen bonding because they have no hydrogen atoms bonded to the oxygen atoms. Ethers therefore boil at relatively low temperatures compared to alcohols; in fact, they boil at approximately the same temperatures as alkanes of comparable molecular weight.

Ethers are only slightly polar and therefore only slightly soluble in water. They are rather inert to most organic reagents and are frequently used as solvents.

Note:

There's no H on the phone "FON"—therefore, no H-bonds and lower boiling points than alcohols.

SYNTHESIS

The Williamson Ether Synthesis produces ethers from the reaction of metal alkoxides with primary alkyl halides or tosylates. The alkoxides behave as nucleophiles, and displace the halide or tosylate via an S_N2 reaction, producing an ether.

Note:

Look! It's S_N2 again! Therefore, you need a strong nucleophile (alkoxides are good) and an unhindered substrate (not bulky or highly substituted).

$$(CH_3)_3C{-}O^-\ Na^+ + CH_3{-}Br \longrightarrow (CH_3)_3C{-}OCH_3 + NaBr$$

It is important to remember that alkoxides will attack only nonhindered halides. Thus, to synthesize a methyl ether, an alkoxide must attack a methyl halide; the reaction cannot be accomplished with methoxide ion attacking a hindered alkyl halide substrate.

The Williamson Ether Synthesis can also be applied to phenols. Relatively mild reaction conditions are sufficient, due to the phenols' acidity.

OH (phenol) + Br (ethyl bromide) —NaOH, H_2O→ O (ethyl phenyl ether)

Note:

More S_N2! This time, however, it's intramolecular—the nucleophile and leaving group are part of the same molecule.

Cyclic ethers are prepared in a number of ways. 107 can be synthesized by means of an internal S_N2 displacement.

HO, H, H, H, H, Br —Base, $-H^+$→ O^-, H, H, H, H, Br → O (oxirane)

Note:

Q: *Why are intramolecular reactions favored?*

A: *Both rate and equilibrium are strongly affected by the reagent concentrations. In intramolecular reactions, the reagents "see" fairly high concentrations of one another—they're basically "tied together."*

Oxidation of an alkene with a **peroxy acid** (general formula RCOOOH) such as MCPBA (*m*-chloroperoxybenzoic acid) will also produce an oxirane.

cyclohexene —MCPBA→ O (cyclohexene oxide)

REACTIONS

A. PEROXIDE FORMATION

Ethers react with the oxygen in air to form highly explosive compounds called **peroxides** (general formula ROOR).

B. CLEAVAGE

Cleavage of straight-chain ethers will take place only under vigorous conditions: usually at high temperatures in the presence of HBr or HI. Cleavage is initiated by protonation of the ether oxygen. The reaction then proceeds by an S_N1 or S_N2 mechanism, depending on the conditions and the structure of the ether. Although not shown below, the alcohol products usually react with a second molecule of hydrogen halide to produce an alkyl halide.

$$CH_3CH_2-\ddot{O}-CH_2CH_3 \underset{}{\overset{H^+}{\rightleftharpoons}} CH_3CH_2-\overset{H}{\overset{|}{O^+}}-CH_2CH_3 \xrightarrow{Br^-} CH_3CH_2Br + CH_3CH_2\ddot{O}H \quad S_N2$$

$$(CH_3)_3C-\ddot{O}-CH_3 \overset{H^+}{\rightleftharpoons} (CH_3)_3C-\overset{H}{\overset{|}{O^+}}-CH_3 \rightleftharpoons (CH_3)_3C^+ + CH_3\ddot{O}H \quad S_N1$$

$$(CH_3)_3C^+ \xrightarrow{Br^-} (CH_3)_3C-Br$$

Note:

Remember, strong bases are poor leaving groups. Without protonation, the leaving group would be an alkoxide (strongly basic), and the reaction would be unlikely to proceed.

In a Nutshell:

Cleavage of straight chain ethers is acid-catalyzed. Cleavage of cyclic ethers can be acid- or base-catalyzed.

Since epoxides are highly-strained cyclic ethers, they are susceptible to S_N2 reactions. Unlike straight-chain ethers, these reactions can be catalyzed by acid or base. In symmetrical epoxides, either carbon can be nucleophilically attacked; but in asymmetrical epoxides, the most substituted carbon is nucleophilically attacked in the presence of acid, and the least substituted carbon is attacked in the presence of base:

$H_2C—CHCH_3$ (epoxide, O)

HBr → $HO—CH_2—CH(CH_3)—Br$

Acid-catalyzed ring opening

Br^- → $Br—CH_2—CH(CH_3)—O^-$ $\xrightarrow{HBr}$ $Br—CH_2—CH(CH_3)—OH$ + Br^-

Base-catalyzed ring opening

Note:

Base-catalyzed cleavage has the most S_N2 character, while acid-catalyzed cleavage seems to have some S_N1 character.

Base-catalyzed cleavage has the most S_N2 character, so it occurs at the least hindered (least substituted) carbon. The basic environment provides the best nucleophile.

In contrast, acid-catalyzed cleavage is thought to have some S_N1 character as well as some S_N2 character. The epoxide O can be protonated, making it a better leaving group. This gives the carbons a bit of positive charge. Since substitution stabilizes this charge (remember, 3° carbons make the best carbocations), the more substituted C becomes a good target for nucleophilic attack.

Don't let epoxides intimidate you; the same basic principles and reaction mechanisms apply, just as we've seen with more simple compounds.

In a Nutshell:

Because they have an –OR instead of an –OH, ethers do not hydrogen bond, and so their physical properties are different from alcohols.

Key reactions include:
1) *epoxide formation;*
2) *S_N1 and S_N2 reactions (i.e., Williamson Ether Synthesis).*

REVIEW PROBLEMS

1. Provide IUPAC names for the following alcohols and classify them as primary, secondary, or tertiary.

a. 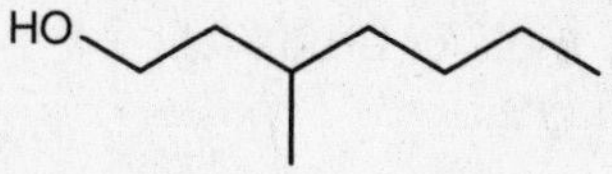

b. 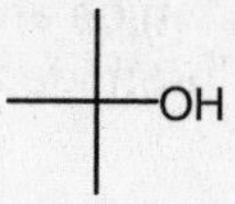

c. 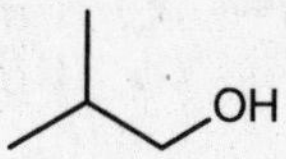

d. 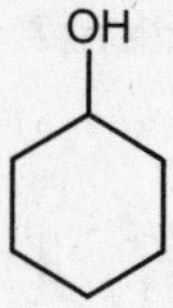

e. 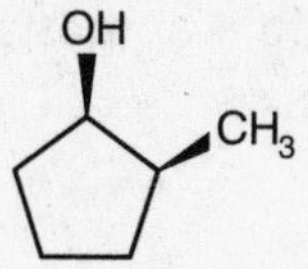

f. 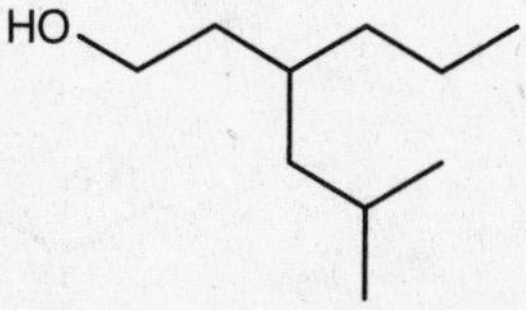

2. Alcohols have higher boiling points than the analogous ethers and hydrocarbons because

A. the oxygen atoms in alcohols have shorter bond lengths.
B. hydrogen bonding is present in alcohols.
C. alcohols are more acidic than the analogous ethers or hydrocarbons.
D. All of the above.

3. Why are alcohols of lower molecular weight more soluble in water than those of higher molecular weight?

4. Tertiary alcohols are oxidized with difficulty because:

A. there is no hydrogen attached to the carbon with the hydroxyl group.
B. there is no hydrogen attached to the α-carbon.
C. tertiary alcohols contain hydroxyl groups with no polarization.
D. they are relatively inert.

5. Which of the following reagents should be used to convert $CH_3(CH_2)_3CH_2OH$ into $CH_3(CH_2)_3CHO$?

A. $KMnO_4$
B. Jones' reagent
C. PCC ($C_5H_6NCrO_3Cl$)
D. $LiAlH_4$

6. The reaction of 1 mole of diethyl ether with hydrobromic acid results in the production of:

   A. 2 moles of ethyl bromide.
   B. 2 moles of ethanol.
   C. 1 mole of ethylbromide and 1 mole of ethanol.
   D. 1 mole of methylbromide and 1 mole of propanol.

7. Which of the following reagents should be used to oxidize the steroid hormone testosterone to 4-androsterone-3,17-dione?

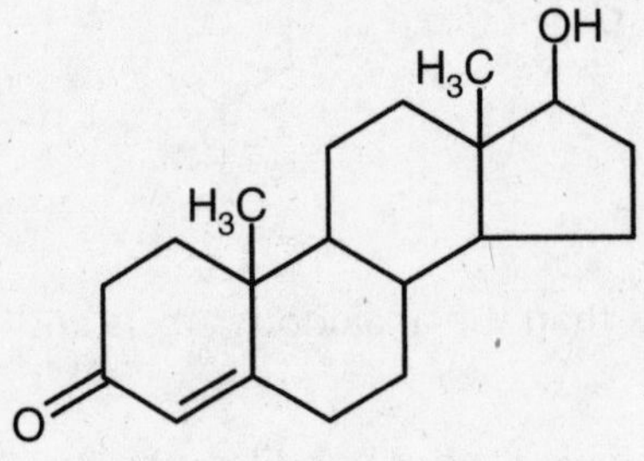

testosterone

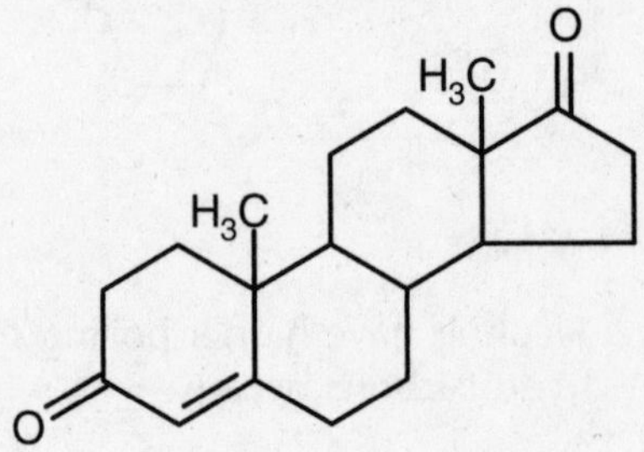

4-androsterone-3, 17-dione

   A. Dilute $KMnO_4$
   B. O_3/CH_2Cl_2; Zn/H_2O
   C. PCC
   D. $LiAlH_4$

SOLUTIONS TO REVIEW PROBLEMS BEGIN ON THE FOLLOWING PAGE.

SOLUTIONS TO REVIEW PROBLEMS

1. **a** **3-methyl-1-heptanol**
a primary alcohol

b ***t*-butyl alcohol or *t*-butanol**
a tertiary alcohol

c **isobutyl alcohol**
a primary alcohol

d **cyclohexanol**
a secondary alcohol

e ***cis*-2-methylcyclopentanol**
a secondary alcohol

f **5-methyl-3-propyl-1-hexanol**
a primary alcohol

2. **B** Alcohols have higher boiling points than their analogous ethers and hydrocarbons because alcohols have a polarized O–H bond where the oxygen is partially negative and the hydrogen is partially positive. This enables the oxygen atoms of other alcohol molecules to be attracted to the hydrogen to form a weak hydrogen bond. These hydrogen bonds make it difficult for the alcohol to vaporize, thereby increasing the boiling point. The analogous hydrocarbons and ethers do not form hydrogen bonds and therefore vaporize at lower temperatures. Choice A is a nonsense choice. The bond length of the oxygen is not a factor in determining the boiling point of a substance. Choice C is incorrect because although alcohols are more acidic than their analogous ethers or hydrocarbons, this property does not affect the boiling point of a substance.

3. Alcohols of lower molecular weight are more soluble in water than larger alcohols because the hydrophilic hydroxyl group can form hydrogen bonds with water. The nonpolar hydrocarbon chain is not solvated by the water because the alkyl group is hydrophobic. As the molecular weight of an alcohol increases, so does the length of the hydrocarbon chain, making the alcohol more hydrophobic. As the molecule becomes more hydrophobic, the alcohol becomes less soluble in water.

4. **A** Tertiary alcohols can be oxidized but only under extreme conditions, since they do not have a hydrogen attached to the carbon with the hydroxyl group. Alcohol oxidation involves the removal of such a hydrogen; if none is present, a carbon-carbon bond must be cleaved instead. This requires a great deal of energy, and will therefore occur only under extreme conditions. Choice B is incorrect because the number of hydrogens attached to the α-carbon is irrelevant to the mechanism of alcohol oxidation. Choice C is incorrect because the hydroxyl group of a tertiary carbon **is** polarized.

5. **C** The best way to prepare aldehydes from primary alcohols is to use PCC (pyridinium chlorochromate, $C_5H_6NCrO_3Cl$), which is choice C. $KMnO_4$, choice A, is a strong oxidizing agent and converts a primary alcohol to a carboxylic acid. Jones' reagent, choice B, also converts a primary alcohol into a carboxylic acid. $LiAlH_4$, choice D, is a reducing agent; it cannot reduce an alcohol further and will certainly not oxidize an alcohol to an
aldehyde.

6. **A** When 1M of an ether reacts with HBr, the initial products are 1M alcohol and 1M alkyl bromide (choice C). However, under these acidic conditions, Br displaces H_2O, resulting ultimately in 1M each of two alkyl bromides. In this case, since the ether is symmetric, the product is 2M ethyl bromide, choice A. Choice B is incorrect because under these conditions, the alcohol is protonated, and H_2O (a good leaving group) is replaced by Br to form ethyl bromide. Choice D is wrong because the ether molecule is split at the oxygen atom; it does not rearrange, as would be required to produce a 3-carbon and a 1-carbon fragment.

7. **C** The best way to oxidize this 2° alcohol to a ketone is with PCC, choice C. $KMnO_4$ (choice A) would oxidize the double bond to a diol, while $LiAlH_4$ (choice D) is a reducing agent, not an oxidizing agent. Choice B is wrong because ozone would cleave the double bond in testosterone.

Aldehydes and Ketones

CHAPTER EIGHTEEN

Aldehydes and **ketones** are compounds that contain the **carbonyl group, C = O,** a double bond between a carbon atom and an oxygen atom. A ketone has two alkyl or aryl groups bonded to the carbonyl, whereas an aldehyde has one alkyl group and one hydrogen (or, in the case of formaldehyde, two hydrogens) bonded to the carbonyl. The carbonyl group is one of the most important functional groups in organic chemistry. In addition to aldehydes and ketones, it is also found in carboxylic acids, esters, amides, and more complicated compounds.

In a Nutshell:

The carbonyl group (C=O) is one of the most important functional groups. It has a dipole moment, with δ+ on C and δ– on O.

NOMENCLATURE

In the IUPAC system, aldehydes are named with the suffix **-al**. The position of the aldehyde group does not need to be specified: it must occupy the terminal (C–1) position. Common names exist for the first five aldehydes: formaldehyde, acetaldehyde, propionaldehyde, butyraldehyde, and valeraldehyde.

Note:

An aldehyde is a terminal functional group; it defines C-1.

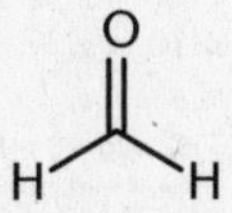

methanal
(formaldehyde)

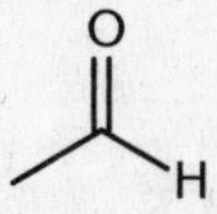

ethanal
(acetaldehyde)

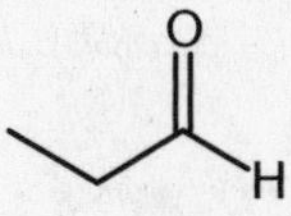

propanal
(propionaldehyde)

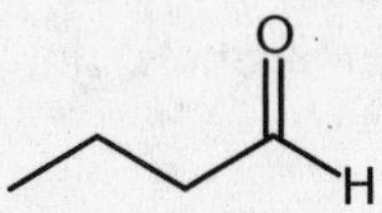

butanal
(butyraldehyde)

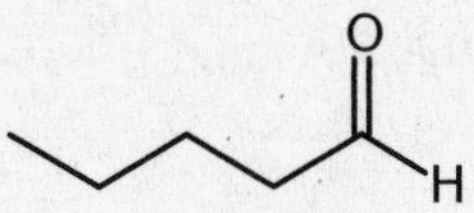

pentanal
(valeraldehyde)

In more complicated molecules, the suffix **-carbaldehyde** can be used. In addition, the aldehyde can be named as a functional group with the prefix **formyl-**.

cyclopentanecarbaldehyde

m-formylbenzoic acid

Ketones are named with the suffix **-one**. The location of the carbonyl group must be specified with a number, except in cyclic ketones, where it is assumed to occupy the number 1 position. The common system of naming **ketones** lists the two alkyl groups followed by the word *ketone.* When it is necessary to name the carbonyl as a substituent, the prefix **oxo-** is used.

2-propanone
(dimethyl ketone)
(acetone)

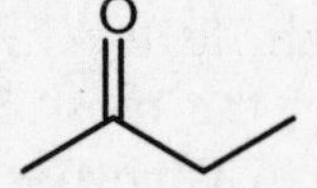

2-butanone
(ethyl methyl ketone)

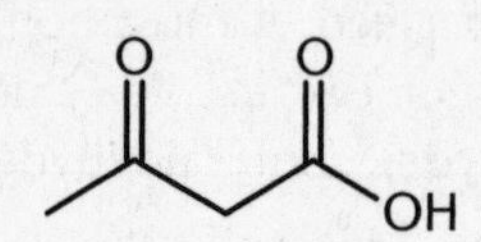

3-oxobutanoic acid

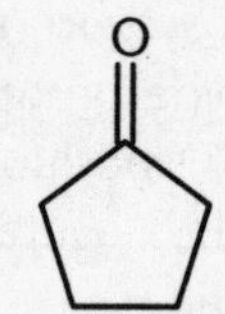

cyclopentanone

PHYSICAL PROPERTIES

The physical properties of aldehydes and ketones are governed by the presence of the carbonyl group. The dipole moments associated with the polar carbonyl groups align, causing an elevation in boiling point relative to the alkanes. This elevation is less than that in alcohols, since no hydrogen bonding is involved.

Note:

The carbonyl group has a dipole moment. Oxygen is more electronegative—it is an "electron hog," pulling the electrons away from the carbon.

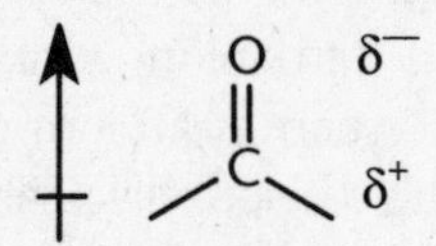

Note:

Dipole–dipole interactions elevate the boiling point, but not as strongly as do hydrogen bonds.

SYNTHESIS

There are numerous methods of preparing aldehydes and ketones; four of the most common are described below.

A. OXIDATION OF ALCOHOLS

An aldehyde can be obtained from the oxidation of a primary alcohol; a ketone can be obtained from a secondary alcohol. As mentioned in Chapter 17, these reactions are usually performed with PCC, sodium or potassium dichromate, or chromium trioxide (Jones' reagent).

Note:

Primary alcohols can be oxidized to aldehydes. Secondary alcohols can be oxidized to ketones.

B. OZONOLYSIS OF ALKENES

Double bonds can be oxidatively cleaved to yield aldehydes and/or ketones, typically with ozone. See Chapter 15 for more details.

C. FRIEDEL-CRAFTS ACYLATION

This reaction, discussed in Chapter 16, produces ketones of the form R–CO–Ar.

REACTIONS

A. ENOLIZATION AND REACTIONS OF ENOLS

Protons alpha to carbonyl groups are relatively acidic ($pK_a \approx 20$), due to resonance stabilization of the conjugate base. A hydrogen atom that detaches itself from the alpha carbon has a finite probability of reattaching itself to the oxygen instead of the carbon. Therefore, aldehydes and ketones exist in solution as a mixture of two isomers, the familiar **keto** form, and the **enol** form, representing the unsaturated alcohol. (**ene** = the double bond, **ol** = the alcohol, so **ene** + **ol** = **enol**). The two isomers, which differ only in the placement of a proton, are called **tautomers**. The equilibrium between the tautomers lies far to the keto side. The process of interconverting from the keto to the enol tautomer is called **enolization**.

In a Nutshell:

Aldehydes and ketones have the traditional keto form (C = O), as well as the less common enol tautomer (enol = ene + ol). The enol form can act as a nucleophile.

O ⇌ OH

Enols are the necessary intermediates in many reactions of aldehydes and ketones. The enolate carbanion, which is nucleophilic, can be created with a strong base such as lithium diisopropyl amide (LDA) or potassium hydride, KH. This nucleophilic carbanion reacts via S_N2 with α,β-unsaturated carbonyl compounds in reactions called **Michael additions.**

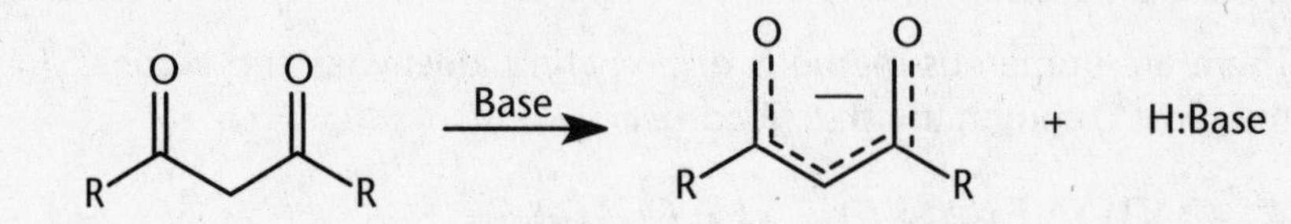

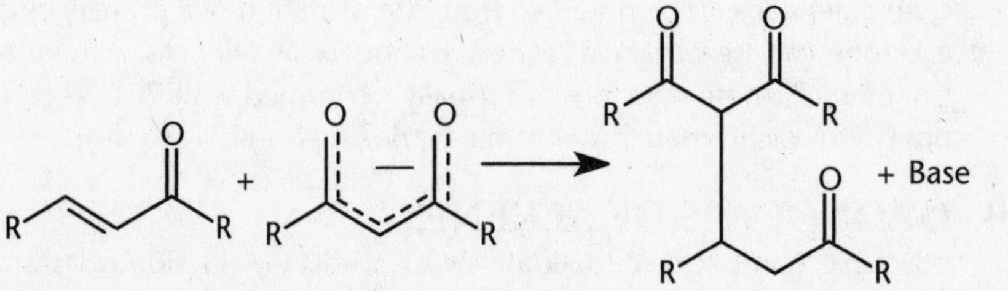

B. ADDITION REACTIONS

General Reaction Mechanism: Nucleophilic Addition To A Carbonyl

Many of the reactions of aldehydes and ketones share this general reaction mechanism. Rather than memorizing them all individually, focus on understanding the basic pattern. Then, you can learn how each reaction exemplifies it.

The C=O bond is polarized, with a partial positive charge on C and a partial negative charge on O. This makes the carbon ripe for nucleophilic attack.

Note:

The carbonyl carbon is a good target for nucleophilic attack.

The nucleophile attacks, forming a bond to the C, which causes the π bond in the C=O to break. This generates a tetrahedral intermediate. If no good leaving group is present, the double bond cannot reform, and so the final product is nearly identical to the intermediate, except that usually the —O^- will accept a proton to become a hydroxyl (–OH).

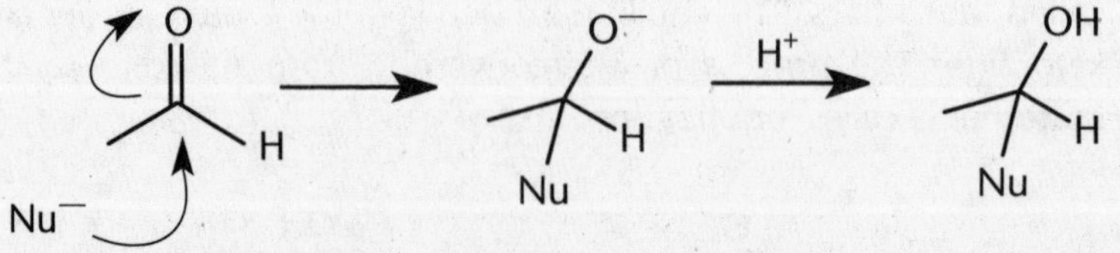

Although the figure only shows nucleophilic addition to an aldehyde, this mechanism applies to ketones as well.

1. Hydration

In the presence of water, aldehydes and ketones react to form *gem*-diols (1,1-diols). In this case, water acts as the nucleophile attacking at the carbonyl carbon. This hydration reaction proceeds slowly; the rate may be increased by the addition of a small amount of acid or base.

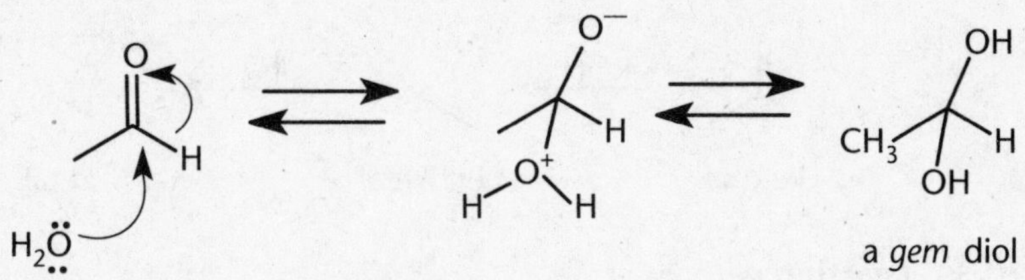

a *gem* diol

Note:

When H_2O is the nucleophile, hydration occurs.

2. Acetal and Ketal Formation

A reaction similar to hydration occurs when aldehydes and ketones are treated with alcohols. When one equivalent of alcohol (the nucleophile in this reaction) is added to an aldehyde or ketone, the product is a **hemiacetal** or a **hemiketal**, respectively. When two equivalents of alcohol are added, the product is an **acetal** or a **ketal,** respectively. The reaction mechanism is the same as for hydration and is catalyzed by anhydrous acid. Acetals and ketals, which are comparatively inert, are frequently used as protecting groups for carbonyl functionalities. They can easily be converted back to the carbonyl with aqueous acid.

aldehyde

hemiacetal

Note:

When ROH is the nucleophile, hemiacetals (or hemiketals) and acetals (or ketals) are formed.

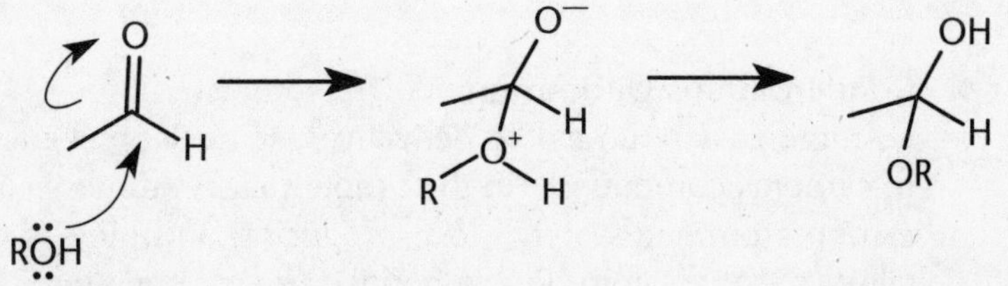

O ROH HO OR ROH RO OR
H H H
aldehyde hemiacetal acetal

O ROH HO OR ROH RO OR
ketone hemiketal ketal

Note:

When CN^- is the nucleophile, cyanohydrins are formed.

3. **Reaction with HCN**

Aldehydes and ketones react with HCN (hydrogen cyanide) to produce stable compounds called **cyanohydrins**. HCN dissociates and the nucleophilic cyanide anion attacks the carbonyl carbon atom. Protonation of the oxygen produces the cyanohydrin. The cyanohydrin gains its stability from the newly formed C–C bond (in contrast, when a carbonyl reacts with HCl, a weak C–Cl bond is formed, and the resulting chlorohydrin is unstable).

$+ CN^-$ O → $H^+ +$ O^- CN → OH CN

4. **Condensations with Ammonia Derivatives**

Ammonia and some of its derivatives are nucleophiles and can add to carbonyl compounds. In the simplest case, ammonia adds to the carbon atom and water is lost, producing an **imine,** a compound with a nitrogen atom double-bonded to a carbon atom. (A reaction in which water is lost between two molecules is called a **condensation reaction.**)

In this case, the first part of the reaction follows the mechanism of nucleophilic addition described above. However, after formation of a tetrahedral intermediate, this reaction proceeds further: the C=O double bond reforms and a leaving group is kicked off. This mechanism is called nucleophilic substitution on a carbonyl and will be described in greater detail in Chapter 19.

Some common ammonia derivatives that react with aldehydes and ketones are hydroxylamine (H_2NOH), hydrazine (H_2NNH_2), and semicarbazide ($H_2NNHCONH_2$); these form oximes, hydrazones, and semicarbazones, respectively.

$$R{-}C({=}O){-}R + \ddot{N}H_3 \longrightarrow R{-}C(O^-)(^+NH_3){-}R \xrightarrow{+H^+} R{-}C(^+OH_2)(:NH_2){-}R \longrightarrow R{-}C({=}^+NH_2){-}R \xrightarrow{-H^+} R_2C{=}NH$$

Don't worry too much about protons coming and going; there should be plenty in the solution, so you can transiently put them where needed to facilitate this reaction.

Examples of other potential nucleophiles and their respective products are shown below.

$$CH_3C(O)CH_3 + NH_3 \longrightarrow CH_3C(NH)CH_3 + H_2O$$

$$CH_3C(O)CH_3 + H_2NOH \longrightarrow CH_3C(NOH)CH_3 + H_2O$$

$$CH_3C(O)CH_3 + H_2NNH_2 \longrightarrow CH_3C(NNH_2)CH_3 + H_2O$$

$$CH_3C(O)CH_3 + H_2NNHCONH_2 \longrightarrow CH_3C(NNHCONH_2)CH_3 + H_2O$$

Note:

Nitrogen-containing compounds can be nucleophiles too.

In a Nutshell:

In the aldol condensation, aldehydes act as nucleophile and target (substrate).

C. THE ALDOL CONDENSATION

The aldol condensation is an important reaction that basically follows the mechanism of nucleophilic addition to a carbonyl that was described above. In this case, an aldehyde acts both as nucleophile (enol form) and target (keto form.) When acetaldehyde (ethanal) is treated with base, an enolate ion is produced. This enolate ion, being nucleophilic, can react with the carbonyl group of another acetaldehyde molecule. The product is 3-hydroxybutanal, which contains both an alcohol and an aldehyde functionality. This type of compound is called an **aldol,** from **alde**hyde and alcoh**ol**. With stronger base and higher temperatures, condensation occurs, producing an α,β-unsaturated aldehyde. This type of condensation reaction has become known as the **aldol condensation.**

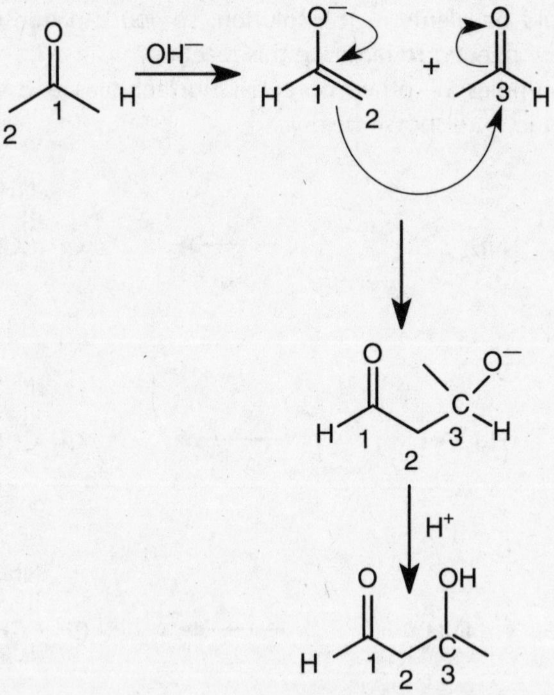

3-hydroxybutanal
(an aldol)

Note:

An aldol is a molecule that is both an aldehyde and an alcohol.

When heated, this molecule can undergo elimination and lose H_2O to form a double bond:

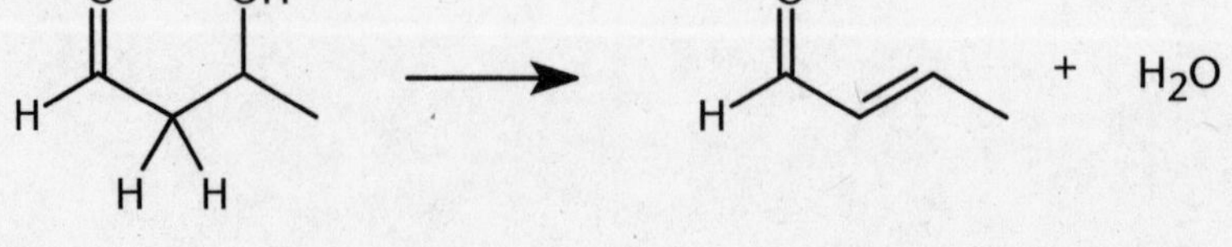

Note:

An aldol can lose H_2O to form a carbon–carbon double bond, in an E1 reaction.

The aldol condensation is most useful when only one type of aldehyde or ketone is present, since mixed condensations usually result in a mixture of products.

D. THE WITTIG REACTION

The **Wittig Reaction** is a method of forming carbon-carbon double bonds by converting aldehydes and ketones into alkenes. The first step involves the formation of a phosphonium salt from the S_N2 reaction of an alkyl halide with the nucleophile triphenylphosphine, $(C_6H_5)_3P$. The phosphonium salt is then deprotonated (losing the proton α to the phosphorus) with a strong base, yielding a neutral compound called an **ylide** (pronounced "ill-id") or **phosphorane.** (The phosphorus atom may be drawn as pentavalent, utilizing the low-lying 3d atomic orbitals.)

In a Nutshell:

The Wittig reaction ultimately converts C = O to C = C.

$$(C_6H_5)_3P + CH_3Br \longrightarrow (C_6H_5)_3PCH_3 + Br^-$$

$$(C_6H_5)_3\overset{+}{P}\text{—}CH_3\ Br^- \xrightarrow{\text{Base}} (C_6H_5)_3P{=}CH_2 \longleftrightarrow (C_6H_5)_3\overset{+}{P}\text{—}\overset{-}{C}H_2$$

phosphonium salt

ylide

Notice that an ylide is a type of carbanion and has nucleophilic properties. When combined with an aldehyde or ketone, an ylide attacks the carbonyl carbon, giving an intermediate called a *betaine*, which forms a four-membered ring intermediate called an oxaphosphetane. This decomposes to yield an alkene and triphenylphosphine oxide.

Note:

The ylide can act as a nucleophile and attack the carbonyl.

$(C_6H_5)_3\overset{+}{P}\text{—}\overset{-}{C}H_2$ + cyclohexanone → $(C_6H_5)_3\overset{+}{P}$—H_2C—(cyclohexyl)—O^- → $(C_6H_5)_3P$—H_2C—(cyclohexyl)—O^- → $O{=}P(C_6H_5)_3$ + methylenecyclohexane (CH_2)

ylide

The decomposition reaction is driven by the strength of the phosphorus-oxygen bond that is formed.

E. OXIDATION AND REDUCTION

Aldehydes and ketones occupy the middle of the oxidation-reduction continuum. They are more oxidized than alcohols but less oxidized than carboxylic acids.

Aldehydes can be oxidized with a number of different reagents, such as $KMnO_4$, CrO_3, Ag_2O, or H_2O_2. The product of oxidation is a carboxylic acid.

O
LAH
or
$NaBH_4$
OH

Note:

Aldehydes can be oxidized to carboxylic acids or reduced to alcohols.

Ketones can be reduced to alcohols.

A number of different reagents will reduce aldehydes and ketones to alcohols. The most common is lithium aluminum hydride (LAH); sodium borohydride ($NaBH_4$) is often used when milder conditions are needed.

O
LAH
or
$NaBH_4$
OH

Aldehydes and ketones can be completely reduced to alkanes by two common methods. In the **Wolff-Kishner** Reduction, the carbonyl is first converted to a hydrazone, which releases molecular nitrogen (N_2) when heated and forms an alkane (the protons being abstracted from the solvent). The Wolff-Kishner reaction is performed in basic solution and therefore is only useful when the product is stable under basic conditions.

Note:

Both aldehydes and ketones can be fully reduced to alkanes:
– Wolff-Kishner (H_2NNH_2)
– Clemmensen (Hg (Zn), HCl)

O
H_2NNH_2
NNH_2
Base
Δ

An alternative reduction not subject to this restriction is the **Clemmensen Reduction**, where an aldehyde or ketone is heated with amalgamated zinc in hydrochloric acid.

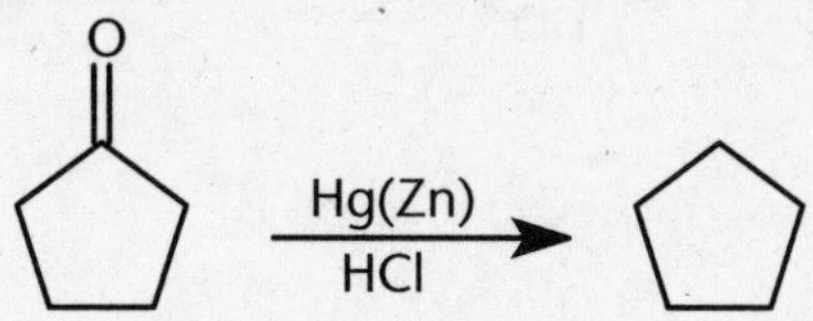

Hg(Zn)
HCl

REVIEW PROBLEMS

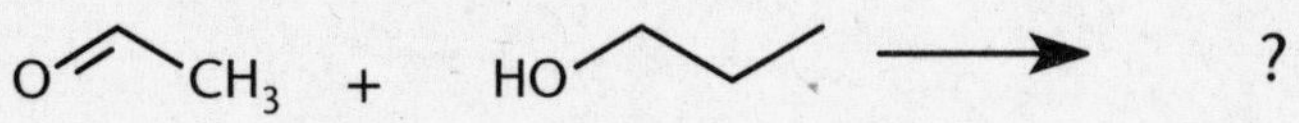

O=⁀CH_3 + HO⁀⁀ ⟶ ?

1. The product of the above reaction is

A.

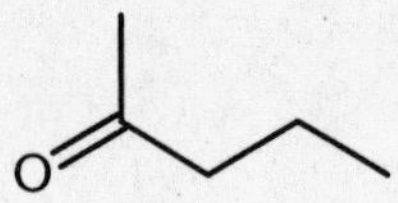

B. 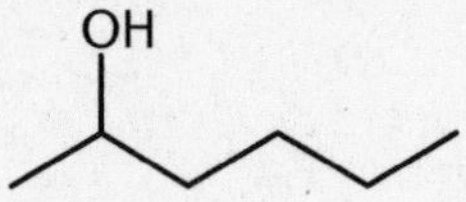 OH

C. 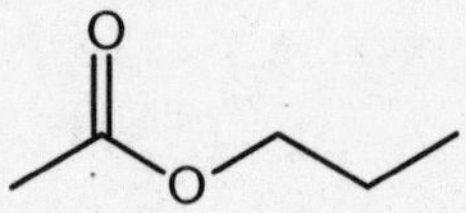 O, O

D. 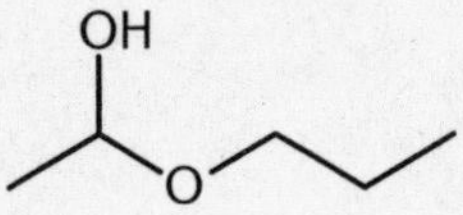 OH, O

O, H_3C, C_2H_5 + ⁀OH ⟶ ?

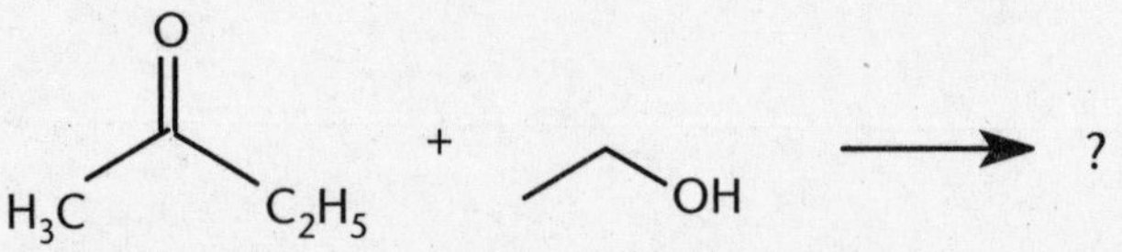

2. The major product of the above reaction is

A. 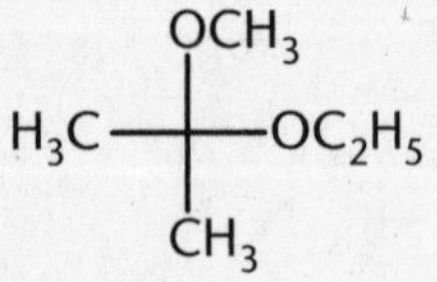 OCH_3, H_3C—|—OC_2H_5, CH_3

B. 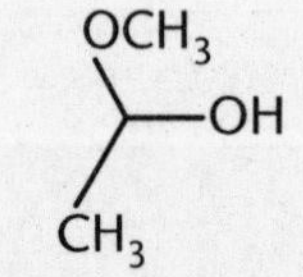 OCH_3, —OH, CH_3

C. 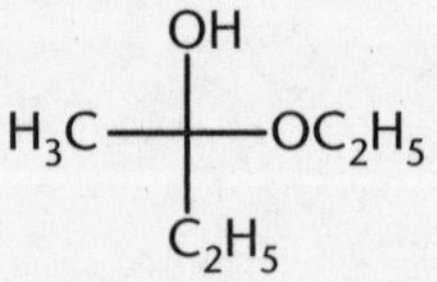 OH, H_3C—|—OC_2H_5, C_2H_5

D. 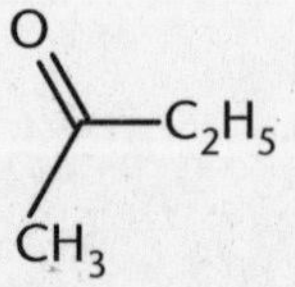 O, —C_2H_5, CH_3

O + NH_2 —ether→ ?

3. The product of the above reaction is

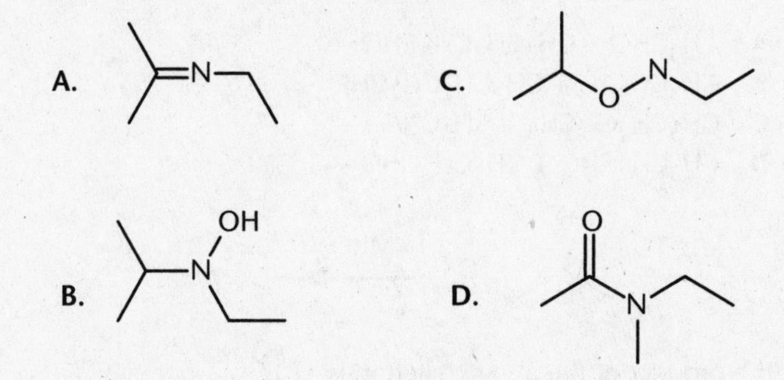

4. All of the following properties are responsible for the reactivity of the carbonyl bond in propanone EXCEPT the fact that

   A. the carbonyl carbon is electrophilic.
   B. the carbonyl oxygen is electron-withdrawing.
   C. a resonance structure of the compound places a positive charge on the carbonyl carbon.
   D. the π electrons are mobile and are pulled towards the carbonyl carbon.

5. The above reaction is an example of

O ⇌ HO

   A. esterification.
   B. tautomerization.
   C. elimination.
   D. dehydration.

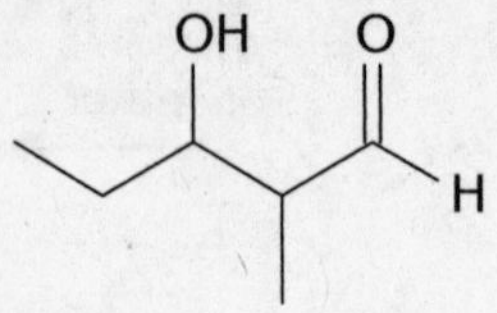

6. Which of the following reactions produces the above compound?

A. $CH_3CHO + CH_3CH_2CH_2CHO \rightarrow$
B. $CH_3COCH_3 + CH_3CH_2CH_2CHO \rightarrow$
C. $CH_3CH_2COCH_3 + CH_3CHO \rightarrow$
D. $CH_3CH_2CHO + CH_3CH_2CHO \rightarrow$

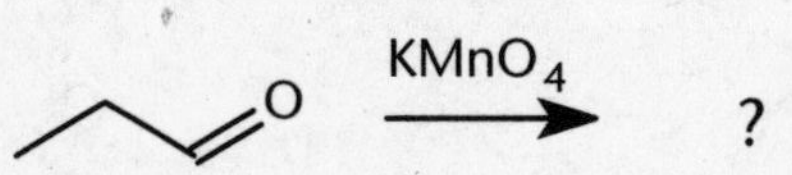

KMnO$_4$?

7. The product of the above reaction is

A. C_3H_7OH
B. C_2H_5COOH
C. C_3H_7CHO
D. CH_3COOH

8. Heating an aldehyde with Zn in HCl produces

A. a ketone.
B. an alkane.
C. an alcohol.
D. a carboxylic acid.

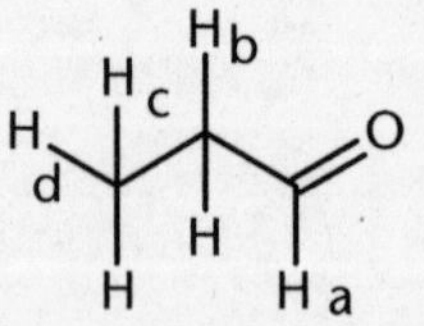

9. Which hydrogen atom in the above compound is the most acidic?

A. a
B. b
C. c
D. d

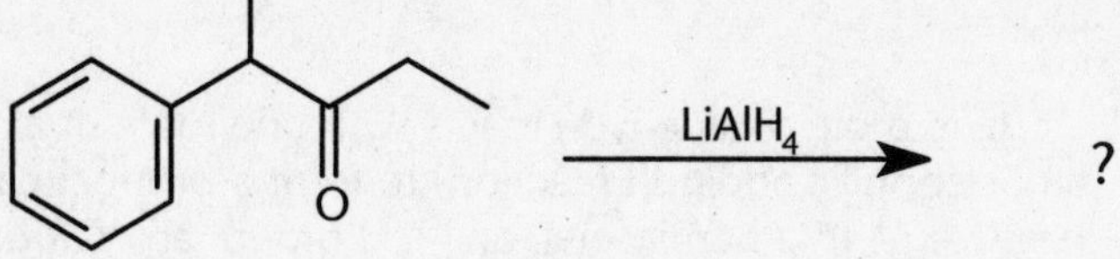

$\xrightarrow{LiAlH_4}$?

10. The product obtained in the above reaction is

A. 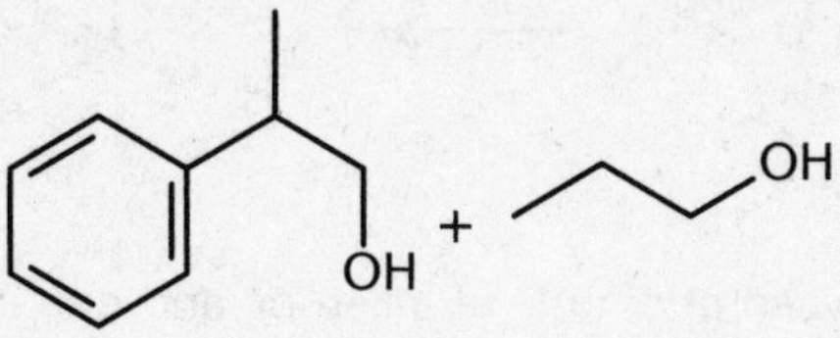 OH + OH C. 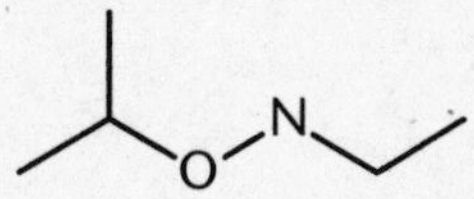 O N

B. 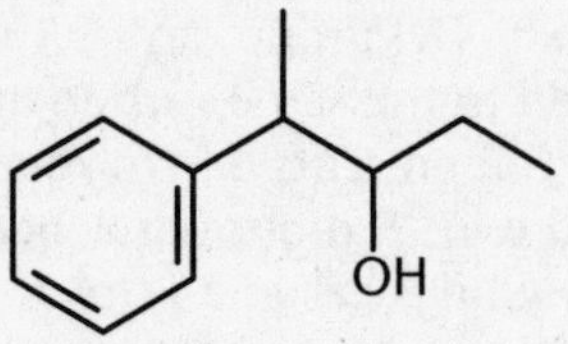 OH D. 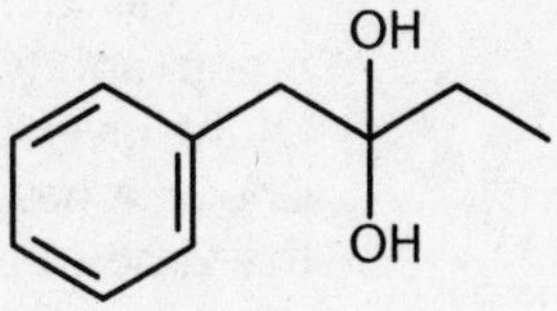 OH OH

11. Draw the following compounds

A. 3-Methyl-2-pentanone
B. 3-Hydroxypentanal
C. Benzyl phenyl ketone
D. Cyclohexane carboaldehyde

12. The product of the reaction between benzaldehyde and an excess of ethanol (C_2H_5OH) in the presence of anhydrous HCl is:

A. 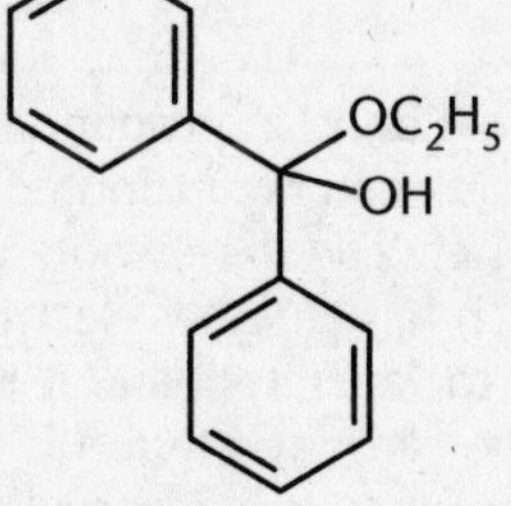 OC_2H_5 OH

B. 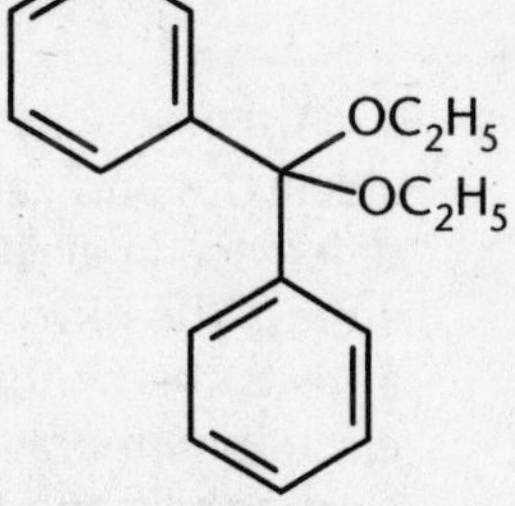 OC_2H_5 OC_2H_5

C. 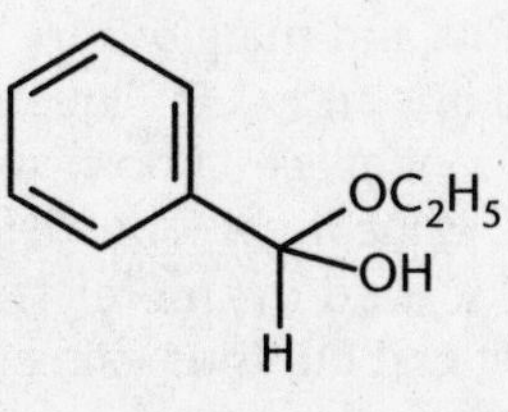 OC_2H_5 OH H

D. 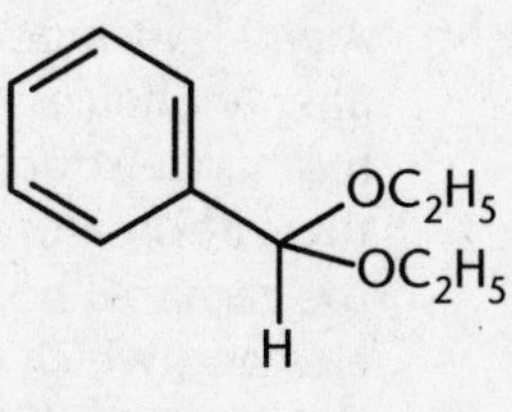 OC_2H_5 OC_2H_5 H

SOLUTIONS TO REVIEW PROBLEMS

1. **D** One mole of aldehyde reacts with one mole of alcohol via a nucleophilic addition reaction, to form a product called a *hemiacetal.* In a hemiacetal, an –OH group, an –OR group, a H atom, and a –R group are attached to the same carbon atom.

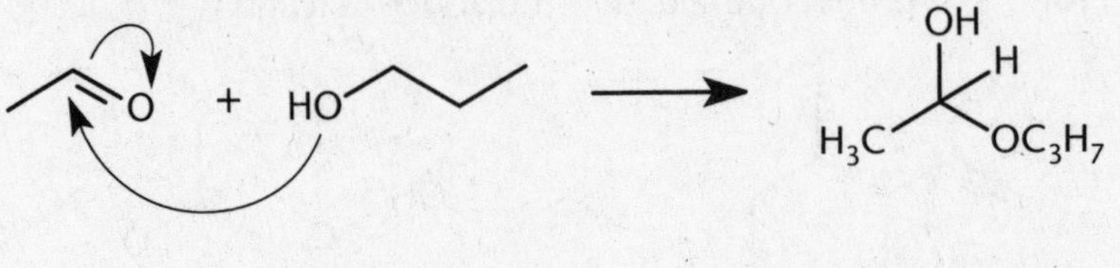

2. **C** The reaction between one mole of a ketone and one mole of an alcohol produces a compound analogous to a hemiacetal that's called a *hemiketal.* This has an –OH group, an –OR group, and two –R groups attached to the same carbon atom. Of the given choices, only choice C represents a hemiketal. Choice A has two –OR groups and two –R groups attached to the same carbon atom; this compound is called a *ketal.* Choice B is a hemiacetal since it has an –OH group, an –OR group, a H atom, and a –R group attached to the same carbon atom. Choice D is a ketone. The correct choice, therefore, is C. Note that a hemiketal is a very unstable compound; it reacts rapidly with a second mole of alcohol to form a ketal.

3. **A** Aldehydes and ketones react with ammonia and primary amines to form imines (also called *Schiff bases*), compounds with a double bond between a carbon atom and a nitrogen atom.

$$(CH_3)_2C = O + H_2N - C_2H_5 \rightarrow (CH_3)_2C = NCH_2CH_3 + H_2O$$

The correct choice is A.

4. **D** The reactivity of the carbonyl bond in propanone, and in aldehydes and ketones in general, can be attributed to the difference in electronegativity between the carbon and oxygen atoms. The oxygen atom, which has a higher electronegativity, attracts the bonding electrons towards itself and is therefore electron-withdrawing. Thus, the carbonyl carbon is electrophilic and the carbonyl oxygen is nucleophilic. Choices A and B include true statements and therefore are incorrect answer choices (remember, this is an EXCEPT question). The resonance structure of propanone places a positive charge on the carbon atom, so choice C is incorrect. The π electrons of the carbonyl bond are pulled toward the more electronegative element, which is oxygen, not carbon; thus, choice D is a false statement and the correct answer choice.

5. **B**

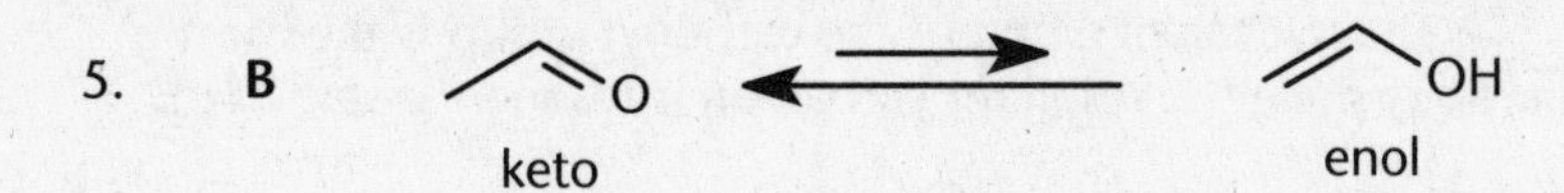

Esterification, choice A, is the formation of esters from carboxylic acids and alcohols. Tautomerization, choice B, is the interconversion of keto and enol forms of a compound. An elimination reaction, choice C, is a reaction in which a part of a reactant is lost and a multiple bond is introduced. A dehydration reaction, choice D, is one in which a molecule of water is eliminated. The above reaction involves an interconversion of keto and enol forms of ethanal. The correct choice is therefore B. Note that equilibrium lies to the left in the above reaction, since the keto form is more stable.

6. **D**

The above reaction is an example of aldol condensation. In the presence of a base, the alpha H is abstracted from an aldehyde, forming an enolate ion, CH_3CH^-CHO. This enolate ion then attacks the carbonyl groups of the other aldehyde molecule, CH_3CH_2CHO, forming the above aldol. The correct choice is D.

7. **B** Aldehydes are easily oxidized to the corresponding carboxylic acids by $KMnO_4$. The –CHO group is converted to –COOH. In this reaction, therefore, C_2H_5CHO is oxidized to C_2H_5COOH, which is choice B. In choice A, the aldehyde has been reduced to an alcohol. In choice C, a $-CH_2$ group has been added. Thus, choices A and C are incorrect. In choice D, the –CHO group has been oxidized to –COOH, but a $-CH_2$ group has been deleted, so choice D is incorrect.

8. **B** Heating an aldehyde or a ketone with amalgamated Zn/HCl converts it to the corresponding alkane; this reaction is called the Clemmensen Reduction. Note that aldehydes and ketones can also be converted to alkanes under basic conditions by reaction with hydrazine (the Wolff-Kishner Reduction).

9. **B** The hydrogen alpha to the carbonyl group is the most acidic, since the resultant carbanion is resonance-stabilized:

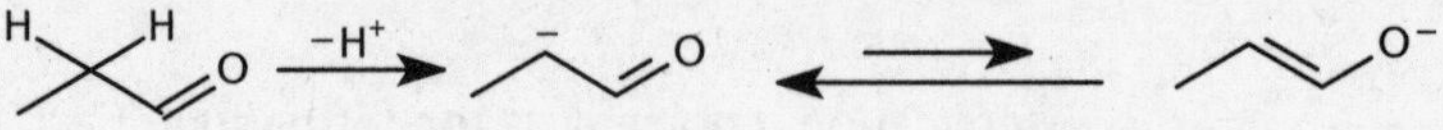

$\xrightarrow{-H^+}$ ⟷ O^-

10. **B** $LiAlH_4$ reduces carboxylic acids, esters, and aldehydes to primary alcohols, and ketones to secondary alcohols. In this reaction, therefore, the ketone is converted to a secondary alcohol. Thus, the correct answer is choice B, $C_6H_5CH(CH_3)CHOHCH_2CH_3$.

11.

A. O

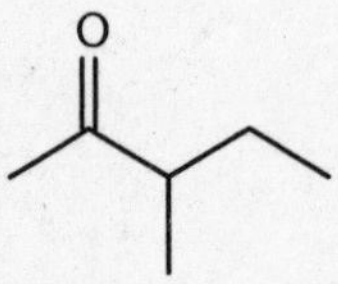

C. O

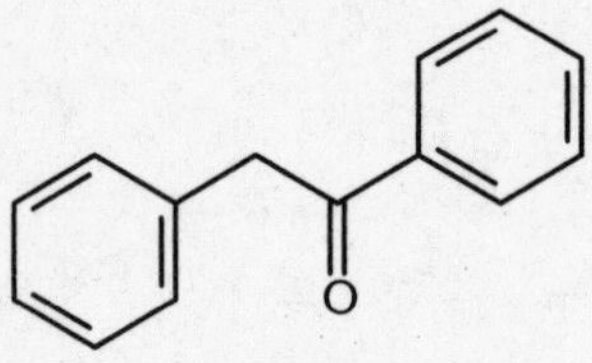

B. OH O

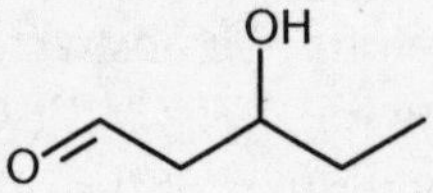

D. O

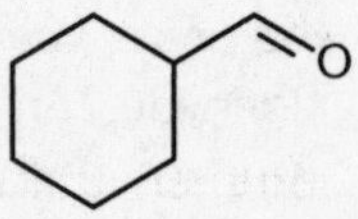

12. **D** This molecule corresponds to an acetal: two alkoxyl functionalities bonded to the same carbon. This question states that an excess of ethanol is present, so benzaldehyde will first be converted to a hemiacetal, having an alkoxyl and a hydroxyl functionality bonded to the same carbon, then an acetal. Choices A and B are wrong because they show the presence of two benzene rings in the final product. Choice C is wrong since this is the hemiacetal that is formed initially, which then goes on to react with excess ethanol to produce the acetal.

Carboxylic Acids

CHAPTER NINETEEN

Carboxylic acids are compounds that contain hydroxyl groups attached to carbonyl groups. This functionality is known as a **carboxyl group**. The hydroxyl hydrogen atoms are acidic, with pK_a values in the general range of 3 to 6. Carboxylic acids occur widely in nature and are synthesized by all living organisms.

Note:

pK_a values generally range from 3 to 6.

NOMENCLATURE

In the IUPAC system of nomenclature, carboxylic acids are named by adding the suffix **-oic acid** to the alkyl root. The chain is numbered so that the carboxyl group receives the lowest possible number. Additional substituents are named in the usual fashion.

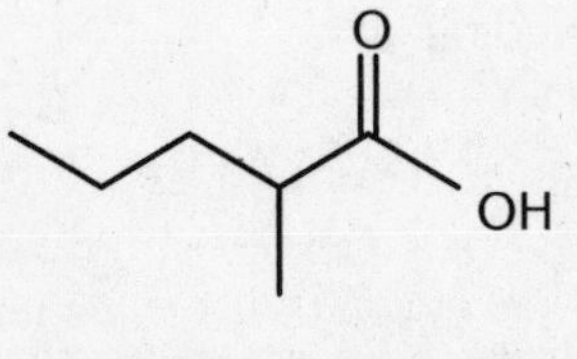

O OH

2-methylpentanoic acid

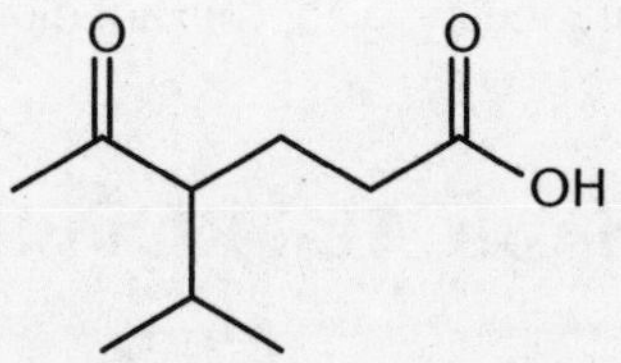

O O OH

4-isopropyl-5-oxohexanoic acid

Note:

This is a very high priority group! It determines C-1 of the carbon backbone, as well as the suffix (-oic acid).

Carboxylic acids were among the first organic compounds discovered. Their original names continue today in the common system of nomenclature. For example, formic acid (from Latin *formica,* meaning ant) was found in ants and butyric acid (from Latin *butyrum,* meaning butter) in rancid butter. The common and IUPAC names of the first three carboxylic acids are listed in the figure below.

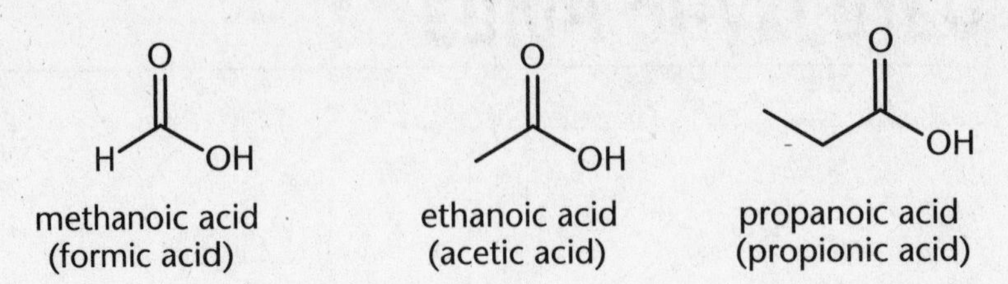

methanoic acid (formic acid) | ethanoic acid (acetic acid) | propanoic acid (propionic acid)

Note:

Salts have the suffix -ate (e.g., sodium hexanoate).

Cyclic carboxylic acids are usually named as cycloalkane carboxylic acids. The carbon atom to which the carboxyl group is attached is numbered 1. Salts of carboxylic acids are named beginning with the cation, followed by the name of the acid with the ending **-ate** replacing **-ic acid**. Typical examples are:

CH_3 Cl COOH

1-chloro-2-methylcyclopentane carboxylic acid

O O^- Na^+

sodium hexanoate

Dicarboxylic acids—compounds with two carboxyl groups—are common in biological systems. The first six straight-chain terminal dicarboxylic acids are oxalic, malonic, succinic, glutaric, adipic, and pimelic acids. Their IUPAC names are ethanedioic acid, propanedioic acid, butanedioic acid, pentanedioic acid, hexanedioic acid, and heptanedioic acid.

PHYSICAL PROPERTIES

A. HYDROGEN BONDING

Carboxylic acids are polar and can form hydrogen bonds. As a result, carboxylic acids can form dimers: pairs of molecules connected by hydrogen bonds. The boiling points of carboxylic acids are therefore even higher than those of the corresponding alcohols. The boiling points follow the usual trend of increasing with molecular weight.

In a Nutshell:

Carboxylic acids are polar and can form hydrogen bonds. Their acidity is due to resonance stabilization and can be enhanced by adding electronegative groups or other potential resonance structures.

B. ACIDITY

The acidity of carboxylic acids is due to the resonance stabilization of the carboxylate anion (the conjugate base). When the hydroxyl proton dissociates from the acid, the negative charge left on the carboxylate group is delocalized between the two oxygen atoms.

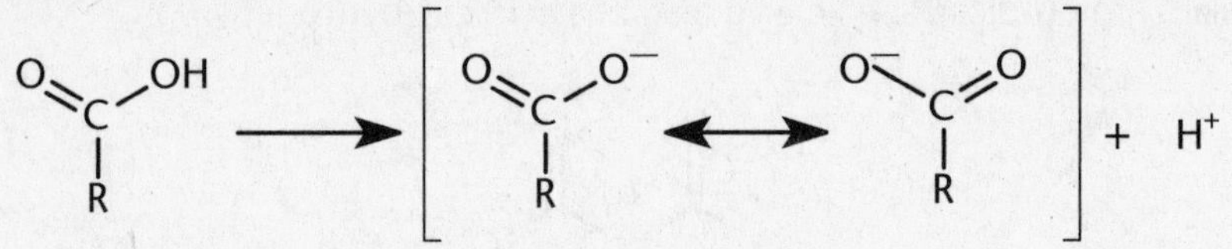

Substituents on carbon atoms adjacent to a carboxyl group can influence acidity. Electron-withdrawing groups such as –Cl or $-NO_2$ further delocalize the negative charge and increase acidity. Electron-donating groups such as $-NH_2$ or $-OCH_3$ destabilize the negative charge, making the compound less acidic.

In dicarboxylic acids, one –COOH group (which is electron-withdrawing) influences the other, making the compound more acidic than the analogous monocarboxylic acid. The second carboxyl group is then influenced by the carboxylate anion. Ionization of the second group will create a doubly charged species, in which the two negative charges repel each other. Since this is unfavorable, the second proton is less acidic than that of a monocarboxylic acid.

β-dicarboxylic acids are notable for the high acidity of the α-hydrogens located between the two carboxyl groups ($pK_a \sim 10$). Loss of this acidic hydrogen atom produces a carbanion that is stabilized by the electron-withdrawing effect of the two carboxyl groups (the same effect seen in β-ketoacids, $RC{=}OCH_2\ COOH$).

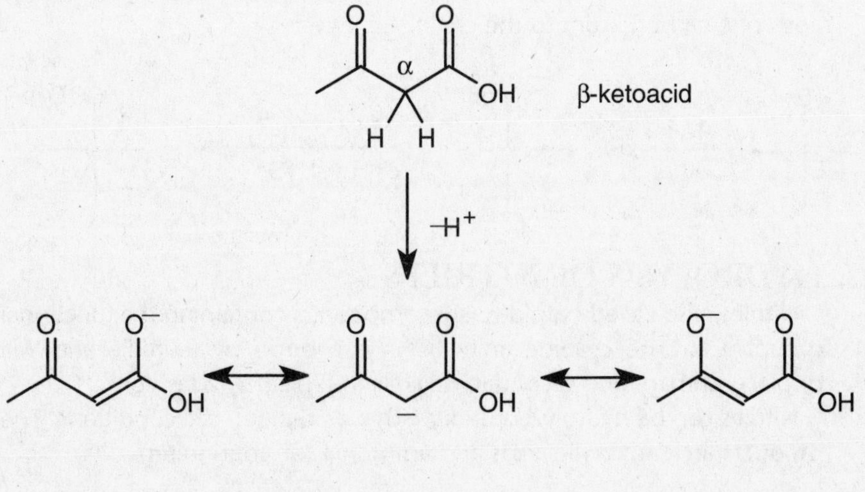

β-ketoacid

$-H^+$

MCAT Favorite:

Other ways to stabilize the negative charge (and thus increase acidity) are:

- *electron-withdrawing groups (e.g., halides);*
- *groups that allow more resonance stabilization (e.g., benzyl or allyl substituents).*

The more of such groups that exist, and the closer to the acid they are, the stronger the acid.

Similarly, the β-dicarboxylic acid also has acidic α hydrogens.

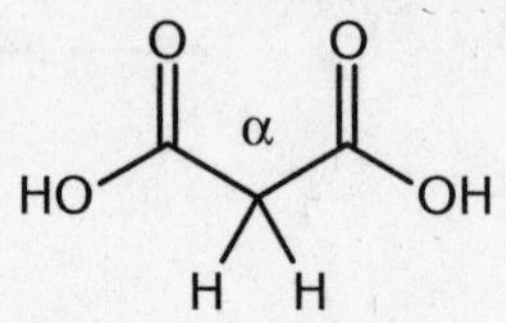

HO–C(=O)–CH₂(α)–C(=O)–OH

Mnemonic:

Carboxylic acids are the most oxidized.

SYNTHESIS

A. OXIDATION REACTIONS

Carboxylic acids can be prepared via oxidation of aldehydes, primary alcohols, and certain alkylbenzenes. The oxidant is usually potassium permanganate, $KMnO_4$. Note that secondary and tertiary alcohols cannot be oxidized to carboxylic acids because of valence limitations.

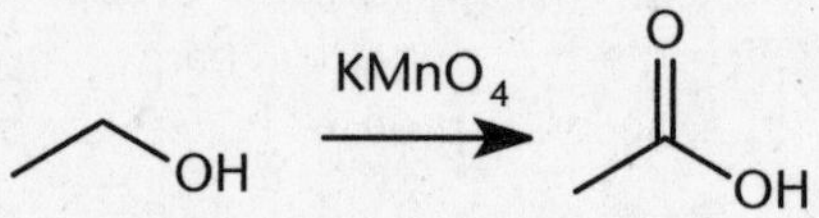

$$\text{CH}_3\text{CH}_2\text{OH} \xrightarrow{KMnO_4} \text{CH}_3\text{COOH}$$

B. CARBONATION OF ORGANOMETALLIC REAGENTS

Organometallic reagents, such as Grignard reagents, react with carbon dioxide (CO_2) to form carboxylic acids. This reaction is useful for the conversion of tertiary alkyl halides into carboxylic acids, which cannot be accomplished through other methods. Note that this reaction adds one carbon atom to the chain.

$$(CH_3)_3C\text{–Br} \xrightarrow[\text{ether}]{\text{Mg}} (CH_3)_3C\text{–MgBr} \xrightarrow[\text{2) } H^+,\ H_2O]{\text{1) } CO_2 \text{ gas}} (CH_3)_3C\text{–COOH}$$

Note:

Note that both carbonation and hydrolysis increase the chain length by one carbon.

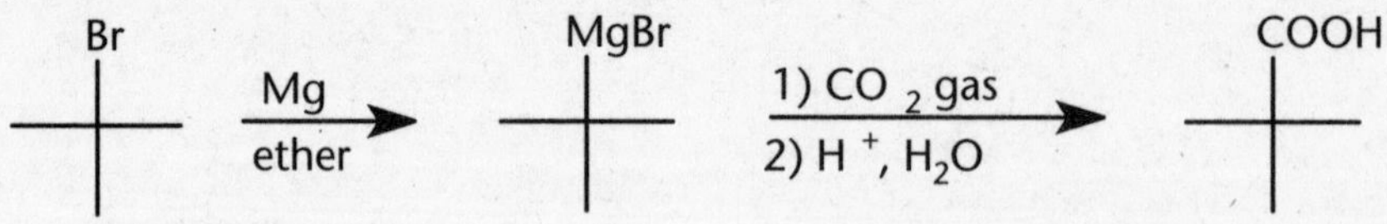

C. HYDROLYSIS OF NITRILES

Nitriles, also called cyanides, are compounds containing the functional group –CN. The cyanide anion CN^- is a good nucleophile and will displace primary and secondary halides in typical S_N2 fashion.

Nitriles can be hydrolyzed under either acidic or basic conditions. The products are carboxylic acids and ammonia (or ammonium salts).

$$CH_3Cl \longrightarrow CH_3CN \longrightarrow CH_3COOH + NH_4^+$$

This allows for the conversion of alkyl halides into carboxylic acids. As in the carbonation reaction, an additional carbon atom is introduced. For instance, if the desired product is acetic acid, a possible starting material would be methyl iodide.

REACTIONS

A. SOAP FORMATION

When long-chain carboxylic acids react with sodium or potassium hydroxide, they form salts. These salts, called soaps, are able to solubilize nonpolar organic compounds in aqueous solutions because they possess both a nonpolar "tail" and a polar carboxylate "head."

O^-Na^+

nonpolar tail polar head

When placed in aqueous solution, soap molecules arrange themselves into spherical structures called **micelles**. The polar heads face outward, where they can be solvated by water molecules, and the nonpolar hydrocarbon chains are inside the sphere, protected from the solvent. Nonpolar molecules such as grease can dissolve in the hydrocarbon interior of the spherical micelle, while the micelle as a whole is soluble in water because of its polar shell.

Note:

RCOOH + NaOH

↓

$RCOO^-Na^+$
(a soap)
+
H_2O

Note:

nonpolar "tail" = hydrophobic.
polar head = hydrophilic.

Real World Analogy:

Notice the similarity between micelle formation and the assembly of the plasma membrane (made of phospholipids).

Clinical Correlate:

In the small intestine, consumed fat is solubilized in micelles, not with detergent but with bile salts, which have a structure similar to soaps—hydrophobic tail and hydrophilic head!

B. NUCLEOPHILIC SUBSTITUTION

Many of the reactions that carboxylic acids (and their derivatives) participate in can be described by a single mechanism: nucleophilic substitution. This mechanism is very similar to nucleophilic addition to a carbonyl, shown in the preceding chapter. The key difference: nucleophilic substitution concludes with re-formation of the C=O double bond and elimination of a leaving group.

$$\text{RC(=O)OH} + \text{Nu}^- \longrightarrow \text{R–C(O}^-\text{)(OH)(Nu)} \xrightarrow{H^+} \text{R–C(O}^-\text{)(}\overset{+}{\text{O}}\text{H}_2\text{)(Nu)} \longrightarrow \text{RC(=O)Nu}$$

1. Reduction

Carboxylic acids occupy the most oxidized side of the oxidation-reduction continuum (see Chapter 17). Carboxylic acids are reduced with lithium aluminum hydride (LAH) to the corresponding alcohols. Aldehyde intermediates that may be formed in the course of the reaction are also reduced to the alcohol. The reaction occurs by nucleophilic addition of hydride to the carbonyl group.

$$\underset{\text{carboxylic acid}}{\text{R–C(=O)–OH}} + \text{H}^- \longrightarrow \text{R–C(O}^-\text{)(H)–OH}_2 \longrightarrow \underset{\text{aldehyde}}{\text{R–C(=O)H}} \xrightarrow{\text{H}^-} \text{R–C(O}^-\text{)(H)–H} \longrightarrow \underset{\text{alcohol}}{\text{R–C(OH)(H)–H}}$$

2. Ester Formation

Carboxylic acids react with alcohols under acidic conditions to form esters and water. In acidic solution, the O on the C=O can become protonated. This accentuates the polarity of the bond, putting even more + charge on the C and making it even more susceptible to nucleophilic attack. This condensation reaction occurs most rapidly with primary alcohols.

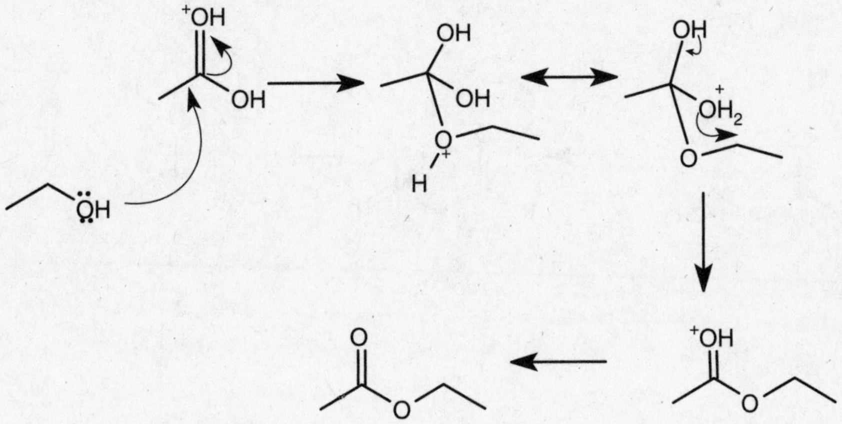

Note:

Protonating the C = O makes the C even more ripe for nucleophilic attack.

3. Acyl Halide Formation

Acyl halides, also called acid halides, are compounds with carbonyl groups bonded to halides. Several different reagents can accomplish this transformation; thionyl chloride, $SOCl_2$, is the most common.

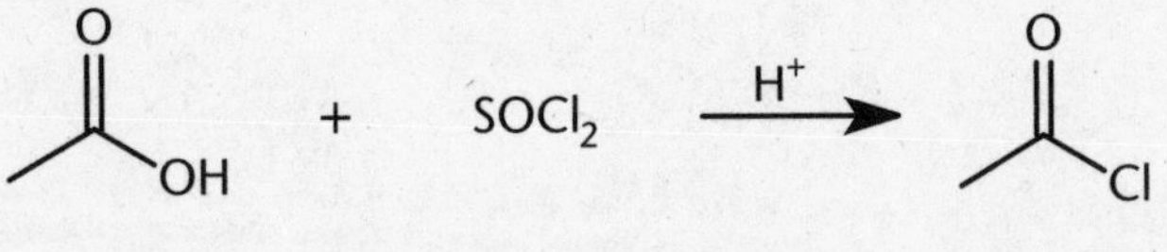

$$\text{CH}_3\text{COOH} + SOCl_2 \xrightarrow{H^+} \text{CH}_3\text{COCl}$$

Note:

Acid chlorides are among the highest energy (least stable and most reactive) members of the carbonyl family.

Acid chlorides are very reactive, as the greater electron-withdrawing power of the Cl^- makes the carbonyl carbon more susceptible to nucleophilic attack than the carbonyl carbon of a carboxylic acid. Thus acid chlorides are frequently used as intermediates in the conversion of carboxylic acids to esters and amides.

In A Nutshell:

Key reactions include:

- *soap formation (e.g., when neutralized by NaOH or KOH);*
- *nucleophilic substitution;*
- *decarboxylation, in which CO_2 is lost.*

C. DECARBOXYLATION

Carboxylic acids can undergo decarboxylation reactions, resulting in the loss of carbon dioxide.

1,3-dicarboxylic acids and other β-keto acids may spontaneously decarboxylate when heated. The carboxyl group is lost and replaced with a hydrogen. The reaction proceeds through a six-membered ring transition state. The enol initially formed tautomerizes to the more stable keto form.

H O O R O → O H O R O → O H R + O C O

enol

keto form
(more stable)

Bridge:

Decarboxylation is very common in biochemical pathways in the body, such as the Krebs cycle (see Biology Chapter 3).

REVIEW PROBLEMS

1. Which compound in each of the following pairs is more acidic?

   A. CH_3COOH or $CH_2ClCOOH$
   B. $HOOCCH_2COOH$ or $HOOCCH_2COO^-$
   C. NH_2CH_2COOH or NO_2CH_2COOH

2. Which of these molecules could be classified as a soap?

   A. $CH_3(CH_2)_{17}CH_2COOH$
   B. CH_3COOH
   C. $CH_3(CH_2)_{19}CH_2COO^-Na^+$
   D. $CH_3COO^-Na^+$

3. Which of these compounds would be expected to decarboxylate when heated?

   A. 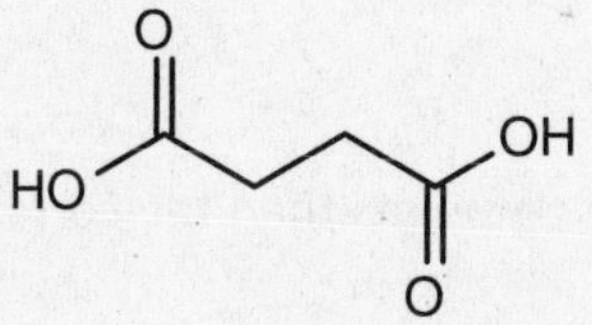 HO, O, OH, O
   C. 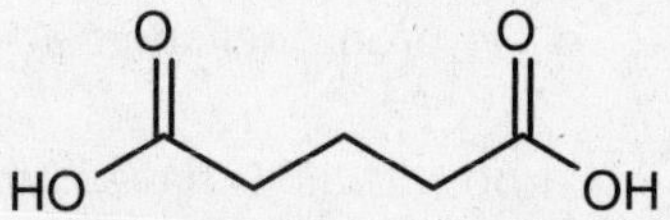 O, O, HO, OH
   B. 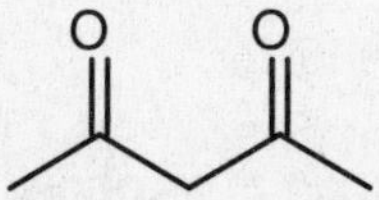 O, O
   D. 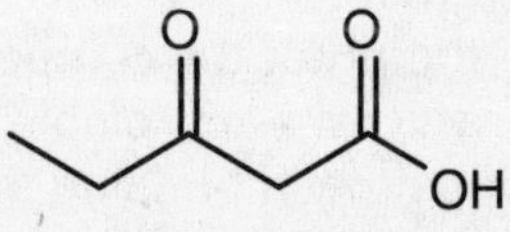 O, O, OH

4. Oxidation of which of the following compounds is most likely to yield a carboxylic acid?

   A. Acetone
   B. Cyclohexanone
   C. 2-Propanol
   D. Methanol

5. Give the IUPAC name for each of the following carboxylic acids:

a. 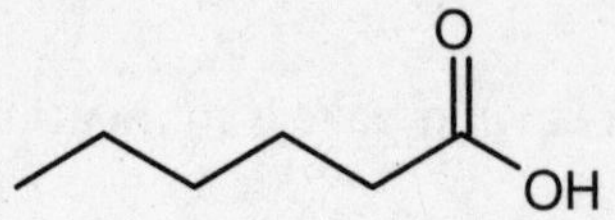 [structure] OH

b. 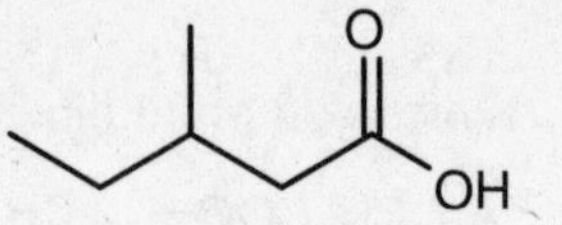 [structure] OH

c. 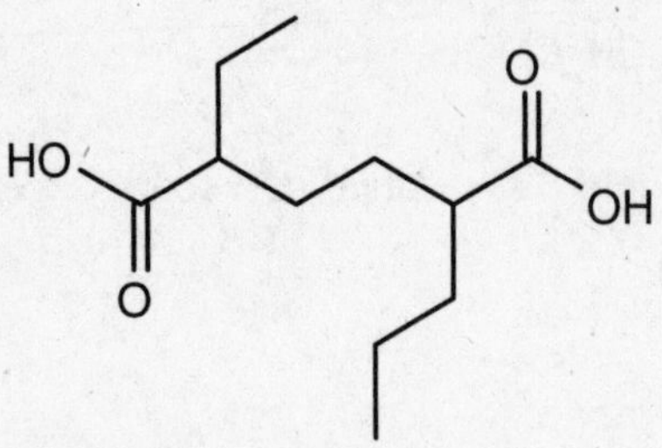 HO [structure] OH

d. 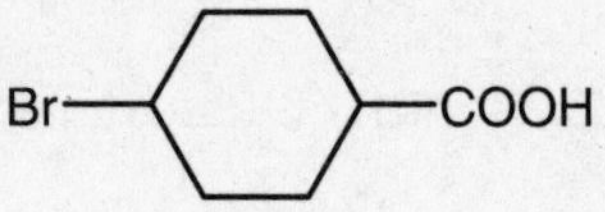 Br—[cyclohexane]—COOH

6. Draw the structures of the following carboxylic acids

a. 2,3-dimethylpentanoic acid
b. 3-butenoic acid
c. *p*-hydroxybenzoic acid
d. 4-bromohexanoic acid

7. Carboxylic acids have higher boiling points than the corresponding alcohols because

A. molecular weight is increased by the additional carboxyl group.
B. the pH of the compound in solution is lower.
C. acid salts are soluble in water.
D. hydrogen bonding is much stronger than in alcohols.

8. Which of the following carboxylic acids will be the most acidic?

A. $CH_3CHClCH_2COOH$
B. $CH_3CH_2CCl_2COOH$
C. $CH_3CH_2CHClCOOH$
D. $CH_3CH_2CH_2COOH$

9. Which of the following substituted benzoic acid compounds will be the least acidic?

A. 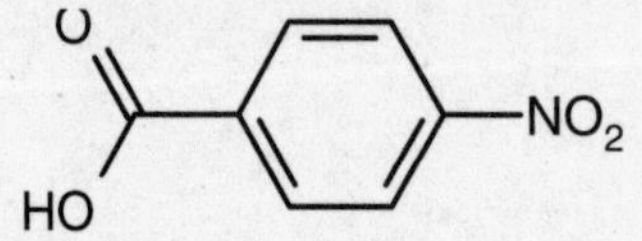 HO(O=)C—C_6H_4—NO_2

C. HO(O=)C—C_6H_4—OH

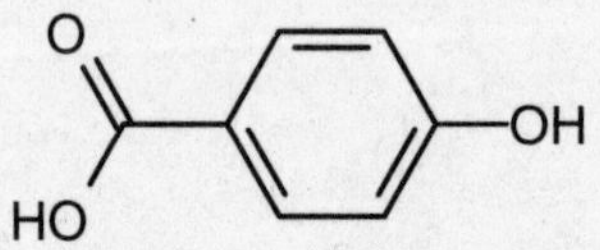

B. 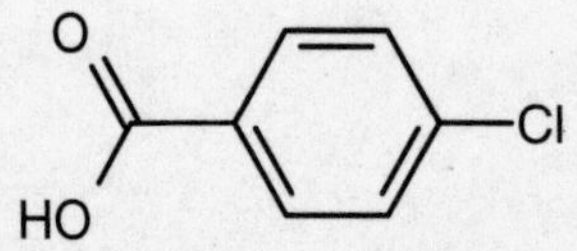 HO(O=)C—C_6H_4—Cl

D. 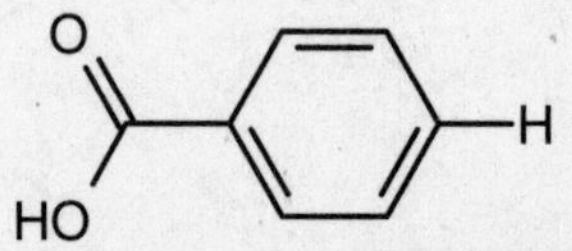 HO(O=)C—C_6H_4—H

10. Rank the following compounds in order of increasing acidity.

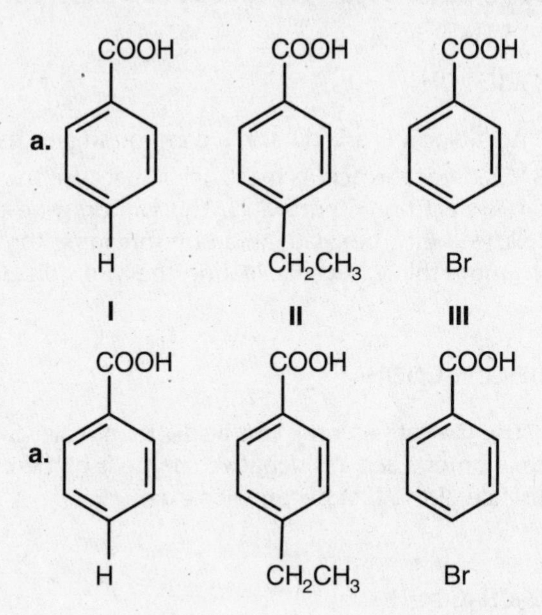

11. Predict the final product of the following reaction

$$CH_3(CH_2)_4CH_2OH \xrightarrow[\text{acetone}]{CrO_3,\ H_2SO_4}$$

1-hexanol

A. $CH_3(CH_2)_4CHO$
B. $CH_3(CH_2)_4COOH$
C. $CH_3(CH_2)_4CH_3$
D. $HOOC(CH_2)_4COOH$

12. Carboxylic acids can be reacted in one step to form all of the following compounds EXCEPT

A. acyl halides.
B. amides.
C. alkenes.
D. alcohols.

13. The reduction of a carboxylic acid by lithium aluminum hydride will yield what final product?

A. An aldehyde
B. An ester
C. A ketone
D. An alcohol

SOLUTIONS TO REVIEW PROBLEMS

1. **A $CH_2ClCOOH$**

Acetic acid is a fairly acidic compound and has a pK_a of 4.8. Chloroacetic acid is more acidic because the chlorine atom is electron-withdrawing. This withdrawing effect stabilizes the carboxylate anion by spreading the charge out even more thinly, thus facilitating the acid's dissociation.

B $HOOCCH_2COOH$

The maleate anion is less acidic than maleic acid because formation of a second negative charge is hindered by the presence of the first negative charge.

C NO_2CH_2COOH

Nitroacetic acid is more acidic than aminoacetic acid, because the nitro group is electron-withdrawing while the amino group is electron-donating. The withdrawing effect delocalizes the negative charge on the anion, making it more stable. In contrast, the donating effect of the amino group concentrates the negative charge, creating a higher-energy state and destabilizing the anion.

2. **C** A soap is a long-chain hydrocarbon with a highly polar end. Generally, this polar end or head is a salt of a carboxylic acid. Choice C fits these criteria and is the correct answer. The remaining choices all fail one or both of the criteria and are therefore wrong. Choice A is not a salt. Choice B is acetic acid, which is not a salt and does not possess a long chain. Choice D is sodium acetate, which is a salt but does not have a long hydrocarbon chain.

3. **D** This compound is a β-keto acid: a keto functionality β to a carboxyl functionality. Decarboxylation occurs with β-keto acids and 1,3 diacids because they can form a cyclic transition state that permits simultaneous transfer of a hydrogen and loss of carbon dioxide. Choice B is a diketone, and this will definitely not decarboxylate. Choices A and C are 1,4 and 1,5 diacids, respectively, and will decarboxylate, but with more difficulty. Again, the correct answer is choice D.

4. **D** Oxidation of methanol, choice D, will yield first formaldehyde and then formic acid; this is the correct answer. Acetone, choice A, cannot be oxidized further unless extremely harsh conditions are used. This is because the carbonyl carbon is bonded to two alkyl groups, and further oxidation would necessitate cleavage of a carbon-carbon bond. Choice B, cyclohexanone, is likewise limited in its options for further oxidation. Choice C, 2-propanol, can be oxidized to acetone, but no further without harsh conditions.

5. a. **Hexanoic acid**

   b. **3-methylpentanoic acid**

   c. **2-ethyl-5-propylhexanedioic acid**

   d. **4-bromocyclohexanecarboxylic acid**

6.

a. 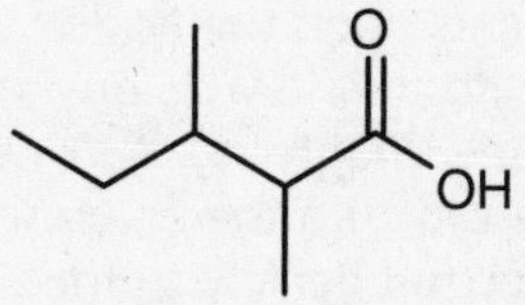 O OH

c. 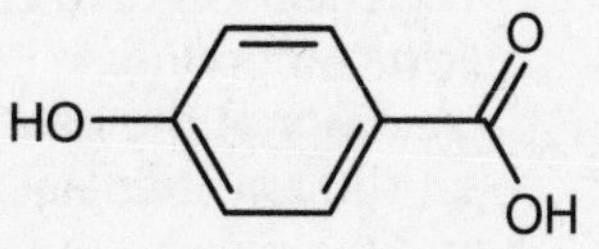 HO O OH

b. 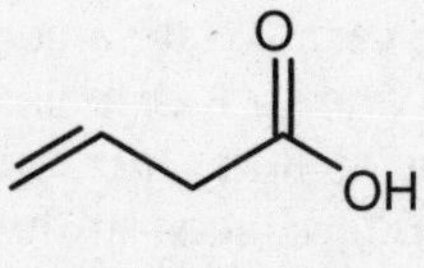 O OH

d. 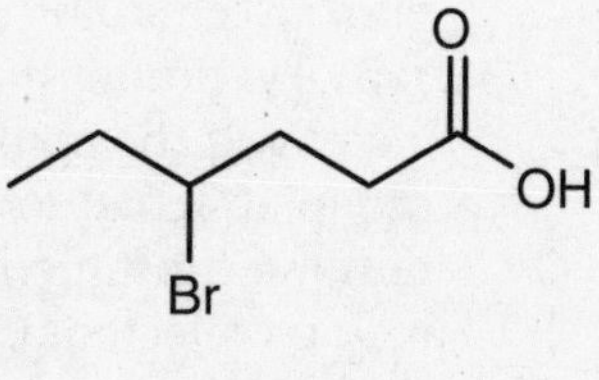 O OH Br

7. **D** The boiling points of compounds depend on the strength of the attractive forces between molecules. In both alcohols and carboxylic acids, the major form of intermolecular attraction is hydrogen bonding; however, the hydrogen bonds of carboxylic acids are much stronger than those of alcohols, since the acids are much more polar than the alcohols. This makes the boiling points of carboxylic acids higher than those of the corresponding alcohols, so choice D is correct. Boiling points are also dependent on molecular weight, choice A, but in this case the influence of the small difference in molecular weight is negligible compared with the effect of hydrogen

bonding. Therefore, choice A is wrong. Choice B describes the behavior of acids in solution; although this is also dependent on intermolecular forces, it is not otherwise related to the behavior of the pure acid, so choice B is wrong. Choice C discusses the behavior of an acid's salt in solution, which is wrong for the same reason.

8. **B** The acidity of carboxylic acids is significantly increased by the substitution of the highly electronegative halogens into the carbon chain. Their electron-withdrawing effect upon the carboxyl group increases the stability of the carboxylate anion, favoring the dissociation of the proton. This effect is especially strong for α-halogenated carboxylic acids. Among the five carboxylic acids listed, choice D is the only unsubstituted acid and therefore must have the lowest acidity. Choice A is β-halogenated, while choices B and C are α-halogenated, so A may be rejected. Finally, choice B contains 2 α-halogens and choice C includes only 1, so the electron-withdrawing effect in choice B is stronger and B is the correct answer.

9. **C** The effects of different substituents upon the acidity of benzoic acid compounds is correlated with their effects on the reactivity of the benzene ring (see Chapter 6). Activating substituents donate electron density into the benzene ring, and the ring in turn donates electron density to the carboxyl group, destabilizing the benzoate ion formed and therefore decreasing a compound's acidity. Deactivating substituents have the opposite effect: they withdraw electrons from the ring, which in turn withdraws negative charge from the carboxyl group, thus stabilizing the carboxylate anion and increasing the compound's acidity. Choice A contains a nitro group attached to the ring, and choice B has a chloride; both of these substituents have deactivating effects, so these choices may be rejected. Choice D is unsubstituted benzoic acid, while choice C has a strongly activating substituent, hydroxyl. Thus, choice C will be the least acidic and is the correct answer.

10. a. An ethyl substituent has an activating effect upon the benzene ring; thus, the acidity of compound II is very low. The bromine substituent, on the other hand, has a deactivating effect on the benzene ring, which makes compound III highly acidic. The acidity of unsubstituted benzoic acid (compound I) is somewhere in the middle. Therefore the order of increasing acidity is II, I, III.

b. In this case, compound I is the strongest acid because the nitro group is a powerful deactivating substituent. On the other hand, II is a stronger acid than III because the methyl group in compound III donates electron density to the carboxyl group, decreasing acidity. Here the order of increasing acidity is III, II, I.

11. **B** Jones' reagent (chromium trioxide in aqueous sulfuric acid) oxidizes primary alcohols directly to monocarboxylic acids, so choice B is correct. This reagent is too strong an oxidizing agent to give an aldehyde (aldehyde will be formed but will immediately be oxidized further), so choice A is wrong. Choice D, a dicarboxylic acid, cannot form because there is no functional group "handle" on the other end of the molecule for the reagent to attack, and it cannot attack the inert alkane. Nor will it produce an alkane such as choice C, so this is also wrong.

12. **C** Carboxylic acids can form all of these types of compounds, except for alkenes, in one step. Acyl halides (choice A) are formed with thionyl chloride. Amides (choice B) are formed by reaction with ammonia. Alcohols (choice D) may be formed using a variety of reducing agents. To form alkenes (choice C), carboxylic acids may be reduced to alcohols, which can then be transformed into alkenes by elimination.

13. **D** Lithium aluminum hydride is a very strong reducing agent. Its reaction with carboxylic acids yields alcohols, choice D. Aldehydes are intermediate products of this reaction, therefore choice A is wrong. Esters are formed from carboxylic acids by reaction with alcohols, so choice B is wrong. Ketones are formed by the Friedel-Crafts Acylation of the acyl chloride derivatives of acids, so choice C is wrong.

Carboxylic Acid Derivatives

CHAPTER TWENTY

Carboxylic acids can be converted into several types of derivatives: **acyl halides, anhydrides, amides,** and **esters.** These are compounds in which the –OH of the carboxyl group has been replaced with **–X, –OCOR, –NH_2,** or **–OR,** respectively. They readily undergo nucleophilic substitution reactions, including hydrolysis (H_2O as nucleophile) which produces the original carboxylic acid. They also undergo other additions and substitutions, including various interconversions between different acid derivatives. In general, the acyl halides are the most reactive of the carboxylic acid derivatives, followed by the anhydrides, the esters, and the amides.

Note:

Order of reactivity: acyl halides > anhydrides > esters > amides

ACYL HALIDES

A. NOMENCLATURE

Acyl halides are also called **acid** or **alkanoyl halides**. (The acyl group is RCO–.) They are the most reactive of the carboxylic acid derivatives. They are named in the IUPAC system by changing the *-ic acid* ending of the carboxylic acid to **-yl halide**. Some typical examples are ethanoyl chloride (also called acetyl chloride), benzoyl chloride, and *n*-butanoyl bromide.

Flashback:

Carboxylic acid nomenclature was discussed in Chapter 19.

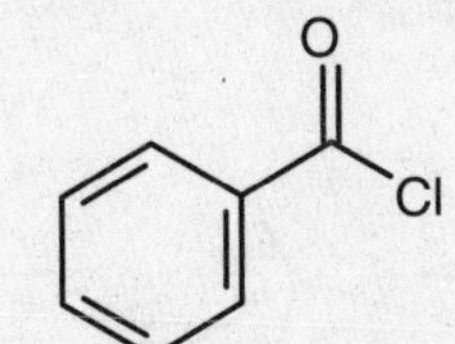

ethanoyl chloride (acetyl chloride) | benzoyl chloride | *n*-butanoyl bromide

B. SYNTHESIS

The most common acyl halides are the acid chlorides, although acid bromides and iodides are occasionally encountered. They are prepared

Note:

The most common acyl halide is acyl chloride, made with RCOOH + $SOCl_2$.

by reaction of the carboxylic acid with thionyl chloride, $SOCl_2$, producing SO_2 and HCl as side products. Alternatively, PCl_3 or PCl_5 (or PBr_3, to make an acid bromide) will accomplish the same transformation.

$$CH_3COOH + SOCl_2 \longrightarrow CH_3COCl + SO_2 + HCl$$

C. REACTIONS: NUCLEOPHILIC ACYL SUBSTITUTION

The following reactions of acyl halides proceed via the mechanism of nucleophilic substitution on a carbonyl, shown in detail in Chapter 19.

1. Hydrolysis

The simplest reaction of acid halides is their conversion back to carboxylic acids. They react very rapidly with water to form the corresponding acid, along with HCl, which is responsible for their irritating odor.

Mnemonic:

Hydrolysis = hydro + lysis, or cleavage by water.

$H_2\ddot{O}$: ; O⁻ ; Cl ; $^+OH_2$; O ; OH ; + HCl

2. Conversion into Esters

Acyl halides can be converted into esters by reaction with alcohols. The same type of nucleophilic attack found in hydrolysis leads to the formation of a tetrahedral intermediate, with the hydroxyl oxygen as the nucleophile. Chloride is displaced and HCl is released as the side-product.

EtÖH ; O ; Cl ; O⁻ ; +O ; H Et ; Cl ; O ; OEt ; + HCl

3. Conversion into Amides

Acyl halides can be converted into amides (compounds of the general formula $RCONR_2$) by an analogous reaction with amines. Nucleophilic amines, such as ammonia, attack the carbonyl group, displacing chloride. The side product is ammonium chloride, formed from excess ammonia and HCl.

$$CH_3COCl + 2\,\ddot{N}H_3 \longrightarrow CH_3CONH_2 + NH_4Cl$$

D. OTHER REACTIONS

1. Friedel-Crafts Acylation

Aromatic rings can be acylated in a Friedel-Crafts reaction. The mechanism is electrophilic aromatic substitution, and the attacking reagent is an acylium ion, formed by reaction of an acid chloride with $AlCl_3$ or another Lewis acid. The product is an alkyl aryl ketone.

$$H_3C\text{–}COCl + C_6H_6 \xrightarrow{AlCl_3} C_6H_5COCH_3$$

Flashback:

Friedel-Crafts reactions were discussed in Chapter 16.

Note:

The mechanism for the Friedel-Crafts acylation is electrophilic aromatic substitution.

2. Reduction

Acid halides can be reduced to alcohols, or selectively reduced to the intermediate aldehydes. Catalytic hydrogenation in the presence of a "poison" like quinoline accomplishes the latter transformation. (Compare with Lindlar's catalyst, Chapter 15.)

$$CH_3COCl \xrightarrow[\text{Pd/BaSO}_4\text{, Quinoline}]{H_2} CH_3CHO$$

Mnemonic:

Anhydride means "without water." Anhydrides are formed by two acid molecules condensing (losing water).

ANHYDRIDES

A. NOMENCLATURE

Anhydrides, also called **acid anhydrides**, are the condensation dimers of carboxylic acids, with the general formula RCOOCOR. They are named by substituting the word **anhydride** for the word *acid* in an alkanoic acid. The most common and important anhydride is acetic anhydride, the dimer of acetic acid. Other common anhydrides, such as succinic, maleic, and phthalic anhydrides, are **cyclic anhydrides** arising from intramolecular condensation or dehydration of diacids.

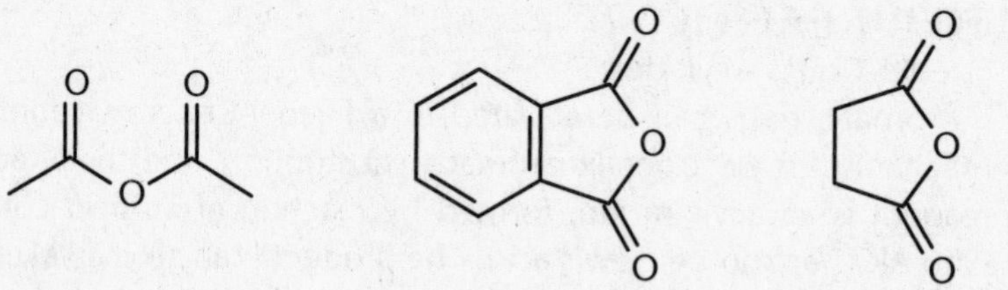

acetic anhydride (ethanoic anhydride) phthalic anhydride succinic anhydride

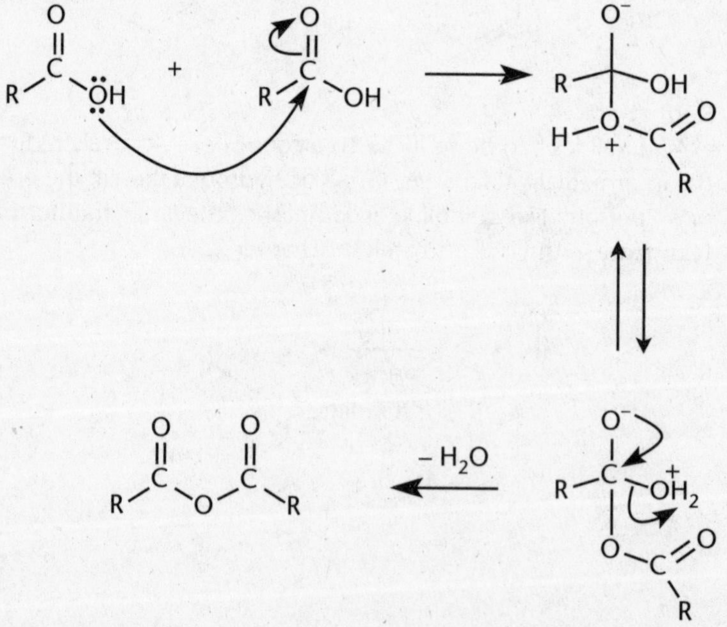

$- H_2O$

Condensation of two carboxylic acid molecules to form an anhydride.

B. SYNTHESIS

Anhydrides can be synthesized by reaction of an acid chloride with a carboxylate salt.

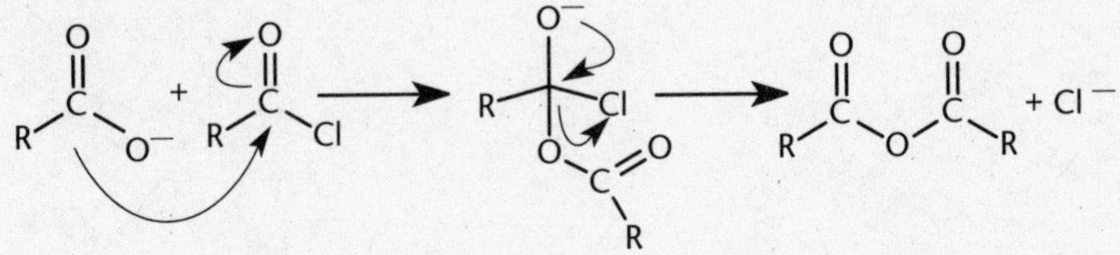

Reaction of acid chloride with carboxylate anion to form an anhydride.

Certain cyclic anhydrides can be formed simply by heating carboxylic acids. The reaction is driven by the increased stability of the newly formed ring; hence, only five- and six-membered ring anhydrides are easily made. In this case, the hydroxyl of one –COOH acts as a nucleophile, attacking the carbonyl on the other –COOH.

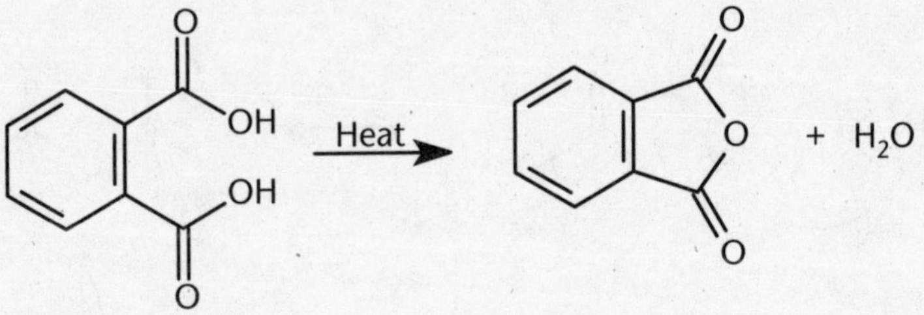

Heat → + H_2O

o-phtalic acid   phthalic anhydride

C. REACTIONS

Anhydrides react under the same conditions as acid chlorides, but since they are somewhat more stable, they are a bit less reactive. The reactions are slower and produce a carboxylic acid as the side product instead of HCl. Cyclic anhydrides are also subject to these reactions, which cause ring-opening at the anhydride group along with formation of the new functional groups.

1. Hydrolysis

Anhydrides are converted into carboxylic acids when exposed to water.

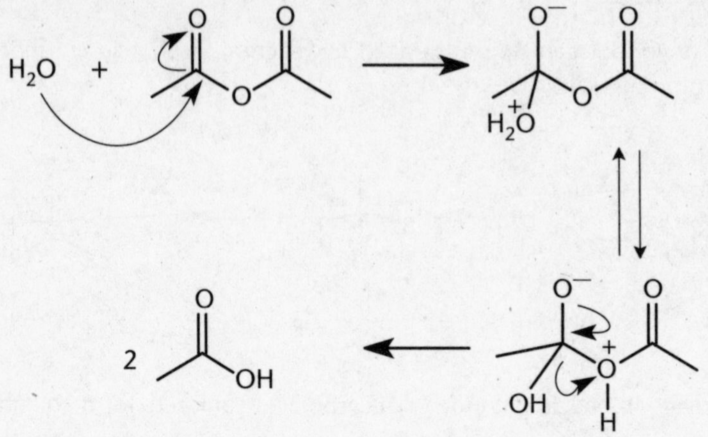

Note that in this reaction, the leaving group is actually a carboxylic acid.

2. **Conversion into Amides**

Anhydrides are cleaved by ammonia, producing amides and ammonium carboxylates.

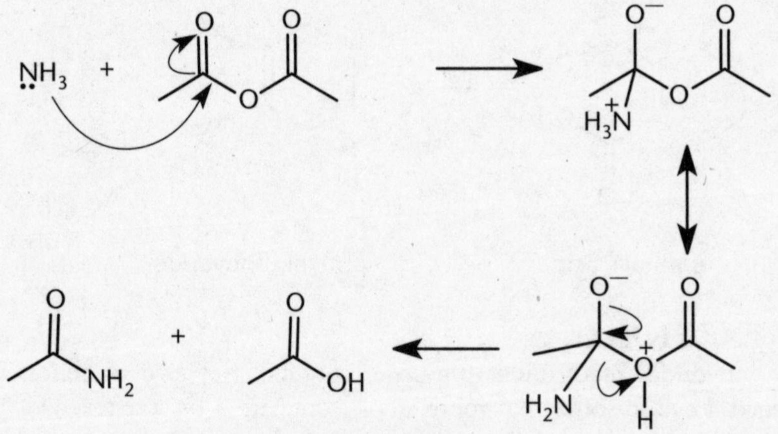

Then:

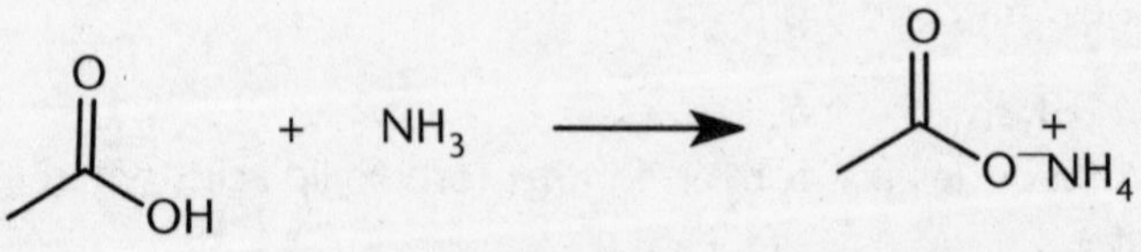

Thus, even though the leaving group is actually a carboxylic acid, the final products are an amide and the ammonium salt of a carboxylate anion.

3. **Conversion into Esters and Carboxylic acids**

Anhydrides react with alcohols to form esters and carboxylic acids.

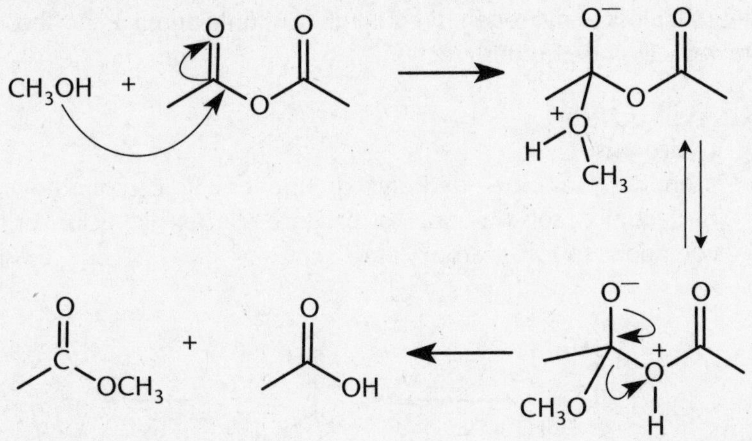

4. **Acylation**

Friedel-Crafts acylation occurs readily with $AlCl_3$ or other Lewis acid catalysts.

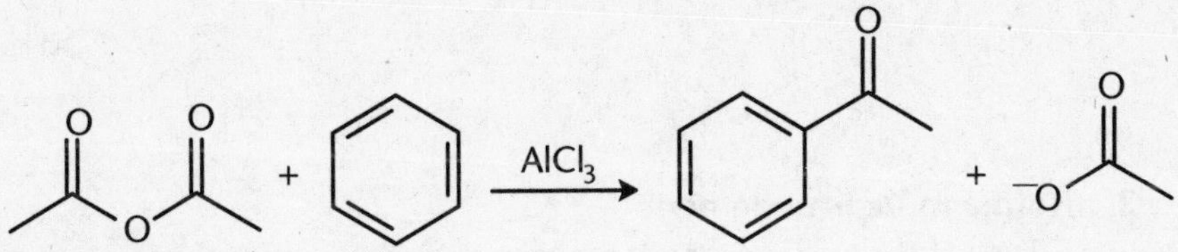

AMIDES

A. NOMENCLATURE

Amides are compounds with the general formula $RCONR_2$. They are named by replacing the *-oic* acid ending with **-amide**. Alkyl substituents on the nitrogen atom are listed as prefixes, and their location is specified with the letter *N*. For example:

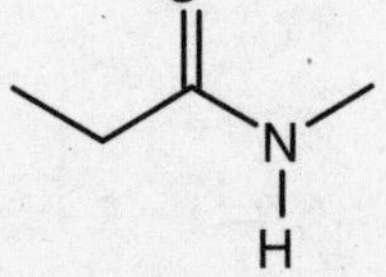

N-methylpropanamide

Note:

This reaction proceeds via electrophilic aromatic substitution.

Bridge:

The peptide bond is actually an amide linkage.

Note:

Only primary and secondary amines can be used to form amides.

B. SYNTHESIS

Amides are generally synthesized by the reaction of acid chlorides with amines or by the reaction of acid anhydrides with ammonia (see above). Note that loss of hydrogen is required; thus, only primary and secondary amines will undergo this reaction.

C. REACTIONS

1. Hydrolysis

Amides can be hydrolyzed under acidic conditions, via nucleophilic substitution, to produce carboxylic acids or basic conditions to form carboxylates:

Note:

Under acidic conditions, the carbonyl oxygen becomes protonated, making the carbon more susceptible to nucleophilic attack.

H_2O + $\overset{+}{O}H$ / NH_2 → OH / NH_2 / $\overset{+}{H_2O}$ ⇌ OH / $\overset{+}{N}H_3$ / OH → $\overset{+}{O}H$ / OH + NH_3 —$-H^+$→ NH_3 + O / OH

2. Hofmann Rearrangement

The **Hofmann rearrangement** converts amides to primary amines with the loss of the carbonyl carbon. The mechanism involves the formation of a **nitrene**, the nitrogen analog of a carbene. The nitrene is attached to the carbonyl group and rearranges to form an **isocyanate,** which, under the reaction condition is hydrolyzed to the amine.

Note:

Hofmann rearrangement: amides → primary amines (with loss of a carbon)

$$RC(=O)NH_2 \xrightarrow{BrO^-} R{-}C(=O){-}\ddot{N}H{-}Br + OH^- \rightleftharpoons R{-}C(=O){-}\ddot{N}{-}Br + H_2O$$

$$R{-}C(=O){-}\ddot{N}{-}Br \longrightarrow O{=}C{=}\ddot{N}{-}R + Br^- \xrightarrow{H_2O} H_2\ddot{N}R + CO_2 + Br^-$$

nitrene isocyanate

3. **Reduction**

Amides can be reduced with LAH to the corresponding amine. Notice that this differs from the product of the Hofmann rearrangement in that no carbon atom is lost.

Note:

Reduction also produces amines, but no carbon is lost.

ESTERS

A. NOMENCLATURE

Esters are the dehydration products of carboxylic acids and alcohols. They are commonly found in many fruits and perfumes. They are named in the IUPAC system as **alkyl** or **aryl alkanoates**. For example, ethyl acetate, derived from the condensation of acetic acid and ethanol, is called ethyl ethanoate according to IUPAC nomenclature.

Note:

An acid + an alcohol = an ester.

B. SYNTHESIS

Mixtures of carboxylic acids and alcohols will condense into esters, liberating water, under acidic conditions. Esters can also be obtained from reaction of acid chlorides or anhydrides with alcohols (see above). Phenolic (aromatic) esters are produced in the same way, although the aromatic acid chlorides are less reactive than aliphatic acid chlorides, so that base must generally be added as a catalyst.

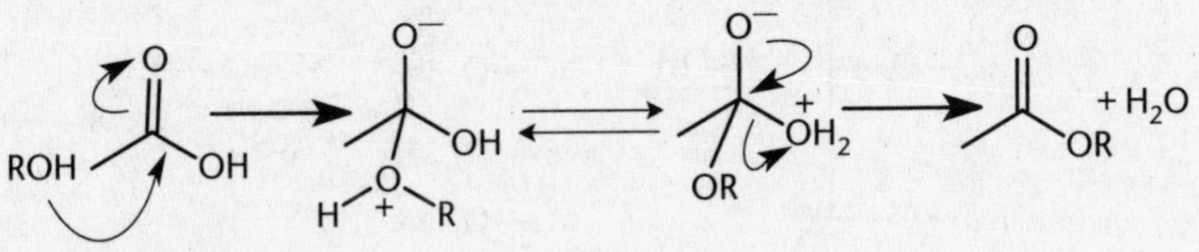

C. REACTIONS

1. **Hydrolysis**

Esters, like the other derivatives of carboxylic acids, can be hydrolyzed, yielding carboxylic acids and alcohols. Hydrolysis can take place under either acidic or basic conditions.

MCAT Favorite:

Hydrolysis, especially of esters, is a popular topic.

Bridge:

Triacylglycerols are actually esters, with glycerol as the alcohol (ROH) and free fatty acids as RCOOH.

Under acidic conditions:

H_2O ^+OH OH O H + H OH OH O + H ^+OH OH + OH

The reaction proceeds similarly under basic conditions, except that the oxygen on the C=O is not protonated, and the nucleophile is OH^-.

Triacylglycerols, also called fats, are esters of long-chain carboxylic acids, often called fatty acids, and glycerol (1,2,3-propanetriol). **Saponification** is the process whereby fats are hydrolyzed under basic conditions to produce soaps. (Note: acidification of the soap retrieves triacylglycerol.)

RC(=O)—O— / RC(=O)—O— / RC(=O)—O— —NaOH→ RC(=O)—O^-Na^+ (×3) + HO— / HO— / HO—

Triacylglycerol Soap Glycerol

2. **Conversion into Amides**

Nitrogen bases such as ammonia will attack the electron-deficient carbonyl carbon atom, displacing alkoxide, to yield an amide and an alcohol side-product. Here, ammonia is the nucleophile.

O + NH_3 → O NH_2 + HO

3. Transesterification

Alcohols can act as nucleophiles and displace the alkoxy groups on esters. This process, which transforms one ester into another, is called **transesterification**.

+ HO ⇌ + $HOCH_3$

4. Grignard Addition

Grignard reagents add to the carbonyl groups of esters to form ketones; however, these ketones are more reactive than the initial esters and are readily attacked by more Grignard reagent. Two equivalents of Grignard reagent can thus be used to produce tertiary alcohols with good yield. (The intermediate ketone can be isolated only if the alkyl groups are sufficiently bulky to prevent further attack.) This reaction proceeds via nucleophilic substitution followed by nucleophilic addition.

CH_3CH_2MgBr → 1. CH_3CH_2MgBr 2. H_2O → OH

3-methyl-3-pentanol

5. Condensation Reactions

An important reaction of esters is the **Claisen condensation.** In the simplest case, two moles of ethyl acetate react under basic conditions to produce a β-keto ester, ethyl 3-oxobutanoate, or acetoacetic ester by its common name. (The Claisen condensation is also called the **acetoacetic ester condensation.**) The reaction proceeds by addition of an enolate anion to the carbonyl group of another ester, followed by displacement of ethoxide ion. This mechanism is analogous to that of the aldol condensation.

Mnemonic:

When a different alcohol attacks the ester, the ester is transformed, and transesterification results.

Note:

Grignard reagents, RMgX, are essentially equivalent to R^- nucleophiles.

Note:

In the Claisen condensation, the enolate ion of one ester acts as a nucleophile, attacking another ester.

Real World Analogy:

The Claisen condensation is the mechanism by which the long hydrocarbon chains of lipids are synthesized in biological systems. Acetyl coenzyme A performs the function of ethyl acetate, and long chains are built up by units of two carbon atoms. This is why long-chain compounds with 14, 16, and 18 carbon atoms are more common in living organisms than those with the odd chain-lengths of 15 and 17.

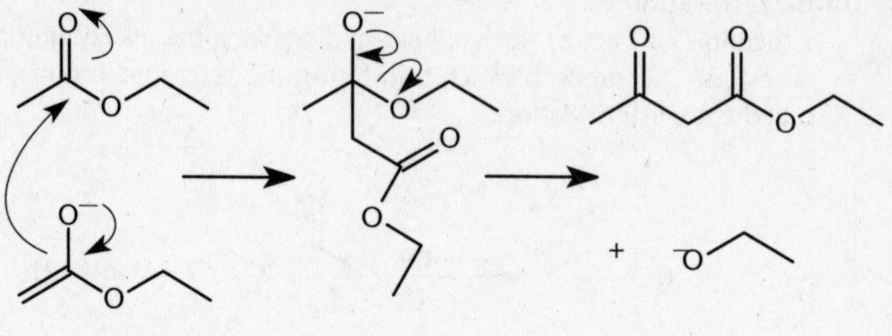

6. **Reduction**

Esters may be reduced to primary alcohols with LAH, but not with $NaBH_4$. This allows for selective reduction in molecules with multiple functional groups.

Note:

LAH is essentially equivalent to a H^- nucleophile.

LAH

OH + OH

D. PHOSPHATE ESTERS

While phosphoric acid derivatives are not carboxylic acid derivatives they form esters similar to those above.

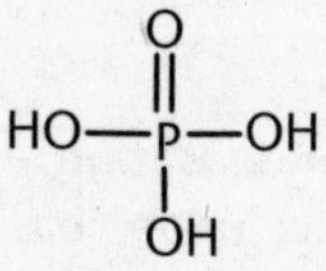

HO—P(=O)(OH)—OH   RO—P(=O)(OR)—OR   where R = H or hydrocarbon

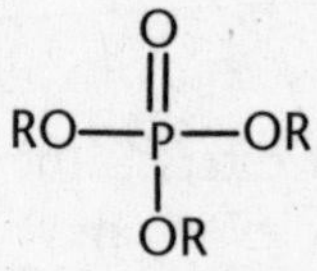

phosphoric acid   phosphoric ester

Phosphoric acid and the mono- and diesters are acidic (more so than carboxylic acids) and usually exist as anions. Like all esters, under acidic conditions they can be cleaved into the parent acid (here, H_3PO_4) and alcohols.

Phosphate esters are found in living systems in the form of **phospholipids** (phosphoglycerides), in which glycerol is attached to two carboxylic acids and one phosphoric acid.

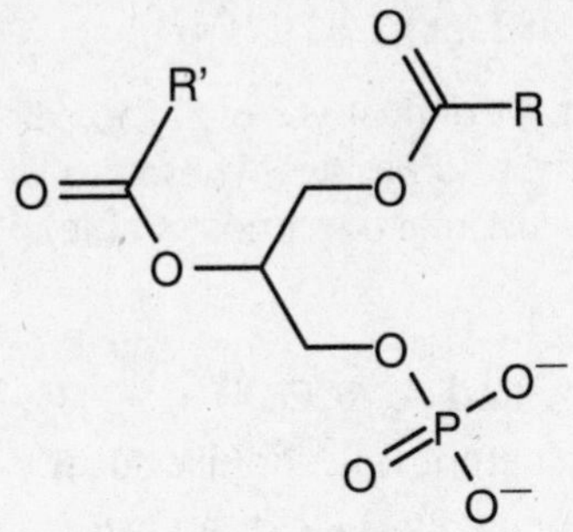

R'
O
O
R
O
O
O
O^-
P
O
O^-

phosphatidic acid
diacylglycerol phosphate
(a phosphoglyceride)

Phospholipids are the main component of cell membranes, and phospholipid/carbohydrate polymers form the backbone of nucleic acids, the hereditary material of life (see Chapter 24). The nucleic acid derivative **adenosine triphosphate (ATP)** can give up and regain one or more phosphate groups. ATP facilitates many biological reactions by releasing phosphate groups to other compounds, thereby increasing their reactivities.

Note:

Remember phosphodiester bonds? They hold the backbone of DNA together, connecting nucleotides with covalent linkages.

SUMMARY OF REACTIONS

- The most important derivatives of carboxylic acid are acyl halides, anhydrides, esters, and amides. These are listed from most reactive (least stable) to least reactive (most stable).

ACYL HALIDES:

- can be formed by adding RCOOH + $SOCl_2$, PCl_3 or PCl_5, or PBr_3;
- undergo many different nucleophilic substitutions; H_2O yields carboxylic acid, while ROH yields an ester, and NH_3 yields an amide;
- can participate in Friedel-Crafts acylation to form an alkyl aryl ketone;
- can be reduced to alcohols or, selectively, to aldehydes.

ANHYDRIDES:

- can be formed by RCOOH + RCOOH (condensation) or $RCOO^-$ + RCOCl (substitution);
- undergo many nucleophilic substitution reactions, forming products that include carboxylic acids, amides, and esters;
- can participate in Friedel-Crafts acylation.

ESTERS:

- formed by RCOOH + ROH or, better, by acid chlorides or anhydrides + ROH;
- hydrolyze to yield acids + alcohols; adding ammonia yields an amide;
- reaction with Grignard reagent (2 moles) produces a tertiary alcohol;
- In Claisen condensation, analogous to the aldol, the ester acts both as nucleophile and target;
- very important in biological processes, particularly phosphate esters, which can be found in membranes, nucleic acids, and metabolic reactions.

AMIDES:

- can be formed by acid chlorides + amines, or acid anhydrides + ammonia;
- hydrolysis yields carboxylic acids or carboxylate anions;
- can be transformed to primary amines via Hofmann rearrangement or reduction.

REVIEW PROBLEMS

1. Name each of the following compounds according to the IUPAC system.

a. 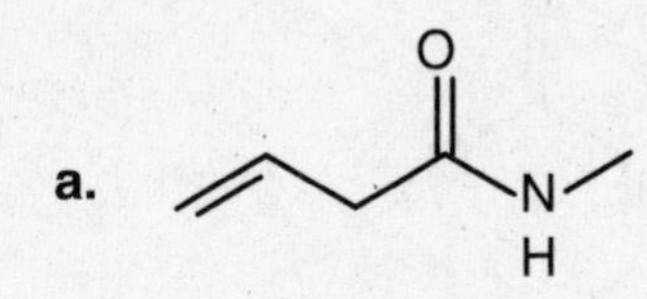

b. 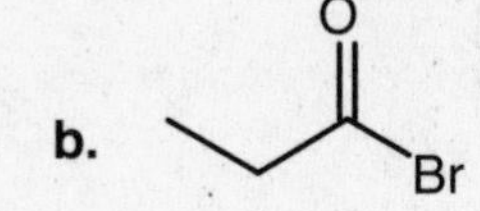

c.

d.

e.

2. What would be the product of the following reaction?

$SOCl_2$ → ?

A.

B.

C.

D.

E.

3. During the hydrolysis of an acid chloride, pyridine (a base) is usually added to the reaction vessel. This is done because

   A. the reaction leads to the production of hydroxide ions.
   B. the acyl chloride is unreactive.

C. the hydrolysis reaction leads to the formation of HCl.
D. the pyridine reacts in a side-reaction with the carboxylic acid product.

4. What would be the primary product of the following reaction?

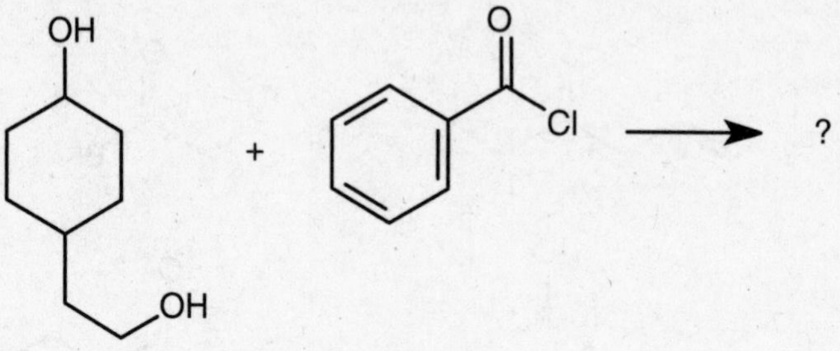

A.

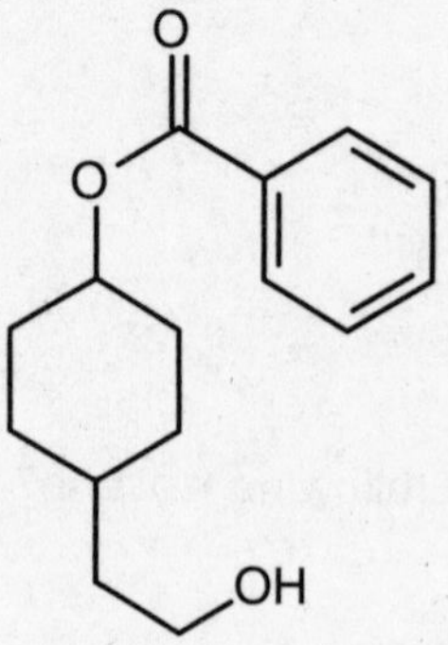

B.

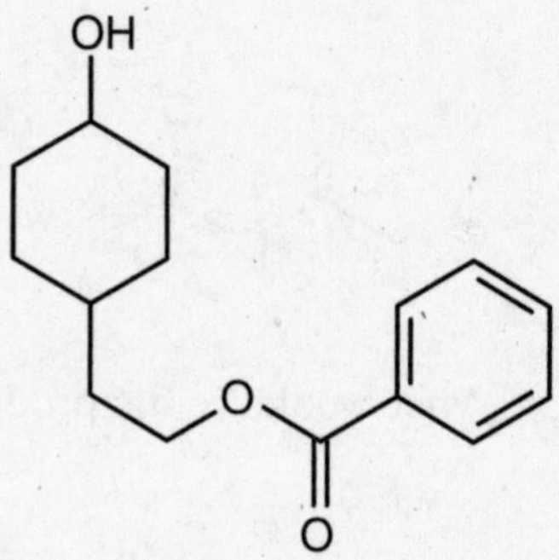

C.

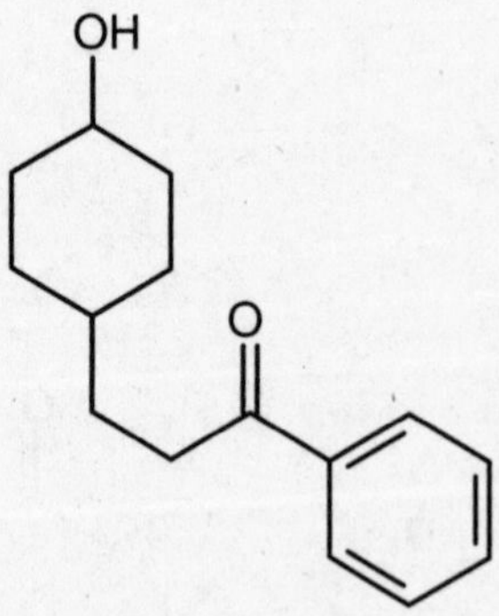

D.

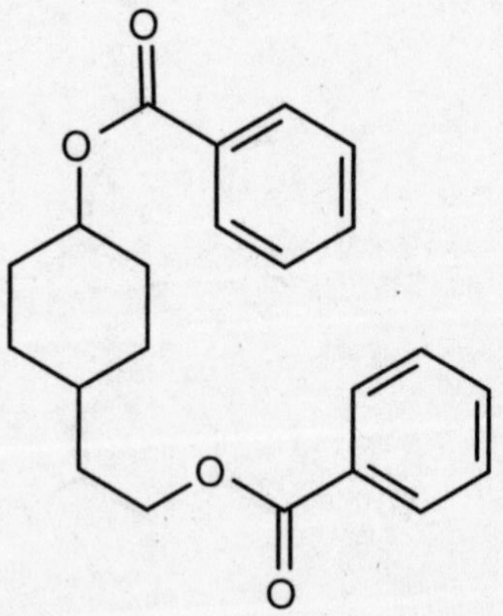

5. In order to produce a primary amide, an acid chloride should be treated with

A. ammonia.
B. an alcohol.
C. a primary amine.
D. a tertiary amine.

6. Which of the following would be the best method of producing methyl propanoate?

   A. Reacting propanoic acid and methanol in the presence of a mineral acid.
   B. Reacting ethanol with propanoyl chloride in the presence of a base.
   C. Reacting propanoyl chloride with an aqueous base.
   D. Reacting propanoic acid with ethanol in the presence of a mineral acid.

7. What would be the product(s) of the following reaction?

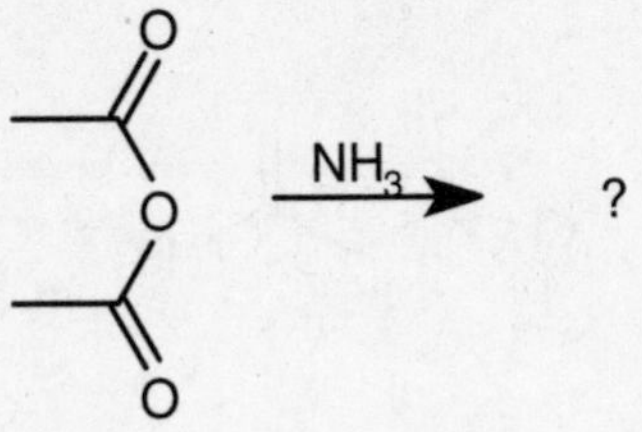

O, O, O, NH_3, ?

A. NH_2, O, NH_2

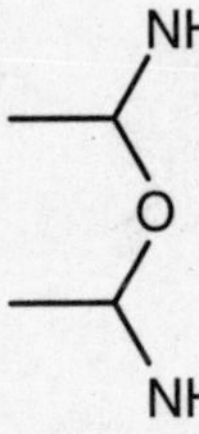

B. 2 NH_2, OH

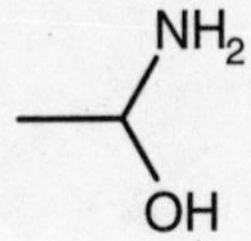

C. NH_2, O, O

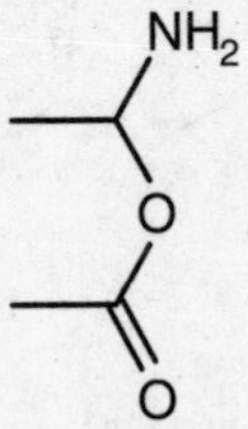

D. NH_2, O + $O^-NH_4^+$, O

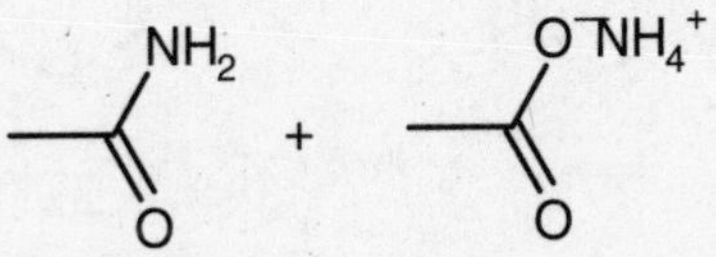

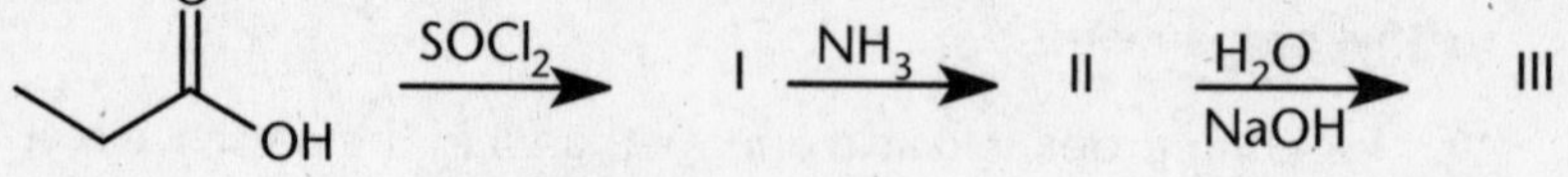

8. Which of the following correctly shows the intermediates and products of the reaction above?

A. 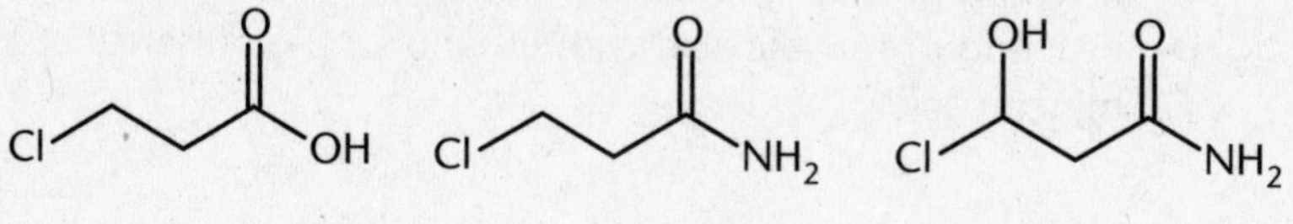 I, II, III

B. 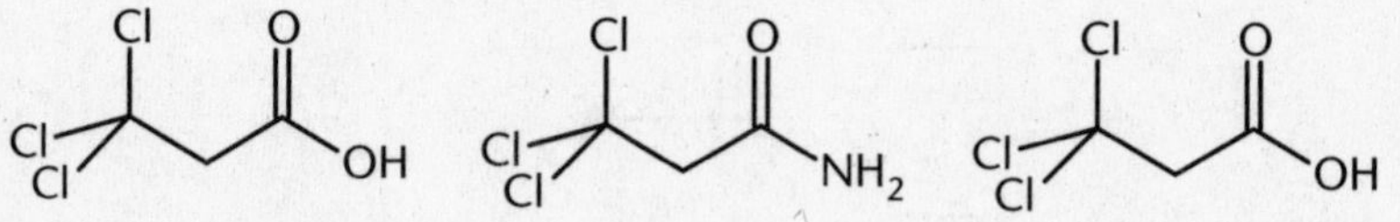 I, II, III

C. 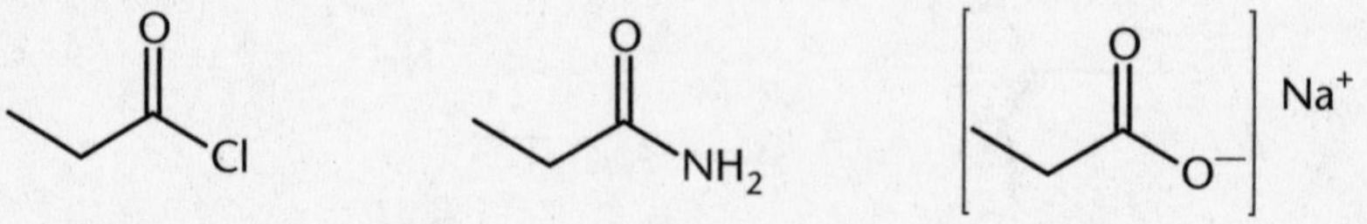 I, II, III Na^+

D. 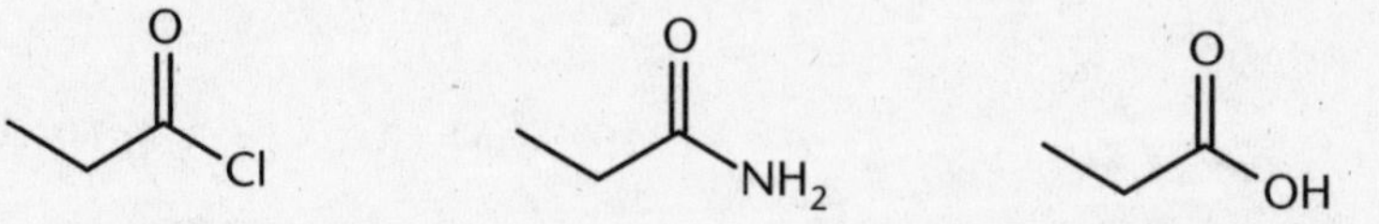 I, II, III

9. Prepare the following compounds from pentanoic acid. Give all reactants and reaction conditions.

a. 1-pentanol
b. Pentanoyl bromide
c. *N*-methylpentanamide
d. Ethyl pentanoate
e. Pentanoic anhydride

10. What is saponification and why is it important?

SOLUTIONS TO REVIEW PROBLEMS BEGIN ON THE FOLLOWING PAGE

SOLUTIONS TO REVIEW PROBLEMS

1. a. ***N*-methyl-3-butenamide**

   b. **Propanoyl bromide**

   c. **Pentanamide**

   d. **Propyl propanoate**

   e. **Ethanoic anhydride**

2. **D** Treating a carboxylic acid with thionyl chloride results in the production of an acyl chloride. In this reaction, butanoic acid is converted to butanoyl chloride, which is choice D. Since none of the other choices are acyl chlorides, they can be eliminated.

3. **C** Hydrolysis of an acid chloride results in the formation of a carboxylic acid and HCl. Therefore, since pyridine can act as a base, it serves to neutralize the HCl that is formed. The reaction does not result in the formation of hydroxide ions, so choice A is wrong. Pyridine does not react with the carboxylic acid product, it reacts with the HCl, so choice D is wrong. Finally, choice B is incorrect because the acyl chloride is very reactive. Again, the correct answer is choice C.

4. **B** In this question, an acid chloride is treated with an alcohol, and the product will be an ester. However, the esterification process is affected by the presence of bulky side-chains on either reactant. It is easier to esterify an unhindered alcohol than a hindered one. In this reaction, the primary hydroxyl group is less hindered and will react with benzoyl chloride more rapidly, so choice B is correct. Choice A is incorrect because the hydroxyl group is a hindered secondary hydroxyl, and the reaction rate will be slower. Choice C is incorrect because it is not an ester. Choice D is incorrect because steric hindrance would prevent this product from being formed.

5. **A** Acid chlorides react with ammonia or other amines to form amides. Since the amine is replacing the hydroxyl group of the carbonyl, there must be at least one hydrogen on the amine. Therefore, only ammonia and primary or secondary amines can

undergo this reaction. In order to obtain a primary amide, ammonia, choice A, must be used. The reaction of an alcohol with an acid chloride produces an ester, so choice B is incorrect. A primary amine reacting with an acid chloride would give an N substituted; thus, choice C is incorrect. Choice D is wrong because tertiary amines will not react with acid chlorides.

6. **A** Methyl propanoate is an ester, it can be synthesized by reacting a carboxylic acid with an alcohol in the presence of acid: choice A. Reacting ethanol with propanoyl chloride (choice B) will also result in the formation of an ester, but because ethanol is used, ethyl propanoate will be formed, not methyl propanoate. This is also the case for choice D, since ethanol is used here as well. Therefore, choices B and D are wrong. Choice C is incorrect because propanoyl chloride will not form an ester in the presence of base alone. Therefore, choice A is the correct response.

7. **D** This question asks for the products when ammonia reacts with acetic anhydride. Recall from the notes that an amide and an ammonium carboxylate will be formed. The only choice showing such a pair is D, acetamide and ammonium acetate.

8. **C** This question gives a reaction scheme for the interconversion of propanoic acid to various derivatives, and asks what intermediate products are formed. The first reaction involves the formation of an acid chloride using thionyl chloride. Acid chlorides are made by replacing the hydroxyl group with chlorine. Thus, choices A and B, which depict intact hydroxyl groups, can be eliminated. The second reaction is an ammonolysis of propanoyl chloride. The product should be propanamide, since ammonia will replace the chloride on the carbonyl carbon. The final reaction involves amide hydrolysis. Hydrolysis leads to carboxylic acid formation. Distinguishing between choices C and D, which both have a carboxylic acid as the third product, involves understanding how carboxylic acids exist in acidic and basic conditions. In acidic solution, the carboxyl group will be protonated, while in basic solution, the carboxyl group will be deprotonated. This reaction involves hydrolysis in the presence of base; therefore, the resulting carboxylic acid will exist in solution as a carboxylate salt. Thus, choice C is the correct answer, since it has sodium propanoate as the product of the third reaction.

9. This question asks about preparation of pentanoic acid derivatives. Where there is more than one way to make the products, the most efficient method will be given.

a. **1-Pentanol**

Carboxylic acids are easily reduced by LAH to produce the corresponding primary alcohol.

b. **Pentanoyl bromide**

To form pentanoyl bromide, pentanoic acid is reacted with PBr_3. The bromide replaces the hydroxyl on the carbonyl carbon.

c. ***N*-methylpentanamide**

N-methylpentanamide can be prepared by first producing the acid chloride using thionyl chloride, and then reacting it with methylamine to yield the amide.

d. **Ethyl pentanoate**

The ethyl ester of pentanoic acid can be formed by reacting it directly with ethanol in the presence of hydrochloric acid.

e. **Pentanoic anhydride**

The most common method of preparing anhydrides is the reaction between an acid chloride and a carboxylate anion. To form pentanoic anhydride, one mole of pentanoic acid must be treated with thionyl chloride to yield pentanoyl chloride. This reacts with one mole of sodium pentanoate to form pentanoic anhydride.

10. Discussed on page 370.

Amines and Nitrogen-Containing Compounds

CHAPTER TWENTY ONE

NOMENCLATURE

Amines are compounds of the general formula NR_3. They are classified according to the number of alkyl (or aryl) groups to which they are bound. A **primary** (**1°**) amine is attached to one alkyl group, a **secondary** (**2°**) amine to two, and a **tertiary** (**3°**) amine to three. A nitrogen atom attached to four alkyl groups is called a **quaternary ammonium compound**. The nitrogen carries a positive charge; thus, these compounds generally exist as salts.

In the common system, amines are generally named as alkylamines. The groups are designated individually or by using the prefixes di- or tri- if they are the same. In the IUPAC system, amines are named by substituting the suffix **-amine** for the final "e" of the name of the alkane to which the nitrogen is attached. *N* is used to label substituents attached to the nitrogen in secondary or tertiary amines. The prefix **amino-** is used for naming compounds containing an OH or a CO_2H group. Aromatic amines are named as derivatives of aniline ($C_6H_5NH_2$), the IUPAC name for which is benzenamine.

Formula:	$CH_3CH_2NH_2$	$CH_3CH_2N(CH_3)_2$	$H_2NCH_2CH_2OH$
IUPAC:	Ethanamine	*N*,*N*-Dimethylethanamine	2-Aminoethanol
Common:	Ethylamine	Dimethylethylamine	————

There are many other nitrogen-containing organic compounds. **Amides** are the condensation products of carboxylic acids and amines, and have already been discussed in Chapter 20. **Carbamates** are compounds with the general formula RNHC(O)OR'. They are also called **urethanes,** and can form polymers called **polyurethanes**. Carbamates are derived from compounds called **isocyanates**, (general formula RNCO) by the addition of an alcohol. **Enamines** are the nitrogen analogs of enols, with an amine group attached to one carbon of a double bond. **Imines** are nitrogen compounds that

Clinical Correlate:

All amino acids, as you might guess, are amines. Peptide bonds between amino acids in proteins are amide bonds. Urea, a molecule used to store ammonia for removal from the body, is also an amide.

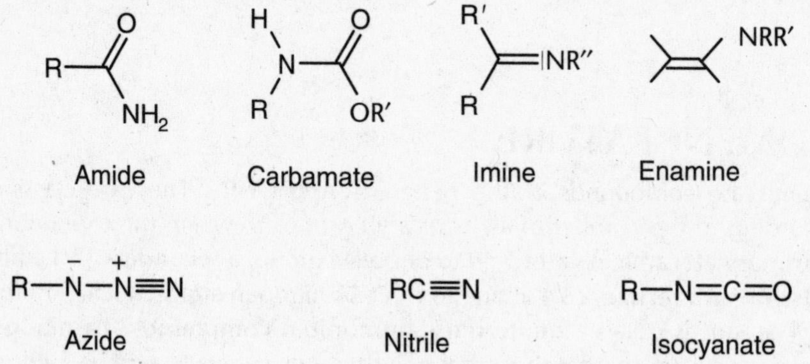

contain nitrogen-carbon double bonds. **Nitriles,** or **cyanides,** are compounds with a triple bond between a carbon atom and a nitrogen atom. They are named with either the prefix **cyano-** or the suffix **-nitrile. Nitro** compounds contain the nitro group, NO_2. **Diazo** compounds contain an N_2 functionality. They tend to lose N_2 to form carbenes. **Azides** are compounds with an N_3 functionality. When azides lose nitrogen (N_2), they form **nitrenes**, the nitrogen analogs of carbenes. Examples of these various compounds are listed below.

R—C(=O)—NH₂ — Amide

H(R)N—C(=O)—OR′ — Carbamate

R′(R)C=NR″ — Imine

Enamine (NRR′)

$R-\overset{-}{N}-\overset{+}{N}\equiv N$ — Azide

$RC\equiv N$ — Nitrile

$R-N=C=O$ — Isocyanate

PROPERTIES

The boiling points of amines are between those of alkanes and alcohols. For example, ammonia boils at –33°C, whereas methane boils at –161°C and methanol at 64.5°C. As molecular weight increases, so do boiling points. Primary and secondary amines can form hydrogen bonds, while tertiary amines cannot; therefore, tertiary amines have lower boiling points. Since nitrogen is not as electronegative as oxygen, the hydrogen bonds of amines are not as strong as those of alcohols.

The nitrogen atom in an amine is approximately sp^3 hybridized. Nitrogen must bond to only three substituents in order to complete its octet; a lone pair occupies the last sp^3 orbital. This lone pair is very important to the chemistry of amines; it is associated with their basic and nucleophilic properties.

Nitrogen atoms bonded to three different substituents are chiral because of the geometry of the orbitals. However, these enantiomers cannot be isolated, because they interconvert rapidly in a process called **nitrogen inversion**: an inversion of the sp^3 orbital occupied by the lone pair. The activation energy for this process is only 6 kcal/mol, and only at very low temperatures is it significantly slowed or stopped.

Note:

Certain amines can be optically active if inversion is hindered by structure.

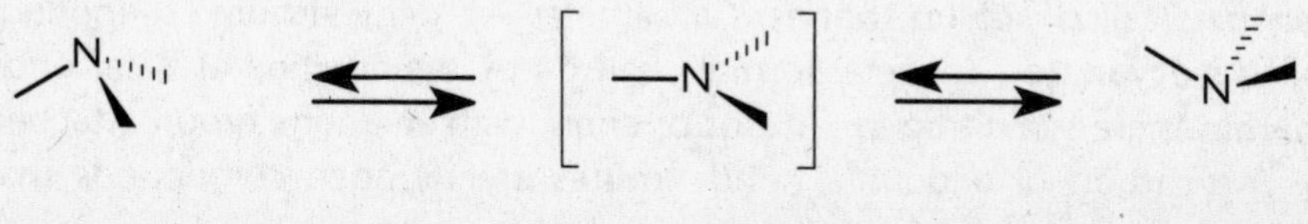

Amines are bases and readily accept protons to form ammonium ions. The pK_b values of alkyl amines are around 4, making them slightly more basic than ammonia (pK_b = 4.76), but less basic than hydroxide (pK_b = –1.7). Aromatic amines such as aniline (pK_b = 9.42) are far less basic than aliphatic amines, because the electron-withdrawing effect of the ring reduces the basicity of the amino group. The presence of other substituents on the ring alters the basicity of anilines: electron-donating groups (such as $-OH$, $-CH_3$, and $-NH_2$) increase basicity, while electron-withdrawing groups (such as NO_2) reduce basicity.

Amines also function as very weak acids. The pK_a's of amines are around 35, and a very strong base is required for deprotonation. For example, the proton of diisopropylamine may be removed with butyllithium, forming the sterically hindered base lithium diisopropylamide, LDA.

Real World Analogy:

An interesting property of several nitrogen-containing compounds, such as nitroglycerin and nitrous oxide, is their ability to act as relaxants. Nitroglycerine is given sublingually to relieve coronary artery spasms in people with chest pain, and nitrous oxide (laughing gas) is a common dental anesthetic. Nitroglycerine has the additional property of rapidly decomposing to form gas and is thus fairly explosive.

SYNTHESIS

A. ALKYLATION OF AMMONIA

1. Direct

Alkyl halides react with ammonia to produce alkylammonium halide salts. Ammonia functions as a nucleophile and displaces the halide atom. When the salt is treated with base, the alkylamine product is formed.

$$CH_3Br + NH_3 \longrightarrow CH_3\overset{+}{N}H_3Br^- \xrightarrow{NaOH} CH_3NH_2 + NaBr + H_2O$$

This reaction often leads to side products, because the alkylamine formed is nucleophilic and can react with the alkyl halide to form more complex products.

Flashback:

Addition of ammonia to an alkyl halide and the Gabriel synthesis are both S_N2 reactions. With its unshared electron pair, ammonia is a very good nucleophile, while the halides are all good leaving groups (see Chapter 14).

2. Gabriel Synthesis

The **Gabriel synthesis** converts a primary alkyl halide to a primary amine. The use of a disguised form of ammonia prevents side-product formation.

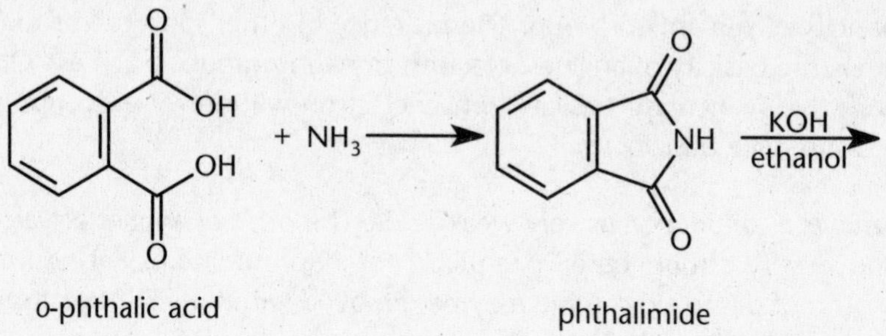

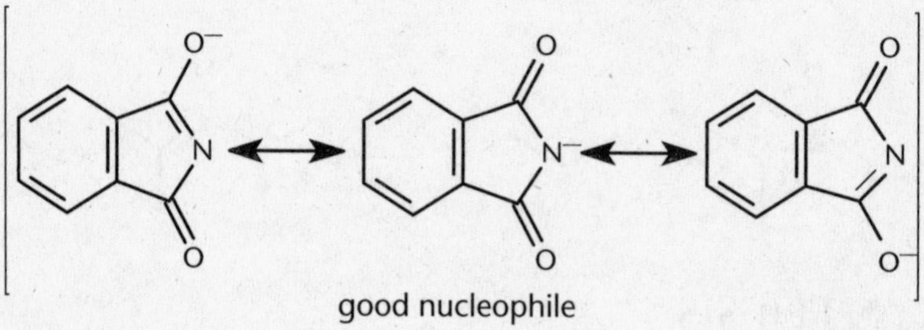

o-phthalic acid + NH_3 → phthalimide $\xrightarrow[\text{ethanol}]{\text{KOH}}$

good nucleophile

Phthalimide, the condensation product of phthalic acid and ammonia, acts as a good nucleophile when deprotonated. It displaces halide ions, forming *N*-alkylphthalimides, which do not react with other alkyl halides. When the reaction is complete, the *N*-alkylphthalimide can be hydrolyzed with aqueous base to produce the alkylamine.

$$\text{N}^-\text{-phthalimide} + \text{CH}_3\text{CH}_2\text{Br} \longrightarrow \text{phthalimide-NCH}_2\text{CH}_3 + \text{Br}^- \xrightarrow[\text{H}_2\text{O}]{\text{NaOH}} \text{phthalate dianion} + \text{H}_2\text{N-CH}_2\text{CH}_3$$

B. REDUCTION

Amines can be obtained from other nitrogen-containing functionalities via reduction reactions.

1. From Nitro Compounds:

Nitro compounds are easily reduced to primary amines. The most common reducing agent is iron or zinc and dilute hydrochloric acid, although many other reagents can be used. This reaction is especially useful for aromatic compounds, because nitration of aromatic rings is facile.

$$\text{o-}CH_3C_6H_4NO_2 \xrightarrow[\text{dil. HCL}]{\text{Zn}} \text{o-}CH_3C_6H_4NH_2$$

2. From Nitriles:

Nitriles can be reduced with hydrogen and a catalyst, or with lithium aluminum hydride (LAH), to produce primary amines.

$$CH_3CH_2C{\equiv}N \xrightarrow{\text{LAH}} CH_3CH_2CH_2NH_2$$

In a Nutshell:

Amines can be formed by:

1) S_N2 reactions
- *ammonia reacting with alkyl halides*
- *Gabriel synthesis*

2) Reduction of:
- *amides*
- *aniline and its derivatives*
- *nitriles*
- *imines*

Amines can be destroyed (converted to alkenes) by exhaustive methylation.

Note:

An imine is a nitrogen double bonded to a carbon and has about the same polarity as a carbonyl functionality.

3. **From Imines:**

Amines can be synthesized by **reductive amination**: a process whereby an aldehyde or ketone is reacted with ammonia, a primary amine, or a secondary amine to form a primary, secondary, or tertiary amine, respectively. When the amine reacts with the aldehyde or the ketone, an imine is produced. Consequently, it will undergo hydride reduction in much the same way that a carbonyl does. When the imine is reduced with hydrogen in the presence of a catalyst, an amine is produced.

O
+ NH_3 →
NH
H_2-Ni
Raney Nickel
NH_2

acetone
imine
isopropylimine
amine
isopropylamine
(aminoisopropane)

4. **From Amides:**

Amides can be reduced with LAH to form amines (see Chapter 20).

O
R NH_2
LAH →
R NH_2

REACTIONS

A. EXHAUSTIVE METHYLATION

Flashback:

Elimination reactions were discussed in Chapter 15.

Exhaustive methylation is also known as **Hofmann elimination.** In this process, an amine is converted to a quaternary ammonium iodide by treatment with excess methyl iodide. Treatment with silver oxide and water converts this to the ammonium hydroxide, which, when heated, undergoes elimination to form an alkene and an amine. The predominant alkene formed is the least substituted, in contrast with normal elimination reactions, where the predominant alkene product is the most substituted.

NH_2
excess MeI →
I^-
$\overset{+}{N}(CH_3)_3$
Ag_2O, H_2O →
^-OH
$\overset{+}{N}(CH_3)_3$
Δ →
+ $N(CH_3)_3$ + H_2O

REVIEW PROBLEMS

1. A compound with the general formula $R_4\ N^+X^-$ is classified as a

   A. secondary amine.
   B. quaternary ammonium salt.
   C. tertiary amine.
   D. primary amine

2. A compound with the structural formula $C_6H_5-\overset{-}{N}-\overset{+}{N}\equiv N$ is called

   A. a urethane.
   B. a diazo compound.
   C. an azide.
   D. a nitrile.

3. Amines have lower boiling points than the corresponding alcohols because

   A. they have higher molecular weights.
   B. they form much stronger hydrogen bonds.
   C. they form weaker hydrogen bonds.
   D. there is no systematic difference between the boiling points of amines and alcohols.

4. Which of the following would be formed if methyl bromide was reacted with phthalimide, and followed by hydrolysis with aqueous base?

   A. $C_2H_5NH_2$
   B. CH_3NH_2
   C. $(C_2H_5)_3N$
   D. $(CH_3)_4N^+Br^-$

5. The reaction of benzamide with $LiAlH_4$ yields which of the following compounds?

   A. Benzoic acid
   B. Benzonitrile
   C. Benzylamine
   D. Ammonium benzoate

6. Suggest a method of converting RCOOH to RNH_2.

7. Which of the following amines has the highest boiling point?

A. CH_3NH_2
B. $CH_3(CH_2)_6NH_2$
C. $CH_3(CH_2)_3NH_2$
D. $(CH_3)_3CNH_2$

8. If 2-amino-3-methylbutane were treated with excess methyl iodide, silver oxide, and water, what would be the major reaction products?

A. Ammonia and 2-methyl-2-butene
B. Trimethylamine and 3-methyl-1-butene
C. Trimethylamine and 2-methyl-2-butene
D. Ammonia and 3-methyl-1-butene

9. Nylon, a polyamide, is produced from hexanediamine and a substance X. This substance X is most probably

A. an amine.
B. a carboxylic acid.
C. a nitrile.
D. an alcohol.

10. A researcher wants to prepare a primary amine (RNH_2). He uses an alkyl halide (RX) and ammonia to produce an alkyl ammonium halide salt ($RNH_3^+X^-$), which he then treats with sodium hydroxide to produce the primary amine. He finds that the product is always contaminated with a secondary amine (R_2NH) and a tertiary amine (R_3N). What has he done wrong and how can he produce only the primary amine he desires?

SOLUTIONS TO REVIEW PROBLEMS BEGIN ON THE FOLLOWING PAGE

SOLUTIONS TO REVIEW PROBLEMS

1. **B** A quaternary ammonium salt has four substituents attached to the nitrogen, resulting in a positive charge on this atom. As a result, this compound forms a salt where X^- is usually a halide. Primary amines have the general formula RNH_2, secondary amines have the general formula R_2NH, and tertiary amines have the general formula R_3N. Therefore, choices A, C, and D are incorrect.

2. **C** This is an azide, which is unstable and readily loses nitrogen to yield a nitrene. A urethane (choice A) has the formula RNHCOOR′. A diazo compound (choice B) has the formula $R-N{\equiv}N$. A nitrile (choice D), also called a *cyanide*, has the formula $R{-}C{\equiv}N$.

3. **C** Amines form weaker hydrogen bonds than alcohols, since nitrogen has a lower electronegativity than oxygen. The molecules are not held together as tightly and are therefore more volatile.

4. **B** The reaction between methyl bromide and phthalimide results in the formation of methyl phthalimide. Subsequent hydrolysis then yields methylamine, so answer choice B is the correct response. Therefore, the overall reaction is the conversion of a primary alkyl halide into a primary amine. Choice A is wrong because this contains an ethyl group, not a methyl group. In order to form this compound, the initial reactant should be ethyl bromide. Choices C and D are incorrect as these are tertiary and quaternary nitrogen compounds respectively and the reaction only converts primary alkyl halides into primary amines. Again, the correct answer is choice B.

5. **C** Lithium aluminum hydride is a good reducing agent and is used to reduce amides to amines. Reduction of benzamide will result in the formation of benzylamine: choice C. Hydrolysis of benzamide would result in the formation of benzoic acid, so choice A is incorrect. Benzonitrile would be formed by the dehydration of amides, so choice B is also wrong. To form ammonium benzoate (choice D) benzamide would first have to be hydrolyzed and then reacted with ammonia, so this answer choice is also incorrect.

6. One method is transformation of the carboxylic acid to an acid chloride, followed by the Curtius rearrangement:

1. $RC(=O)OH \xrightarrow{SOCl_2} RC(=O)Cl$

2. $RC(=O)Cl \xrightarrow{NaN_3} RC(=O){-}N{-}\overset{+}{N}{\equiv}N \xrightarrow{-N_2} O{=}C{=}N{-}R$

Curtius rearrangement

3. $O{=}C{=}N{-}R \xrightarrow[-CO_2]{H_2O} H_2N{-}R$

7. **B** As the molecular weights of amines increase, so do their boiling points. Of the choices given, choice B, heptylamine, has the highest molecular weight and therefore the highest boiling point, 142–144°C. Choice A, methylamine, has a boiling point of –6.3°C. Butylamine, choice C, has a boiling point of 77.5°C, and *t*-butylamine, choice D, has a boiling point of 44.4°C.

8. **B** Treatment of an amine with excess methyl iodide, silver oxide, and water is called exhaustive methylation or Hofmann elimination. The products formed are a trisubstituted amine and an alkene. Since 2-amino-3-methylbutane is a primary amine, it will take up three methyl groups; the trisubstituted amine produced will be trimethylamine. The predominant alkene product will be the least substituted alkene, because removal of a secondary hydrogen is sterically hindered. Therefore, this reaction will produce 3-methyl-1-butene, plus trimethylamine, choice B. Choices A and D are incorrect; ammonia cannot be a product of this reaction since the mechanism involves the addition of methyl groups. Choice C is incorrect because 2-methyl-2-butene, the more substituted alkene, would not be the predominant product.

9. **B** An amide is formed from an amine and a carboxyl group or its acyl derivatives. In this question, an amine is already given; the compound to be identified must be an acyl compound. The only acyl compound among the choices given is a carboxylic acid, choice B.

10. **B** The researcher gets a contaminated product because he is using a direct alkylation process, which does not stop cleanly after the first alkylation. The primary amine produced by direct alkylation is an excellent nucleophile and will react with other alkyl halides to produce secondary and tertiary amines.

$$NH_3 + R-X \rightarrow RNH_2 \quad (1°)$$

$$RNH_2 + R-X \rightarrow R_2NH \quad (2°)$$

$$R_2NH + R-X \rightarrow R_3N \quad (3°)$$

Instead he should use the Gabriel synthesis, which converts an alkyl halide to a primary amine without unwanted side products. In this reaction, an imide, not ammonia, is used as the nucleophile. The product is an *N*-alkylimide, which cannot react further with alkyl halides. After the reaction is complete, the *N*-alkylimide is hydrolyzed with base to produce the primary alkylamine.

Purification and Separation

CHAPTER TWENTY TWO

Much of organic chemistry is concerned with the isolation and purification of the desired reaction product. A reaction itself may be completed in a matter of minutes, but separating the product from the reaction mixture is often a difficult and rather time-consuming process. Many different techniques have been developed to accomplish this objective: to obtain a pure compound separated from solvents, reagents, and other products.

BASIC TECHNIQUES

A. EXTRACTION

One way of separating out a desired product is through **extraction**, the transfer of a dissolved compound (here, the desired product) from one solvent into another in which it is more soluble. Most impurities will be left behind in the first solvent. The two solvents should be immiscible (form two layers that do not mix because of mutual insolubility). The two layers are temporarily mixed together so that solute can pass from one to the other. For example, a solution of isobutyric acid in diethyl ether can be extracted with water. Isobutyric acid is more soluble in water than in ether, and so when the two solvents are placed together, isobutyric acid transfers to the water phase.

The water (aqueous) and ether (organic) phases are separated in a specialized piece of glassware called a separatory funnel. Once separated, the isobutyric acid can be isolated from the aqueous phase in pure form. Some isobutyric acid will remain dissolved in the ether phase, so the extraction should be repeated several times with fresh solvent (water). More product can be obtained with successive extractions; i.e., it is more effective to perform three successive extractions of 10 mL each than to perform one extraction of 30 mL. Once the compound has been isolated in its purified form in a solvent, it can then be obtained by evaporation of the solvent.

Note:

Think of the aqueous and organic layers as being like oil and water in salad dressing: you can shake the mixture to increase their interaction , but ultimately they will separate again.

Bridge:

Extraction depends on the rules of solubility—"like dissolves like"! Remember the three intermolecular forces that affect solubility:

1. *hydrogen bonding: compounds that can do this, such as alcohols or acids, will move most easily into the aqueous layer.*
2. *dipole-dipole interactions: these compounds are less likely to move into the aqueous layer.*
3. *Van der Waals (London) forces: with only these interactions, compounds are least likely to move into the aqueous layer.*

Bridge:

You can use the properties of acids and bases to your advantage in extraction:

$HA + base \longrightarrow A^- + Base{:}H^+$

When the acid dissociates, the anion formed will be more soluble in the aqueous layer (because it is charged) than was the original form.

Thus, adding a base will help you extract an acid.

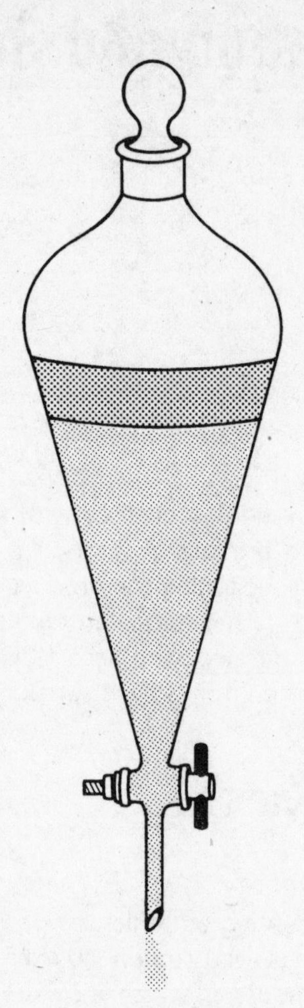

Separatory funnel

An extraction carried out to remove unwanted impurities rather than to isolate a pure product is called a **wash**.

B. FILTRATION

Filtration is used to isolate a solid from a liquid. In this technique, a liquid/solid mixture is poured onto a paper filter that allows only the solvent to pass through. The result of this process is the separation of the solid (often referred to as the residue) from the liquid or **filtrate**. The two basic types of filtration are **gravity filtration** and **vacuum filtration**. In gravity filtration, the solvent's own weight pulls it through the filter. Frequently, however, the pores of the filter become clogged with solid, slowing the rate of filtration. For this reason, in gravity filtration it is generally desirable for the substance of interest to be in solution (dissolved in the solvent), while impurities remain undissolved and can be filtered out. This allows the desired product to flow more easily and rapidly through the apparatus. To ensure that the product remains dissolved, gravity filtration is usually carried out with hot solvent.

In vacuum filtration, the solvent is forced through the filter by a vacuum on the other side. Vacuum filtration is used to isolate relatively large quantities of solid, usually when the solid is the desired product.

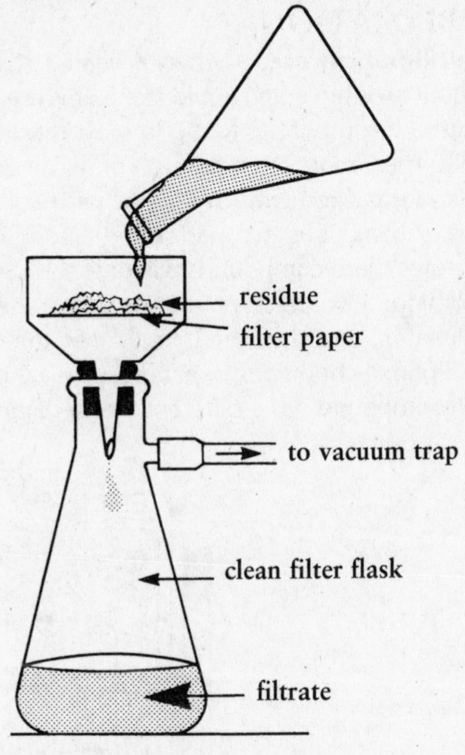

Vacuum filtration

C. RECRYSTALLIZATION

Recrystallization is a process in which impure crystals are dissolved in a minimum amount of hot solvent. As the solvent is cooled, the crystals reform, leaving the impurities in solution. In order for recrystallization to be effective, the solvent must be chosen carefully. It must dissolve the solid while it is hot, but not while it is cold. In addition, it must dissolve the impurities at both temperatures, so that they remain in solution. Solvent choice is usually a matter of trial and error, although some generalizations can be made. An estimate of polarity is useful, since polar solvents dissolve polar compounds while nonpolar solvents dissolve nonpolar compounds. A solvent with intermediate polarity is generally desirable in recrystallization. In addition, the solvent should have a low enough freezing point that the solution may be sufficiently cooled.

In some instances, a mixed solvent system may be used. Here the

Note:

Ideally, the desired product should have solubility that depends on temperature—it should be more soluble at high temperature, less so at low. In contrast, impurities should be equally soluble at various temperatures.

Bridge:

Remember the phase diagram?

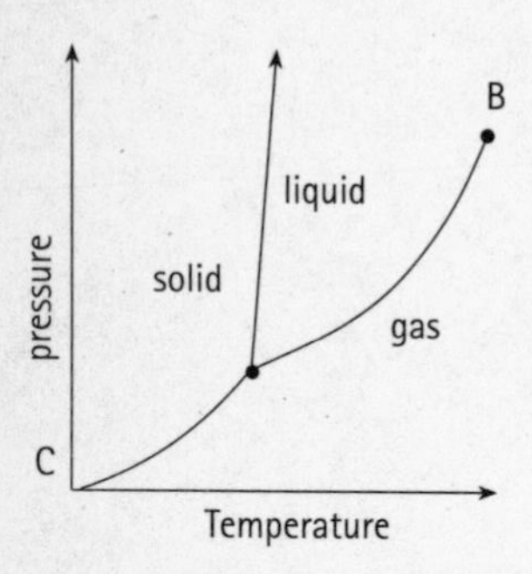

To make a solid sublime, you must either
- *a) raise the temperature at a low enough pressure; or*
- *b) lower the pressure at a very cold temperature.*

Ever wonder why the ice cubes in your freezer seem to be shrinking?

Clinical Correlate:

Centrifugation is generally used to separate big things from each other. For example, you can:

- *centrifuge blood to separate cells (RBC, WBC, platelets) from plasma;*
- *centrifuge cell debris to separate out organelles of interest, such as mitochondria;*
- *centrifuge (at ultra-high speeds) to separate big DNA molecules, such as bacterial chromosomes, from smaller ones, such as plasmids.*

crude compound is dissolved in a solvent in which it is highly soluble. Another solvent, in which the compound is less soluble, is then added in drops, just until solid begins to precipitate. The solution is heated a bit more to redissolve the precipitate, and then slowly cooled to induce crystal formation.

D. SUBLIMATION

Sublimation occurs when a heated solid turns directly into a gas, without an intervening liquid stage. It is used as a method of purification because the impurities found in most reaction mixtures will not sublime easily. The vapors are made to condense on a **cold finger**, a piece of glassware packed with dry ice or with cold water running through it. Most sublimations are performed under vacuum, because at higher pressures more compounds will pass through a liquid phase rather than subliming; low pressure also reduces the temperature required for sublimation and thus the danger that the compound will decompose. The optimal conditions depend on the compound to be purified, since each compound has a different phase diagram.

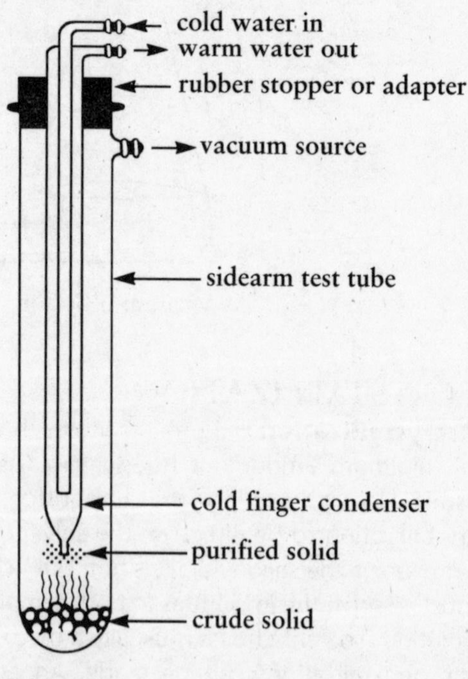

Sublimation

E. CENTRIFUGATION

Particles in a solution settle, or **sediment**, at different rates depending upon their mass, their density, and their shape. Sedimentation can be accelerated by **centrifuging** the solution. A centrifuge is an apparatus in which test tubes containing the solution are spun at high speed, which subjects them to centrifugal force. Compounds of greater mass and density settle toward the bottom of the test tubes, while lighter compounds remain near the top. This method of separation is effective for many different types of compounds, and is frequently used in biochemistry to separate cells, organelles, and biological macromolecules.

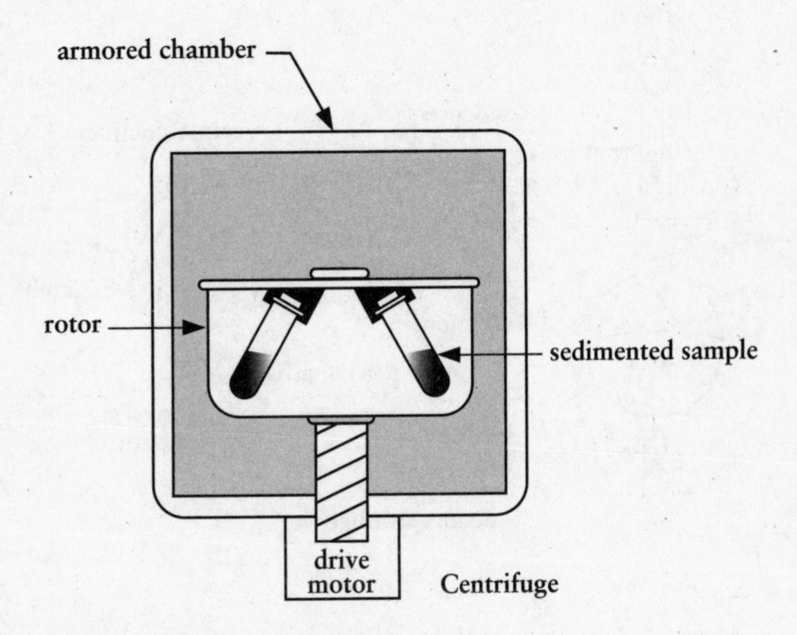

Centrifuge

Mnemonic:

<u>S</u>e<u>D</u>imentation depends on <u>S</u>ize (mass) and <u>D</u>ensity.

DISTILLATION

Distillation is the separation of one liquid from another through vaporization and condensation. A mixture of two (or more) miscible liquids is slowly heated; the compound with the lowest boiling point is preferentially vaporized, condenses on a water-cooled distillation column, and is separated from the other, higher-boiling compound(s). (Immiscible liquids can be separated in a separatory funnel and thus do not require distillation.)

A. SIMPLE

Simple distillation is used to separate liquids that boil *below* 150°C and at least 25°C apart. The apparatus consists of a distilling flask containing the two liquids, a distillation column consisting of a thermometer and a condenser, and a receiving flask to collect the distillate.

Bridge:

Boiling point is strongly affected by intermolecular forces:

- *H bonds (this is why H_2O has such a high bp);*
- *dipole-dipole interactions;*
- *dispersion (London) forces (the reason that longer molecules tend to have higher bp than shorter ones).*

B. VACUUM

Vacuum distillation is used to separate liquids that boil *above* 150°C and at least 25°C apart. The entire system is operated under reduced pressure, lowering the boiling points of the liquids and thus preventing their decomposition due to excessive temperature.

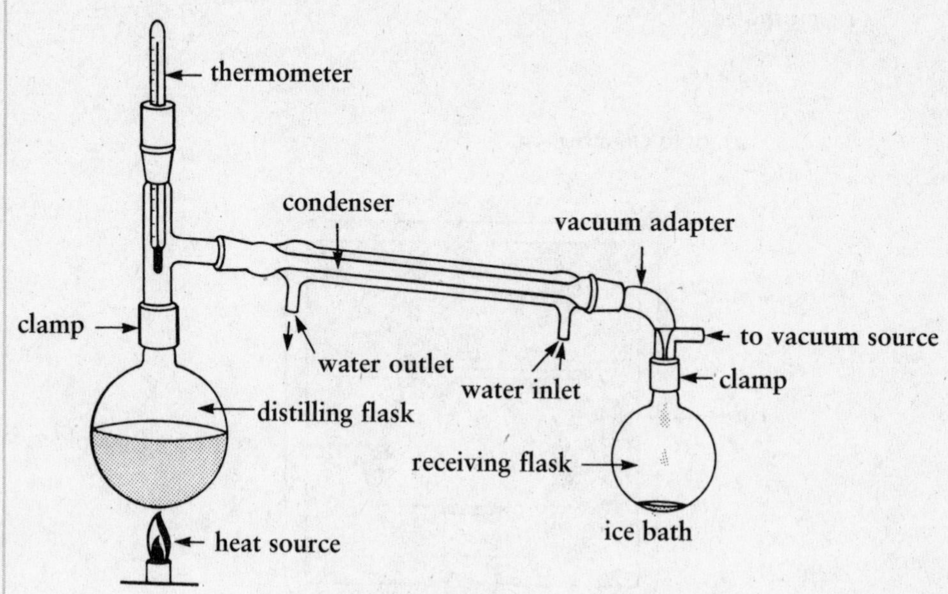

Vacuum distillation

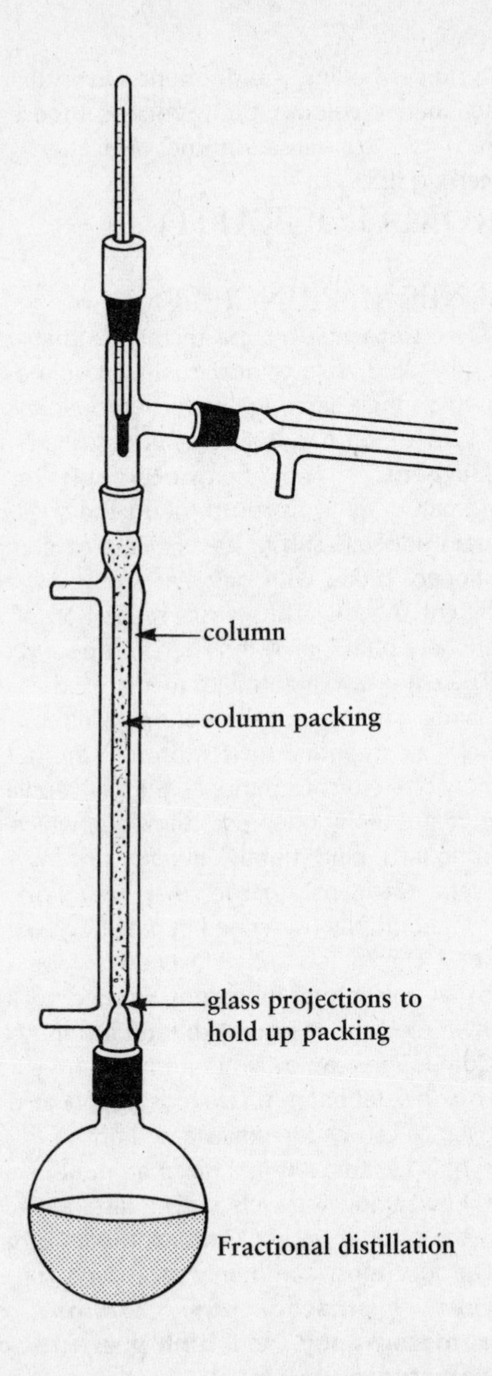

Fractional distillation

C. FRACTIONAL

Fractional distillation is used to separate liquids that boil less than 25°C apart. A fractionating column is used to connect the distilling flask to the distillation column. It is filled with inert objects, such as glass beads, which have a large surface area. The vapors condense on these surfaces, reevaporate, and then condense further up the column. Each time the liquid evaporates, the vapors contain a greater proportion of

Note:

Fractional distillation can be thought of as repeated distillation: it's like distilling the compounds over and over again.

the lower-boiling component. Eventually, near the top of the fractionating column, the vapor is composed solely of one component, which will condense on the distillation column and collect in the receiving flask.

CHROMATOGRAPHY

A. GENERAL PRINCIPLES

Chromatography is a technique that allows scientists to separate, identify, and isolate individual compounds from a complex mixture based on their differing chemical properties. First, the sample is placed, or loaded, onto a solid medium called the **stationary phase** or **adsorbant**. Then, the **mobile phase**, a liquid (or gas for gas chromatography), is run through the stationary phase, to displace (or **elute**) adhered substances. Different compounds will adhere to the stationary phase with different strengths, and therefore migrate with different speeds. This causes separation of the compounds within the stationary phase, allowing each compound to be isolated.

There are several forms of media used as the stationary phase, which separate compounds based on different chemical properties. How quickly a compound travels through the stationary phase depends on a variety of factors. Commonly, the key is polarity. For instance, thin layer chromatography often uses silica gel, which is highly polar. Thus, polar compounds bind tightly, eluting poorly into the less polar organic solvent. Size or charge may also play a role, as in column chromatography (described in detail below). Newer techniques, such as affinity chromatography, take advantage of unique properties of a substance (such as its strong binding to a specific antibody or to a known receptor or ligand) to bind it tightly to the stationary phase.

Compounds can be distinguished from each other because they travel across the stationary phase (adsorbant) at different rates. In practice, a substance can be identified based on:

- how far it travels in a given amount of time (as in TLC); or
- how rapidly it travels a given distance, e.g., how quickly it elutes off the column (as in GC or column chromatography.)

The four most commonly used types of chromatography are **thin-layer chromatography**, **column chromatography**, **gas chromatography**, and **high-pressure** (or **performance**) **liquid chromatography**.

Note:

Key idea: Chromatography separates compounds based on how strongly they adhere to the solid, or stationary, phase (or, in other words, how easily they come off into the mobile phase).

B. THIN-LAYER CHROMATOGRAPHY

The adsorbant in thin-layer chromatography (TLC) is either a piece of paper or a thin layer of silica gel or alumina on a plastic or glass sheet. The mixture to be separated is placed on the adsorbant; this is called **spotting**, because a small, well-defined spot is desirable. The TLC plate is then **developed**—placed upright in a developing chamber (usually a beaker with a lid or a wide-mouthed jar), containing **eluant** (solvent) approximately 1/4 inch deep (this value depends on the size of the plate). It is imperative that the initial spots on the plate be above the level of the solvent, or else they will simply elute off the plate into the solvent rather than moving neatly up the plate itself. The solvent creeps up the plate by capillary action, moving different compounds at different rates. When the **solvent front** nears the top of the plate, the plate is removed from the chamber and allowed to dry.

Chromatography is often done with silica gel, which is very polar and hydrophilic. The mobile phase, usually an organic solvent of weak to moderate polarity, is then used to "run" the sample through the gel. Non-polar compounds move very quickly, while polar molecules are stuck tightly to the gel. The more polar the solvent, the faster the sample will migrate. Reverse-phase chromatography is just the opposite. Here the stationary phase is very non-polar, so polar molecules run very quickly, while non-polar molecules stick more tightly.

The spots of individual compounds (usually white) are not usually visible on the white TLC plate. They are **visualized** by placing the TLC plate under UV light, which will show any compounds that are UV-sensitive (see Chapter 23, Spectroscopy); or by allowing iodine, I_2, to stain the spots. Other chemical staining agents include phosphomolybdic acid and vanillin. Note that these compounds destroy the product (usually by oxidation), so that it cannot be recovered for further study.

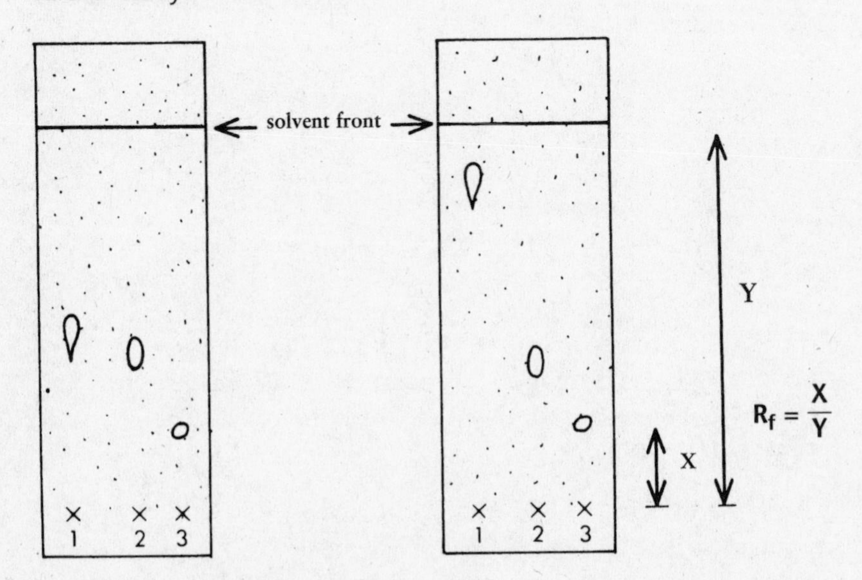

Thin-layer chromatograms

The distance a compound travels, divided by the distance the solvent travels, is called the **R_f value**. This value is relatively constant for a particular compound in a particular solvent, and can therefore be used for identification.

TLC is most frequently used for qualitative identification (i.e., determining the identity of a compound). It can also be used on a larger scale, as a means of purification. **Preparative** or **prep TLC** uses a large TLC plate upon which a sizeable streak of a mixture is placed. As the plate develops, the streak splits into bands of individual compounds, which can then be scraped off. Rinsing with a polar solvent will recover the pure compounds from the silica.

C. COLUMN CHROMATOGRAPHY

The principle behind column chromatography is the same as for TLC. Column chromatography, however, uses silica gel or alumina as an adsorbant (not paper), and this adsorbant is in the form of a column (not a layer), allowing much more separation. In TLC the solvent and compounds move up the plate (by capillary action), whereas in column chromatography they move down the column (by gravity). Sometimes the solvent is forced through the column with nitrogen gas; this is called **flash column chromatography**.

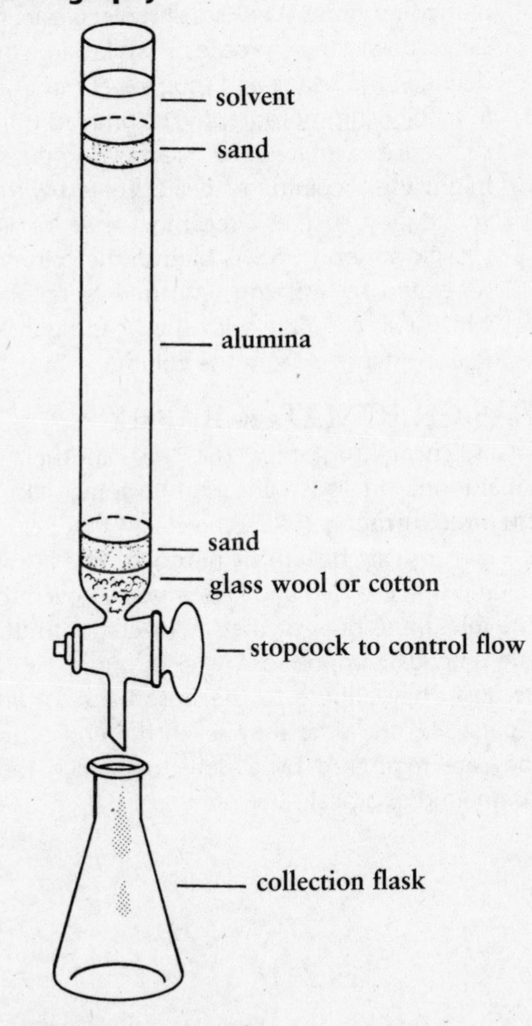

Column chromatography

The solvent drips out the end of the column and fractions are collected in flasks or test tubes. These fractions contain bands corresponding to the different compounds, and when the solvents are evaporated, the compounds can be isolated.

Column chromatography is particularly useful in biochemistry, because it can be used to separate macromolecules such as proteins or nucleic acids. Several techniques exist:

1) In ion exchange chromatography, the beads in the column are coated with charged substances, and so they will attract or bind compounds with an opposing charge. For instance, a positively-charged column will attract and hold negative substances while letting those with positive charge pass through.
2) In size-exclusion chromatography, the column contains beads with many tiny pores. Very small molecules can enter the beads, which slows down their progress, while large molecules move around/between the beads and thus travel through the column faster.
3) In affinity chromotography, columns can be "customized" to bind a substance of interest. For example, to purify substance A, a scientist might use a column of beads coated with something that binds A very tightly, such as a receptor for A, A's biological target, or even a specific antibody. A will bind to the column very tightly. It can later be eluted by washing with free receptor (or target or antibody), which will compete with the bead-bound receptor and ultimately free substance A from the column.

D. GAS CHROMATOGRAPHY

Gas chromatography (GC) is another method of qualitative separation. In gas chromatography, also called **vapor-phase chromatography** (VPC), the eluant that passes through the adsorbant is a gas, usually helium or nitrogen. The adsorbant is inside a 30-foot column that is coiled and kept inside an oven to control its temperature. The mixture to be separated is injected into the column and vaporized. The gaseous compounds travel through the column at different rates, because they adhere to the adsorbant to different degrees, and will separate by the time they reach the end of the column. At this point they are registered by a detector, which records the presence of a compound as a peak.

Note:

Again, note that there is a stationary phase (here, a 30-foot column) and a mobile phase or eluant (here, a gas).

Note:

To identify a compound or distinguish 2 different compounds, look at their "retention times"—that is, how long it took for each to travel through the column.

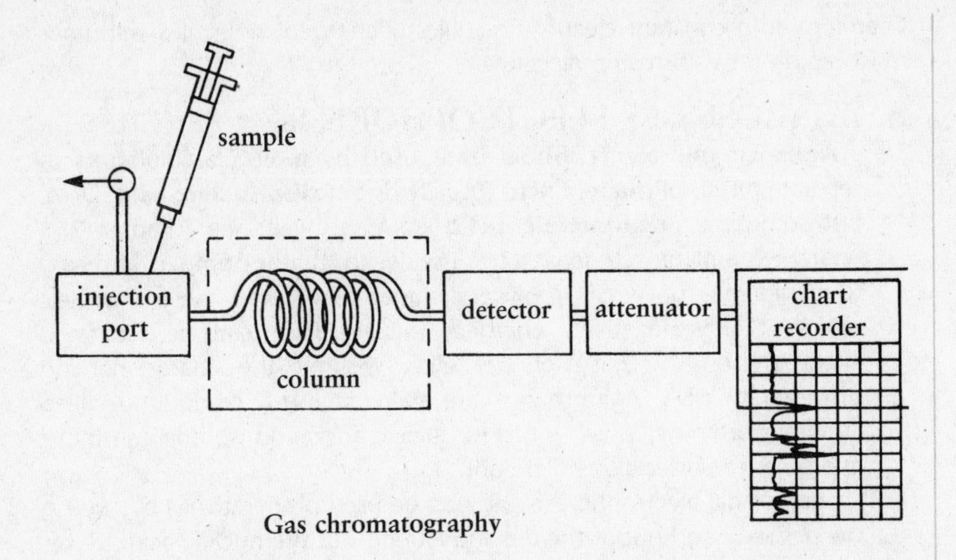

Gas chromatography

GC can be used on a larger scale for quantitative separation, and is then called preparative or prep GC. This is, however, very tedious and difficult to perform.

E. HPLC

HPLC stands for either high-pressure or high-performance liquid chromatography. The eluant is a liquid that travels through a column similar to a GC column, but under pressure. In the past, very high pressures were used; now they are much lower, hence the change from high *pressure* to high *performance*.

In HPLC, a sample is injected into the column and separation occurs as it flows through. The compounds pass through a detector and are collected as the solvent flows out the end of the apparatus. The eluant may vary, as in thin-layer or column chromatography.

Mnemonic:

In electrophoresis:
Anions are attracted to the Anode
while
Cations are attracted to the Cathode.

Note:

In most forms of electrophoresis, the size of a macromolecule is usually the most important factor—small molecules move faster, while large ones move more slowly and may in fact take hours to leave the well.

ELECTROPHORESIS

When a molecule is placed in an electric field, it will move towards either the cathode or the anode depending on its size and charge. **Electrophoresis** employs this phenomenon to separate macromolecules (usually biological macromolecules) such as proteins or DNA. The migration velocity, v, of a molecule is directly proportional to the electric field strength, E, and to the net charge on the molecule, z, and is inversely proportional to a frictional coefficient, f, which depends on the mass and shape of the migrating molecules.

$$v = \frac{Ez}{f}$$

Real World Example:

To make your gel run faster:
- *increase the voltage (↑ E)*
- *use a lower percent of agarose or acrylamide (↓ f)*

Note:

SDS–PAGE and agarose gel electrophoresis separate molecules based on size.

Note:

When pH = pI, a protein stops moving.

Note:

Isoelectric focusing separates proteins based on charge.

Therefore, in a constant electric field, highly charged molecules will move most rapidly, as will small molecules.

A. AGAROSE GEL ELECTROPHORESIS

Agarose gel electrophoresis is used by molecular biologists to separate pieces of **nucleic acid** (usually **deoxyribonucleic acid**, DNA, but sometimes **ribonucleic acid**, RNA, as well; see Chapter 25). Agarose is a plant gel, derived from seaweed, that is nontoxic and easy to manipulate (unlike SDS/polyacrylamide). Since every piece of nucleic acid is highly negatively charged, nucleic acids can be separated effectively on the basis of size even without the charge-masking provided by SDS. Agarose gels are stained with a compound called ethidium bromide, which binds to nucleic acids and is visualized by its fluorescence under ultraviolet light.

Agarose gel electrophoresis can also be used preparatively, by cutting the desired band out of the gel and eluting out the nucleic acid.

B. SDS-POLYACRYLAMIDE GEL ELECTROPHORESIS

SDS-polyacrylamide gel electrophoresis separates proteins on the basis of mass, not charge. Polyacrylamide gel is the standard medium for electrophoresis. SDS is sodium dodecyl sulfate, which disrupts noncovalent interactions. It binds to proteins and creates large negative net charges, neutralizing the protein's original net charge. As proteins move through the gel, the only variable affecting their velocity is f, the frictional coefficient, which is dependent on mass. After separation, the gel is stained so that the protein bands can be visualized.

C. ISOELECTRIC FOCUSING

A protein may be characterized by its **isoelectric point** (pI), which is the pH at which its net charge (the sum of the charges on all of its component amino acids; see Chapter 25) is zero. If a mixture of proteins is placed in an electric field in a gel with a pH gradient, the proteins will move until they reach the point at which the pH is equal to their pI. At this location, the protein will be uncharged and will no longer move in the field. Molecules differing by as little as one charge can be separated in this manner, which is called **isoelectric focusing**.

SUMMARY OF PURIFICATION METHODS

Method	Use
Extraction	separates dissolved substances based on differential solubility in aqueous vs. organic solvents
Filtration	separates solids from liquids
Recrystallization	separates solids based on differential solubility; temperature is important here
Sublimation	separates solids based on their ability to sublime
Centrifugation	separates large things (like cells, organelles and macromolecules) based on mass and density
Distillation	separates liquids based on boiling point, which in turn depends on intermolecular forces
Chromatography	uses a stationary phase and a mobile phase to separate compounds based on how tightly they adhere (generally due to polarity but sometimes size as well)
Electrophoresis	used to separate biological macromolecules (such as proteins or nucleic acids) based on size and sometimes charge

Bridge:

Since amino acids and proteins are organic molecules, the fundamental principles of acid-base chemistry apply to them as well.

- *At a low pH, [H+] is relatively high. Thus, at a pH < pI proteins will tend to be protonated and, as a result, positively charged.*
- *At a relatively high (basic) pH, [H+] is fairly low and proteins will tend to be de-pronated—thus carrying a negative charge.*

REVIEW PROBLEMS

1. A mixture of sand, benzoic acid, and naphthalene in ether is best separated by

- **A.** filtration, followed by acidic extraction, followed by recrystallization.
- **B.** filtration, followed by basic extraction, followed by evaporation.
- **C.** extraction, followed by sublimation, followed by GC.
- **D.** filtration, followed by electrophoresis, followed by extraction.

2. Fractional distillation would most likely be used to separate which of the following compounds?

- **A.** methylene chloride (bp 41°C) and water (bp 100°C)
- **B.** ethyl acetate (bp 77°C) and ethanol (bp 80°C)
- **C.** aniline (bp 184°C) and benzyl alcohol (bp 22°C)
- **D.** aniline (bp 184°C) and water (bp 100°C)

3. Which of the following compounds would be the most effective in extracting benzoic acid from a diethyl ether solution?

- **A.** Tetrahydrofuran
- **B.** Aqueous hydrochloric acid
- **C.** Aqueous sodium hydroxide
- **D.** Water

4. Which of the following techniques would best separate red blood cells from blood plasma?

- **A.** gel electrophoresis
- **B.** centrifugation
- **C.** isoelectric focusing
- **D.** HPLC

Questions 5, 6, and 7 refer to the following table:

Compound	Distance Travelled
benzyl alcohol	1.0 cm
benzyl acetate	2.6 cm
p-nitrophenol	2.3 cm
naphthalene	4.0 cm

5. Calculate the R_f values for the compounds listed above, which were developed on silica gel with ether as the eluant. Assume that the solvent front travelled a distance of 10 cm. Rank the compounds in decreasing order of polarity.

6. What would be the effect on the R_f values if the TLC described above were run with hexane rather than ether as the eluant?

A. No effect
B. Increase tenfold
C. Double
D. Decrease

7. If benzyl alcohol, benzyl acetate, *p*-nitrophenol, and naphthalene were separated by column chromatography with ether on silica gel, which compound would elute first?

A. Benzyl alcohol
B. Benzyl acetate
C. *p*-Nitrophenol
D. Naphthalene

8. Which of the following would be the best procedure for extracting acetaldehyde from an aqueous solution?

A. A single extraction with 100 mL of ether
B. Two successive extractions with 50 mL portions of ether
C. Three successive extractions with 33.3 mL portions of ether
D. Four successive extractions with 25 mL portions of ether

Questions 9 and 10 refer to the following table.

	pI	*MW*
Protein A	4.5	25,000
Protein B	6.0	10,000
Protein C	9.5	12,000

9. At what pH can the above proteins be separated by isoelectric focusing?

A. 4.5
B. 6.0
C. 9.5
D. 7.0

10. The proteins were treated with SDS and separated by gel electrophoresis. Predict the order of migration towards the anode.

SOLUTIONS TO REVIEW PROBLEMS

1. **B** In this question, three substances must be separated using a combination of techniques. The first step should be the most obvious: removal of the sand by filtration. Sand is an insoluble impurity and cannot be extracted. Thus, choice C is wrong. After filtration, the remaining compounds are still dissolved in ether. If the solution is extracted with aqueous base, the benzoate anion is formed and becomes dissolved in the aqueous layer, while naphthalene, a nonpolar compound, remains in the ether. The compounds are now separated, and evaporation of the ether will yield purified naphthalene. (The benzoic acid can be isolated by acidifying the aqueous layer and filtering the precipitate.) This combination of techniques is given in choice B, and this is the correct answer. Choice A is wrong because the benzoic acid will not dissociate in an acidic solution, and so will not become water-soluble (on the other hand, if the solvent were basic, it *would* separate benzoic acid). Choice D is incorrect because electrophoresis will not separate the two organic compounds, since they have the same charge.

2. **B** Fractional distillation is the most effective procedure for separating two liquids that boil within a few degrees of each other. Ethyl acetate and ethanol, choice B, boil within three degrees of each other and thus would be good candidates for fractional distillation. In the other three answer choices, the compounds have widely separated boiling points. Fractional distillation could be used, but is not needed; simple distillation or vacuum distillation, which are easier to do, would be effective.

3. **C** By extracting with sodium hydroxide, benzoic acid will be converted to its sodium salt, sodium benzoate. Sodium benzoate, unlike its acid counterpart, will dissolve in an aqueous solution. The aqueous layer simply has to be acidified in order to retrieve benzoic acid. Choice A is wrong because diethyl ether and tetrahydrofuran are miscible; this answer choice can be discarded. Hydrochloric acid will not transform benzoic acid into a soluble salt, so choice B is incorrect. Finally, choice D is wrong because benzoic acid is insoluble in water. Again, choice C is the correct answer.

4. **B** Red blood cells suspended in blood plasma are best separated by centrifugation. In this technique, the blood cells would be forced to the bottom of the tube which could then be separated from the plasma. Because this is essentially a phase difference, a solid and a liquid, electrophoresis does not have to be employed, so choices A and C can be eliminated. Choice D is wrong because HPLC is used to separate smaller organic compounds.

5. The R_f values for the compounds are as follows: benzyl alcohol = 0.1; benzyl acetate = 0.26; *p*-nitrophenol = 0.23; naphthalene = 0.4. The more polar compounds have smaller R_f values, because they adhere to the adsorbant more strongly. Therefore, the compounds in decreasing order of polarity are: benzyl alcohol (polar), *p*-nitrophenol, benzyl acetate, naphthalene (nonpolar).

6. **D** Hexane is less polar than ether, and therefore is less likely to displace compounds adsorbed to the silica gel. This would decrease the distance the compounds would travel, decreasing the R_f values.

7. **D** In column chromatography, as in TLC, the least polar compound travels most rapidly. This means that naphthalene, with an R_f value of 0.4, would travel most rapidly and would be the first to elute from the column.

8. **D** It is more effective to perform four successive extractions with small amounts of ether than to perform one extraction with a large amount of ether.

9. **B** The three proteins may be separated when the isoelectric point of one of these proteins is equal to the pH of the solution. In this problem, protein B has the intermediate pI (6.0), and so the proteins can best be separated when the pH of the solution equals the pI of B, or 6.0. Since protein A is neutral at pH = 4.5, it becomes basic (negatively charged) at pH = 6. Thus, at this pH, protein A migrates to the anode. On the other hand, protein C is neutral at pH = 9.5. Therefore, at pH = 6 it becomes acidic (positively charged) and migrates to the cathode. Finally, protein B remains in the solution because

at pH equal to its isoelectric point its molecules are uncharged and therefore are not affected by the potential difference. Thus, the three proteins are separated.

10. The order of migration in the SDS-polyacrylamide gel electrophoresis is inversely proportional to the molecular weight. From the molecular weights given, it is clear that the order of the increasing migration rate is A, C, B.

Spectroscopy

TWENTY THREE

Once an organic compound is isolated, it must be characterized and identified. If it is a known compound, identification can often be made from elemental analysis or determination of the melting point. With new or more complex compounds, other methods must be used. **Spectroscopy** is the process of measuring the energy differences between the possible states of a molecular system by determining the frequencies of electromagnetic radiation (light) absorbed by the molecules. The possible states are quantized energy levels associated with different types of molecular motion, including molecular rotation, vibration of bonds, and electron movement. Different types of spectroscopy measure these different types of molecular motion, identifying specific functional groups and how they are connected.

Spectroscopy is useful because only a very small quantity of sample is needed. In addition, the sample may be reused after an IR, NMR, or UV spectrum is obtained.

INFRARED

A. BASIC THEORY

Infrared (IR) spectroscopy measures molecular vibrations, which include bond **stretching**, **bending**, and **rotation**. The useful absorptions of infrared light occur in the 3,000–30,000 nm region, which corresponds to 3,500–300 cm^{-1} (called **wavenumbers**). When light of these wavelengths/wavenumbers is absorbed, the molecules enter higher (excited) vibrational states.

Bond stretching (which can be of two types: symmetric or asymmetric) involves the largest change in energy and is observed in the region 1,500–4,000 cm^{-1}. Bending vibrations are observed in the region 400–1500 cm^{-1}. Four different types of vibration that can occur are shown below.

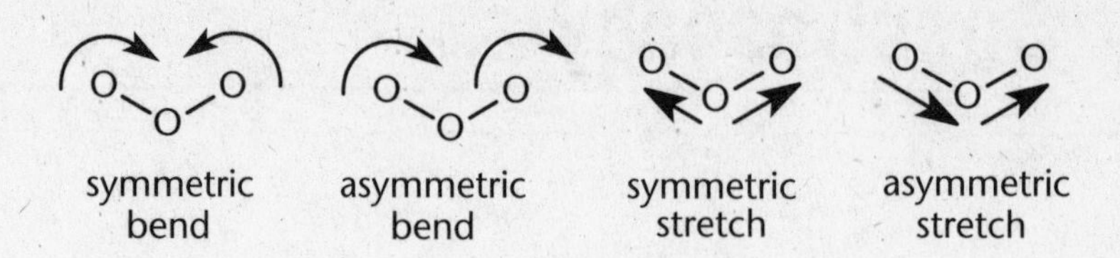

In addition to bending and stretching vibrations, more complex vibrations may occur. These can be combinations of bending, stretching, and rotation frequencies or complex frequency patterns caused by the motion of the whole molecule. Absorptions of these types are seen in the region 1,500–400 cm^{-1}. This region of the spectrum is known as the **fingerprint region** and is characteristic of a molecule; it is, therefore, frequently used to identify a substance.

Note:

Symmetric stretches do not show up in IR spectra since they involve no net change in dipole moment.

In order for an absorption to be recorded, the motion must result in a change in a bond dipole moment. Molecules comprised of atoms with the same electronegativity, as well as symmetrical molecules, do not experience a changing dipole moment and therefore do not exhibit absorption. For example, O_2 and Br_2 do not absorb, but HCl and CO do.

A typical spectrum is obtained by passing infrared light (of frequencies from approximately 4,000–400 cm^{-1}) through a sample, and recording the absorption pattern. Percent transmittance is plotted versus frequency, where percent transmittance = absorption^{-1} (%T = A^{-1}); absorptions appear as valleys on the spectrum.

Note:

Wave numbers (cm^{-1}) are not the same as frequency.

$\nu = \frac{c}{\lambda}$, *while wave number* $= \frac{1}{\lambda}$

B. Characteristic Absorptions

Particular functional groups absorb at localized frequencies. For example, alcohols absorb around 3,300 cm^{-1}, carbonyl groups around 1,700 cm^{-1}, and ethers around 1,100 cm^{-1}. The table below lists the specific absorptions of key functional groups and the vibrations with which they are associated.

Common Infrared Absorption Peaks

Functional Group	Frequency (cm^{-1})	Vibration
Alkanes	2800 — 3000	C—H
	1200	C—C
Alkenes	3080 — 3140	=C—H
	1645	C=C
Alkynes	2200	C≡C
	3300	≡C—H
Aromatic	2900 — 3100	C—H
	1475 — 1625	C—C
Alcohols	3100 — 3500	O—H (broad)
Ethers	1050 — 1150	C—O
Aldehydes	2700 — 2900	(O)C—H
	1725 — 1750	C=O
Ketones	1700 — 1750	C=O
Acids	1700 — 1750	C=O
	2900 — 3300	O—H (broad)
Amines	3100 — 3500	N—H (sharp)

In a Nutshell:

IR spectroscopy is best used for identification of functional groups. The most important peaks to know are those for alcohols (don't forget this is a BROAD peak), acids (BROADEST peak), ketones, and amines (sharp peak). If you know nothing else here, know these!

C. APPLICATION

A great deal of information can be obtained from an IR spectrum. Most of the useful functional group information is found between 1,400 and 4,000 cm^{-1}.

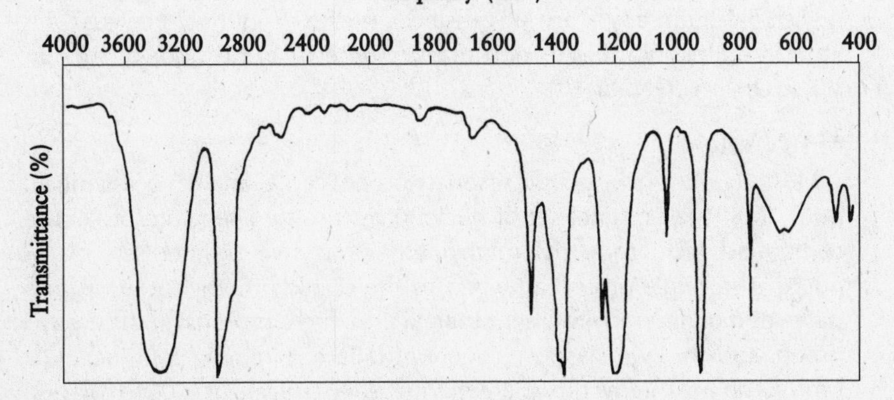

The figure above shows the IR spectrum of an aliphatic alcohol. The large peak at 3,300 cm^{-1} is due to the presence of the hydroxyl group, while the peak at 3,000 cm^{-1} can be attributed to the alkane portion of the molecule.

NUCLEAR MAGNETIC RESONANCE

A. BASIC THEORY

Nuclear Magnetic Resonance (**NMR**) spectroscopy is one of the most widely used spectroscopic tools in organic chemistry. NMR is based on the fact that certain nuclei have magnetic moments which are normally oriented at random. When such nuclei are placed in a magnetic field, their magnetic moments tend to align either with or against the direction of this applied field. Nuclei whose magnetic moments are aligned with the field are said to be in the α **state** (lower energy), while those whose moments are aligned against the field are said to be in the β **state** (higher energy). If the nuclei are then irradiated with electromagnetic radiation, some will be excited into the β state.The absorption corresponding to this excitation occurs at different frequencies depending on an atom's environment. The nuclear magnetic moments are affected by other nearby atoms that also possess magnetic moments. Hence, a compound may contain many nuclei that resonate at different frequencies, producing a very complex spectrum.

A typical NMR spectrum is a plot of frequency versus absorption of energy during resonance. Frequency *decreases* toward the right. Alternatively, varying magnetic field may be plotted on the x axis, *increasing* toward the right. Because different NMR spectrometers operate at different magnetic field strengths, a standardized method of plotting the NMR spectrum has been adopted. An arbitrary variable, called **chemical shift** (represented by the symbol δ), with units of **parts per million (ppm)** of spectrometer frequency, is plotted on the x axis.

NMR is most commonly used to study ^{1}H nuclei (protons) and ^{13}C nuclei, although any atom possessing a nuclear spin (any nucleus with an odd atomic number or odd mass number) can be studied, such as ^{19}F, ^{17}O, ^{14}N, ^{15}N, or ^{31}P.

B. ^{1}H NMR

Most ^{1}H nuclei come into resonance between 0 and 10 δ downfield from TMS. Each distinct set of nuclei gives rise to a separate peak. The compound dichloromethyl methyl ether has two distinct sets of ^{1}H nuclei. The single proton attached to the dichloromethyl group is in a different magnetic environment than are the three protons on the methyl group, and the two classes resonate at different frequencies. The three protons on the methyl group are magnetically equivalent, due to rotation about the oxygen-carbon single bond, and resonate at the same frequency. Thus, two separate peaks are expected, as shown below.

Clinical Correlate:

Magnetic Resonance Imaging (MRI) is a noninvasive medical diagnostic tool which utilizes proton NMR. Multiple cross-sectional scans of the patient's body are taken, and the various chemical shifts of absorbing protons are translated into specific colors. A picture is thus produced which shows the relative density of specific types of protons; for instance, a dark blue area on the MRI output screen indicates a high concentration of the types of protons giving rise to the range of resonances which correlate to dark blue. Comparison to normal MRI then allows the diagnostician to detect abnormalities in the scanned region.

Bridge:

Nuclei are positively charged spheres. Rotation of a charged sphere produces a magnetic moment.

Note:

TMS provides a reference peak. The signal for its H atoms is assigned a $\delta = 0$.

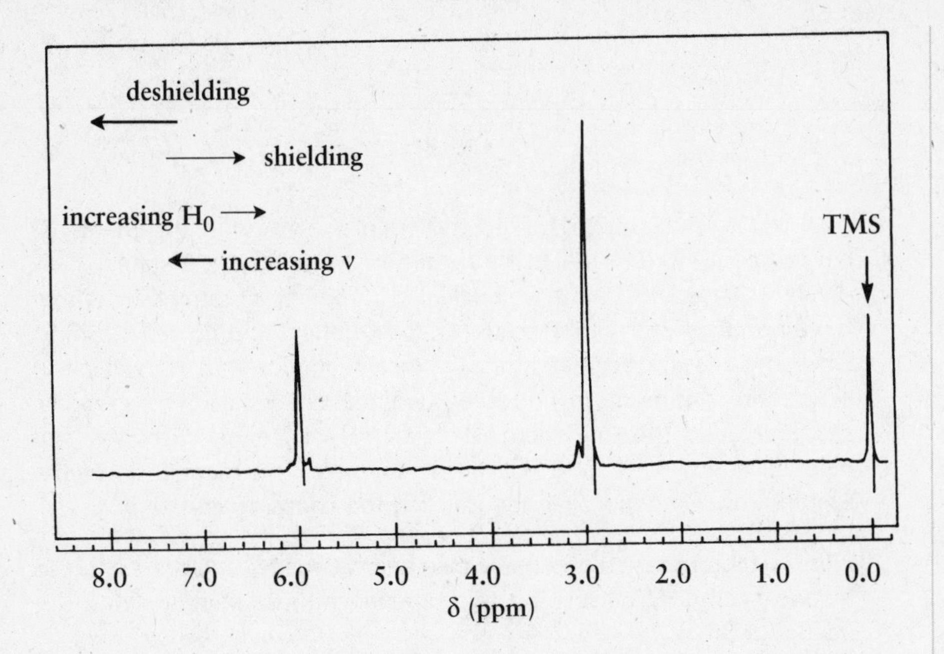

The left-hand peak corresponds to the single dichloromethyl proton and the middle peak to the three methyl protons (the one on the far right is the TMS reference peak). Notice that if the areas under the peaks are integrated, the ratio between them is 3:1, corresponding to the number of protons producing each peak.

The single proton comes into resonance downfield from the methyl protons. This phenomenon is due to the electron-withdrawing effect of the chlorine atoms. The electron cloud that surrounds the ^1H nucleus ordinarily screens the nucleus somewhat from the applied magnetic field. The chlorine atoms pull away the electron cloud and **deshield** the nucleus. Thus, the nucleus resonates in a lower field than it would otherwise. By the same rationale, electron-donating atoms, such as the silicon atoms in TMS, **shield** the ^1H nuclei, causing them to come into resonance at a higher field.

If two magnetically different protons are within three bonds of each other, a phenomenon known as **coupling**, or **splitting** occurs. Consider two protons, H_a and H_b, on the molecule 1,1-dibromo-2,2-dichloroethane (see below).

In A Nutshell:

- *Each peak represents a single proton or a group of equivalent protons. (The number of peaks represents the number of groups of non-equivalent protons.)*
- *The relative area of each peak reflects the ratio of the protons producing each peak.*
- *The position of the peak (upfield or downfield) due to shielding or deshielding effects reflects the chemical environment of the protons.*

Mnemonic:

Downfield is Deshielded.

$$\mathrm{H_a{-}C(Cl)_2{-}C(Br)_2{-}H_b}$$

Note:

The splitting of the peak represents the number of adjacent hydrogens. A peak will be split into n + 1 peaks where n is the number of adjacent hydrogens.

At any given time, H_a can experience two different magnetic environments, since H_b can be in either the α or the β state. These different states of H_b influence nucleus H_a (if the two H atoms are within three bonds of each other), causing slight upfield and downfield shifts. Since there is approximately a 50 percent chance that H_b will be in either state, this results in a **doublet**, two peaks of equal intensity equally spaced around the true chemical shift of H_a. H_b experiences the two different states of H_a and is likewise coupled. The magnitude of the splitting, usually denoted in Hz, is called the **coupling constant, J.**

In 1,1-dibromo-2-chloroethane (see below), the H_a nucleus is affected by two nearby H_b nuclei, and can experience four different states: $\alpha\alpha$, $\alpha\beta$, $\beta\alpha$, or $\beta\beta$.

$$\mathrm{H_b{-}C(Cl)(H_b){-}C(Br)_2{-}H_a}$$

The $\alpha\beta$ and $\beta\alpha$ states have the same net effect on the H_a nucleus, and the resonances occur at the same frequency. The $\alpha\alpha$ and $\beta\beta$ states resonate at frequencies different from each other and from the $\alpha\beta/\beta\alpha$ frequency. The result is three peaks centered around the true chemical shift, with an area ratio of 1:2:1. In general, n hydrogen atoms couple to give n + 1 peaks, whose area ratios are given by Pascal's triangle, shown in the table below.

Pascal's Triangle

Number of Adjacent Hydrogens	Total Number of Peaks	Area Ratios
0	1	1
1	2	1:1
2	3	1:2:1
3	4	1:3:3:1
4	5	1:4:6:4:1
5	6	1:5:10:10:5:1
6	7	1:6:15:20:15:6:1
7	8	1:7:21:35:35:21:7:1

The following table indicates the chemical shift ranges of several different types of protons:

Chemical Shifts

Type of Proton	Approximate Chemical Shift δ (ppm) Downfield from TMS
RCH_3	0.9
RCH_2	1.25
R_3CH	1.5
–CH=CH	4.6–6
–C≡CH	2–3
Ar–H	6–8.5
–CHX	2–4.5
–CHOH/–CHOR	3.4–4
RCHO	9–10
RCHCO–	2–2.5
–CHCOOH/–CHCOOR	2–2.6
–CHOH–CH_2OH	1–5.5
ArOH	4–12
–COOH	10.5–12
–NH_2	1–5

C. ^{13}C NMR

^{13}C NMR is very similar to ^{1}H NMR. Most ^{13}C NMR signals, however, occur 0–200 δ downfield from the carbon peak of TMS. Another significant difference is that only 1.1 percent of carbon atoms are ^{13}C atoms. This has two effects: first, a much larger sample is needed to run a ^{13}C spectrum (about 50 mg compared with 1 mg for ^{1}H NMR), and second, coupling between carbon atoms is generally not observed.

Coupling *is* observed, however, between carbon atoms and the protons directly attached to them. This one-bond coupling is analogous to the three-bond coupling in ^{1}H NMR. For example, if a carbon atom is attached to two protons, it can experience four different states of those protons (αα, αβ, βα, and ββ), and the carbon signal is split into a triplet with the area ratio 1:2:1.

An additional feature of ^{13}C NMR is the ability to record a spectrum *without* the coupling of adjacent protons. This is called **spin decoupling**, and produces a spectrum of **singlets**, each corresponding to a separate, magnetically equivalent, carbon atom. For example, compare the following spectra of 1,1,2-trichloropropane. One is a typical **spin-decoupled spectrum**, and the other (see below) is spin-coupled.

In a Nutshell:

Proton NMR is good for

1. *Determining the relative number of protons and their relative chemical environments.*
2. *Showing how many adjacent protons there are by splitting patterns.*
3. *Showing certain functional groups.*

Spin-Decoupled Spectrum of 1,1,2-Trichloropropane

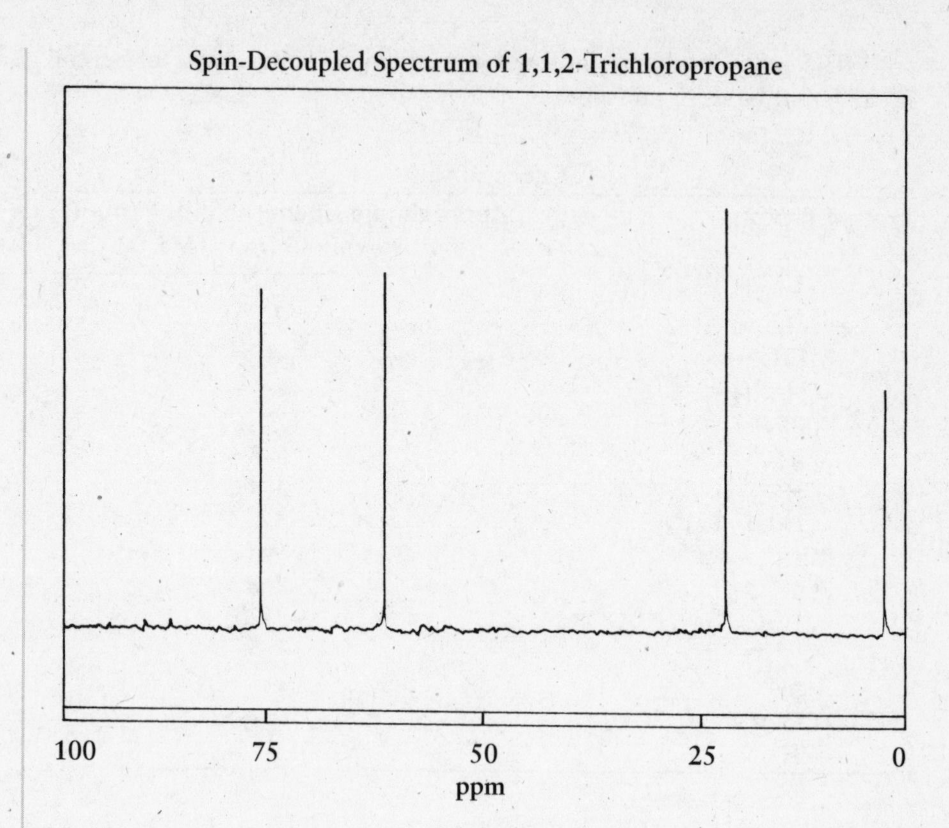

In a Nutshell:

Carbon NMR can show

1. *The number of different carbons with their relative chemical environments.*
2. *Their number of hydrogens (spin-coupled NMR only).*

Spin-Coupled Spectrum of 1,1,2-Trichloropropane

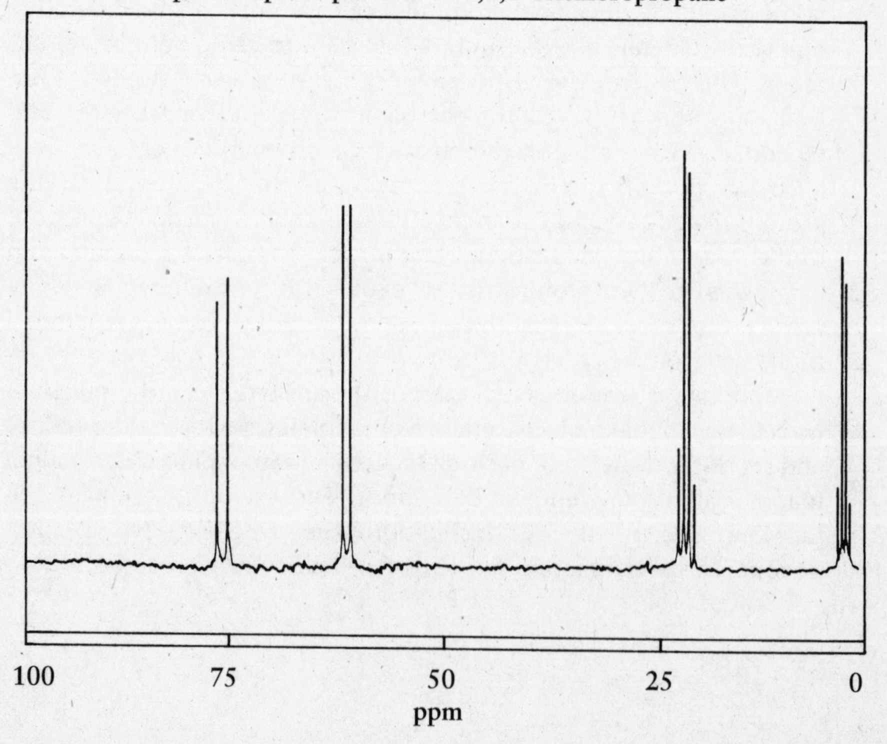

In general, NMR spectroscopy provides information about the carbon skeleton of a compound, along with some suggestion of its functional groups. Specifically, NMR can provide the following types of information:

1. The number of nonequivalent nuclei, determined from the number of peaks.
2. The magnetic environment of a nucleus, determined by the chemical shift.
3. The relative numbers of nuclei, determined by integrating the peak areas.
4. The number of neighboring nuclei, determined by the splitting pattern observed (except for ^{13}C in the spin-decoupled mode).

ULTRAVIOLET SPECTROSCOPY

A. BASIC THEORY

Ultraviolet spectra are obtained by passing ultraviolet light through a chemical sample (usually dissolved in an inert, non-absorbing solvent) and plotting absorbance versus wavelength. The wavelength of maximum absorbance provides information on the extent of the conjugated system as well as other structural and compositional information.

Flashback:

Bonding and antibonding orbitals were discussed in Chapter 13.

In a Nutshell:

UV spectroscopy is most useful for studying compounds containing double bonds, and/or hetero atoms with lone pairs.

MASS SPECTROMETRY

A. BASIC THEORY

Mass spectrometry differs from the methods thus far discussed in that it is not true spectroscopy, i.e., no absorption of electromagnetic radiation is involved, and in that it is a destructive technique—mass spectrometry, does not allow for reuse of the sample once the analysis is complete. Most commonly used mass spectrometers utilize a high speed beam of electrons to ionize the sample to be analyzed, a particle accelerator to put the charged particles in flight, a magnetic field to deflect the accelerated cationic fragments, and a detector which records the number of particles of each mass exiting the deflector area. The initially formed ion is the molecular cation-radical (M^+) resulting from a single electron being removed from a molecule of the sample. This unstable species usually decomposes rapidly into a cationic fragment and a radical fragment. Since there are many molecules in the sample and (usually) more than one way for the initially-formed cation-radical to decompose into fragments, a typical mass spectrum is composed of many lines, each corresponding to a specific mass/charge ratio (m/e). The spectrum itself plots mass/charge on the horizontal axis and relative abundance of the various cationic fragments on the vertical axis.

Note:

UV spectroscopy can be applied quantitatively by using Beer's law:

$$A = \varepsilon bc$$

where A is absorbance (measured by UV spectroscopy), ε is a constant for the substance at a given wavelength, b is path length (usually equal to one), and c is concentration.

Note:

The initial ion formed in a mass spectrometry is a cation radical, which breaks up into cations and radicals. Only cations are deflected by the magnetic field and thus only cations show up on the spectrum.

B. CHARACTERISTICS

The tallest peak, belonging to the most common ion, is called the **base peak**, and is assigned the relative abundance value of 100 percent. The peak with the highest m/e ratio (see figure below) is generally the **molecular ion peak (parent ion peak)**, M^+, from which the molecular weight, M, can be obtained. The charge value is usually 1; hence the m/e ratio can usually be read as the mass of the fragment.

C. APPLICATION

Fragmentation patterns often provide information that helps identify or distinguish certain compounds. In particular, the fragmentation pattern provides clues to the compound's structure. For example, while IR spectroscopy would be of little use in distinguishing between propionaldehyde and butyraldehyde, a mass spectrum would allow unambiguous identification.

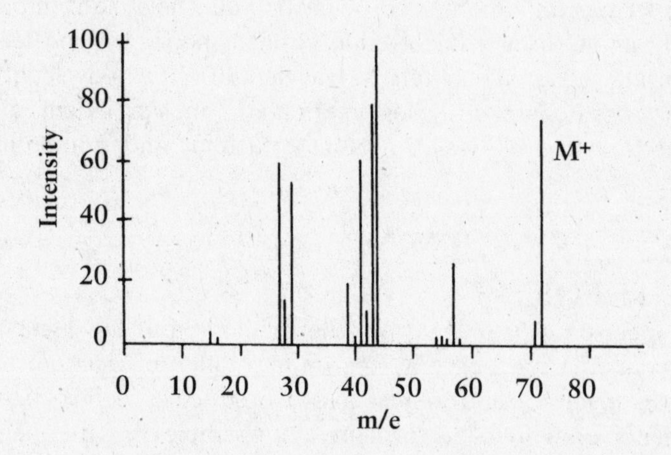

The figure above shows the mass spectrum of butyraldehyde. The peak at m/e = 72 corresponds to the molecular cation-radical, M^+, while the base peak at m/e = 44 corresponds to the cationic fragment resulting from the loss of a C_2H_4 neutral fragment (M – 28 = 44). Other peaks of note include those at 57(M – 15, loss of CH_3 radical), 43(M – 29, loss of C_2H_5 radical), and at 29 (M – 43, loss of C_3H_7 radical). The small peak at m/e = 15 can be attributed to the unstabled (and therefore not abundant) methyl cation.

REVIEW PROBLEMS

1. IR spectroscopy is most useful for distinguishing

   A. double and triple bonds.
   B. C–H bonds.
   C. chirality of molecules.
   D. composition of racemic mixtures.

2. Oxygen (O_2) does not exhibit an IR spectrum because

   A. it has no molecular motions.
   B. it is not possible to record IR spectra of a gaseous molecule.
   C. molecular vibrations do not result in a change in the dipole moment of the molecules.
   D. None of the above.

3. If IR spectroscopy were employed to monitor the oxidation of benzyl alcohol to benzaldehyde, which of the following would provide the best evidence that the reaction was proceeding as planned?

   A. Comparing the fingerprint region of the spectra of starting material and product.
   B. Noting the change in intensity of the peaks corresponding to the phenyl ring.
   C. Noting the appearance of a broad absorption peak in the region of 3100–3500 cm^{-1}.
   D. Noting the appearance of a strong absorption in the region of 1700 cm^{-1}.

4. Which of the following chemical shifts would correspond to an aldehyde proton signal in a ^{1}H NMR spectrum?

   A. 9.5 ppm
   B. 7.0 ppm
   C. 11.0 ppm
   D. 1.0 ppm

5. The isotope ^{12}C is not useful for NMR because

   A. it is not abundant in nature.
   B. its resonances are not sensitive to the presence of neighboring atoms.
   C. it has no magnetic moment.
   D. the signal-to-noise ratio in the spectrum is too low.

6. The NMR spectrums of ethanol in the presence and absence of water are shown below. Explain the additional lines that appear in the spectrum of anhydrous CH_3CH_2OH.

NMR spectrum of ethanol

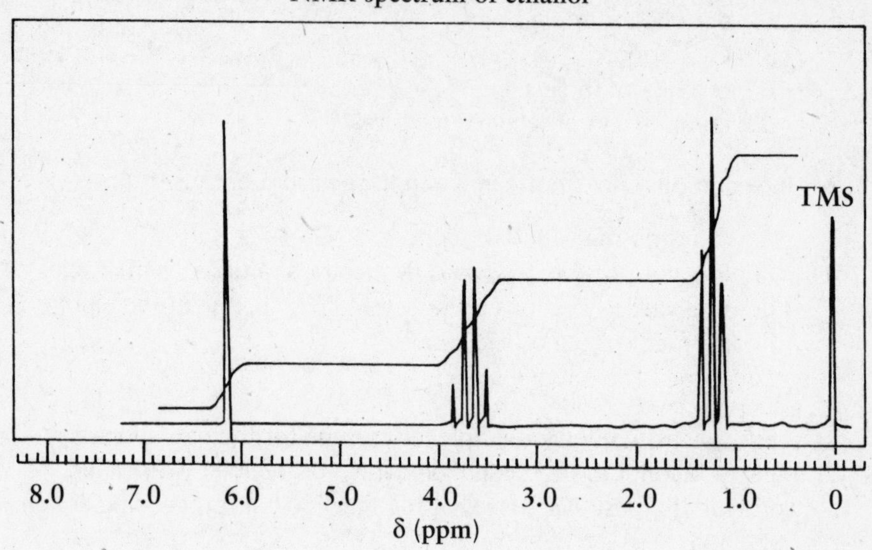

NMR spectrum of anhydrous ethanol

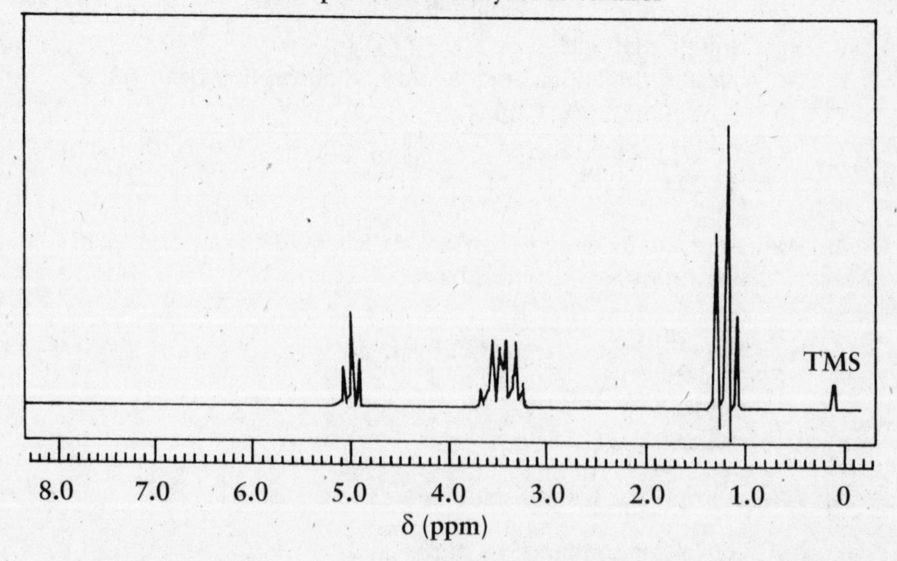

7. In ^{13}C NMR, splitting of spectral lines is due to

   A. coupling between a carbon atom and protons attached to that carbon atom.
   B. coupling between a carbon atom and protons attached to adjacent carbon atoms.
   C. coupling between adjacent carbon atoms.
   D. coupling between two adjacent protons.

8. In the ^{13}C NMR spectrum of methyl phenyl ketone, which carbon atom will appear the farthest downfield?

9. UV spectroscopy is most useful for detecting

   A. aldehydes and ketones
   B. unconjugated alkenes
   C. conjugated alkenes
   D. aliphatic acids and amines

10. Mass spectroscopy results in the separation of fragments according to

    A. atomic mass
    B. mass-to-charge ratio
    C. viscosity
    D. absorption wavelength

SOLUTIONS TO REVIEW PROBLEMS

1. **A** IR is most useful for distinguishing between different functional groups. Almost all organic compounds have C–H bonds, so that except for "fingerprinting" a compound, these absorptions are not useful. Very little information about the optical properties of a compound, such as choices C and D, can be obtained by IR spectroscopy.

2. **C** Since molecular oxygen is homonuclear and diatomic, there is no net change in its dipole moment during vibration or rotation; in other words, the compound does not absorb in the infrared. Diatomic nitrogen and chlorine exhibit similar behavior. IR spectroscopy is based on the principle that when the molecule vibrates or rotates, there is a change in dipole moment; therefore, choice C is the correct answer. Choice A is wrong because oxygen does have molecular motions. Choice B is wrong because it is possible to record the IR of a gaseous molecule, as long as it shows a change in its dipole moment when it vibrates.

3. **D** In this reaction, the functional group is changing from a hydroxyl to an aldehyde. This means that a sharp stretching peak will appear at around 1700 cm^{-1}, which corresponds to the carbonyl functionality. Therefore, choice D is the correct response. choice C is wrong because the reaction will be characterized by the disappearance of a peak at 3100–3500 cm^{-1}, not the appearance of one (this peak corresponds to the hydroxyl functionality). Choice A is certainly useful, but is not as good a method as choice D. Choice B is the least useful, as it is the C = O and –OH stretches that need to be considered.

4. **A** The peak at 9.5 ppm corresponds to an aldehyde proton. This signal lies downfield because the carbonyl oxygen is electron withdrawing and deshields the proton. Choice C corresponds to a carboxyl proton, and is even further downfield, because the acidic proton is deshielded to a greater degree than the aldehyde proton. Choice B is wrong because this chemical shift corresponds to aromatic protons. Choice D is wrong because this upfield signal is characteristic of an alkyl proton.

5. **C** This isotope has no magnetic moment and will therefore not exhibit resonance. Nuclei that possess a magnetic moment are typically nuclei with odd-numbered masses (^{1}H, ^{11}B, ^{13}C, ^{15}N, ^{19}F, etc.) or those with an even mass but an odd atomic number (^{2}H, ^{10}B). Note that ^{12}C is very abundant in nature.

6. The normal spectrum of CH_3CH_2OH has a CH_3 group split into a triplet by the CH_2 group protons, a CH_2 group split into a quartet by the CH_3 group protons, and a hydroxyl proton singlet. This proton exchanges rapidly with the protons of water and is not able to participate in splitting. Under anhydrous conditions, this proton can couple, and it splits the neighboring quartet into a doublet, so that an octet is observed. The –OH signal itself appears as a triplet due to splitting by the protons of the neighboring CH_2 group.

7. **A** Coupling between adjacent carbon atoms (choice C) is rarely seen due to the low abundance of ^{13}C. Coupling between carbon and protons on the adjacent carbon (choice B) is never observed. Since proton coupling is only relevant to ^{1}H NMR, choice D is incorrect.

8. The carbonyl carbon atom will be farthest downfield because of the electron-withdrawing ability of the oxygen atom. ^{13}C NMR follows the same rules of chemical shift as ^{1}H NMR.

9. **C** Most conjugated alkenes give an intense UV absorption. Aldehydes, ketones, acids, and amines all absorb in the UV. However, other forms of spectroscopy (mainly IR and NMR) are more useful for precise identification. Isolated alkenes, choice B, can rarely be identified by UV.

10. **B** In mass spectrometry, a molecule is broken down into smaller charged fragments. These fragments are passed through a magnetic field and are identified according to their mass-to-charge ratios; therefore, choice B is the correct answer. Choice D is the basis for IR and NMR, not mass spectrometry, so this is incorrect. Viscosity, choice C, doesn't form the basis for any of the spectroscopic techniques discussed, so it is also wrong. Finally, choice A is incorrect because the separation of fragments does not depend solely on mass, but on charge as well, and the fragments are mostly polyatomic.

Carbohydrates

CHAPTER TWENTY FOUR

Carbohydrates are compounds containing carbon, hydrogen, and oxygen in the form of polyhydroxylated aldehydes or ketones. They have the general formula $C_m(H_2O)_n$ and serve many functions in biological systems, most notably as the chemical energy source for most organisms. A single carbohydrate unit is a **monosaccharide** (simple sugar), and a molecule with two sugars is a **disaccharide**. **Oligosaccharides** are short carbohydrate chains, while **polysaccharides** are long carbohydrate chains.

In a Nutshell:

Carbohydrates are aldehydes or ketones with many hydroxyl groups.

MONOSACCHARIDES

Monosaccharides, the simplest carbohydrate units, are classified according to the number of carbons they possess. For example, **trioses**, **tetroses**, **pentoses**, and **hexoses** have 3, 4, 5, and 6 carbons, respectively. The basic structure of monosaccharides is exemplified by the simplest, glyceraldehyde.

Note:

Monosaccharides are the simplest units and are classified by the number of carbons.

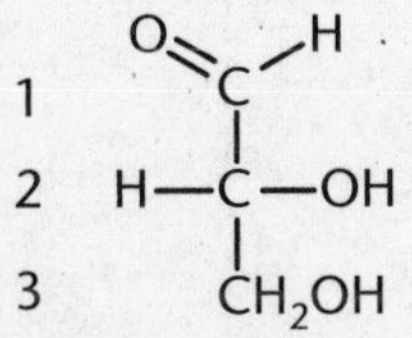

```
          O    H
1          ⟍C⟋
            |
2      H—C—OH
            |
3          CH₂OH
```

Glyceraldehyde

Glyceraldehyde is a polyhydroxylated aldehyde or **aldose** (aldehyde sugar). A polyhydroxylated ketone is called a **ketose** (ketone sugar). The numbering of the carbon atoms in a monosaccharide begins with the end closest to the carbonyl group.

Note:

As usual, number the molecule from the end closest to the carbonyl.

A. STEREOCHEMISTRY

The stereochemistry of monosaccharides can be understood by studying the enantiomeric configurations of glyceraldehyde.

Note:

The D and L designations are based on the sterochemistry of glyceraldehyde.

mirror

CHO		CHO
H—C—OH		HO—C—H
CH_2OH		CH_2OH
D-Glyceraldehyde		L-Glyceraldehyde

Mnemonic:

If Lowest -OH is on the Left, the molecule is L. If the -OH is on the Right, it's D (from the Latin root dextro, meaning "right").

The D and L configurations of glyceraldehyde were assigned early in this century (before the *R* and *S* configurations were used) to designate the optical rotation of each enantiomer. D-glyceraldehyde was later determined to exhibit a positive rotation (designated as D-(+)-glyceraldehyde) and L-glyceraldehyde a negative rotation (designated as L-(–)-glyceraldehyde. However, other monosaccharides are assigned the D or L configuration depending on their relationship to glyceraldehyde: a molecule whose highest numbered chiral center (the chiral center farthest from the carbonyl) has the same configuration as D-(+)-glyceraldehyde is classed as a D sugar. A molecule that has its highest numbered chiral center in the same configuration as L-(–)-glyceraldehyde is classed as an L sugar. This is illustrated below:

mirror

CHO		CHO
H—OH		HO—H
HO—H		H—OH
H—OH		HO—H
H—OH		HO—H
CH_2OH		CH_2OH
D-Glucose		L-Glucose

Note:

Epimers differ in configuration at only one carbon.

Monosaccharide stereoisomers are divided into two optical families, D and L; the stereoisomers within one family are known as **diastereomers**. Aldose diastereomers which differ only about the configuration of one carbon are known as **epimers**. For instance, D-ribose and D-arabinose are pentose epimers. They differ in configuration only at C–2.

CHO
H—OH
H—OH
H—OH
CH_2OH

D-Ribose

CHO
HO—H
H—OH
H—OH
CH_2OH

D-Arabinose

Note:

Fructose is a ketose while glucose, galactose, and mannose are all aldoses.

Some important monosaccharides are shown below.

CH_2OH
C=O
HO—H
H—OH
H—OH
CH_2OH

D-Fructose

CHO
H—OH
HO—H
H—OH
H—OH
CH_2OH

D-Glucose

CHO
H—OH
HO—H
HO—H
H—OH
CH_2OH

D-Galactose

CHO
HO—H
HO—H
H—OH
H—OH
CH_2OH

D-Mannose

B. RING PROPERTIES

Because monosaccharides contain both a hydroxyl group and a carbonyl group, they can undergo intramolecular reactions to form cyclic hemiacetals (or hemiketals, in the case of ketoses). These cyclic molecules are stable in solution and may exist as six-membered **pyranose** rings (as in glucose) or five-membered **furanose** rings. Like cyclohexane, the pyranose rings adopt a chairlike configuration, and the substituents assume axial or equatorial positions so as to minimize steric hindrance. When converting the monosaccharide from its straight-chain Fischer projection to the Haworth projection (shown in the figure below), it is important to remember that any group on the right of the Fischer projection will be pointing down, while any group on the left side of the Fischer projection will be pointing up. The following reaction scheme depicts the formation of a cyclic hemiacetal from D-glucose.

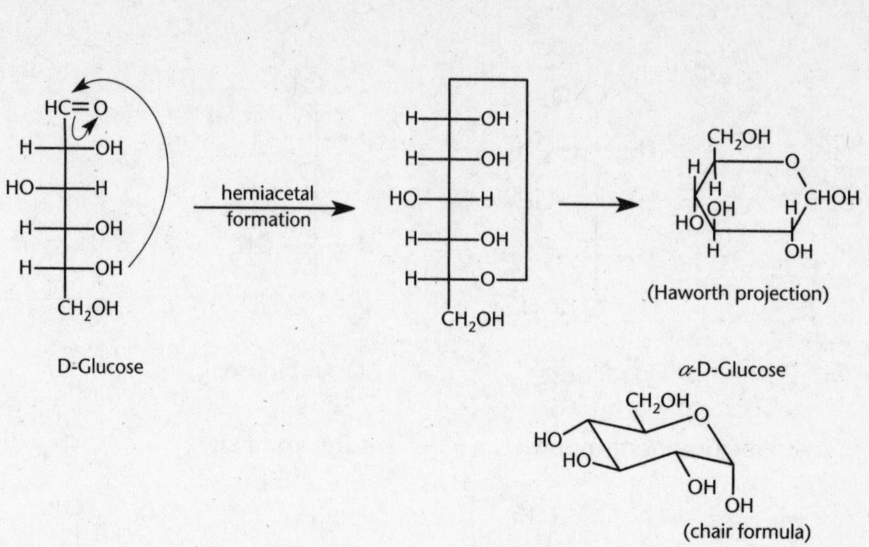

D-Glucose

hemiacetal formation

(Haworth projection)

α-D-Glucose

(chair formula)

Note:

Anomers differ in configuration only at the newly formed chiral center. α = down. β = up.

When a straight-chain monosaccharide is converted to its cyclic form, the carbonyl carbon (C–1 for glucose) becomes chiral. Cyclic stereoisomers differing about the new chiral carbon are known as **anomers**. In glucose, the alpha anomer has the –OH group of C–1 *trans* to the CH_2OH substituent (down), while the beta anomer has the –OH group of C–1 *cis* to the CH_2OH substituent (up).

When exposed to water, hemiacetal rings spontaneously open and then reform. Because of bond rotation between C–1 and C–2, either the alpha or beta anomer may be formed. The reaction is more rapid when catalyzed by acid or base. The spontaneous change of configuration about C–1 is known as **mutarotation**, and results in a mixture containing both anomers in their equilibrium concentrations (for glucose, 36 percent alpha:64 percent beta). The alpha configuration is less favored because the hydroxyl group of C–1 is axial, making the molecule more sterically strained (beta is equatorial).

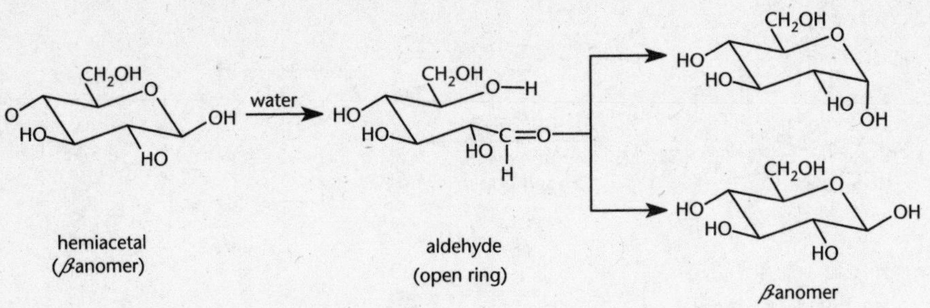

hemiacetal (βanomer) —water→ aldehyde (open ring) → αanomer / βanomer

C. MONOSACCHARIDE REACTIONS

1. Ester Formation

Monosaccharides contain hydroxyl groups and can undergo many of the same reactions as simple alcohols. Therefore, they may be converted to either esters or ethers. In the presence of acid anhydride and base, all of the hydroxyl groups will be esterified. The following reaction is an example of glucose esterification.

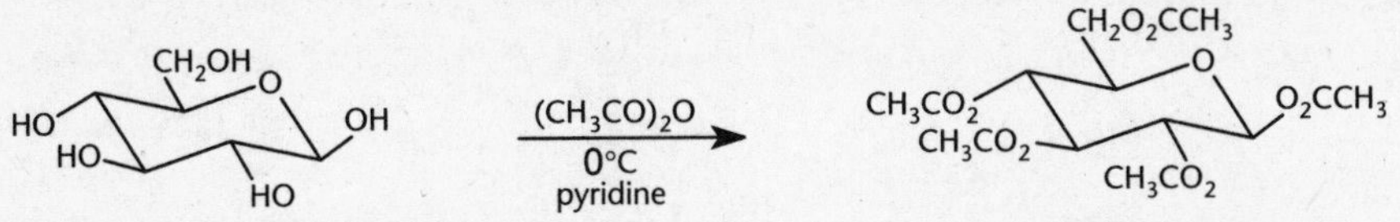

βD-Glucose —$(CH_3CO)_2O$, 0°C, pyridine→ Penta-O-acetyl- βD-glucose

2. Oxidation of Monosaccharides

As they switch between anomeric configurations, the hemiacetal rings spend a short period of time in the open-chain aldehyde form. Like all aldehydes, these can be oxidized to carboxylic acids called **aldonic acids**. Thus, the aldoses are reducing agents. Any monosaccharide with a hemiacetal ring (–OH on C–1) is considered a **reducing sugar** and can be oxidized. Both Tollens' reagent and Benedict's reagent can be used to detect the presence of reducing sugars. A positive Tollens' test involves the reduction of Ag^+ to form metallic silver. When Benedict's reagent is used, a red precipitate of Cu_2O indicates the presence of a reducing sugar. Ketose sugars are also reducing sugars and give positive Tollens' and Benedict's tests, because they can isomerize to aldoses via keto-enol shifts.

Note:

Reaction types are determined by the functional groups that are present. Think of alcohols and carbonyls.

In a Nutshell:

Key reactions of monosaccharides include:
- *ester formation*
- *oxidation*
- *glycosidic reactions*

Clinical Correlate:

Benedict's test is used to detect glucose in the urine of diabetics.

COOH
H—OH
HO—H
H—OH
H—OH
CH_2OH

β-D-Glucose $\xrightarrow[\text{Benedict's Solution}]{Cu(OH)_2}$ D-Gluconic Acid (an aldonic acid) + $Cu_2O(s)$ (red solid)

Note:

Glycosidic reactions:

hemiacetal + alcohol → acetal.

3. Glycosidic Reactions

Hemiacetal monosaccharides will react with alcohols under acidic conditions. The anomeric hydroxyl group is transformed into an alkoxy group, yielding a mixture of the alpha and beta acetals. The resulting bond is called a **glycosidic linkage**, and the acetal is known as a **glycoside**. An example is the reaction of glucose with ethanol.

β-D-Glucose $\xrightarrow[HCl]{C_2H_5OH}$ Ethyl-α-D-glucoside (an acetal) + Ethyl-β-D-glucoside (an acetal) + H_2O

Glycosides do not mutarotate and are stable in water.

DISACCHARIDES

As discussed above, a monosaccharide may react with alcohols to give acetals. When that alcohol is another monosaccharide, the product is called a **disaccharide**. The formation of a disaccharide is shown below.

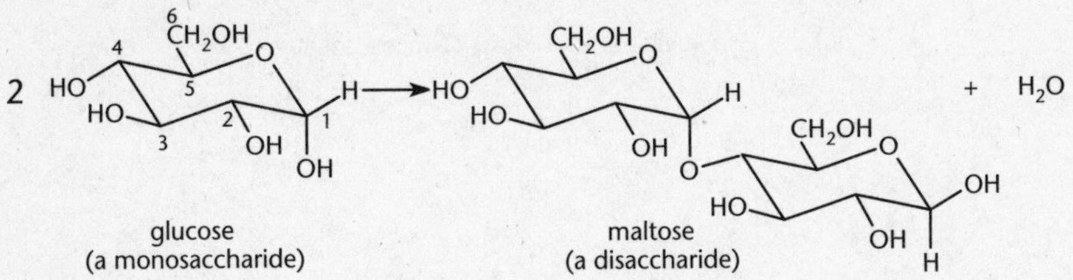

glucose
(a monosaccharide)

maltose
(a disaccharide)

The most common glycosidic linkage occurs between C–1 of the first sugar and C–4 of the second, and is designated as a 1,4′ link. 1,6′ and 1,2′ bonds are also observed. The glycosidic bonds may be either alpha or beta, depending on the orientation of the hydroxyl group on the anomeric carbon.

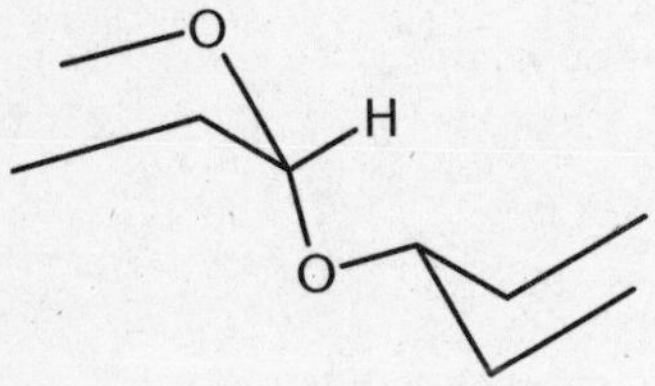

α-glycosidic linkage

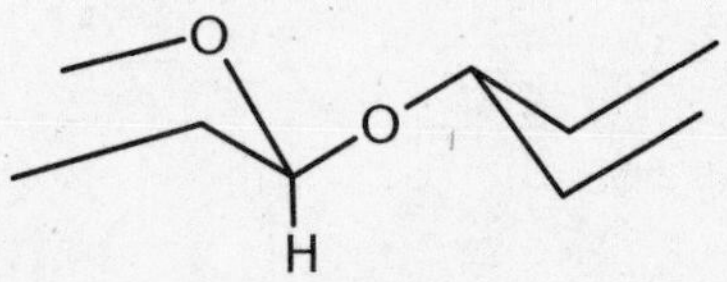

β-glycosidic linkage

These glycosidic linkages can often be cleaved in the presence of aqueous acid. For example, the glycosidic linkage of maltose, a disaccharide, can be cleaved to yield two molecules of glucose.

Note:

In the body, enzymes are needed to ensure that the correct glycosidic linkages form. Without enzymes, the reactions are nonspecific and tend to keep going, never stopping at the disaccharide level.

POLYSACCHARIDES

Polysaccharides are formed via linkage of monosaccharide units with glycosidic bonds. The three most important biological polysaccharides are **cellulose**, **starch**, and **glycogen**. Cellulose is comprised of D-glucose linked by 1,4′-beta-glycosidic bonds. Cellulose is the structural component of plants. Starch stores energy in plants and glycogen stores energy in animals; both are formed by linking glucose units in 1,4′-alpha-glycosidic bonds, with occasional 1,6′-alpha-glycosidic bonds creating branches. While all three are composed of glucose subunits, the orientation about the anomeric carbon gives them biological differences. Cellulose cannot be digested by humans, while starch and glycogen can, and are important energy sources for living organisms.

Note:

Key biological polysaccharides:

cellulose (1,4′ beta);
starch and glycogen (mostly 1,4′ alpha; some 1,6′ alpha).

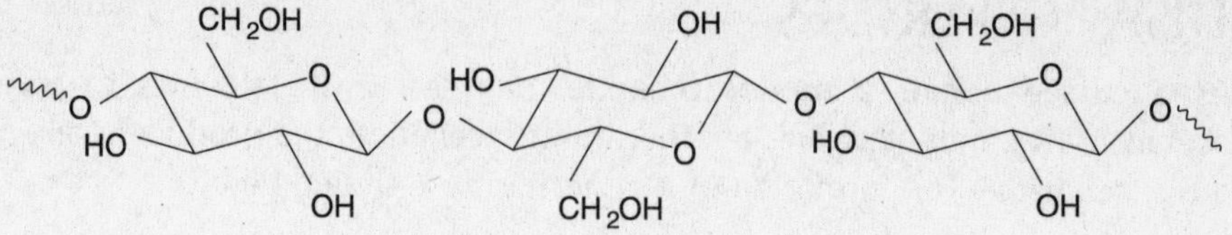

Cellulose, a 1,4′-β-D-Glucose polymer

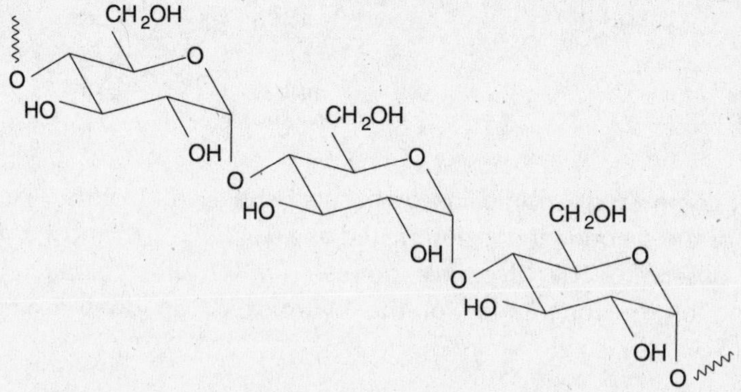

Starch, a 1,4′-α-D-Glucose polymer

REVIEW PROBLEMS

1. When glucose is in a straight-chain formation, it

   A. is an aldoketose.
   B. is a pentose.
   C. has five chiral carbons.
   D. is one of sixteen stereoisomers.

2. All of the following are true of epimers EXCEPT

   A. they differ in configuration about only one carbon.
   B. they usually have slightly different chemical and physical properties.
   C. they are diastereomers (with the exception of glyceraldehyde).
   D. they always have equal but opposite optical activities.

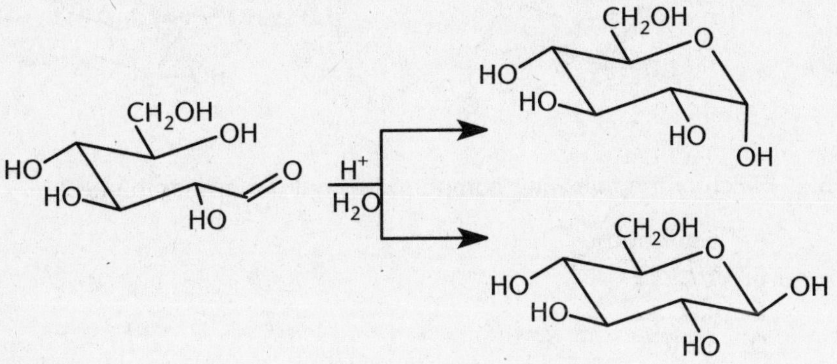

CH_2OH OH HO HO HO O H^+ H_2O → CH_2OH O HO HO HO OH ; CH_2OH O HO HO OH HO

3. The above reaction is an example of one step in

   A. aldehyde formation.
   B. hemiketal formation.
   C. mutarotation.
   D. glycosidic bond cleavage.

4. What is the product of the following reaction?

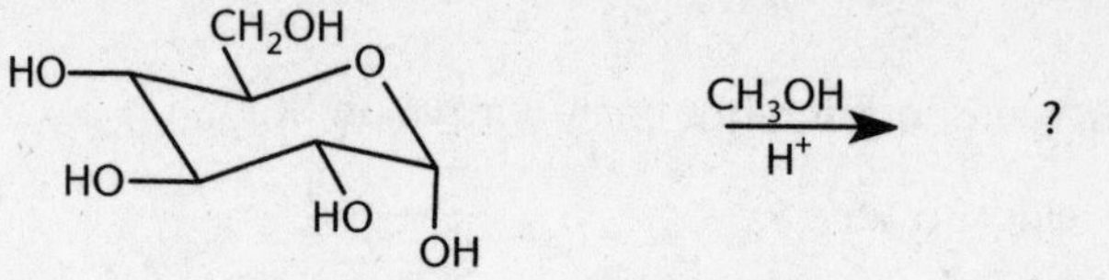

CH_2OH / HO / HO / HO / OH ring structure $\xrightarrow[H^+]{CH_3OH}$?

A. ring with CH_2OCH_3, H_3CO, H_3CO, H_3CO, OCH_3

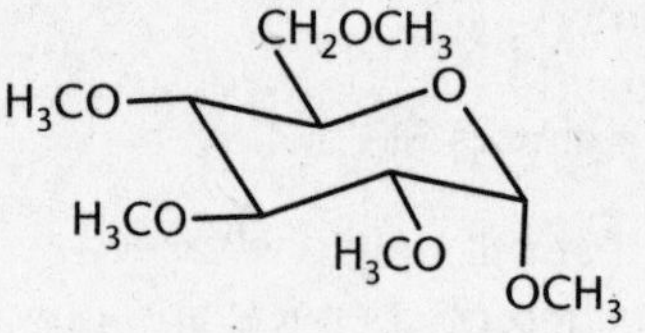

B. ring with CH_2OH, HO, HO, HO, OCH_3

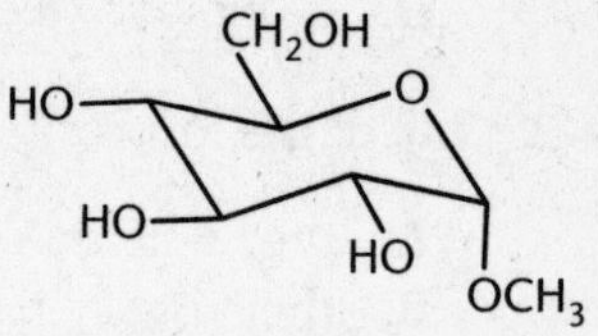

C. ring with CH_2OH, H_3CO, HO, HO, OH

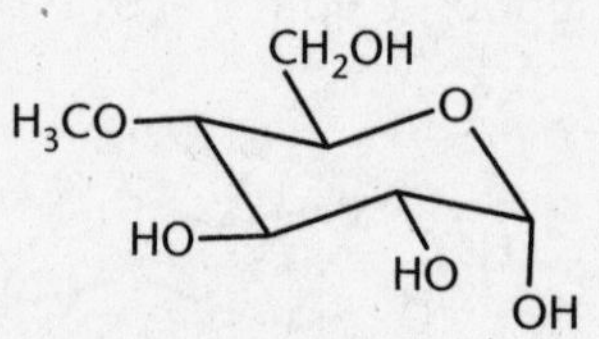

D. ring with CH_2OCH_3, HO, HO, HO, OH

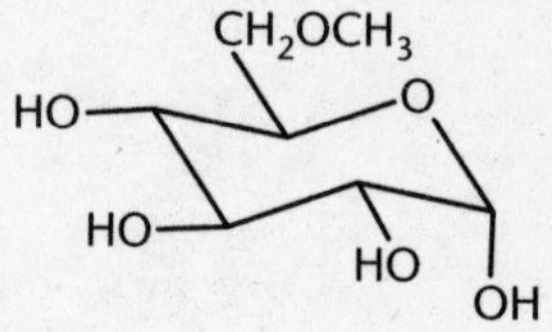

5. Which of the following compounds is not a monosaccharide?

A. deoxyribose
B. fructose
C. glucose
D. maltose

6. The cyclic forms of monosaccharides are:

I. hemiacetals
II. hemiketals
III. acetals

A. I only
B. III only
C. I and II
D. I, II and III

7. Which of the following sugars represents an aldohexose?

A.

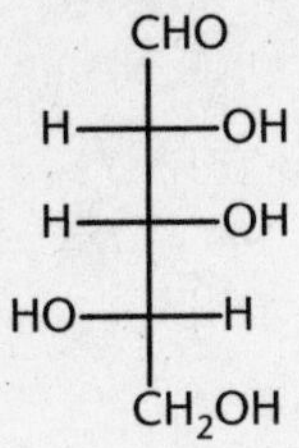

B.

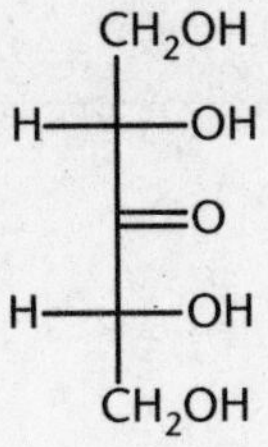

C.

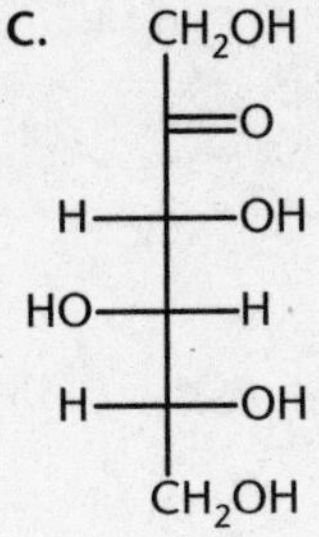

D.

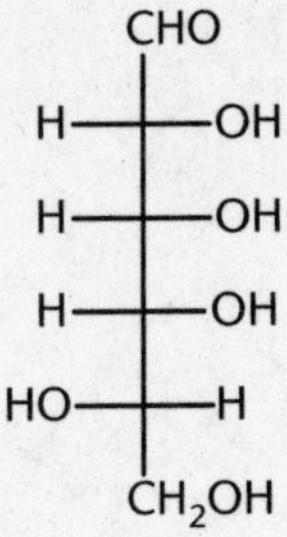

8. When the following straight-chain Fischer projection is converted to a chair or ring conformation, what will be its structure?

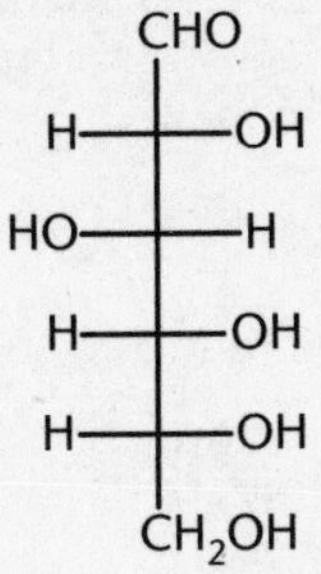

A.

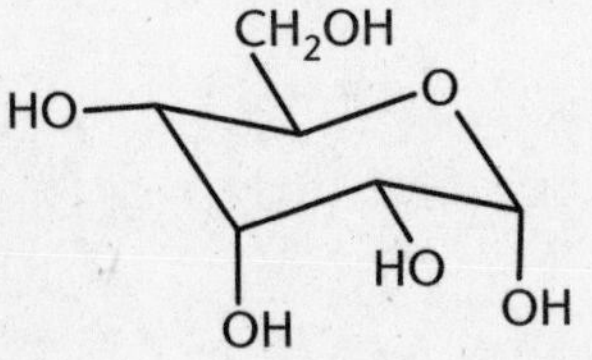

B.

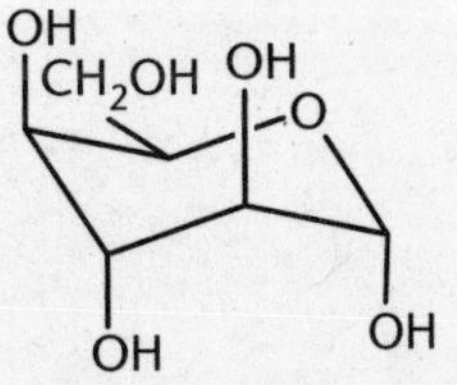

C.

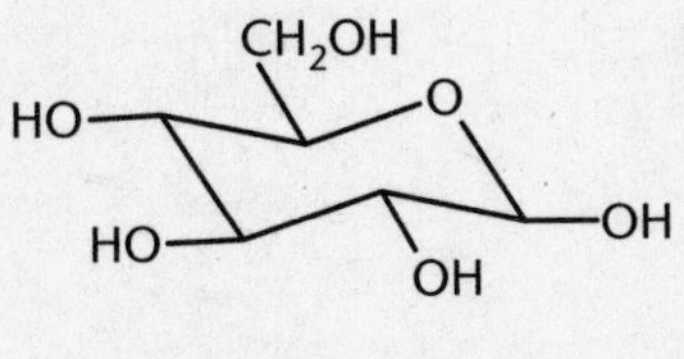

D.

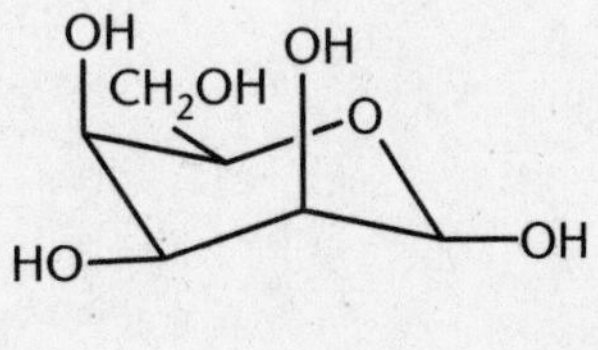

9. What would be the product of the following reaction?

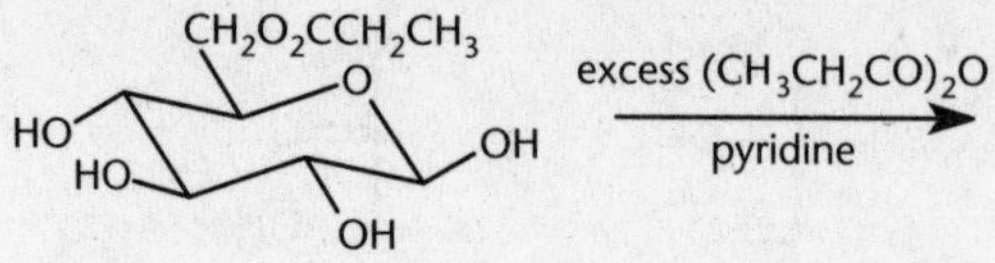

$CH_2O_2CCH_2CH_3$, HO, HO, OH, OH (ring O) — excess $(CH_3CH_2CO)_2O$, pyridine →

A. $CH_2O_2CCH_2CH_3$; HO, HO, OH, OH

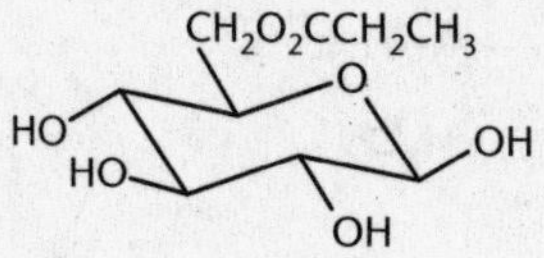

B. CH_2OH; HO, $H_3CH_2CCO_2$, OH, OH

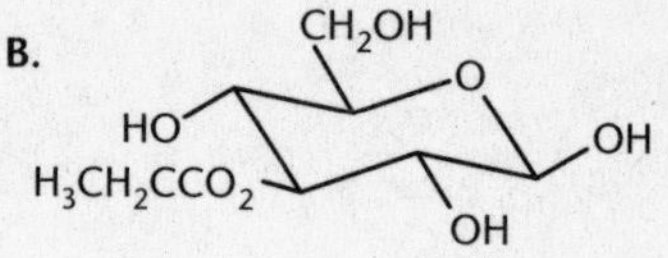

C. $CH_2O_2CCH_2CH_3$; $H_3CH_2CCO_2$, $H_3CH_2CCO_2$, $O_2CCH_2CH_3$, $O_2CCH_2CH_3$

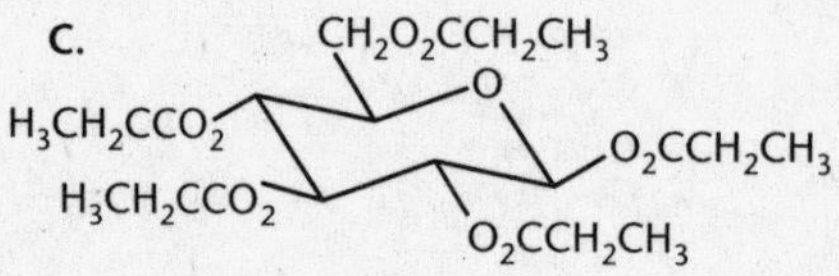

D. CH_2OH; HO, HO, $O_2CCH_2CH_3$, OH

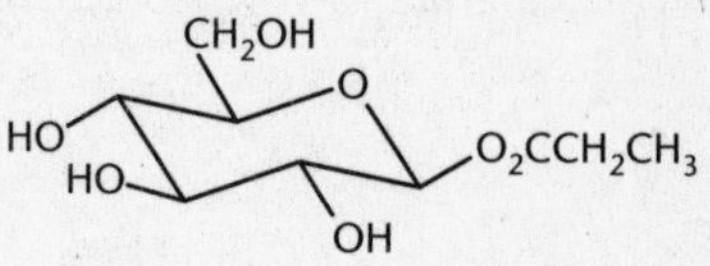

10. Which of the following are reducing sugars?

A. fructose
B. galactose
C. glucose
D. All of the above

SOLUTIONS TO REVIEW PROBLEMS BEGIN ON THE FOLLOWING PAGE

SOLUTIONS TO REVIEW PROBLEMS

1. **D** Glucose is an aldohexose, meaning that it has one aldehyde group and six carbons. Given this information, choices A and B can be eliminated. In aldose sugars, each nonterminal carbon is chiral. Therefore, glucose has four chiral centers, not five. The number of stereoisomers possible for a chiral molecule is 2^n, where n is the number of chiral carbons. Since glucose has four chiral centers, there are sixteen stereoisomers possible. Thus, choice D is correct.

2. **D** Epimers are monosaccharide diastereomers that differ in their configuration about only one carbon. As with all diastereomers, epimers have different chemical and physical properties, and their optical activities have no relationship to each other. Therefore, choice D is the only statement that does not apply to epimers.

3. **C** In solution, the hemiacetal ring of glucose will break open spontaneously and then reform. When the ring is broken, bond rotation occurs between C–1 and C–2 to produce either the alpha or the beta anomer. The reaction given in this question depicts the mutarotation of glucose.

4. **B** When glucose is reacted with ethanol under acid catalysis, the hemiacetal is converted to an acetal via replacement of the anomeric hydroxyl group with an alkoxy group. The result is a type of acetal known as a *glycoside.*

5. **D** Maltose is a disaccharide made of two joined glucoses. All the other choices are monosaccharides.

6. **C** Monosaccharides can exists as hemiacetals or hemiketals depending upon whether they are aldoses (monosaccharides that contain an aldehyde functionality in their open chain forms) or ketoses (monosaccharides that contain a ketone functionality in their open chain forms). When a monosaccharide is in its cyclic form, it is either a hemiacetal or hemiketal. Here, the anomeric carbon is attached to the oxygen in the ring (which constitutes the acetal or ketal group), and is also attached to a hydroxyl functionality (hence it is only a hemiacetal or hemiketal, since a full acetal or ketal would involve the conversion of this functionality to a alkoxy group). Therefore, choices I and II are true, making choice C the correct response.

7. **D** Aldohexoses are sugars that consist of six carbons and contain an aldehyde group. Glucose is an aldohexose.

8. **C**

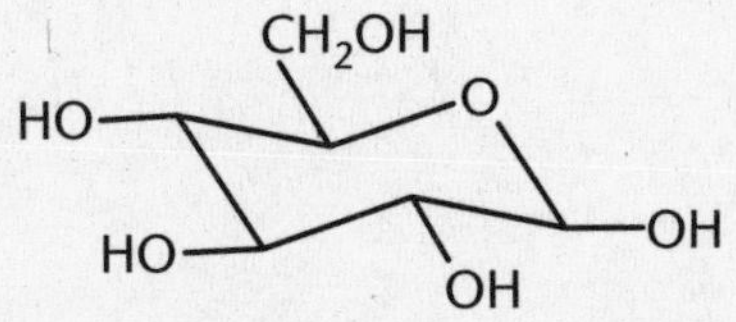

CH_2OH
HO O
HO OH
OH

9. **C** Glucose, because it has hydroxyl groups, can react like any alcohol and be converted to an ester. Therefore, the reaction of glucose with an anhydride in the presence of base, shown in this question, creates an ester group at each hydroxyl position. Choice C is the only answer which shows every hydroxyl group esterified.

10. **D** All aldose sugars are considered reducing sugars, since they are easily oxidized to carboxylic acids by such reagents as Tollens' and Benedict's solutions. Galactose and glucose are aldoses; thus, they are reducing sugars. Although fructose is not an aldose, it can be oxidized because it can isomerize to an aldose via a few keto-enol shifts in a basic solution. Therefore, all three monosaccharides are reducing sugars, and choice D is correct.

Amino Acids, Peptides, and Proteins

CHAPTER TWENTY FIVE

Proteins are large polymers composed of many amino acid subunits. Proteins have diverse biological roles; for example, they provide structure (keratin, collagen), regulate body metabolism via hormonal control (insulin), and serve as catalysts (enzymes).

AMINO ACIDS

Amino acids contain an amine group and a carboxyl group attached to a single carbon atom (the alpha carbon atom). The other two substituents of the alpha carbon are usually a hydrogen atom and a variable side-chain referred to as the **R-group**.

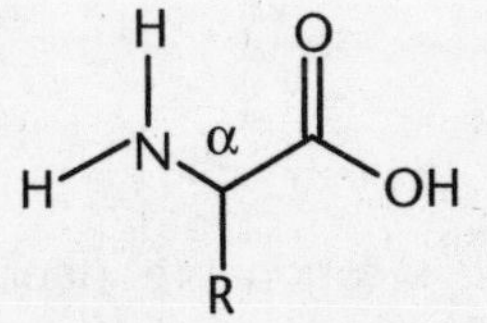

H, N, α, O, H, OH, R

The alpha carbon is a chiral center (except in glycine, the simplest amino acid, where R = H), and thus all amino acids (except for glycine) are optically active. Naturally-occurring amino acids (of which there are 20) are L-enantiomers (see Chapters 12 and 24).

Note:

Except for glycine, all amino acids are chiral.

By convention, the Fischer projection for an amino acid is drawn with the amino group on the left.

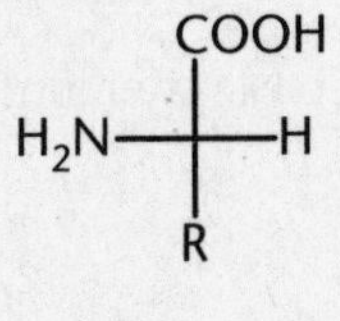

COOH, H_2N, H, R — L-amino acid

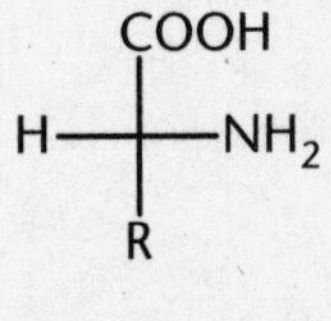

COOH, H, NH_2, R — D-amino acid

A. ACID-BASE CHARACTERISTICS

Amino acids have an acidic carboxyl group and a basic amino group on the same molecule. As a result, when they are in solution, amino acids sometimes take the form of dipolar ions, or **zwitterions** (from German *zwitter,* hybrid). The two halves of the molecules neutralize each other, so that at neutral pH they exist in the form of internal salts.

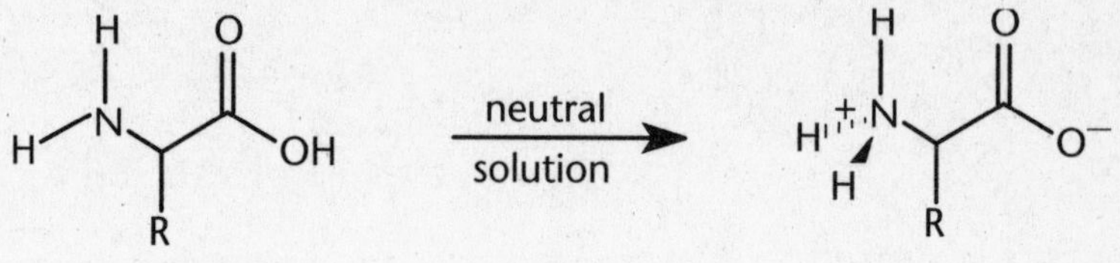

neutral solution

amino acid zwitterion

Amino acids are **amphoteric**; i.e., they may act as either acids or bases, depending on their environment. Amino acids in acidic solution are fully protonated. Since they have two protons that can dissociate—one from the carboxyl group and one from the amino group—amino acids have at least two dissociation constants, K_{a1} and K_{a2}.

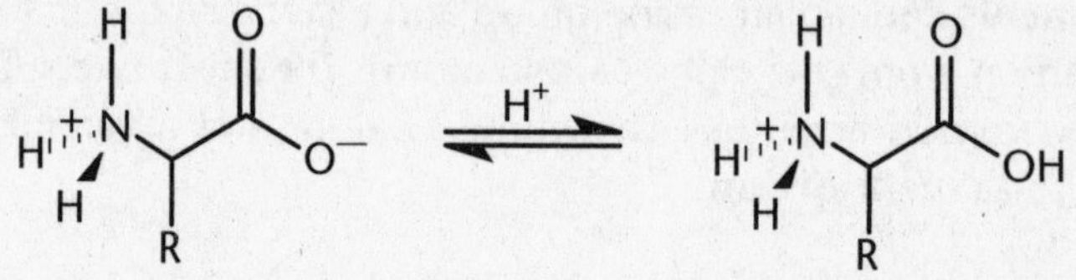

H^+

[neutral] [acidic solution]

Amino acids in basic solution are deprotonated. They have two proton-accepting groups and, therefore, at least two dissociation constants, K_{b1} and K_{b2}.

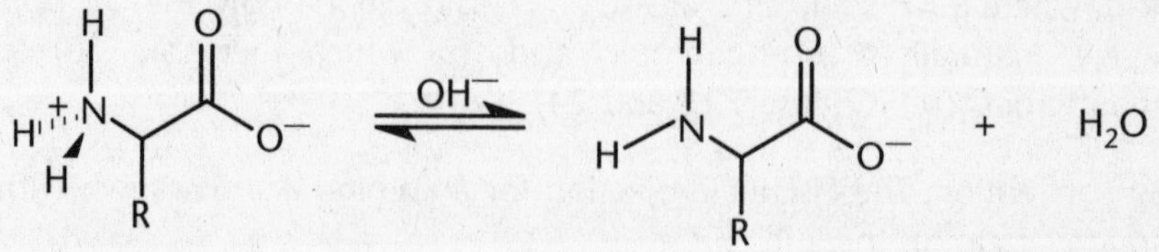

OH^- + H_2O

[neutral] [basic solution]

Real World Analogy:

Isoelectric focusing distinguishes proteins based on their pI values. In isoelectric focusing, a charged protein migrates through a pH gradient in the electric field until it reaches a pH region equal to its pI value, at which point the protein stops moving and may be visualized or extracted. Proteins which differ only slightly in their pI values can be separated by isoelectric focusing.

At low pH, the amino acid carries an excess positive charge, and at high pH, the amino acid carries an excess negative charge. The intermediate pH, at which the amino acid is electrically neutral and exists as a zwitterion, is the **isoelectric point (pI)**, or **isoelectric pH**, of the amino acid.

The isoelectric pH lies between pK_{a1} and pK_{a2}.

Note:

At its isoelectric point, an amino acid is uncharged.

B. TITRATION OF AMINO ACIDS

Because of their acidic and basic properties, amino acids can be titrated. The titration of each proton occurs as a distinct step resembling that of a simple monoprotic acid. The titration curve of glycine is shown below.

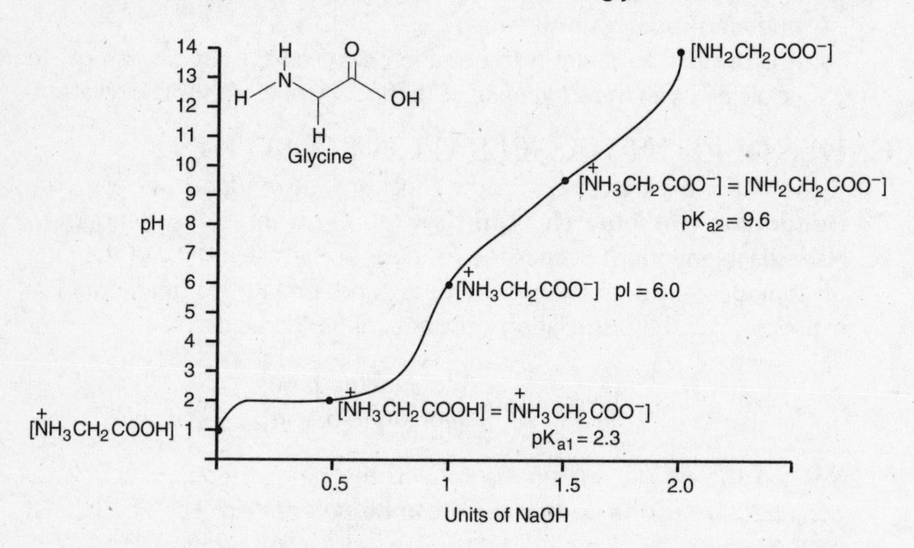

A 1M glycine solution is acidic; the glycine exists predominantly as $^+NH_3CH_2COOH$. The amino acid is fully protonated and carries a positive charge. As the solution is titrated with NaOH, carboxyl groups lose a proton. During this stage, the amino acid acts as a buffer and the pH changes very slowly. When 0.5 mol of base has been added to the amino acid solution, the concentrations of $^+NH_3CH_2COOH$ and $^+NH_3CH_2COO^-$ (its zwitterion) are equimolar. At this point the pH is equal to the pK_{a1}, and the solution is buffered against pH changes.

As more base is added, all of the carboxyl groups are deprotonated. The amino acid loses buffering capacity, and thus the pH rises more rapidly. When 1 mol of base has been added, glycine exists predominantly as $^+NH_3CH_2COO^-$. The amino acid is now electrically neutral; the pH is equal to glycine's pI.

Note:

Titration with base: first the carboxyl group is deprotonated, then the amino group.

Glycine passes through a second buffering stage during which pH change is slow because continued titration deprotonates amino groups. When 1.5 mol of base have been added, the concentrations of $^+NH_3CH_2COO^-$ and $NH_2CH_2COO^-$ are equimolar, and the pH is equal to pK_{a2}.

Note:

Amino acids pass through at least two buffering stages, one at each pK_a.

As another 0.5 mol of base is added, all of the amino groups are deprotonated to $NH_2CH_2COO^-$; glycine is now completely deprotonated.

Certain things should be noted about the titration of amino acids:

1. When adding base, the carboxyl group loses its proton first; after all of the carboxyl groups are fully deprotonated, the amino group loses its acidic proton.
2. Two moles of base must be added in order to deprotonate one mole of most amino acids. The first mole deprotonates the carboxyl group, while the second mole deprotonates the amino group.
3. The buffering capacity of the amino acid is greatest at or near the two dissociation constants, K_{a1} and K_{a2}. At the isoelectric point, its buffering capacity is minimal.
4. It is possible to perform the titration in reverse, from alkaline pH to acidic pH, with the addition of acid; the sequence of events is reversed.

C. HENDERSON-HASSELBALCH EQUATION

The ratio of an amino acid's ions are dependent on pH. The **Henderson-Hasselbalch equation** (also known as the Henderson-Hasselbach equation) defines the relationship between pH and the ratio of conjugate acid to conjugate base, and provides a mathematical expression for the dissociation constants of amino acids.

$$pH = pK_a + \log \frac{[\text{conjugated base}]}{[\text{conjugated acid}]}$$

When the pK_{a1} of glycine is known, the ratio of conjugate acid to conjugate base for a particular pH can be determined. For example, at pH 3.3, glycine, which has a pK_a of 2.3, will have the ratios:

$$3.3 = 2.3 + \log \frac{[H_3N^+CH_2COO^-]}{[H_3N^+CH_2COOH]}$$

By subtraction: $\log \frac{[H_3N^+CH_2COO^-]}{[H_3N^+CH_2COOH]} = 1$

The antilog of 1 = 10, thus: $\frac{[H_3N^+CH_2COO^-]}{[H_3N^+CH_2COOH]} = \frac{10}{1}$

So, in this example, there are ten times as many zwitterions as there are of the fully protonated form.

The Henderson-Hasselbalch equation can be used experimentally to prepare buffer solutions of amino acids. The best buffering regions of amino acids occur within one pH unit of the pK_a or pK_b. For example, the carboxyl group of glycine, which has a pK_a of 2.3, shows high buffering capacity between pH 1.3 and 3.3.

Note:

The best buffering occurs within one pH unit of the pK_a.

D. AMINO ACID SIDE-CHAINS

Amino acid side-chains (R-groups) give chemical diversity to the backbone of the amino acid molecule. They also give proteins some distinguishing features. The twenty amino acids are classified according to whether their side chains are **nonpolar**, **polar** (but uncharged), **acidic**, or **basic**.

1. Nonpolar Amino Acids

Nonpolar amino acids have R-groups that are saturated hydrocarbons. The R-groups are hydrophobic and decrease the solubility of the amino acid in water. Amino acids with nonpolar side-chains are usually found buried within protein molecules, away from the aqueous cellular environment.

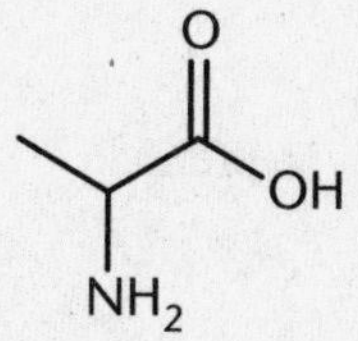

O OH NH_2

Alanine

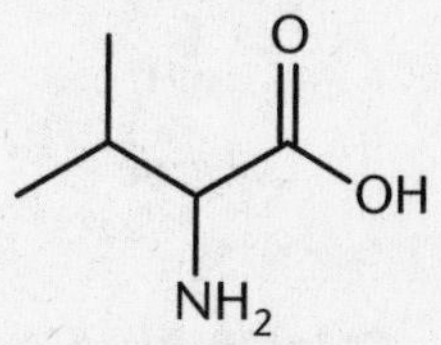

O OH NH_2

Valine

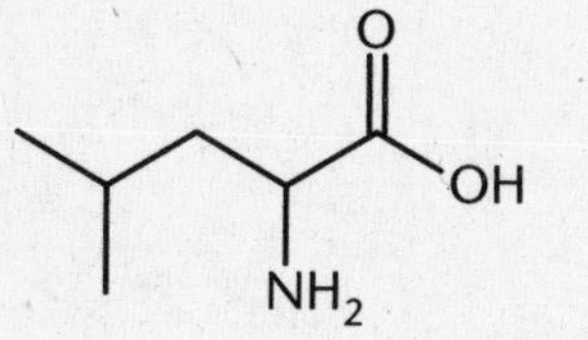

O OH NH_2

Leucine

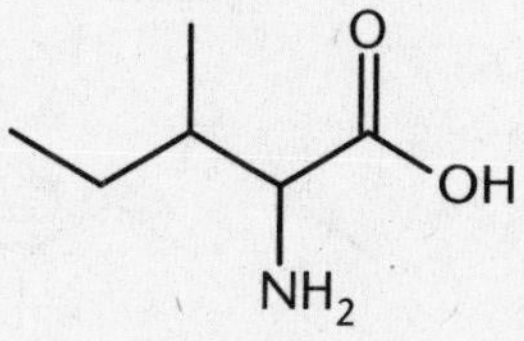

O OH NH_2

Isoleucine

Note:

Nonpolar amino acids are often found at the core of globular proteins or in transmembrane regions of proteins that are in contact with the hydrophobic portion of the phospholipid membrane.

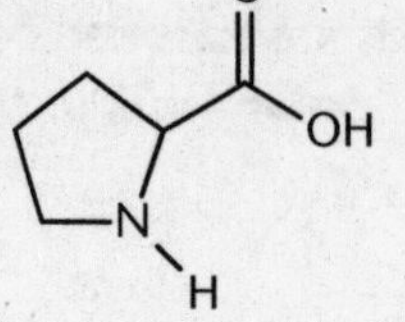

Proline

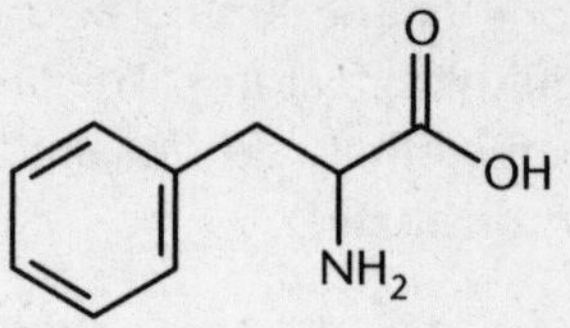

Phenylalanine

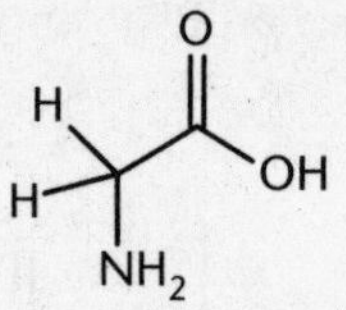

Glycine

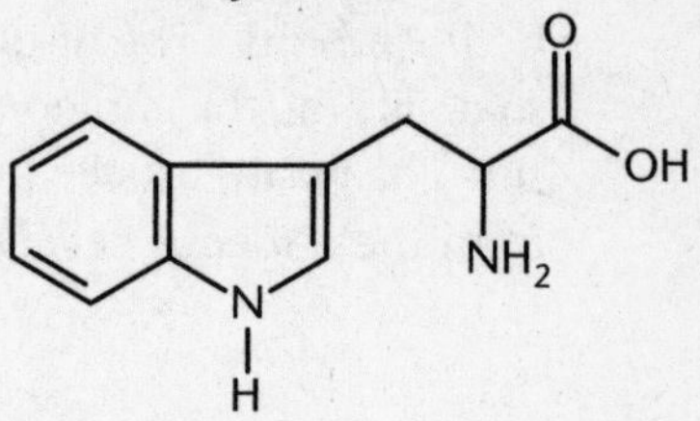

Tryptophan

Note:

The other types of amino acids (polar, acidic and basic) are found in regions of proteins that are exposed to the aqueous, polar environment.

2. Polar Amino Acids

Polar amino acids have polar, uncharged R-groups that are hydrophilic, increasing the solubility of the amino acid in water. They are usually found on protein surfaces.

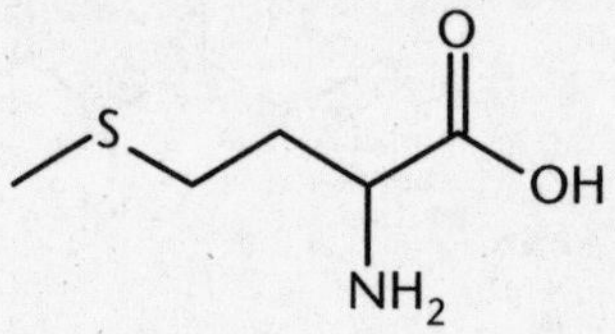

Methionine

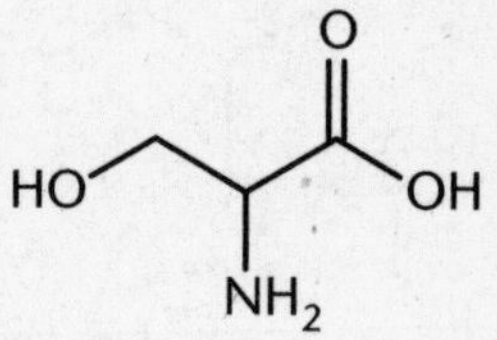

Serine

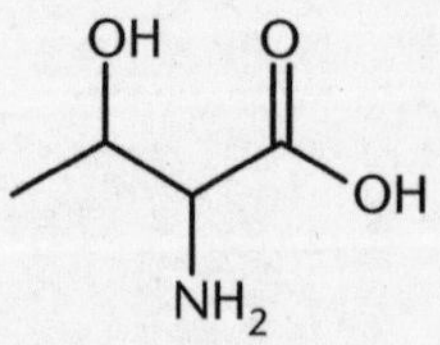

Threonine

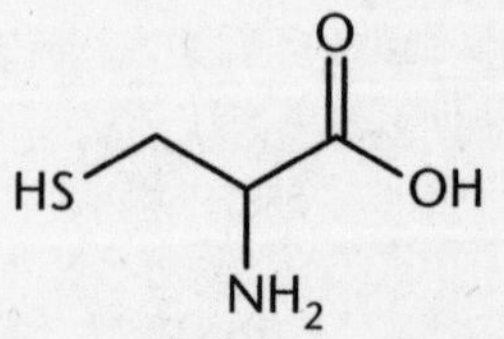

Cysteine

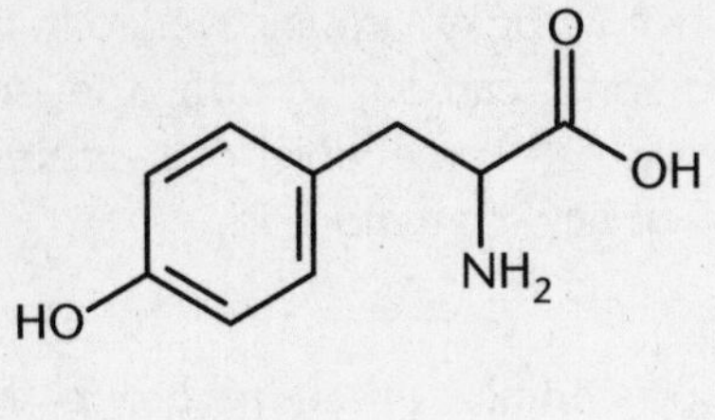

Tyrosine

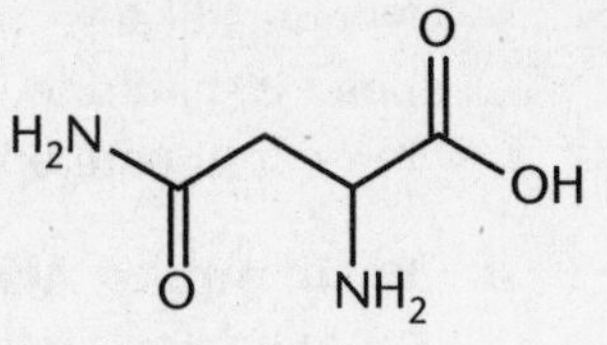

Asparagine

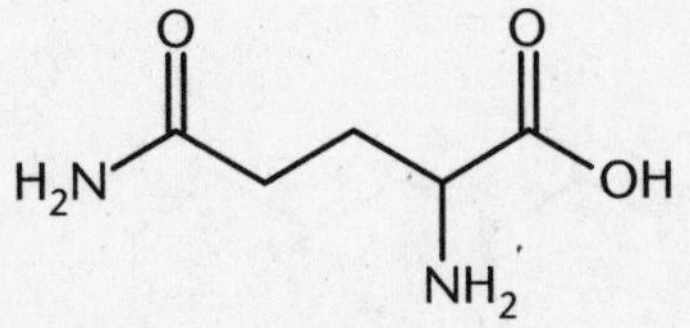

Glutamine

Note:

Acidic amino acids have a negative charge at physiological pH (pH 7.4).

3. Acidic Amino Acids

Amino acids whose R-group contains a carboxyl group are called acidic amino acids. They have a net negative charge at physiological pH (pH 7.4), and exist in salt form in the body. They often play important roles in the substrate-binding sites of enzymes.

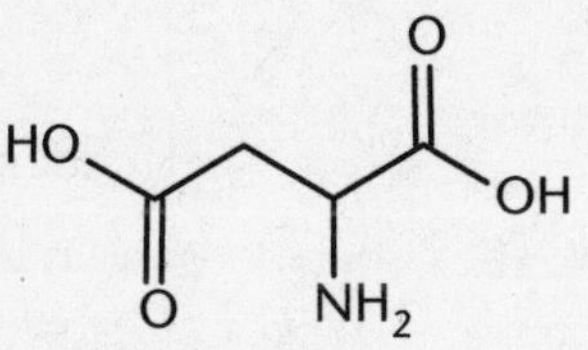

Aspartic Acid

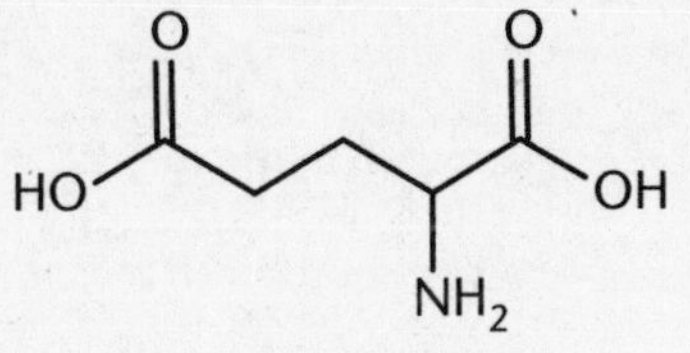

Glutamic Acid

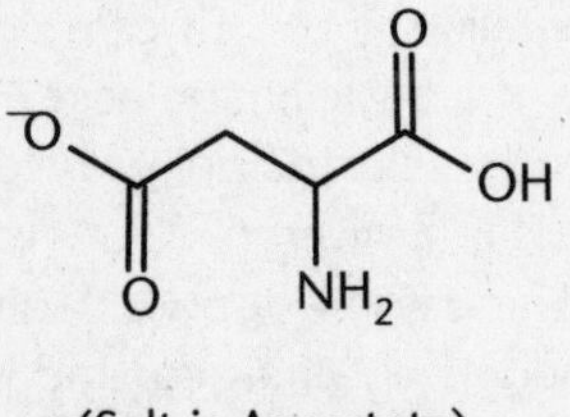

(Salt is Aspartate)

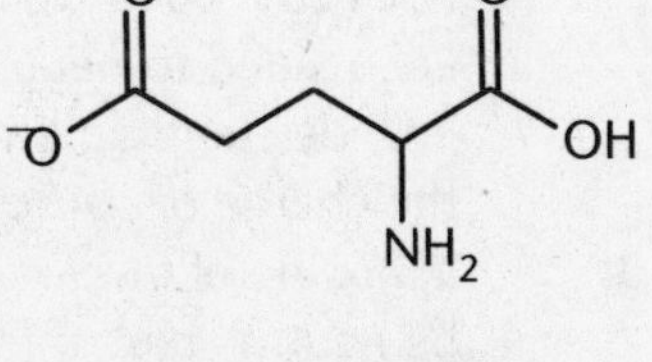

(Salt is Glutamate)

Note:

Acidic amino acids have three distinct pK_a's.

Aspartic acid and glutamic acid each have three groups that must be neutralized during titration (two –COOH and one $–NH_3^+$). Therefore, their titration curve is different from the standard curve for amino acids (exemplified by glycine). The molecule has three distinct dissociation constants—pK_{a1}, pK_{a2}, and pK_{a3}—although the

Note:

At physiological pH, basic amino acids have a net positive charge. They also have three dissociation constants.

Pavlov's Dog:

You can use pI to predict an amino acid's charge at a given pH.
If pH < pI, think positive charge.
If pH > pI, think negative charge.

neutralization curves of the two carboxyl groups overlap to a certain extent. Because of the additional carboxyl group, the isoelectric point is shifted toward an acidic pH. Three moles of base are needed to deprotonate one mole of an acidic amino acid.

4. Basic Amino Acids

Amino acids whose R-group contains an amino group are called basic amino acids and carry a net positive charge at physiological pH.

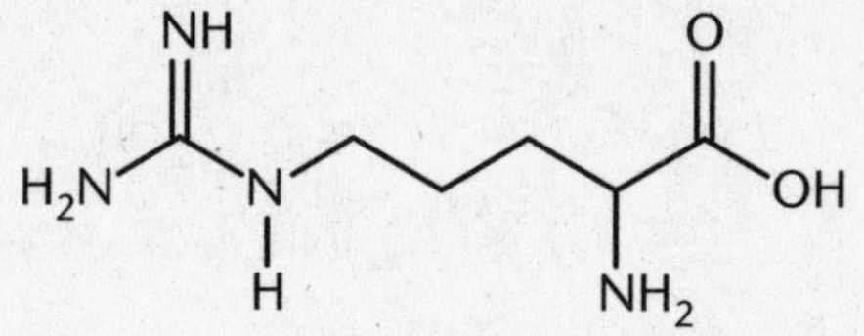

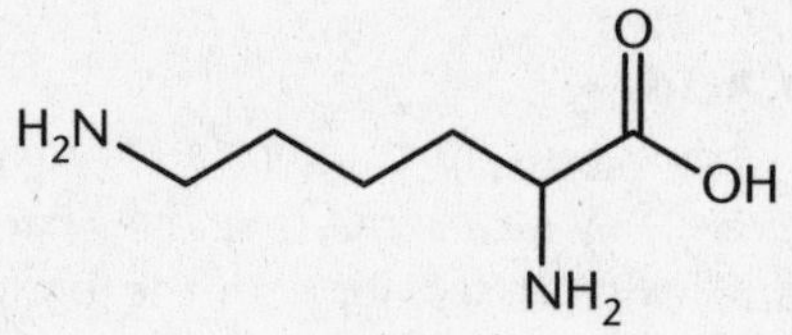

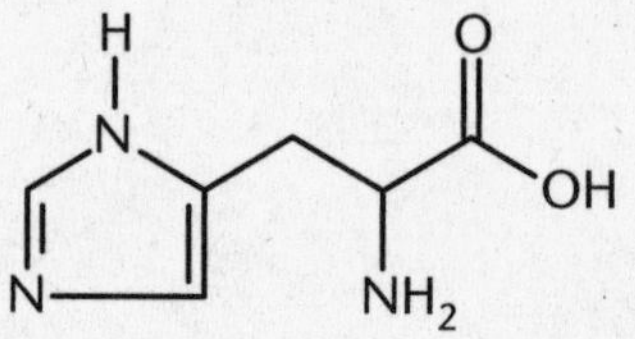

NH, O, H_2N, N, H, OH, NH_2 — Arginine

O, H_2N, OH, NH_2 — Lysine

H, N, O, OH, N, NH_2 — Histidine

The titration curve of amino acids with basic R-groups is modified by the additional amino group that must be neutralized. Although basic amino acids have three dissociation constants, the neutralization curves for the two amino groups overlap. The isoelectric point is shifted toward an alkaline pH. Three moles of acid are needed to neutralize one mole of a basic amino acid.

Understanding titration curves and isoelectric points helps predict the charge of particular amino acids at a given pH. For example, in a mixture of glycine, glutamic acid, and lysine at pH 6.0, glycine will be neutral, glutamic acid will be negatively-charged, and lysine will be positively-charged.

PEPTIDES

Peptides are composed of amino acid subunits, sometimes called **residues**, linked by **peptide bonds**. Peptides are small proteins (the distinction between a peptide and protein is vague). Two amino acids joined together form a **dipeptide**, three form a **tripeptide**, and many amino acids linked together form a **polypeptide**.

Note:

A peptide bond is an amide bond.

A. REACTIONS

Amino acids are joined by **peptide bonds** (amide bonds) between the carboxyl group of one amino acid and the amino group of another. This bond is formed via a condensation reaction (a reaction in which water is lost). The reverse reaction, hydrolysis (cleavage with the addition of water) of the peptide bond, is catalyzed by an acid or base.

Certain enzymes digest the chain at specific peptide linkages. For example, **trypsin** cleaves at the carboxyl end of arginine and lysine; chymotrypsin cleaves at the carboxyl end of phenylalanine, tyrosine, and tryptophan.

Bridge:

Refer to Chapter 21 for the reactions concerning nitrogen-containing compounds.

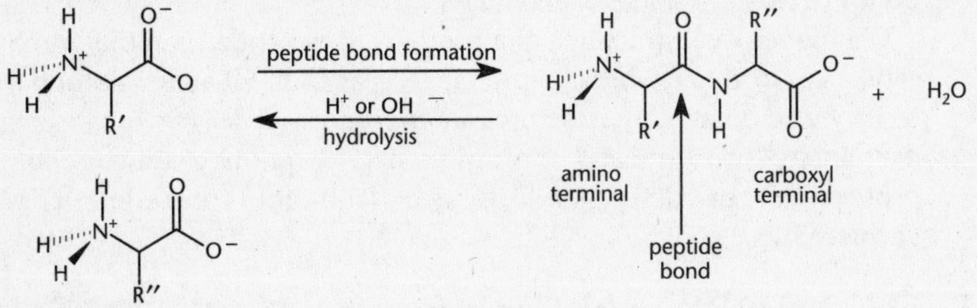

peptide bond formation
H^+ or OH^-
hydrolysis
+ H_2O
amino terminal
carboxyl terminal
peptide bond

B. PROPERTIES

The terminal amino acid with a free alpha-amino group is known as the **amino-terminal** or **N-terminal** residue, while the terminal residue with a free carboxyl group is called the **carboxy-terminal** or **C-terminal** residue. By convention, peptides are drawn with the N-terminal end on the left and the C-terminal end on the right.

Amides have two resonance structures, and the true structure is a hybrid with partial double-bond character. As a result, rotation about the C–N bond is restricted. The bonds on either side of the peptide unit, however, have a great deal of rotational freedom.

Note:

Rotation is limited around the peptide bond because resonance gives the C–N bond partial double-bond character.

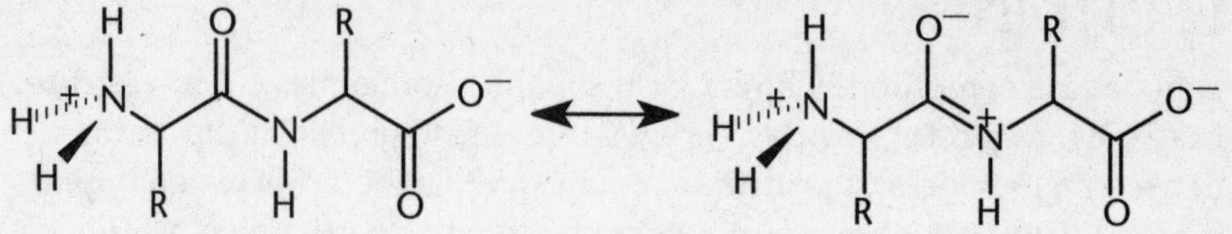

PROTEINS

Proteins are polypeptides that can range from only a few up to more than a thousand amino acids in length. Proteins serve many diverse functions in biological systems, acting as enzymes, hormones, membrane pores, receptors, and elements of cell structure. Four structural levels of protein structure—**primary**, **secondary**, **tertiary**, and **quaternary**—are described below.

A. PRIMARY STRUCTURE

The primary structure of the protein refers to the sequence of amino acids, listed from the N-terminus to the C-terminus, liked by covalent bonds between residues in the chain.

The higher-level structures of a protein are dependent on the primary sequence; in other words, a protein will assume whatever secondary, tertiary, and quaternary structures are most energetically favorable given its primary structure and environment. The primary structure of a protein can be determined using a laboratory procedure called **sequencing**.

B. SECONDARY STRUCTURE

The secondary structure of a protein refers to the local structure of neighboring amino acids, governed mostly by hydrogen bond interactions within and between peptide chains. The two most common types of secondary structures are the α-**helix** and the β-**pleated sheet**.

1. α-Helix

α-helix is a rod-like structure in which the peptide chain coils clockwise about a central axis. The helix is stabilized by intramolecular hydrogen bonds between carbonyl oxygen atoms and amine hydrogen atoms four residues away. The side-chains point away from the structure's core and interact with the cellular environment. A typical protein with this structure is **keratin**, which is found in feathers and hair.

2. β-Pleated Sheet

In β-pleated sheets, the peptide chains lie alongside each other in rows. The chains are held together by intramolecular hydrogen bonds between carbonyl oxygen atoms on one peptide chain and amine hydrogen atoms on another. In order to accommodate the maximum number of hydrogen bonds, the β-pleated sheet assumes a rippled, or pleated, shape. The R-groups of the amino residues point above and below the plane of the β-pleated sheet. Silk fibers are composed of β-pleated sheets.

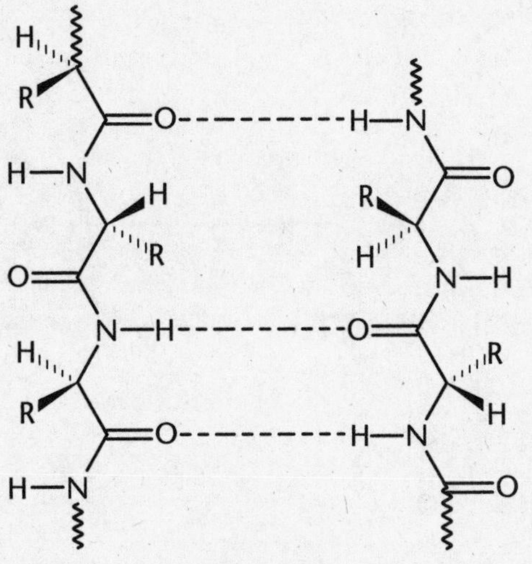

β-pleated sheet

Note:

Secondary structure describes hydrogen bonding. Common secondary structures include the α-helix and the β-pleated sheet.

In a Nutshell:

- *Primary structure consists of the amino acid sequence and covalent bonds.*
- *Secondary structure refers to the local structure of a protein as largely determined by hydrogen bond interactions.*
- *Tertiary structure is the three- dimensional shape of the protein.*
- *Quaternary structure is the arrangement of polypeptide subunits.*
- *Conjugated proteins have prosthetic groups.*

C. TERTIARY STRUCTURE

Tertiary structure refers to the three-dimensional shape of the protein, as determined by hydrophilic and hydrophobic interactions between the R-groups of amino acids that are far apart on the chain, and by the distribution of disulfide bonds. In a disulfide bond, two **cysteine** molecules become oxidized to form **cystine**. Disulfide bonds create loops in the protein chain.

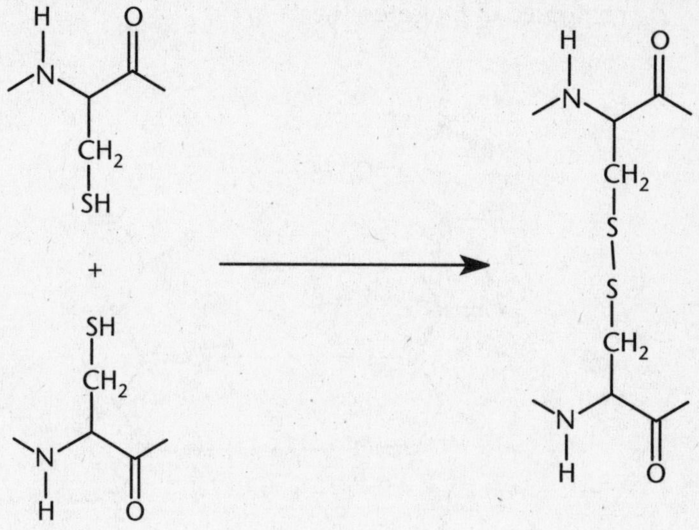

H O N CH$_2$ SH + SH CH$_2$ N H O → H O N CH$_2$ S S CH$_2$ N H O

cysteine cystine

Other amino acids have significant effects on tertiary structures as well. For instance, proline, because of its shape, cannot fit into an α-helix, thereby causing a kink in the chain.

Amino acids with hydrophilic (polar and charged) R-groups tend to arrange themselves toward the outside of the protein, where they interact with the aqueous cellular environment. Amino acids with hydrophobic R-groups tend to be found close together, protected from the aqueous environment by polar amino and carboxyl groups.

Proteins are divided into two major classifications on the basis of tertiary structure. **Fibrous proteins**, such as **collagen**, are found as sheets or long strands, while **globular proteins**, such as **myoglobin**, are spherical in shape.

D. QUATERNARY STRUCTURE

Some proteins contain more than one polypeptide subunit. The quaternary structure refers to the way in which these subunits arrange themselves to yield a functional protein molecule. **Hemoglobin**, which is composed of four polypeptide chains, possesses quaternary structure.

E. CONJUGATED PROTEINS

Certain proteins, known as **conjugated proteins**, derive part of their function from covalently attached molecules called **prosthetic groups**. Prosthetic groups may be organic molecules or metal ions. Many vitamins are prosthetic groups. Proteins with lipid, carbohydrate, and nucleic acid prosthetic groups are referred to as **lipoproteins**, **glycoproteins**, and **nucleoproteins**, respectively. Prosthetic groups play major roles in determining the function of the proteins with which they are associated. For example, the **heme group** carries oxygen in both myoglobin and hemoglobin. The heme is composed of an organic porphyrin ring with an iron atom bound in the center. Hemoglobin is inactive without the heme group.

F. DENATURATION OF PROTEINS

Denaturation, or **melting**, is a process in which proteins lose their three-dimensional structure and revert to a **random-coil** state. Denaturation can be caused by detergent, or by changes in pH, temperature, or solute concentration. The weak intermolecular forces keeping the protein stable and functional are disrupted. When a protein denatures, the damage is usually permanent. However, certain gentle denaturing agents do not permanently disrupt the protein. Removing the reagent might allow the protein to **renature** (regain its structure and function).

Note:

Denaturation is the loss of three-dimensional structure.

Real-World Analogy:

Permanent denaturation occurs when cooking egg whites. They denature and form a solid, rubbery mass that cannot be transformed back to its clear liquid form.

REVIEW PROBLEMS

1. If a mixture of alanine (pI=6) and aspartic acid (pI=3) is subjected to electrophoresis at pH 3, which of the following would you expect to occur?

A. Alanine will migrate to the cathode while aspartic acid migrates to the anode.
B. Alanine will not move while aspartic acid migrates to the cathode.
C. Aspartic acid will not move while alanine migrates to the cathode.
D. Alanine will migrate to the anode while aspartic acid migrates to the cathode.

2. In a neutral solution, most amino acids exist as

A. positively-charged compounds.
B. zwitterions.
C. negatively-charged compounds.
D. hydrophobic molecules.

3. What would be the charge of the following amino acid at pH 7?

$$HOOCCH_2CH_2CH(NH_2)COOH$$

A. neutral
B. negative
C. positive
D. None of the above

4. If the following amino acid (pI=9.74) in acidic solution is completely titrated with sodium hydroxide, what will be its charge at pH 3, 7, and 11?

$$H_2NCH_2CH_2CH_2CH_2CH(NH_2)COOH$$

A. positive, neutral, negative
B. negative, neutral, positive
C. neutral, positive, positive
D. positive, positive, negative

5. Amino acids with nonpolar R-groups have which of the following characteristics in aqueous solution?

A. They are hydrophilic and found buried within proteins.
B. They are hydrophobic and found buried within proteins.
C. They are hydrophobic and found on protein surfaces.
D. They are hydrophilic and found on protein surfaces.

6. All of the following statements concerning peptide bonds are true EXCEPT:

A. their formation involves a reaction between an amine group and a carboxyl group.
B. they are the primary bonds found in proteins.
C. they have partial double-bond character.
D. their formation involves hydration reactions.

7. How many different tripeptides can be formed that contain one valine, one alanine, and one leucine?

A. 5
B. 6
C. 7
D. 8

8. Beside peptide bonds, what other covalent bonds are commonly found in peptides?

A. Hydrogen bonds
B. Ether bonds
C. Disulfide bonds
D. Hydrophobic bonds

9. Discuss primary, secondary, tertiary, and quaternary structures in proteins. What are the defining characteristics of each category?

10. α-helices are secondary structures characterized by

A. intramolecular hydrogen bonds.
B. disulfide bonds.
C. a rippled effect.
D. intermolecular hydrogen bonds.

11. Denaturation involves the loss of what types of structure?

A. Primary
B. Secondary
C. Tertiary
D. Both B and C

SOLUTIONS TO REVIEW PROBLEMS BEGIN ON THE FOLLOWING PAGE

SOLUTIONS TO REVIEW PROBLEMS

1. **C** At pH 6, alanine will exist as a neutral, dipolar ion: the amino group will be protonated while the carboxyl group will be deprotonated. Lowering the pH to 3 will result in protonation of the carboxyl group, so the molecule will assume an overall positive charge. Alanine will, therefore, migrate to the cathode. On the other hand, aspartic acid will exist as a neutral dipolar ion at pH 3 since this is equivalent to its isoelectric point. Therefore, when it is subjected to electrophoresis, it will not move. In summary, alanine will migrate to the cathode while aspartic acid will not move, making choice C the correct response.

2. **B** Discussed on page 451.

3. **B** The amino acid in question is glutamic acid, which is an acidic amino acid because it contains an extra carboxyl group. At neutral pH, both of the carboxyl groups are ionized, so there are two negative charges on the molecule. Only one of the charges is neutralized by the positive charge on the amino group, so the molecule has an overall negative charge. Thus, the answer is choice B.

4. **D** At pH 3, the amine and carboxyl groups will be protonated to give a net positive charge. As the pH rises to 7, the proton will first dissociate from the carboxyl, but both amine groups will still be fully protonated, so the charge will still be positive. At pH 11, the molecule is above its isoelectric point and will be fully deprotonated, resulting in two neutral amine groups and a negatively-charged carboxylate group, so the charge at pH 11 will be negative. Therefore, the correct sequence of charges is positive, positive, negative corresponding to choice D.

5. **B** Nonpolar molecules or groups are those whose negative and positive centers of charge coincide. They are not soluble in water and are thus hydrophobic. Amino acids with hydrophobic R-groups are considered hydrophobic molecules; they tend to be found buried within the protein molecules where they do not have to interact with the aqueous cellular environment. Choices A and D are incorrect because nonpolar R-groups cannot be hydrophilic. Choice C is incorrect because nonpolar molecules are seldom located on the surface of proteins, where they would interact unfavorably with the aqueous cellular environment.

6. **D** Formation of a peptide bond, which is the primary covalent bond found in proteins, involves a condensation reaction between the amine group of one amino acid and the carboxyl group of an adjacent amino acid. As a result of the carbonyl group present at the bond, the double bond resonates between C=O and C=N. This resonance gives the peptide bond a partial double-bond character, and limits rotation about the bond. From this information, it can be seen that choices A, B, and C are all characteristics of the peptide bond. Choice D is false because the formation of the peptide bond is a condensation reaction involving the loss of water, rather than a hydration reaction, which involves the addition of water.

7. **B** The 6 tripeptides that can be formed are:

Val-Ala-Leu, Val-Leu-Ala,

Ala-Val-Leu, Ala-Leu-Val,

Leu-Val-Ala, and Leu-Ala-Val.

8. **C** The key word in this question is "covalent." While hydrogen bonds and hydrophobic bonds are involved in peptide structure, they are not considered covalent bonds, since they do not involve sharing electrons. Therefore, choices A and D are incorrect. Ether bonds are covalent bonds, but they are not found in peptides. The correct answer is disulfide bonds, choice C. Disulfide bonds are covalent bonds forming between the sulfur-bearing R-groups of cysteines. The resulting cystine molecule constitutes a disulfide bridge and often causes a loop in the peptide chain.

9. Discussed on page 458–460.

10. **A** When discussing secondary structure, the most important bond is the hydrogen bond. The rigid α-helices are held together by hydrogen bonds between the carbonyl oxygen of one peptide bond and the amine hydrogen of a peptide bond four residues removed. This hydrogen bond is intramolecular, so choice A is correct. Disulfide bonds are covalent bonds usually associated with primary and tertiary structure, therefore, choice B is incorrect. Choices C and D are incorrect since the rippled effect and intermolecular hydrogen bonds are both characteristic of β-pleated sheets.

11. **D** Protein denaturation involves the loss of three-dimensional structure and function. Since the three-dimensional shape of a protein is conferred by secondary and tertiary structures, denaturation disrupts these structures. Therefore, both choices B and C are correct. Denaturation does not cause a loss of primary structure since it does not cause peptide bonds to break; thus, choice A is incorrect.

INDEX

A

B

C

D

E

F

G

H

I

J

K

L

M

N

O

P

Q

R

S

T

U

V

W

Y

Z

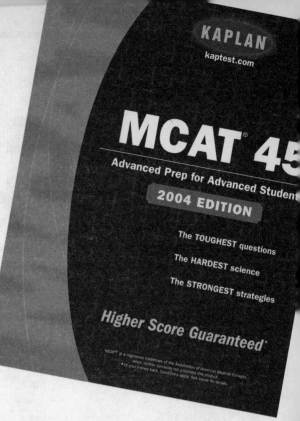